Proceedings

of the 5th International Yellow River Forum on Ensuring Water Right of the River's Demand and Healthy River Basin Maintenance

Volume V

Yellow River Conservancy Press

图书在版编目(CIP)数据

第五届黄河国际论坛论文集/尚宏琦,骆向新主编. —郑州:
黄河水利出版社,2015.9
ISBN 978 – 7 – 5509 – 0399 – 9

I. ①第… Ⅱ.①尚… ②骆… Ⅲ.①黄河 – 河道整治 –
国际学术会议 – 文集 Ⅳ.①TV882.1 – 53

中国版本图书馆 CIP 数据核字(2012)第 314288 号

出 版 社:黄河水利出版社
　　　　地址:河南省郑州市顺河路黄委会综合楼 14 层　　　　邮政编码:450003
发行单位:黄河水利出版社
　　　　发行部电话:0371 – 66026940、66020550、66028024、66022620(传真)
　　　　E-mail:hhslcbs@ 126. com
承印单位:河南省瑞光印务股份有限公司
开本:787 mm × 1 092 mm　　1/16
印张:149.75
印数:1—1 000
版次:2015 年 9 月第 1 版　　　　　　　　　　印次:2015 年 9 月第 1 次印刷

定价(全五册):960.00 元(US＄155.00)

Under the Auspices of

Ministry of Water Resources, People's Republic of China

Sponsored & Hosted by

Yellow River Conservancy Commission(YRCC), Ministry of Water Resources, P. R. China
China Yellow River Foundation(CYRF)

Editing Committee of Proceedings of the 5th International Yellow River Forum on Ensuring Water Right of the River's Demand and Healthy River Basin Maintenance

Welcome

(preface)

The 5th International Yellow River Forum (IYRF) is sponsored by Yellow River Conservancy Commission (YRCC) and China Yellow River Foundation (CYRF). On behalf of the Organizing Committee of the conference, I warmly welcome you from over the world to Zhengzhou to attend the 5th IYRF. I sincerely appreciate the valuable contributions of all the delegates.

As an international academic conference, IYRF aims to set up a platform of wide exchange and cooperation for global experts, scholars, managers and stakeholders in water and related fields. Since the initiation in 2003, IYRF has been hosted for four times successfully, which shows new concepts and achievements of the Yellow River management and water management in China, demonstrates the new scientific results in nowadays world water and related fields, and promotes water knowledge sharing and cooperation in the world.

The central theme of the 5th IYRF is "Ensuring water right of the river's demand and healthy river basin maintenance". The Organizing Committee of the 5th IYRF has received near one thousand paper abstracts. Reviewed by the Technical Committee, part of the abstracts are finally collected into the Technical Paper Abstracts of the 5th IYRF.

An ambience of collaboration, respect, and innovation will once again define the forum environment, as experts researchers, representatives from national and local governments, international organizations, universities, research institutions and civil communities gather to discuss, express and listen to the opportunities, challenges and solutions to ensure the sustainable water resources management.

We appreciate the generous supports from the co – sponsors, including domestic and abroad governments and organizations. We also would like to thank the members of the Organizing Committee and the Technical Committee for their great supports and the hard work of the secretariat, as well as all the experts and authors for their outstanding contributions to the 5th IYRF.

Finally, I would like to present my best wishes to the success of the 5th IYRF, and hope every participant to have a good memory about the forum!

Chen Xiaojiang
Chairman of the Organizing Committee, IYRF
Commissioner of YRCC, MWR, China
Zhengzhou, September 2012

Contents

J. Application of Experiences and New Technologies of Water Resources Management(II)

K. Others

4

J. Application of Experiences and New Technologies of Water Resources Management

(II)

Ecohydrological Assessment Tool (EcoHAT) System and Its New Application Cases

Song Wenlong, Yang Shengtian, Wang Xuelei, Wang Yujuan,
Zeng Hongjuan, Wang Mingcheng and Zhang Jing

State Key Laboratory of Remote Sensing Science, Beijing Key Laboratory for Remote
Sensing of Environment and Digital Cities, School of Geography,
Beijing Normal University, Beijing, 100875, China

Abstract: Ecohydrological process model driven by remote sensing data has been international research hotspots in the fields of water resources and water environment, along with the technology development in RS, GIS, and computer. Ecohydrological Assessment Tool (EcoHAT) system is one for simulations of ecohydrological processes. This paper introduces some new application cases of EcoHAT system, discussing the coupling of remote sensing data and the models with physical mechanism. To understand the main issues in fields of water resources and water environment, including water resource evaluation in an ungauged basin, lower soil water simulation at a large scale, simulation of pollution loads from non - point sources, and impact analysis of riparian basin. Spatially, simulation and analysis on water resources and water environment at a large scale has been done through modifying structure of physical models to realize their coupling with remote sensing data. Thus, key parameters of the models will be obtained or determined from public remote sensing products, to reduce the dependence on field monitoring data in traditional methods. For ungauged basin without suitable hydrological models or abundant data, we have constructed a remote sensing data driven distributed hydrological model (RS - DTVGM) coupling with the TRMM and FY - 2 precipitation products and the MODIS LST products, to evaluate water resource. A soil water model with physical mechanism has been coupled with remote sensing data, including the FY - 2 precipitation product, MODIS data and GLDAS data from public platform, to simulate lower layer soil water movement at a large scale in Weihe River. Pollutant loads from non - point sources in a large - scale basin (ENPS - LSB) is estimated and the impacts of riparian is analyzed in Hainan province from 2003 to 2007 by modifying of the model and constructing a riparian model (RIPAM). By comparison with the experiment data, the simulation results are found to be in acceptable differences. Future breakthroughs in scale transformation and interpolation techniques of remote sensing data from multi - source will help improve our understanding in ecohydrological processes.

Key words: ecohydrology, ecohydrological process, remote sensing, EcoHAT, water resources and water environment.

1 Introduction

Ecohydrology is a new frontier branch of hydrology in the context of, water crisis has been a global problem, and the sustainable development and utilization of water resource has been an important and urgent way for the human society development (Huang, Y. L., Fu, B. J., et al., 2003; Yang, S. Y., Yan, D. H., et al., 2009). Ecohydrology, an interdisciplinary science of hydrology and ecology, has been an international research hotspot driven by a series of large - scale international research programs (Wang, G. X., Qian, J., et al., 2008; Yang, Y. G., Xiao, H. L., et al., 2011). Of ecohydrology, the purposes are to realize the sustainable development and utilization of water resource and the healthy and sustainable development of the ecosystem; the study basis has been the relationship of water cycle and vegetation; the cores of study have been soil water and evapotranspiration; the key points have been the coupling and the scale effect of ecological process and hydrological process; the study objects have involved the areas such as wetland ecosystem, arid ecosystem, forest ecosystem, river ecosystem and lake ecosystem, and the interaction of ecological process, hydrological process and the human activities.

To extend the HIMS (Liu, C. M. , Wang, Z. G. , et al. , 2009), making full use of remote sensing data sources, coupling with the physical and chemical mechanisms of the process of ecohydrological model, simulating interaction between hydrological cycle, nutrient cycle and plant growth in the soil – plant – atmosphere continuum (SPAC), a regional scale distributed simulation system, EcoHAT, has been developed. The EcoHAT has integrated the processes of hydrological cycle, nutrient cycle and plant growth in ecosystems, and coupled the remote sensing data with the ecohydrological models and the physical and chemical mechanisms, to solve problems in water resource and water environment. This paper will introduce some new applications of EcoHAT.

2   The development of EcoHAT system

The paper titled " Development of ecohydrological assessment tool and its application" introduced the theoretical basis and the development of EcoHAT system in detail (Liu, C. M. , Yang, S. T. , et al. , 2009). And main achievements and applications can refer to some published papers (Wang, Mannaerts, et al. , 2010; Wang, Wang, et al. , 2011; Yang, Dong, et al. , 2011).

3   New applications and analysis of EcoHAT system

3.1   Water source evaluation in an ungauged basin

Combining remote sensing (RS) technique and Distributed Time Variant Gain Model (DTVGM), a RS driven distributed hydrological model has been constructed, which is called RS – DTVGM. Most variables and parameters of RS – DTVGM could be derived from RS directly or indirectly, which could reduce the dependence of hydrological simulation on observation data and be suitable for application in ungauged basins. Choosing the most popular RS precipitation products, TRMM and FY – 2, it estimated the products' precision in the high latitude region (north of 40°N) at different time scales. Combined with simple and klemen's advanced statistic methods, an integrated method is put forward to estimate air temperature based on the MODIS LST (Land Surface Temperature). Coupled with NCEP/NCAR temperature data, the instantaneous air temperature was converted to daily average temperature. A relatively simple interpolation method is also designed to interpolate lost RS data caused by cloud, which took the elevation as spatial control factor and time interval as temporal control factor. This method could provide continuous temperature data for hydrological model both at spatial and temporal scales. Taking MODIS as main data source, it combined with energy balance theory and Priestley – Taylor and put forward a RS method to estimate instantaneous potential evapotranspiration in sunny days. Daily potential evapotranspiration was calculated from the instantaneous one using the cosine curve and provide input data for evapotranspiration model. It could receive a good estimation precision, but the simulation results of the extreme (high and low) values were not good enough. Took Tekesi River Basin, which is located in northwestern China and has scarce observation data, as a case study, RS – DTVGM was calibrated and validated. Coupled with different kinds of public RS and GIS data and products, most of the hydrological variables and parameters of the RS – DTVGM were extracted. A hydrological information database was constructed and the hydrological simulation was conducted to calibrate and validate the model, the results showed that the RS – DTVGM could be run with a little surface observation data, and received a Nash – Sutcliffe efficiency of $0.5 \sim 0.6$, a water balance efficiency of $0.9 \sim 1.1$ and a relative efficiency R^2 of $0.5 \sim 0.6$.

3.2　Lower soil water simulation at a large scale

Traditional methods for monitoring soil moisture can be used for thick soil layers, but not for regional scale and continuous monitoring. The remote sensing simulation models can meet the needs of regional scale, but the soil layer is confined only to the surface depth, the research on the thick layer for soil moisture inversion is still scarce. This study focuses on the two issues, aiming to explore a remote sensing driven soil water monitoring model for Weihe River Basins to simulate the soil water movement during layers of soil lower to1 m. Combined with remote sensing (RS) technique and Richards equation, a RS driven soil moisture model is constructed. Most variables and parameters could be derived from RS directly or indirectly, which could reduce the dependence of hydrological simulation on observation data and be suitable for application in Weihe River Basins. Firstly, analyses of the effectiveness of remote sensing data from different sources have been implemented. The effectiveness of FY – 2 precipitation data and the GLDAS data on daily, monthly and annual scales have been analyzed respectively. Then, we made full use of public platforms for remote sensing and GIS data products, provided the model input parameters directly, such as precipitation, temperature, pressure, wind speed, relative humidity, surface temperature, leaf area index, land cover, etc. Based on the building model of soil moisture, using the public platform of RS and GIS data products as the main data source (including FY – 2 precipitation estimation products, MODIS land product data, GLDAS assimilation product data, including SRTM DEM Data, Meris Land cover data and soil database HWSD), soil moisture simulations for 1 m soil layer were carried out in Weihe River Basin from 2006 to 2009.

3.3　Simulation of pollution loads from non – point sources and the effect analysis of riparian basin

Songtao Reservoir is the most important water source in Hainan Province of China, playing an important role in 2.05×10^6 mu farmland irrigation and life water usage. In recent years, with the increase of emissions from human life and livestock waste as well as the illegal logging of natural forests and deforestation, non – point source (NPS) pollution is getting worse. Therefore, to evaluate the effects of management and strategy on NPS pollution with the NPS pollution model means a lot for water quality assurance, drinking water safety, ecological environment protection and international tourism island construction in Hainan. This study made an improvement of large – scale NPS pollution model, and applied the model to run a simulation of NPS pollution characteristics on Songtao reservoir region from 2003 ~ 2007. The results show that: ① the slope, slope length and digital streams were extracted from DEM, as well as the basin was divided into 65 sub – watershed basin; ② the land use data, vegetation cover, LAI, albedo and surface temperature were extracted from remote sensing data, and the method to calculate the above parameters were discussed when remote sensing data were unavailable; ③ the method to calculate the solar radiation, net radiation and root depth were presented; ④ the model was calibrated and validated with runoff and water quality data from 2001 to 2005, and the results demonstrated that: ① the correlation coefficient (R^2) and Nash – Suttchiffe coefficient (ENS) of runoff were higher than 0.8 during calibration and verification, with average error of less than 20%; ② the R^2 of nitrate was 0.907 during calibration, and ENS was 0.646, with an average error of 29.3%; the R^2 and ENS of nitrate were higher than 0.88 during verification, with an average error of 4.85%; ③ the R^2 and ENS of ammonia were higher than 0.6 during calibration, with an average error of 4%; the R^2 and ENS of ammonia were higher than 0.84 during verification, with an average error of 28.9%; ④ the R^2 and ENS of TP were higher than 0.88 during calibration and verification, with average error of less than 15%. So, it could be concluded that the simulation accuracy of the model was higher.

A riparian model (RIPAM) is built and used for modeling on different scales, from an important riparian zone to a watershed area in Guanting reservoir, Beijing, China. The nitrogen and phosphorous removal rates are simulated and the effects of land use on nonpoint source pollution are analyzed. Riparian function is also assessed in Guanting reservoir. RIPAM model consists of eight sub – modules, they are soil heat and moisture module, nitrification module, denitrification module, ammonia volatilization module and NPP module, products allocation module and plant uptake module. Remote sensing data including SPOT5 and Landsat TM5 are used in the study to increase the data accuracy and to derive parameters such as vegetation types, community structure, ecological pattern, some background spatial data such as soil type and water, and some environmental factors such as precipitation, soil water, and etc. By comparison with the experiment results, it is found that the simulated results are acceptable on watershed scale.

4 Discussions

A series of remote sensing products, with different accuracies and spatio – temporal resolution, were used directly in the ecohydrological process simulations. This may increase the uncertainty of the models without accuracy calibration and scale transformation. For the poor spatio – temporal continuity of remote sensing products, simple interpolation was unsuitable and would reduce the accuracy of evaluated parameters. And for the restriction in time, technology and data source, modified models and methods for parameters evaluation through remote sensing data have not been validated at higher levels. All of the works need be improved in the future.

Acknowledgements
Financial support came from the Public Project from the Ministry of Water Resources (Grant No. 200901022 – 01).

References

Wang X, C Mannaerts, et al. Evaluation of Soil Nitrogen Emissions from Riparian Zones Coupling Simple Process – oriented Models with Remote Sensing Data [J]. Science of the Total Environment, 2010, 408(16): 3310 – 3318.

Wang X, Q Wang, et al. Evaluating Nitrogen Removal by Vegetation Uptake Using Satellite Image Time Series in Riparian Catchments[J]. Science of the Total Environment, 2011.

Yang S, G Dong, et al. Coupling Xin' Anjiang Model and SWAT to Simulate Agricultural Non – point Source Pollution in Songtao watershed of Hainan, China [J]. Ecological Modelling, 2011, 222(20): 3701 – 3717.

Wang G X, Qian J, et al. Current Situation and Prospect of the Ecological Hydrology Advance in Earth Sciences, 2008 (3):314 – 323.

Liu C M, Wang Z G, et al. Development of Hydro – informatic Modelling System and Its Application [J]. Science in China Series E – technological Sciences, 2009, 38 (3): 350 – 360.

Liu C M, Yang S T, et al. Development of Ecohydrological Assessment Tool and Its Application [J]. Science in China Series E – technological Sciences, 2009(6): 1112 – 1121.

Yang Y G, Xiao H L, et al. Research Advances in Ecohydrological Process and Ecohydrological Function[J]. Journal of Desert Research, 2011, 31(5): 1242 – 1246.

Yang S Y, Yan D H, et al. Progress of Eco – hydrological Coupling Research [J]. Water Resources and Hydropower Engineering, 2009, 40(002):1 – 8.

Huang Y L, Fu B J, et al. Advances in Ecohydrological Process Research[J]. Acta Ecologica Sinica, 2003, 23(3): 580 – 587.

The Application of Muskingum Segmentation Algorithm Forecasting Wubao, Tongguan Flood Process in the Midstream of the Yellow River

Sun Meiyun, *Liu Jun*, *Zuo Jun* and *Liu Huazhen*

College of Hydrology and Water Resources, Hohai University, Nanjing, 210098, China

Abstract: Real – time forecasting the flood in Wubao, Tongguan reach in the Yellow River, understanding the hydrological situation in downstream can contribute to formulate the scheme of water resources allocation, prevent the phenomenon of wasting the water in flood and drying up in drought downstream, which can alleviate the contradiction between the suppy and demand of water resources effectively; meanwhile the accurate flood forecast can lay the solid foundation for the research of sediment in the Yellow River, provide the reasonable basis for managing the Yellow River. Aiming at the character of longer river, the more tributaries and the flood influenced seriously by the tributaries in Wubao, Tongguan interzone of the Yellow River, this paper puts forward the method of parameters calibration in Wu Long and Long Tong interzone respectively forecasting flood process by using Muskingum segmentation algorithm, forecasting and simulating the flood process by using the calibrated parameters, identifing the initial flow in every segment nodes, establishing the scheme of dealing the tributaries based on first combination and then evolvement method. The results from the model validation has shown that the simulated and the observed flood process fit very well. Moreover, the simulated flood volume is fairly satisfactory, which explaines that this model is not only feasible but also has the higher precision.

Key words: Muskingum segmentation algorithm, parameters calibration, flood process forecasting, Wubao, Tongguan interzone of the Yellow River

1 The outline

The flood routing in natural river is a significant research project in academic and application level. There are hydrology and hydraulics routing for river channel flood routing. The latter needs detailed information and calculation cumbersome for solving the St. Venant equations; the former often simplifies the St. Venant equations by using water balance equation instead of the continuity equation and channel storage equation instead of the dynamic equations in the condition of lacking information or never need the hydrological regime in the river section. Muskingum is a method widely used in flood routing in hydrology, its model is just as followed:

$$Q_2 = C_0 I_2 + C_1 I_1 + C_2 Q_1$$
$$C_0 = (-Kx + 0.5\Delta t)/(K - Kx + 0.5\Delta t)$$
$$C_1 = (Kx + 0.5\Delta t)/(K - Kx + 0.5\Delta t) \qquad (1)$$
$$C_2 = (K - Kx - 0.5\Delta t)/(K - Kx + 0.5\Delta t)$$
$$C_0 + C_1 + C_2 = 1$$

where, C_0, C_1, C_2 is the coefficient of flow routing; I_1, I_2 is the inflow at the start and end of the interval respectively, m^3/s; Q_1, Q_2 is the outflow at the start and end of the interval respectively; K is a storage coefficient (dimensions of time); x is a weighting factor; Δt is the interval of time.

Muskingum is linear finite method simplified by channel flow routing equation, which requires K, x are constant and the flow is linear distribution in the period of calculated time and along the way. Therefore, the interval of time Δt can not be too large or too small. The flow changes non – linear in the period of Δt if it is too large; if it is too small the flow can not fit linear distribution along the way. Generally speaking, Δt should equal or close to K.

2 Muskingum segmentation continuous algorithm

Due to the quick rise, sudden fall, short duration of flood and more tributaries existing in Wubao, Tongguan interzone of the Yellow River, it is difficult to forecast the flood in this interzone and the accuracy is not so well. This paper put forward the method of Muskingum segmentation continuous algorithm, improving the forecast accuracy effectively by adding the tributaries into the flood. Muskingum segmentation continuous algorithm was put forward by professor Zhao Renjun in 1962, deviding the channel length L into n segments, the K of each segment is equal, choosing Δt, giving the reach of K, x:

$$n = \frac{K}{K_l} = \frac{K}{\Delta t}$$

$$x_l = \frac{1}{2} - \frac{n(1-2x)}{2} \tag{2}$$

where, n is the number of segments; K_l is storage coefficient in every segment; x_l weighting factor in every segment.

The key of the method is identifing the initial flow in each segment. Considering the time of flood calculation is at primary stage of rising flood and the channel storage is not full enough, the article identify the initial flow in segment nodes using linear interpolation method.

The standard of evaluating the forecast accuracy is the similarity between the predicted and observed flood results. Owing to the more uncertainty factors existing in the flood, therefore, this article just compares the shape between the forecast and the measured flood and the flood volume and time difference of flood peak emergence Δh. The evaluation index of shape is the coefficient of determination DC and the evaluation index of flood volume is the flood volume relative error δ.

$$DC = 1 - S_c^2/\sigma_y^2$$

$$S_c = \sqrt{\left[\sum_{i=1}^{n} (y - y_i)^2\right]/n}$$

$$\sigma_y = \sqrt{\left[\sum_{i=1}^{n} (y_i - \bar{y})^2\right]/n}$$

$$\delta = \left(\sum_{i=1}^{n} y - \sum_{i=1}^{n} y_i\right) \times 100/\sum_{i=1}^{n} y_i$$

$$\Delta h = h - h'$$

where, S_c is mean variance of forecasting error; σ_y is mean variance of forecasting factor; y_i, y is testing flood and forecasting flood respectively; $\bar{y}$ is the mean of the testing flood; δ is the flood total relative error; h is the emerge time of measured flood peak; h' is the emerge time of forecast flood peak.

3 Examples application

3.1 The outline in the study basin

The channel from Wubao to Longmen in the midstream of the Yellow River located in the downstream of dabei main stream, which is a canyon river, The length is 246 km. There are more than 240 tributaries adding to the mainsream. the major ones are Sanchuang River, Qunchan River, Wuding River, Qingjian Rive, Qishui River, Yanshui, Fenchuang River, Shiwangchuan and Zhouchuan River and the hydrological stations on the 9 tributaries respectively are Houdacheng, Pengou, Baijiachuan, Yanchuan, Daning, Ganguyi, Xinshihe, Dacun and Jixian, mentioned in Fig. 1.

The channel from Longmen to Tongguan located up to the Sanmenjia reservoir, that is the channel from Longmen, Hejin, and Huaxian to Tongguan. It is called Xiaobei mainstream because of its characteristics of wandering river different from canyon river (Dabei mainstream) the bigger tributaries in the interzone include Fenhe River, Shushui River, Beiluo River, Weihe River and Jing River, among which Weihe River and Fenhe River are the biggest ones in the Yellow River, seen in Fig. 1.

Fig. 1　Schematic diagram of the Wubao—Tongguan reach in the Yellow River

3.2　The tributaries reach dealing

There are two methods of first combination then evolvement and first evolvement then combination in hydrology when it comes to flood routing with tributaries. Firstly combination and then evolvement is the method of linear superposition the tributaries orderly in the mainstream first, then establishing the Muskingum model deriving the outflow process in the outflow section. Due to the flood influenced seriously by the tributaries, this paper choose the former method adding the tributaries to flood routing.

Given the Δt and the range of K, x, identifying the number of segments n and the initial flow in every segment corresponding to every K, Muskingum algorithm for flood routing is used from Wubao and Longmen respectively to decided whether the tributaries before the next calculation need to be added. Firstly the location of the segement nodes should be identified according to the number of segments n. Then the location of the tributaries is determined based on the distance from the tributaries to Wubao, Longmen and from the segement points to Wubao, Longmen. The assumption that the tributarie x was just between the segement point y and $y + 1$ is made, compared the ralatively distance from the tributaries x to the segement y and to the segement $y + 1$. If it is closer to the segement point y, then the tributaries x is added as the the inflow $y + 1$ after the flood routing y; else, added as the inflow $y + 2$; if the tributarie x is coincided with the segement y, the tributaries x is added as the the inflow $y+1$, then the next period of flood is caculated.

Though the analysis of the flood from 1980 to 2000, the propagation time of flood peak in Wulong and Longtong are 10 ~ 18 h, 15 ~ 23 h respectively. According the meaning of K, preliminary identified K is 10 ~ 18 and 15 ~ 23. Δt is 1 h. n is 10 ~ 18 and 15 ~ 23, x is anyone never greater than 0.5. The tributaries dealing can been seen in Tab. 1 and Tab. 2.

Tab. 1　Statistics of Tributary flood dealing in Wulong interzone

The number of segements	Houdacheng	Peigou	Baijiachuang	Yanchuang	Daning	Ganguyi	Xinshi River	Da Villiage	Ji Courty
10	0	1	3	5	6	6	7	8	8
11	0	1	3	5	7	7	8	9	9
12	0	2	3	6	8	8	9	10	10
13	0	2	3	6	8	8	10	10	11
14	0	2	4	7	9	9	10	11	12
15	0	2	4	7	9	10	11	12	13
16	0	2	4	8	10	10	12	13	13
17	0	2	4	8	11	11	13	14	14
18	0	2	5	9	11	12	13	14	15

Tab. 2　Statistics of Tributary flood dealing in Long tong interzone

The number of segments	Hejin	Hua Courty	Tongguan
15	2	14	15
16	2	15	16
17	2	16	17
18	3	17	18
19	3	18	19
20	3	19	20
21	3	20	21
22	3	21	22
23	3	22	23

3.3　Parameters calibration in model

Firstly the flood peak of Wubao and Longmen is taken respectively as a standard. Then the flood which the flood peak is over 3,000 is chosen. At the time, the flood of corresponding tributaries is extracted. it is respectively 25 and 15 flood satisfing the above conditions in Wulong and Longtong interzone in 1980 ~ 2000. After that, the parameters is calibrated by using the method of test for true. The parameters interval which is much closer to the tested flood is chosen after flood simulation of different K, x combination by establishing model. The satisfing parameters interval are listed in Tab. 3, Tab. 4.

Tab. 3　Statistics of parameter calibration in Wulong interzone

Names of floods	Flood peak of Wubao(m³/s)	K	x	Coefficient of determination	Flood volume relative error(%)	Time difference of flood peak emergence(h)
19800329	3,370.0	16 ~ 18	0.36 ~ 0.47	0.70	− 9.33	1.0
19801007	3,420.0	17 ~ 18	0.23 ~ 0.36	0.74	0.33	− 2.0
19810322	4,140.0	16 ~ 17	0.17 ~ 0.48	0.72	− 15.20	− 2.0
19810702	4,143.3	16 ~ 17	0.35 ~ 0.49	0.60	− 13.44	− 1.0
19810722	6,036.6	13 ~ 16	0.17 ~ 0.37	0.84	− 14.72	− 2.0
19810727	6,810.0	13 ~ 15	0.24 ~ 0.41	0.86	− 0.31	− 1.0
19810806	4,870.0	15 ~ 16	0.26 ~ 0.44	0.94	3.94	− 1.0
19820730	4,730.0	15 ~ 16	0.25 ~ 0.49	0.89	3.62	+ 1.0
19830804	5,460.0	14 ~ 16	0.10 ~ 0.45	0.72	0.69	− 3.0
19880723	4,000.0	16 ~ 17	0.39 ~ 0.49	0.89	3.96	+ 1.0
19890721	12,400.0	13 ~ 14	0.10 ~ 0.36	0.83	− 2.87	− 1.0
19890911	3,300.0	15 ~ 17	0.22 ~ 0.44	0.83	− 11.26	− 1.0
19900317	3,000.0	15 ~ 17	0.10 ~ 0.40	0.84	− 3.91	− 2.0
19910721	4,440.0	15 ~ 17	0.30 ~ 0.45	0.90	0.13	− 1.0
19920725	3,960.0	16 ~ 18	0.20 ~ 0.45	0.76	− 7.14	− 1.0
19920808	9,340.0	13 ~ 14	0.35 ~ 0.45	0.79	− 6.49	+ 1.0
19930318	3,220.0	13 ~ 17	0.10 ~ 0.40	0.94	11.43	− 3.0
19940707	4,270.0	16 ~ 17	0.35 ~ 0.45	0.77	− 9.92	+ 1.0
19940804	6,310.0	13 ~ 14	0.40 ~ 0.50	0.91	− 1.24	− 3.0
19950728	7,600.0	13 ~ 14	0.40 ~ 0.50	0.86	− 12.60	0.0
19960330	3,750.0	15 ~ 16	0.25 ~ 0.40	0.88	− 0.66	− 3.0
19960809	8,640.0	12 ~ 13	0.45 ~ 0.50	0.92	3.13	− 1.0
19970319	3,550.0	14 ~ 16	0.30 ~ 0.50	0.82	− 3.24	− 3.0
19980312	3,030.0	17 ~ 18	0.10 ~ 0.50	0.78	− 3.64	+ 4.0
19980713	6,000.0	13 ~ 14	0.30 ~ 0.45	0.91	− 7.21	0.0

Tab. 4　Statistics of parameter calibration in Longtong interzone

Names of floods	Flood peak of Longmen(m³/s)	K	x	Coefficient of determination	Flood volume relative error(%)	Time difference of flood peak emergence(h)
19810702	6,400	15 ~ 19	0.1 ~ 0.5	0.822,388	− 1.412,55	− 3.0
19820729	5,050	15 ~ 21	0.1 ~ 0.32	0.881,63	− 4.908,27	− 1.0
19810720	5,200	15 ~ 19	0.1 ~ 0.3	0.736,708	− 6.601,61	− 1.0
19870823	6,840	16 ~ 20	0.1 ~ 0.5	0.910,795	− 9.198,64	0
19880804	10,200	15 ~ 17	0.1 ~ 0.3	0.781,384	3.839,361	− 4.0

Continued to Tab. 4

Names of floods	Flood peak of Longmen(m³/s)	K	x	Coefficient of determination	Flood volume relative error(%)	Time difference of flood peak emergence(h)
19890722	8,300	15 ~ 19	0.1 ~ 0.3	0.886,406	5.986,043	+1.0
19910722	4,590	19 ~ 23	0.1 ~ 0.26	0.893,583	7.188,083	0
19940707	4,780	17 ~ 20	0.1 ~ 0.5	0.887,978	0.887,978	0
19940805	10,600	17 ~ 18	0.1 ~ 0.3	0.745,77	6.857,192	0
19950717	7,860	18 ~ 20	0.1 ~ 0.25	0.723,979	3.153,475	0
19960801	11,100	17 ~ 21	0.1 ~ 0.3	0.886,548	− 0.446,26	− 1.0
19970730	5,750	17 ~ 20	0.1 ~ 0.3	0.829,797	− 8.04,498	+2.0
19980313	3,200	15 ~ 23	0.1 ~ 0.5	0.952,685	4.852,57	− 3.0
19850806	6,720	16 ~ 19	0.1 ~ 0.35	0.799,217	− 3.604,92	− 3.0
19860704	3,520	18 ~ 23	0.1 ~ 0.5	0.907,488	− 5.834,56	− 2.0

In order to facilitate forecast statistics, the flood results are classified into different levels according to the flood peak of Wubao and Longmen respectively, and the results are just listed in Tab. 5, Tab. 6.

Tab. 5   Statistics of parameter in classification in Wulong interzone

Flood peak of Wubao(m³/s)	K	x
3,000 ~ 4,000	17	0.30
4,000 ~ 4,500	16	0.40
4,500 ~ 5,000	15	0.40
5,000 ~ 7,000	14	0.40
7,000 ~ 9,000	13	0.40
>9,000	13	0.35

Tab. 6   Statistics of parameter in classification in Longtong interzone

Flood peak of Longmen(m³/s)	K	x
3,000 ~ 4,000	21	0.1
4,000 ~ 5,000	20	0.1
5,000 ~ 7,000	19	0.1
7,000 ~ 9,000	18	0.1
>9,000	17	0.1

3.4 Parameters validation and flood simulation

The measured flood in 2001 ~ 2007 is taked to validate the calibrated parameters. During this period, there are 3 and 4 flood whose flood peak is over 3,000 in Wubao and Longmen station respectively. The rationality of the parameters is validated by forecasting these floods. The flood process is simulated by using the calibrated parameters and the the predicted accuracy index results can been seen in Tab. 7, Tab. 8. The simulated flood process in Wulong interzone can been seen in Fig. 2 ~ Fig. 4 and the simulated results in Longtong interzone can been seen in Fig. 5 ~ Fig. 8.

Tab. 7　Statistics of parameter in prediction model in Wulong interzone

Names of floods	Flood peak of Wubao(m^3/s)	K	x	Coefficient of determination	Flood volume relative error(%)	Time difference of flood peak emergence(h)
20010321	3,000	17	0.40	0.84	-5.4	-1.0
20030730	9,400	13	0.35	0.80	4.8	0
20070320	3,080	17	0.40	0.89	6.3	-6.0

Tab. 8　Statistics of parameter in prediction model in Longtong interzone

Names of floods	Flood peak of Longmen(m^3/s)	K	x	Coefficient of determination	Flood volume relative error(%)	Time difference of flood peak emergence(h)
20010819	3,400	21	0.10	0.92	0.45	+4.0
20020704	4,580	19	0.10	0.89	9.13	-4.0
20030824	3,170	21	0.10	0.85	-3.48	+5.0
20060825	3,220	21	0.10	0.86	-6.15	+2.0

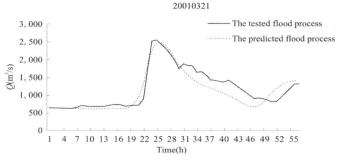

Fig. 2　Comparison graph in measured and forecasting flood process of Longmen in 2001

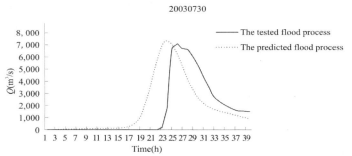

Fig. 3　Comparison graph in measured and forecasting flood process of Longmen in 2003

14

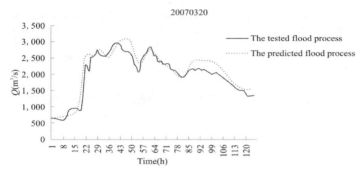

Fig. 4　Comparison graph in measured and forecasting flood process of Longmen in 2007

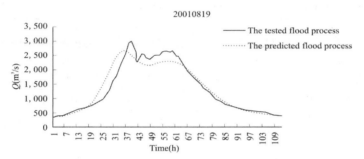

Fig. 5　Comparison graph in measured and forecasting flood process of Tongguan in 2001

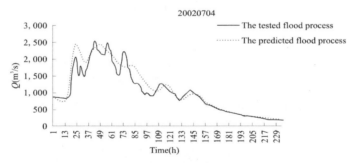

Fig. 6　Comparison graph in measured and forecasting flood process of Tongguan in 2002

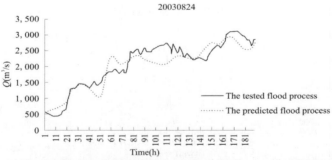

Fig. 7　Comparison graph in measured and forecasting flood process of Tongguan in 2003

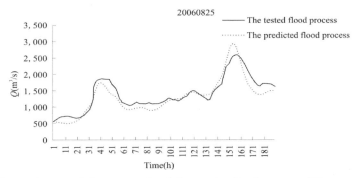

Fig. 8   Comparison graph in measured and forecasting flood process of Tongguan in 2006

As seen from Tab. 7, Tab. 8 and Fig. 2 ~ Fig. 8, though the flood is calculated by using the calibrated parameters, the deduction process can reflect the runoff from the upstream and tributaris accurately, which means the method is feasible, applicable and worth to promote.

3.5   Analysis and discussions

The evaluation index of accuracy including the coefficient of determination, the total relative error and time difference of the peak emergence. The flood whose coefficient of determination above 0.7 and total relative error less than 20% can been recognized as qualified. As seen from Tab. 5, Tab. 6, the coefficient of determinationin in 7 simulated flood are over 0.8. The total relative error is below 10% and the time difference of the peak emergence is sort of small. All in all, the accuracy in simulated flood process is satisfied, which can fit the measured runoff from the upstream and tributaries well.

The flood satisfing the condition in parameters calibrated period is much more, which makes the results of parameters calibrated credibility improved greatly, and lays a good foundation for the simulated and predicted flood. It shows that the parameters change obviously with the flood peak in Wubao and Longmen in Tab. 3 and Tab. 4. Therefore, this paper classifies the flood into different levels according to the flood peak. The parameters in each level are relatively stable, which makes it convenient to forecast; the flood predicted process are not sensitive to parameter x, which can be restricted in a certain range to meet the accuracy requirements. The flood in parameters validation is much less, but the results accuracy is sort of high and satisfied, which shows that the calibrated parameters is of good practicability in the reach.

The precondition of Muskingum segmentation algorithm is assuming that the parameters in each section are the same. Actually the regulation and storage of each section reach is different. Therefore, the actual segement river – length must be different if the segement parameters are the same. It can not fit the actual situation, which inevitably causes the errors.

4   Conclusions

Aiming at the character of more tributaries and the quick rise and sudden fall of flood, this paper puts forward the method of Muskingum segmentation algorithm to forecast the flood process and reduc the time step, and solves the problems of less duration of flood routing. Due to the flood influenced seriously by the tributaries, this article formulates the scheme of dealing the tributaries, fills the blank of traditional Muskingum algorithm applied to the multi – branch reach, improves the accuracy of simulation. The scheme of parameters calibration is established by dividing the flood into different levels according to the flood peak of Wubao, Longmen respectively. The relationship between K, x and the flood is identified. So that every flood corresponds to the unique parameters and the problem of parameters instablity is solved. In a word, this method is simple to be operated and easy to be understood. The simulated results show that the predicted and the tested flood match well. This method has good practicability and high precision.

References

Yuan X H, Zhang S Q, et al. Parameter Estamation of Nonlinear Muskingum Model Using Mixed Genetic Algorithm[J]. Journal of Hydraulic Engineering,2001,5:77 –81.

Zhao Z G, Wang W D. Study on Discharge Calculation Coefficient with Muskingun Method[J]. Water Resources & Hydropower of Northeast China,2006,264:21 –23.

Di J Z. Analysis on Several Different Application with Muskingun Method[J]. Yellow River,1994, 4:5 –7.

He L, Fu X D. Distribution of Flood Travel Time Along the Wubao—Tongguan Reach of Yellow River[J]. South – to – North Water Diversion and Water Science &Technology, 2012, 1: 18 –26.

Zhang X S, Sun J C, et al. Application of Improved Muskingum Method to the Flood Routing of Weihe River[J]. Yellow River,2010,11:36 –38.

Kong F Z, Wang X Z. Method Estimating Muskingum Model Parameters Based on Physical Characteristics of a River Reach[J]. Jortnal of China University of Mining & Technology, 2008,4:756 –758.

Hydrogen and Oxygen's Stable Isotope Composition's Differential Equation Model of Phreatic Water at Several Ideal Conditions

Tong Haibin

Institute of Numerical Simulation of Water Resource and Environment, Henan University, Kaifeng, 475004, China

Abstract: The significance and progress of the meteoric water line and evaporation line was briefly reviewed. Based on the isotope mass conservation principle and Rayleigh fractionation principle, the differential equation models of coupled flow field – stable isotope concentration field in phreatic water system are derived and the relationship between oxygen stable isotope and hydrogen stable isotope in phreatic water system at steady state is discussed. The models of two cases are deduced: ① for the variation of hydrogen and oxygen stable isotope concentration of unconfined steady flow along the flow direction recharging only by precipitation, the corresponding differential equation is founded and the analytic solution is presented. Then the relationship between hydrogen stable isotope and oxygen stable isotope at the same conditions is deduced; ② for the variation of hydrogen and oxygen stable isotope concentration of unconfined steady flow along the flow direction when evaporation exists, the corresponding differential equation is founded and the analytic solution is presented. Then the relationship between hydrogen stable isotope and oxygen isotope (Evaporation Line) at the same conditions is deduced. The presentation of the models mention above gives a help to interpret more quantitative and more objective the stable isotope data and $\delta^2 H - \delta^{18} O$ relation line.

Key words: stable isotope composition of hydrogen and oxygen, phreatic water, steady state, differential equation model

1 Introduction

Precipitation Line and evaporation line is a very important analytic tool in the field of isotope hydrology, widely being used in determining groundwater recharge sources and in dividing the groundwater type and so on. Its advantage is simple and convenient; but its drawback is that only reflects the statistical laws of the specific spatial and temporal scales within the meaning of average probability, and difficult to quantify, easy to cover up the dynamic characteristic of water stable isotope changes. Traditionally, due to the complexity of the hydrological phenomena and limitations of observation methods, test methods that focus on input and output to avoid the middle process, in fact, is a black box method. It can only be an analysis (guess) of the system and can not reveal the nature of the system; and it is the way of experience and difficult to get the new concept. To resolve this problem, in addition to the field and laboratory basic experiment with more in – depth, more systematic and more control, establishing the corresponding white – box models with clear physical meaning is needed.

Based on the Rayleigh fractionation principle and the Saint – Venant equations, Tong H B et al. (2007) presents hydrogen and oxygen stable isotope concentration field – fluid coupled differential equations of flowing river water during the evaporation process. Based on the radioactive decay principle and mass conservation principle, Tong H B et al. (2011) derived radioisotope concentration field – flow field coupling lumped parameter model in the groundwater system. A preliminary exploration in establishing white – box model with clear physical and mathematical meaning is presented in the above study. Based on the thought of above study, according to the isotope mass conservation principle and groundwater dynamics theory, a differential equation model of the hydrogen and oxygen stable isotope changes in one – dimensional phreatic Water system is derived, and then the analytic solution of the stable isotope relationship line between hydrogen and oxygen is derived, a more precise description of the dynamical feature and temporal – spatial variation law of the stable isotope migration of the system is expected from this differential equation model.

Due to limited space, only the hydrogen and oxygen stable isotope variation in the steady state is discussed. Two cases are discussed at the following paragraph: ① accepting the precipitation recharge; ② with phreatic Water's evaporating process.

2 Hydrogen and oxygen stable isotope composition's changes of phreatic water system along the flow direction in steady state

2. 1 Hydrogen and oxygen stable isotope composition's changes of phreatic Water system along the flow direction when recharging from rainfall under the steady – state

2. 1. 1 The model's assumption

Assume that the flow and isotope composition in the phreatic Water system are as follows: ① unconfined stream keeps in a constant flow state; ② flowing water's stable isotope is completely mixed in the vertical direction and is no proliferation in the longitudinal direction; ③ water hydroxide isotopic composition is in a steady state; ④ the isotope composition of precipitation does not change with time and space; ⑤ precipitation intensity is a constant; ⑥ phreatic Water's velocity is so higher that stable isotopes' change caused by water – rock interaction can be ignored.

2. 1. 2 Differential model's establishing and solving

Take the micro water body located in $\left[x - \dfrac{\Delta x}{2}, x + \dfrac{\Delta x}{2} \right]$ as the research object, establishing heavy isotopes' mass conservation equation of the micro water body considering the inflow and outflow of heavy isotopes:

$$(q - \frac{dq}{dx}\frac{\Delta r}{2})(R - \frac{dR}{dx}\frac{\Delta x}{2})dt + \mu H(x)R(x)\Delta x - (q + \frac{dq}{dx}\frac{\Delta x}{2}) \cdot$$
$$(R + \frac{dR}{dx}\frac{\Delta r}{2})dt + wR_p\Delta xdt = \mu HR\Delta r \tag{1}$$

where, w is the precipitation intensity.

Eliminate $\mu H(x)R(x)\Delta x$ on both sides of Eq. (1):

$$(q - \frac{dq}{dx}\frac{\Delta r}{2})(R - \frac{dR}{dx}\frac{\Delta x}{2})dt - (q + \frac{dq}{dx}\frac{\Delta x}{2})(R + \frac{dR}{dx}\frac{\Delta x}{2})dt +$$
$$wR_p\Delta xdt = 0 \tag{2}$$

Both sides of Eq. (2) is divided by dt:

$$(q - \frac{dq}{dx}\frac{\Delta r}{2})(R - \frac{dR}{dx}\frac{\Delta x}{2}) - (q + \frac{dq}{dx}\frac{\Delta x}{2})(R + \frac{dR}{dx}\frac{\Delta x}{2}) + wR_p\Delta x = 0 \tag{3}$$

Expanding and simplifying:

$$-\frac{d(qR)}{dx}\Delta x + wR_p\Delta x = 0 \tag{4}$$

Both sides of Eq. (4) is divided by Δx:

$$\frac{d(qR)}{dx} = wR_p \tag{5}$$

To unconfined flow field in steady state (assuming aquifer bottom plate is horizontal), following equation is presented according to groundwater dynamics:

$$q = -K\frac{dh}{dx}h \tag{6}$$

$$\frac{dq}{dx} = w \tag{7}$$

Assume that the boundary conditions are as follows:

$$h \big|_{x=0} = h_1 \qquad (8)$$

$$h \big|_{x=l} = h_2 \qquad (9)$$

Then the following definite problem is presented:

$$\begin{cases} \dfrac{d\left(-K \dfrac{dh}{dx} h \right)}{dx} = w \\ h \big|_{x=0} = h_1 \\ h \big|_{x=l} = h_2 \end{cases} \qquad (10)$$

The Solution of the definite problem:

$$h = \sqrt{ h_1{}^2 + \dfrac{h_2{}^2 - h_1{}^2}{l} x + \dfrac{w}{K}(lx - x^2) } \qquad (11)$$

$$q = K \dfrac{h_1{}^2 - h_2{}^2}{2l} - \dfrac{1}{2}wl + wx \qquad (12)$$

Expand Eq. (5), we obtain:

$$q \dfrac{dR}{dx} + R \dfrac{dq}{dx} = wR_p \qquad (13)$$

Substituting Eq. (7) into Eq. (13), finishing:

$$q \dfrac{dR}{dx} = w(R_p - R) \qquad (14)$$

Assume the isotope ratio of the standard water samples is R_s:

$$R = (1 + \delta) R_s \qquad (15)$$

Substituting Eq. (15) into Eq. (14) and finishing:

$$q \dfrac{d\delta}{dx} = w(\delta_p - \delta) \qquad (16)$$

Order

$$M = K \dfrac{h_1{}^2 - h_2{}^2}{2l} - \dfrac{1}{2}wl \qquad (17)$$

Then Eq. (12) can be written as:

$$q = M + wx \qquad (18)$$

Substituting Eq. (18) into Eq. (16) and finishing:

$$\dfrac{d\delta}{\delta_p - \delta} = \dfrac{d(M + wx)}{M + wx} \qquad (19)$$

Assume that the boundary conditions of isotope concentration field are as follows:

$$\delta \big|_{x=0} = \delta_0 \qquad (20)$$

There are definite problem:

$$\begin{cases} \dfrac{d\delta}{\delta_p - \delta} = \dfrac{d(M + wx)}{M + wx} \\ \delta \big|_{x=0} = \delta_0 \end{cases} \qquad (21)$$

The solution of the definite problem Eq. (21) is:

$$\delta = (\delta_0 - \delta_p)\left(1 + \dfrac{wx}{M} \right) + \delta_p \qquad (22)$$

Applied the solution Eq. (22) in ^{2}H and ^{18}O, there are:

$$^2\delta = (^2\delta_0 - {}^2\delta_p)\left(1 + \dfrac{wx}{M} \right) + {}^2\delta_p \quad \left(M = K\dfrac{h_1{}^2 - h_2{}^2}{2l} - \dfrac{1}{2}wl \right) \qquad (23)$$

$$^{18}\delta = (^{18}\delta_0 - {}^{18}\delta_p)\left(1 + \dfrac{wx}{M} \right) + {}^{18}\delta_p \quad \left(M = K\dfrac{h_1{}^2 - h_2{}^2}{2l} - \dfrac{1}{2}wl \right) \qquad (24)$$

Eq. (23) and Eq. (24) are the analytic expression of hydrogen and oxygen stable isotope composition's variation of unconfined steady flow system along the flow direction when recharging from rainfall at vertical direction.

2.1.3　Establishment of relation line of $^2\delta - {}^{18}\delta$

Apply Eq. (20) to ^2H and ^{18}O:

$$\frac{d^2\delta}{({}^2\delta_p - {}^2\delta)} = \frac{w}{q}dx \tag{25}$$

$$\frac{d^{18}\delta}{({}^{18}\delta_p - {}^{18}\delta)} = \frac{w}{q}dx \tag{26}$$

Combine Eq. (25) with Eq. (26), we have:

$$\frac{d^2\delta}{({}^2\delta_p - {}^2\delta)} = \frac{d^{18}\delta}{({}^{18}\delta_p - {}^{18}\delta)} \tag{27}$$

Set the definite conditions as following:

$$^2\delta \big|_{18\delta = 18\delta_0} = {}^2\delta_0 \tag{28}$$

Eq. (27) and Eq. (28) composed definite problem:

$$\begin{cases} \dfrac{d^2\delta}{({}^2\delta_p - {}^2\delta)} = \dfrac{d^{18}\delta}{({}^{18}\delta_p - {}^{18}\delta)} \\ {}^2\delta \big|_{18\delta = 18\delta_0} = {}^2\delta_0 \end{cases} \tag{29}$$

Solved the problem Eq. (29), we obtained:

$$^2\delta = \left(\frac{{}^2\delta_0 - {}^2\delta_p}{{}^{18}\delta_0 - {}^{18}\delta_p}\right)({}^{18}\delta - {}^{18}\delta_p) + {}^2\delta_p \tag{30}$$

Eq. (30) is the analytic expression of hydrogen and oxygen stable isotope relation line of unconfined steady flow system when recharging from rainfall at vertical direction.

2.2　Hydrogen and oxygen stable isotope composition's variation of unconfined steady flow system along the flow direction when evaporating of phreatic water exist

2.2.1　Assumptions of the model

Assume that the flow and the isotopic composition in the unconfined aquifer are as follows: ①unconfined flow is in a steady state (constant flow); ② flowing water's stable isotope is completely mixed in the vertical direction and is no proliferation in the longitudinal direction; ③water hydroxide isotopic composition is in a steady state; ④the fractionation factor of hydrogen and oxygen stable isotope is a constant during the water's evaporating process; ⑤the evaporation rate is a constant; ⑥phreatic Water's velocity is so high that stable isotopes' change caused by water – rock interaction can be ignored.

2.2.2　The model's derivation and solution

Take the micro water body located in $\left[x - \dfrac{\Delta x}{2}, x + \dfrac{\Delta x}{2}\right]$ as the research object, establishing heavy isotopes' mass conservation equation of the micro water body considering the inflow and outflow of heavy isotopes:

$$\mu H(x)\Delta x R(x) + \left[q(x) - \frac{dq}{dx}\frac{\Delta x}{2}\right]\left[R(x) - \frac{dR\Delta x}{dx\,2}\right]\Delta t -$$
$$\left[q(x) + \frac{dq}{dx}\frac{\Delta x}{2}\right]\left[R(x) + \frac{dR}{dx} \cdot \frac{\Delta x}{2}\right]\Delta t - \varepsilon\Delta x\Delta t\alpha R(x)$$
$$= \mu H(x)\Delta x R(x) \tag{31}$$

where, ε is the rate of evaporation.

Both sides of Eq. (31) is divided by $\Delta x \Delta t$, finishing:

$$\frac{\mathrm{d}(qR)}{\mathrm{d}x} + \varepsilon \alpha R = 0 \tag{32}$$

Expand Eq. (32):

$$q\frac{\mathrm{d}R}{\mathrm{d}x} + R\frac{\mathrm{d}q}{\mathrm{d}x} + \varepsilon \alpha R = 0 \tag{33}$$

For unconfined steady flow system with evaporation of phreatic Water, there is the continuity equation:

$$\frac{\mathrm{d}q}{\mathrm{d}x} + \varepsilon = 0 \tag{34}$$

Substituting Eq. (34) into Eq. (33):

$$q\frac{\mathrm{d}R}{\mathrm{d}x} + \varepsilon(\alpha - 1)R = 0 \tag{35}$$

Substituting Eq. (15) into Eq. (35) and finishing:

$$q\frac{\mathrm{d}\delta}{\mathrm{d}x} + \varepsilon(\alpha - 1)(1 + \delta) = 0 \tag{36}$$

Assume boundary conditions of flow field are the same as Eq. (8) and Eq. (9) and aquifer bottom plate is horizontal, solved the following definite problem:

$$\begin{cases} q = -K\dfrac{\mathrm{d}h}{\mathrm{d}x}h \\ \dfrac{\mathrm{d}q}{\mathrm{d}x} + \varepsilon = 0 \\ h\big|_{x=0} = h_1 \\ h\big|_{x=l} = h_2 \end{cases} \tag{37}$$

We obtain:

$$h = \sqrt{h_1^{\,2} + \frac{h_2^{\,2} - h_1^{\,2}}{l}x + \frac{\varepsilon}{K}(x^2 - lx)} \tag{38}$$

$$q = K\frac{h_1^{\,2} - h_2^{\,2}}{2l} + \frac{1}{2}\varepsilon l - \varepsilon x \tag{39}$$

Order:

$$M = K\frac{h_1^{\,2} - h_2^{\,2}}{2l} + \frac{1}{2}\varepsilon l \tag{40}$$

Then Eq. (39) can be written as:

$$q = M - \varepsilon x \tag{41}$$

Then substituting Eq. (41) into Eq. (36), we have:

$$(M - \varepsilon x)\frac{\mathrm{d}\delta}{\mathrm{d}x} + \varepsilon(\alpha - 1)(1 + \delta) = 0 \tag{42}$$

Assume isotope concentration field's boundary condition is the same as Eq. (20), there is a definite problem:

$$\begin{cases} (M - \varepsilon x)\dfrac{\mathrm{d}\delta}{\mathrm{d}x} + \varepsilon(\alpha - 1)(1 + \delta) = 0 \\ \delta\big|_{x=0} = \delta_0 \end{cases} \tag{43}$$

The solution of the problem Eq. (43) is:

$$\delta(x) = (1 + \delta_0)\left(1 - \frac{\varepsilon}{M}x\right)^{\alpha - 1} - 1 \tag{44}$$

Apply Eq. (44) to hydrogen and oxygen stable isotope:

$$^2\delta(x) = (1 + {}^2\delta_0)\left(1 - \frac{\varepsilon}{M}x\right)^{2\alpha - 1} - 1 \quad (M = K\frac{h_1^{\ 2} - h_2^{\ 2}}{2l} + \frac{1}{2}\varepsilon l) \tag{45}$$

$$^{18}\delta(x) = (1 + {}^{18}\delta_0)\left(1 - \frac{\varepsilon}{M}x\right)^{18\alpha - 1} - 1 \quad (M = K\frac{h_1^{\ 2} - h_2^{\ 2}}{2l} + \frac{1}{2}\varepsilon l) \tag{46}$$

Eq. (45) and Eq. (46) are the analytic expression of hydrogen and oxygen stable isotope composition's variation of unconfined steady flow system along the flow direction when evaporation of phreatic water exists.

2.2.3 Hydrogen and oxygen stable isotope's relationship of unconfined steady flow system when evaporation of phreatic Water exists (analytic expression of evaporation line)

Used Eq. (36) in oxyhydrogen stable isotope respectively:

$$d^2\delta = -\frac{\varepsilon}{q}({}^2\alpha - 1)(1 + {}^2\delta)dx \tag{47}$$

$$d^{18}\delta = -\frac{\varepsilon}{q}({}^{18}\alpha - 1)(1 + {}^{18}\delta)dx \tag{48}$$

Eq. (47) /Eq. (48), we have:

$$\frac{d^2\delta}{d^{18}\delta} = \left(\frac{{}^2\alpha - 1}{{}^{18}\alpha - 1}\right)\frac{1 + {}^2\delta}{1 + {}^{18}\delta} = K_\alpha\frac{1 + {}^2\delta}{1 + {}^{18}\delta} \quad \left(K_\alpha = \frac{{}^2\alpha - 1}{{}^{18}\alpha - 1}\right) \tag{49}$$

Assume isotope concentration field's definite condition is the same as Eq. (28), there is the definite problem:

$$\begin{cases} \dfrac{d^2\delta}{d^{18}\delta} = K_\alpha\dfrac{1 + {}^2\delta}{1 + {}^{18}\delta} \quad \left(K_\alpha = \dfrac{{}^2\alpha - 1}{{}^{18}\alpha - 1}\right) \\ {}^2\delta\big|_{{}^{18}\delta = {}^{18}\delta_0} = {}^2\delta_0 \end{cases} \tag{50}$$

The solution of problem Eq. (50) is:

$$^2\delta = (1 + {}^2\delta_0)\left(\frac{1 + {}^{18}\delta}{1 + {}^{18}\delta_0}\right)^{K_\alpha} - 1 \quad \left(K_\alpha = \frac{{}^2\alpha - 1}{{}^{18}\alpha - 1}\right) \tag{51}$$

From above analytic expression of the $^2\delta - {}^{18}\delta$ relationship line, we can see that the function form of the relationship line's analytic expression is different when phreatic Water system is in the conditions of precipitation recharge and of evaporation discharge . The former is linear, the latter is nonlinear. This reflects the complexity of the relationship between the hydrogen and oxygen stable isotope composition.

Typically, no matter the surface water (such as lakes, rivers) or groundwater (such as phreatic Water, confined water) , no matter evaporated water or precipitation its relationship is expressed by the linear expression $^2\delta = s \times {}^{18}\delta + d$, the reason for that maybe were : based on the observation data of precipitation in many parts of the world, the linear expression is a linear regression equation that derived by Craig (1961) and the Dansgaard (1964) using statistical methods. But its correctness is not supported by a strict mathematical proof.

Confined water's $^2H - {}^{18}O$ relationship Eq. (29) ~ Eq. (30) and Eq. (50) ~ Eq. (51) derived in several ideal conditions based on the isotopic mass conservation principle and groundwater dynamics theory showed that their relationship is strongly influenced by the initial conditions. It reveals that the linear relationship expression $^2\delta = s \times {}^{18}\delta + d$ is not the exact expression of the phreatic Water system's stable isotope relations even in the ideal case. So that the reasonable infer is that for the real natural phreatic Water system with more complexity, its $^2\delta - {}^{18}\delta$ relationship can not obey this simple linear rule.

The defects of the linear relationship expression $^2\delta = s \times {}^{18}\delta + d$ in the phreatic Water system at least show that the linear relationship $^2\delta = s \times {}^{18}\delta + d$ as expression of Meteoric Water Line (MWL) and Evaporation Line (EL) for different kinds of water body is not necessarily a clear and reasonable formula in the field of isotope hydrology.

Although in the model's derivation process, a number of secondary factors are simplified or ignored. Such as: in model (2.2), fractionation factors are set to be a constant, but in fact they are variables that related to temperature and humidity. In addition, isotope concentration field and flow field are limited to be in steady state in this discussion, some other factors are also simplified or ignored correspondingly: ①flowing water's stable isotope is completely mixed in the vertical direction and is no proliferation in the longitudinal direction; ②phreatic Water's velocity is so high that stable isotopes' change caused by water – rock interaction can be ignored. However, this does not affect the basic conclusions of this paper.

The differential equation model's establishment and the analytic solution's exportation of hydrogen and oxygen stable isotopic composition in unconfined aquifer provides a theoretical basis for a more objective and quantitatively interpretation of hydrogen and oxygen stable isotope data in the phreatic Water system, for understanding $^2H - ^{18}O$ relationship line's physical basis and influencing factors, for revealing information that the $^2H - ^{18}O$ relationship line contained.

References

Friedman I. Deuterium Content of Natural Waters and Other Substances [J]. Geochim. Cosmochim. Acta, 1953,4: 89 – 103.

Craig H. Isotopic Variations in Meteoric Waters[J]. Science, 1961, 133: 1702 – 1708.

Dansgaard W. Stable Isotopes in Precipitation[J]. Tellus, 1964, 16(4): 436 – 468.

Yurtsever Y. Worldwide Survey of Isotopes in Precipitation[R]. Vienna: IAEA report, 1975.

Chen J S. Wang J Y, Zhao X, et al. Study of Groundwater Supply of the Confined Aquifers in the Ejin Basin Based on Isotopic Methods[J]. Geological Review, 2004, 50(6): 649 – 658.

Gu W Z, Lu J J, Xie M, et al. Environmental Isotope Study of the Groundwater Resources in the North Wulan – Buhe desert, Inner Mongolia[J]. Advances In Water Science, 2002,13(3): 326 – 332.

Gu W Z, Lu J J, Tang H X, et al. Challenges of Basin Study to Traditional Hydrological Conceptions: the 50 Years Anniversary of Hydrological Bas in Study of PRC and the 20 Year's Anniversary of Chuzhou Hydrological Laboratory[J]. Advances In Water Science, 2003, 14 (3): 368 – 378.

Tong H B, Chen J S, Wang J Y. Differential Equation Model for River Water with Hydrogen and Oxygen's Stable Isotope Composition [J]. Advances In Water Science, 2007 (04): 552 – 557.

Tong H B, Fu X D, Chen J S. A Panel Model of Isotope Concentration – flow Coupled Field in Groundwater Systems[J]. Journal of Chongqing University, 2011, In press.

Willem G Mook. Environmental Isotopes in the Hydrological Cycle: Theory, Methods, Review [C]// Environmental Isotopes in the Hydrological Cycle. Vienna: IHS of IAEA, 2001.

Xue Y Q. Principles of Groundwater dynamics [M]. Beijing: The Geological Publishing House,1986.

Rui X F. Principles of hydrology[M]. Beijing: China WaterPower Press, 2004.

Saxena R K. Oxygen – 18 Fractionation in Nature and Estimation of Groundwater Recharge[R]. Uppsala: Fyris – Tryck AB, 1987.

Application of 3G Video Monitoring Technology to the Yellow River Basin Management

Wang Huailing, *Wang Peng*, *Qin Wenhai*, *Sun Jian* and *Zhang Yanli*

Information Center of Yellow River Conservancy Commission of Ministry of Water Resources, Zhengzhou, 450003, China

Abstract: 3G video monitoring systems is to realize the visualization management in efficient way, especially in the flood control work, making accurate command and flat conductor become possible. This paper focuses on the 3G video monitoring visualization platform realization of transmission and video system technology, and application introduction of 3G video monitoring technology in the Yellow River. Finally, this paper forecasts the application prospect of video monitoring technology in the Yellow River.

Key words: 3G, wireless video monitoring technology, visual management, basin management, the Yellow River communication network

1 Summary

The Yellow River management and development has gained huge success since 1949, to promote economic and social development of the basin. However, the Yellow River is particularly difficult to be managed, the work is still facing some problems. Construction of water conservancy is obviously backward, security level is apparently lower, compared with the requirements of economic and social development. Reform development task is very formidable. Therefore, in the early stage on the basis of information construction, to strengthen information construction and management of the basin is very important, considering the basin's business development needs and characteristics of water conservancy information technology.

Ice condition information is an important basis of command and decision in the Yellow River. The river reaches of Inner Mongolia have occurred 9 times ice flood burst since 1986, the Yellow River ice has become a top priority of China's winter and spring flood prevention work. Ice condition – prone areas in the Yellow River Basin are located in the Ningxia and Inner Mongolia upper reaches of the Yellow River, also Henan and Shandong reaches. Main flood control units are mostly located in the two sides of the rivers and floodplain thoset have characteristics of scattered, isolated, multi – point and wide range with small amount of information of the single point of communication. Before 2010, the Yellow River ice condition information collection mainly depended on artificial forecast means those were backward of information transmission and worse real – time mobility. When faced an emergency situation, it can not be pass site information go to higher command authorities at the first time, unable to meet multi – point real – time monitoring requirements during the flood control. According to ice changes of the Yellow River flood and the new demands of the ice control, it is necessary to carry out the Yellow River flood control capacity building.

The 3G (3rd – generation) is the third generation of mobile communication technology, which belongs to the cellular mobile communication technology, can support data, images, audio, video and other multimedia business data transmission in high speed. In the 3G business coverage area, the 3G technology could be used to realize the key river and important projects of the remote video monitoring as the visualization management platform.

In the June 2010, Yellow River Conservancy Commission (YRCC) of the Ministry of Water Resources (MWR) issued a notification that "developing flood control information transmission test of 3G technology for the 3G technology application of flood control". A lot of works about early research and field survey have been done in the past time. And the summary of the past video monitoring system construction management experience has been studied. From the November 2010 to the January 2012, the video surveillance system of the Yellow River in the Inner Mongolia section had been launched, including the emergency building of video surveillance system in the Inner Mongolia section of the Yellow River, improving the forecast in advance and the accuracy of monitoring. The Yellow River 3G video monitoring system contains seven fixed monitoring points and one mobile monitoring point, which is the important part of the initial construction of a three – dimensional monitoring system at the Ningxia and Inner Mongolia section of the Yellow River, including to use the satellite remote sensing, UAV and artificial monitoring.

2   The Yellow River 3G video monitoring system

In order to meet the needs of high – definition Remote monitoring video in the Yellow River ice flood control, using advanced 3G mobile communications technology and video compression technology to achieve real – time monitoring of the key river reaches, to protect the information of flood control and video fast, safe, reliable delivery, to realize flood control and monitoring information ' standardized coding, diversification of collection, network transmission and integrated management'.

Real – time information of wireless video site has been sent to the monitoring center, the various command of monitor reach the wireless video site through transmission network. Remote video monitoring system is composed of wireless video surveillance, transmission network and video monitoring center.

Video system uses a two – monitor mode to realize that the bureau of hydrology, flood control office and levels of management of the YRCC share information, see Fig. 1.

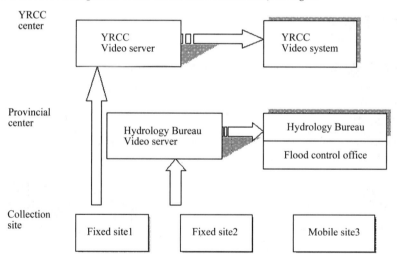

Fig. 1   The data flow diagram of the Yellow River 3G video system

2.1   The choice of video surveillance sites

The choice of video surveillance site should be able to fully reflect the Yellow River flood ice features; also it includes geographical location, geographical features, the conditions of observation, communication conditions, safety conditions and other conditions. The principle of selected sites is as follows:

(1) The video points should prefer which is easy to form great ice.

(2) The video points should prefer which is closed to the main river channel, in order to observe the whole picture of the main channel.

(3) The video points should prefer which is closed to hydrological station and river management segment, in order to facilitate maintenance and management in the future.

According to the above principles, we had selected seven fixed monitoring points, including Sanshenggong, etc.

2.2　The realization of the transmission subsystem

Transmission channel is using the latest 3G wireless access. The business of remote video monitoring system mainly concentrated in the uplink channel, one channel image (in CIF format) need about 300 kbps bandwidth. The theoretical bandwidth of China Mobile TD – SCDMA uplink channel is 384 kbps, practical bandwidth is about 128 kbps or so. It can not meet the transmission requirement. Therefore, according to situation of the site's signal coverage, 3G channel should prefer CDMA2000 of China Telecom and WCDMA of China Unicom. As shown in Tab. 1, China Unicom WCDMA uplink bandwidth is maximum, on the basis of field test, WCDMA is priority preferred. Also the theory uplink bandwidth of China Telecom CDMA2000 can meet the requirements, but its actual bandwidth is far lower than the theoretical value because of its many users. We have used CDMA2000 double card binding mode to increase bandwidth in order to meet the requirements.

Tab. 1　Transmission system comparison

Telecom operators	The type of 3G	Upstream bandwidth	Downstream bandwidth
中国移动通信 CHINA MOBILE	TD – SCDMA	128 kbps	2.8 Mbps
中国电信 CHINA TELECOM	CDMA2000 EVDO	1.8 Mbps	3.1 Mbps
中国联通 CHINA UNICOM	WCDMA HSPA	5.75 Mbps	14.4 Mbps

Communication subsystem includes access and transmission (Fig. 2).

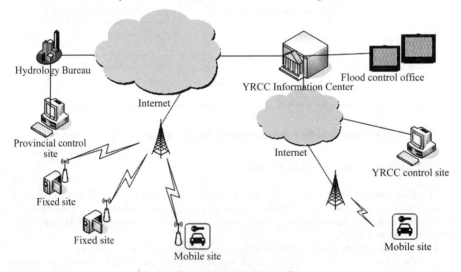

Fig. 2　Transmission scheme diagram

2.3　The realization of the video subsystem

The video subsystem is based on the YRCC video platforms, so it needs to upgrade the original platform. The original platform of the YRCC is based on network video surveillance systems using the video server, its encoding is H. 264. In order to guarantee the smooth upgrade of the platform, the video platform technology route remains unchanged.

2.3.1　Video surveillance technology

Video surveillance technology has been the focus of people's attention, is widely used in various control situations with its intuitive, easy and informative. Video surveillance systems can be broadly divided into three types to see Tab. 2 below.

Tab. 2　Comparison table of the video surveillance systems

System name	System classification	Monitor range	Bandwidth demand	Access mechanism	System stability	System open	System security
Local analog video system	Analog	Small	Higher	Direct access mode	Good	No	Medium
PC – based multimedia video system	Digital	Smaller	High	Direct access mode	Worse	Worse	Medium
Video server – based network video system	Digital	Wide	Lower	Client – server mode	Good	Good	High

The video proposed site is located in the levee of the Yellow River, its network access means is limited, its transmission distance is far away, its monitoring environment is poor and security is higher. To compare the advantages and disadvantages of three types of image monitoring system, technology route of video surveillance system is still selected network video system that based on the video server.

2.3.2　Video compression technology

Video compression technology is an important pivotal factor that is aimed at definition and bandwidth of video system, so its choice is very important.

Video compression technology is also known as video encoding technology. In video surveillance applications, complex algorithms may produce a higher compression ratio, but it can also lead to huge computational overhead, thereby it affects the real – time monitor system. Wireless access technology affected by the environment, transmission parameters dynamic change, stability of the transmission channel is poor. Network transmission that based on TCP / IP has many uncertainties, these problems require that video stream is provided with formidable function of flow control and error detection and correction.

In many video coding algorithms, H. 264 apply to network remote monitoring system in wireless environment. It has a high compression and high – quality smoother image.

For considering the current situation of communication and computer network in the Yellow River, YRCC still use H. 264 video system with high – definition, high – transmission performance, small – occupied bandwidth, in older to use minimum bandwidth to get the best video effects. The research finds that the products used varies widely, mainly due to the technical level of equipment providers, and that there is no direct relationship between compression formats. When we select equipment that should need to be tested (in remote transmission conditions).

2.4   Center video platform

Center video platform realizes compatibility of the old and new systems and enhances system functions by expanding and upgrading the original ice – video system platform. By center management service system and the online video service system, it completes all levels of user authentication, video management, alarm management and remote management of video equipment, completes video and audio access, forwarding, control, conversion display of analog video. By development of humane user interface, you can use the client or web browser to access the video management server, to realize exchange, remote control, video forwarding in wireless monitoring resources at all levels of platform.

Center video platform includes video management server, the video decoder, streaming media server, video management client, the system software and the display unit.

2.5   System compatibility mode

The YRCC video system belongs to remote video monitoring system. Research finds that current mainstream remote video system is in three ways: DVR, network video server, network camera.

Different implementation will result in not compatible between different systems, to realize old and new systems compatibility, to docket two or more video platform systems. For example, the docking is that the new platform dockets platform has been built, both platforms are docking. After docking, the new platform realizes the seamless connection of both platforms, manages and monitors the built video points through the built platforms. Platform for docking in three ways: direct docking, gateway docking and SDK docking.

2.5.1   The way of direct docking platform

Platform for direct docking requires that both of platforms abide by uniform standard, and the standard must have maneuverability.

Platform direct access way is applicable to apply the new system; there are conditions to upgrade the system platform, in which the new system should become mandatory requirements.

2.5.2   The way of docking gateway

Video imaging system of non – standard protocol and stream can use the video gateway to realize conversion, the function of video gateway is the analysis of video stream, networking protocols and realization of conversion function for stream and protocol.

The video gateway can be deployed in front video access centers, and in center video management platform. The premise of this way is that need to open stream format and networking protocols of this video system.

Video gateway can use third – party devices, or under the premise of open stream standard and protocol of network, to realize the conversion of factory private stream format and protocol into standard stream and protocol.

The advantages of using video gateway are that video image can access standard stream and the protocol, in favor of strengthen the reliability of video management platform.

2.5.3   The way of docking SDK

Equipment agreement is non – standard protocol, it needs equipment SDK and device – side protocol stream conversion, then access to the platform through the protocol stream.

The agreement is varied, so the equipment manufacturers need to provide the following:

(1) The official version of the device SDK development Kit and decoding library.

(2) The Demo program for test.

When the manufacturers of built platform provided equipment SDK, who should ensure the availability of the SDK, and access to platform by the protocol stream conversion after the testing and inspection. The access way has the advantage of that this providers do not need to work platform development, only to provide a device SDK and coordinate the new platform development.

3 Promotion and application

In November 2010, the first time ice regular meeting of YRCC was held in Zhengzhou, 3G video monitoring system returned ice condition pictures of Sanhuhekou and Huajiangyingzi to Zhengzhou at the first time, completed large screen – display for flood control by the hardware decoder in the lobby of flood control. The picture is smooth, clear and vision during monitoring period. The YRCC leadership fully endorsed the 3G remote video systems that play an important role in ice work, and make further instructions: on the basis of the full research, the construction of video surveillance systems on the Inner Mongolia reach should be further carried out at the appropriate site selected.

The 3G video surveillance technology is launched step by step, it promotes the use in the Yellow River. We have built successively seven ice video points in 3G coverage area of the Inner Mongolia reach, built six mobile video surveillance in Henan, Shandong, Shanxi, Shaanxi and Inner Mongolia , and five video surveillance in four reservoirs of the Yellow River. The video system gradually established has become of an important means of all – weather and real – time monitoring for the flood and ground inspections, to provide effective information support for flood control, decision – making, disaster prevention and mitigation.

On the basis of the 2009 UAV monitoring, using new technologies of satellite remote sensing, UAV and 3G, we launched research and construction applications for the ice condition of the Yellow River. UAV technology have the advantages of long duration, real – time image transmission, high – risk areas detection, low – cost, flexible, using recycled, avoiding casualties, etc. It is a strong complement to satellite remote sensing and manned aviation remote sensing, further meets the needs of the ice work for timeliness and effectiveness. In 2011, the ice condition of the Inner Mongolia reach was overseen at the first time by using of UAV and 3G technology, using the characteristics of UAV keep watch over the target in Sanhuhekou, and ensuring the key sections and key parts monitoring work. The UAV surveillance video was provided to the YRCC. The image is shown in Fig. 3, Fig. 4.

Fig. 3   UAV&3G video surveillance terminal

Fig. 4　Video image of UAV

The 3G video surveillance technology in the Yellow River has been successfully applied, but most of the ice flood control section not covered by 3G signal, can not use 3G technology to achieve real – time video surveillance. We had been built microwave communications networks and broadband wireless access system in the lower reaches of the Yellow River in Henan and Shandong Provinces, also will build the YRCC satellite communication system in 2012. This will provide an effective information transmission channel for remote video. In the future remote video surveillance system construction will mainly rely on the Yellow River communication network, as an effective means of access, the 3G is a beneficial supplement to the Yellow River communications network. Communication network of YRCC and 3G not only could be used in video monitoring business and flood reporting station as access channel, but also in checking risk situation, water politics, engineering maintenance and other fields. It could provide wireless network access means for the hydrological station along the Yellow River, monitoring station, and engineering maintenance units, solving the last kilometer problem and improving the automation level of the Yellow River basin management.

References

Li Guoying. Digitalization of the Yellow River[J]. Yellow River, 2001(11).
Wang Jiayao, Zou Haiyan. Remarks on Digitalization of Yellow River[J]. Yellow River, 2002.
Zhang Jianhua, Wang Ying. WCDMA Wireless Network Technology [M]. Beijing: Posts & Telecom Press, 2007.
Jack k. Video Demystified[M]. Firth Edition. Beijing: Posts&Telecom Press, 2009.
Jiang Wei, Meng Limin. Design of H. 264 Video Monitoring System Based on 3G Network Transmission [J]. Journal of Hangzhou University of Electronic Science and technology, 2011 (5).

Research on Pipe Line Flowmeter Measurement and Calibration Technology[①]

Wu Xinsheng, *Chen Jin*, *Wang Li*, *Feng Yuan*, *Liao Xiaoyong*
and *Wei Guoyuan*

Yangtze River Scientific Research Institute, Changjiang Water Resources Commission,
Wuhan, 430010, China

Abstract: Water metering is an important measure of fundamental technique for regulation, control and management of water resources. Since the difficulties so far of the offline measurement and calibration for the water meters on the pipelines taking water from rivers and lakes, the basic principles and characteristics of the field calibration method and devices were researched on, by investigating the flowmeter online calibration method in other fields, and the research status and application prospects in China and abroad, the tests were carried out with portable ultrasonic flowmeter in river model and sites, and compared with other flow measurement devices by measurement analysis, a flowmeter calibration technique has been presented for water conservation industry based on the comparison to apply to water industry site piping. It is focusing on the technical requirements, measurement process and calculation methods of online calibration with portable ultrasonic flowmeter in the actural flow. The result provides technical support to effectively carry out water measurement traceability and pipe flowmeter field calibration for the water conservation industry.

Key words: flow measurement, on – line calibration, field comparison method, portable ultrasonic flowmeter

1 Overview

Water metering is an important foundation for technical measures for regulation, control and management of water resources, and the accuracy of flowmeter measurement is closely linked to energy conservation, cost accounting and experimental data, which is included in the catalog of mandatory measuring instruments by China. There are a large number of flow measurement instruments in China, and a general lack of site online test or calibration measures, especially a blank for pipeline flowmeter used in the water concervancy industry. It involves the relevant technical standards for water metering, standardized equipment, values delivery and traceability, and is a very prominent problem. How to carry out the flowmeter field online calibration in actual flow and cycle test is a technical problem to be solved urgently of water resources management.

There are two flowmeter calibration methods of direct and indirect measurement, and calibration modes can be divided into offline and on line two ways (Cai Wuchang, 2006). The direct measurement method is also known as actual flow calibration method, that is, the actual fluid flows through the calibrated instrument, and then the flow rates are measured by the other standard equipment (standard flow meter or other standard flow measuring instruments) and the calibrated instrument for comparison , it is also known as wet calibration. The flow values obtained by this calibration method is both reliable and accurate, is the main method of flowmeter testing standards. From the sense of metrology, on – line actual flow calibration is the most consistent with the metering characteristics of accuracy, consistency, traceability and experimentatation.

The value of flow rate of flowmeter is usually calibrated and transferred in the manufacturer, however, for those measurement equipments of large diameter flowmeter, electromagnetic flowmeter

---

① Fund: the public welfare industry special research of Water Resources Ministry (201001049 – 02)

(including other types of turbine, vortex and ultrasonic flow meter and so on), which are on the pipelines taking water from rivers and lakes, due to the impact of the conditions or unconvenient equipment disassembly, such as water supply is not allowed to be stoped in large transmission pipes, there are many difficulties to offline test. So using field on – site calibration to ensure the accuracy of flow metering is particularly important and urgent.

A big additional errors of a flowmeter even certified by manufacturer will be caused by the appling environment and state at a site, this additional error can only be revealed at the site, and is difficult to determine even if taking a great effort to remove the flowmeter to test, while only online calibration can modify to the inherent error and the additional error together, it shows that even if the online calibration accuracy is not as high as by the laboratory test, but it has more practical significance. For this purpose, this article has researched and discussed in the technical requirement, measurement and calculation methods of portable ultrasonic flowmeter used for pipe flowmeter on – line actual flow calibration.

2　On – line testing and calibration status in China and abroad

Flowmeter on – line calibration service has been early in accordance with the international systems for conformity assessment of products and system certification, the most representative is the ISO9000, it is explicitly requested in the production process, from design, development, production, testing to use, it must be implemented "measurement assurance" required by measurement or calibration traceable to the international (SI) unit system. The unity of the value of flow amount is achieved by international device comparison.

2.1　Field calibration method

Field on – line calibration in actual flow includes the flow standard device method and the tracer method. The former is under the measurement conditions of site, let the fluid, which usally flows through the calibrated instrument, temporarily flows though a fixed (or mobile) standard device or a standard measurement instrument for comparison. The latter is to inject the tracer transiently into the fluid, and then measuring the distance required for tracer flowing after a period of time to calculate the flow amount. The application of tracer method on field flowmeter calibration in the world has reached a matureness, the international standard ISO2975 "Water Flow Measurement in Closed Conduits – Tracer Method" was presented in the 1970s, in China the tracer method with chemical tracer and isotope tracer have been explored in lab by the Industrial Metrology Department of Daqing Petroleum Institute, to verify the feasibility of implementation, it proves that the isotopic tracer is better. However, as environmental pollution, destruction of water quality by chemical reagents and isotopic tracers, it is rarely used in the water conservancy industry.

There is a field on – line dry calibration as well, originated in the 1970s, to solve the difficulty that the calibration in actual flow is unable to realize for large diameter electromagnetic flowmeter. It has been served as the content of industrial standard JIS "Flow measurement method of application of electromagnetic flowmeter" in Japan. From mid – 1990s, famous electromagnetic flowmeter manufacturers in the world have developed special testing equipment, the application of special instruments has been from the water conservancy industry extending to the petroleum, chemical and other process industries.

Dry calibration has been researched in China, such as the measurement principle, characteristics and implementation of the vortex electric field measurement method and the surface weight function method were analyzed in Zhejiang University, and so on. But since one standard is only applicable to flowmeters of one manufacturer, dry calibration method has greater limitations.

Besides, dry calibration is an indirect calibration method (Fu Xing, Hu Liang, Xie Haibo, et al., 2007) based on measuring the circulation area, the structure size and magnetic flux density B of the electromagnetic flow sensor parameters to calculate the flow rate value, this method is not intuitive, not a direct reflection of the water in the pipe.

2.2　Common standard equipment

Flow standard device generally comprises a set of tubes of standard volume or standard flow meters and the corresponding switching valves, you can implement the timing on line calibration of the flowmeter by it. When closing valve 1 and opening valve 2, the fluid through the calibrated flowmeter into the pipe of standard size to two pipe flowmeters rotational calibration, shown in Fig. 1. This approach has been basically implemented to the trade measurement of crude oil in China.

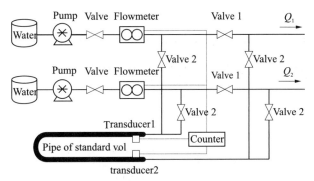

Fig. 1　Online calibration device with pipe of standard volume

While the mobile standard device is composed of special vehicle or trailer loading pipe of standard volume or the standard flowmeter, temporary access to the pipeline for measurement calibration. The truck – mounted pipe of piston type and of standard volume has been made by the U. S. company CSI, and the technology is also used in more than 20 years of experience in China. The product of truck – mounted pipe of spherical volume had been stereotyped by Shanghai Automation Instrumentation Factory; Air Force Oil Research Institute has developed a gas flowmeter calibration on airport "JLC – A type on – line inspection device", it has turbine flowmeter with high precision. Mobile vehicle flow standard device is of flexibility and convenience to test on site, but the cost of equipment is higher.

2.3　Comparison method of portable ultrasonic flowmeter

When using a flow metering standard device for calibration, the additional section of pipe for connecting with, and a switching control valve and other auxiliary equipment are required to be reserved, in fact, it is difficult to meet in most working sites of water supply. Therefore, according to their actual conditions, all water companies have explored a wide range of on – line comparision method and the indirect calibration methods: ① using weighing instrument comparision method in the process such as some container, pool or weighing apparatus, can obtain higher matching and precised result; ② according to the head – flow characteristics of the lift pump to estimate water transport quantity, but the accuracy is too low in this method; and ③ using the comparison method of portable ultrasonic flowmeter.

The comparison method of portable ultrasonic flowmeter is the most convenient among the above methods. The process of the comparison method of standard device is: taking the portable ultrasonic flowmeter with high accuracy as the standard, install it on the water pipe and test it synchronously with the original calibrated flowmeter, and then compare the two measured results , to calculate the relative error for the calibrated flowmeter according to the results, the relative error = (measured value – standard value) / standard value, and then adjust the coefficient of calibrated flowmeter by the calculated relative error, to make the data be consistent both to the calibrated flowmeter and the portable ultrasound flowmeter, and let the calibrated flowmeter be adjusted. The method is currently being used most by water supply enterprises.

However, on – line calibration faces many difficulties. Firstly, the calibrated objects are complex, the measurement conditions of the field are influenced by many factors, environment and time constraints are also very prominent problems to be solved. Secondly, new technology and new equipment is widely used in modern production, a variety of measuring instruments, not only high precision, and multi – parameter, intelligent, and constitute the system in most cases, and even joined the network. All these need to be addressed through a variety of experimental research and practical application.

3　Portable ultrasonic flowmeter calibration method

3.1　Principle and characteristics of portable ultrasonic flowmeter measurement

The ultrasonic flowmeter is a non – contact instrument, it will not change the fluid flow condition, no additional resistance, and can solve problems such as measurement difficulties for other types of instrument to measure corrosive, non – conductive, radioactive and flammable and explosive medium flow.

General flowmeter with the measurement of the diameter increasing will cause the difficulties in manufacturing and transportation, increasing cost, energy loss and the inconvenience of installation, and the cost of portable ultrasonic flowmeter is basically unrelated to the size of the measuring pipe diameter, the scope of application can be applied from a variety of pipes, narrow canals to natural rivers, is unmatched by other instruments. Especially some of the excellent features of portable clamp – on ultrasonic flowmeter is even more obvious: ① the sensor is bundled in the pipeline wall, does not affect the process, can be removable at any time, fast and simple, easy to carry to the sites, the instrument measurement and maintenance does not affect the operation of the pipeline system, cost – effective. The performance – price ratio is high. ② A sensor can be used for a variety of liquid media, measurment diameter of pipe can be from tens of millimeters to a few thousand mm.

Most of the time difference ultrasonic flowmeter uses the time difference by the sound waves spreaded downstream propagation and upstream propagation to measure the fluid flow. The principle shown in Fig. 2, The time difference of the launch acoustic wave spreading from the sensor P_1 to P_2: $\Delta t = t_1 - t_2 = 2LV\cos\theta/c^2$, namely $V = (c^2/2L\cos\theta) \times \Delta t$, where c is the propagation velocity of the ultrasound in stationary fluid, by the calculation of time difference Δt, the flow rate of V can be got, and then, the volumetric flow rate Q can be obtained.

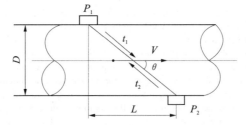

Fig. 2　Diagram of ultrasonic flowmeter measurement principle

3.2　The feasibility of online calibration by portable ultrasonic flowmeter

As we all know, on – site calibration by pipeline flow meter will generally meet difficulties in large diameter and large flow, since the method of liquid flow meter calibration has not been worked out yet in China, and there is no standard flowmeter with accuracy more than three times higher than the online flowmeter, the simple practice is to use a portable ultrasonic flowmeter as a standard device to do online calibration comparison, it is practical approach to large diameter flowmeter online calibration at the current technical conditions.

The requirements of measurement error for many water flow measurement of engineering is to reach as long as ±5% of the standard flow, and the accuracy of portable ultrasonic flow meter measurement is better than ±1%, so is conducive to the site on – line calibration. compare to other industrial measurement, the appling conditions of water conservancy industry is no harsh conditions such as corrosive, radioactive and flammable and explosive medium, when appling the method to calibrate line to the drainage flow meter on line, only need a right operation, reduce random error and additive error as far as possible, and only at a specific point in the flow measurement, the use of the approach of fixing specific points to improve efficiently the accuracy of the standard device (Miao Yusheng, 2009), can achieve the accuracy requirements by the national authorities for the industrial and residential water metering.

3.3 Technical requirements of the standard flowmeter

Portable ultrasonic flow meter used for on – line calibration should be tested in the national legal metrology institutes, and with a valid certification. A flowmeter to be calibrated should be accompanied by an instruction manual, a flowmeter calibrated periodicly should also have the last calibration certificate. The other auxiliary equipments for online calibration, such as tape measure, thickness gauge, stopwatch all should have a valid test certificate.

The state and parameter settings of the flowmeter to be calibrated should be aware before calibration, and make sure clearly that the measured pipeline operating parameters and the temperature, humidity, external magnetic field, the mechanical vibration and ambient conditions of the site, should be consistent with the working conditions of field calibration.

3.4 Measurement and calculation methods

(1) The diameter measurement. Measur the perimeter outer the pipe with gauge near the location where sensor installed several times, and then get the average one as the outer perimeter.

(2) Measurement of wall thickness. Measur the thick of pipe with thickness gauge at and near the location where sensor installed, then get the average one.

(3) The standard meter installation. The operation should be strictly according to the manual, at the first to determine the measurement location on the pipe section to be installed a meter and clean up the pipe wall, and then coated with a coupling agent in the sensor and the pipe measuring point, to make the emitting surface in close contact with the pipe wall. After meter is installated, turn the power on, enter the pipe parameters (material, roughness, temperature, etc.) in, to get the accurate installation distance for sensor to be adjusted and tighten.

(4) Zero point check. If the flow in pipeline can be stopped, the zero flow check is carried out for the flowmeter; if not, dynamic zero check for the standart flowmeter may be conducted before calibration.

(5) Calibration of measurement performance. Read the flow values shown by the flowmeter to be calibrated and the standard device respectively within a certain time, then calculate by Eq. (1) ~ Eq. (3) to get the flowmeter indication error and repeatability (Li Changwu, Zhang Dongfei, Yuan Ming, et al., 2009):

$$E_{ij} = \frac{q_{ij} - (q_s)_{ij}}{(q_s)} \times 100\% \tag{1}$$

$$E_i = \frac{1}{n} \sum_{i=1}^{n} E_{ij} \tag{2}$$

$$(E_r)_i = \sqrt{\frac{1}{(n-1)} \sum_{j=1}^{n} (E_{ij} - E_i)^2} \times 100\% \tag{3}$$

where, E_{ij} is the relative indication error of each flowmeter at each flow point of each calibration time; q_{ij} is the value showed by flowmeter (instantaneous value or cumulative value) at the $i-$th flow point of the j – th calibration; $(q_s)_{ij}$ is the standard value (instantaneous value or cumulative value) at the i – th flow point of the j – th calibration; E_i is the indication error of flowmeter at the i – th flow point; $(E_r)_i$ is repeatability of flowmeter at the i – th flow point.

The measurement deviations of the calibrated flowmeter is as follows:

$$E = \pm \mid E_i \mid_{max} \qquad (4)$$

where, $\mid E_i \mid_{max}$ is the maximum deviation value of flowmeter among all values measured at each calibration point.

It is nessecery to determine the number of calibrated points based on the actual situation at the site, generally opt for one to three, and calibrating three times for each flow point, the largest changes in the instantaneous flow does not exceed 5%. At each calibration, read and record the values showed by flowmeter and the standard device at the same time, if the reading values is the instantaneous value, at least 20 values is needed, then get the mean value; if it is the accumulated value, you should ensure that it is more than 1,000 times the minimum reading, or read the cumulative value for 20 minimum at least.

(6) Calibration results treatment. The measurement error $\mid E \mid$ and repeatability E_r of calibrated flowmeter should satisfy the Eq. (5) and Eq. (6), otherwise it should be to repair the meter or let it off – line to be sent to the laboratory for test or calibration by standard device.

$$\mid E \mid < \sqrt{\sigma_s^2 + \sigma_m^2} \qquad (5)$$

$$E_r < \sqrt{\sigma_s^2 + \sigma_m^2}/3 \qquad (6)$$

where, σ_s^2 is the maximum permissible error of standard flowmeter; σ_m^2 is the maximum permissible error of calibrated flowmeter.

4　Measurement and data analysis

4.1　Ways and means of comparison measurment

4.1.1　Test scheme arrangement

Measurement test is mainly in view of the accuracy and applicability of portable ultrasonic flowmeter. First of all we choose four sets of water pipe for flow controlling of the river model in the Yangtze River Scientific Research Institute, Changjiang Water Resources Comission to carry out, in order to control the test conditions and compare on a variety of test means. We selected six carbon steel pipes of diameters of $\Phi(100 \sim 400)$ mm in the experiment, in order to understand the effects of different diameter on the measurement accuracy possibly. In addition, we adjust the valve opening of the measured pipeline controlling system, in order to understand the impact of different flow on measurement accuracy, and use the method – V and method – Z for observation and analysis of measuring reliability.

Common installation of the sensor includes method – Z (direct type) and method – V (reflection type), shown in Fig. 3. Which installation method will be chosen depends on pipe size and properties of medium. It should be to observe the signal strength of indicator before the formal readings and records of the instrument measuring, generally higher signal strength is better. Compare the results of flow rate (or flow quantity) by installation method – Z and the method – V in test process, if the values are quite different between two methods, there may be significant lateral flow, that is rotating flow phenomena should be noted, and take nessecery measures.

Flow quantities were compared and measured in water pipe using a portable ultrasonic flowmeter, an electromagnetic flowmeter and weir devices, in comparison measurment test, and shown in Fig. 4. The measurement procedure is for continuously supply water, when the values indecated by weir measuring needle and flowmeter indecatar are stability, then start the measurement synchronously to get readings of the three water metering facilities, to verify the accuracy of the portable ultrasonic flowmeter.

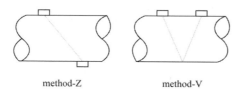

method-Z   method-V

Fig. 3   The two common installation of
ultrasonic transducer

Fig. 4   Three flow measuring device in
the pipeline of the model in
comparison measurment test

4.1.2   Test equipment

The new type mounted F601 portable ultrasonic flowmeter made in Germany is used, the measuring diameter range of a single set of sensors is: $\Phi(50 \sim 3\,400)$ mm; flow rate range: $0.01 \sim 25$ m/s; resolution: 0.025 cm/s; repeatability: 0.15% (depending on the application); the the volumetric flow rate accuracy (flow field is fully developed and is radial symmetry): $\pm 1\%$, $\pm 0.5\%$ (after calibration).

TT100 Ultrasonic Thickness Gauge is used for wall thickness measurement, the measuring range is: $0.8 \sim 300$ mm; resolution: 0.1 mm. There are 5 types of electromagnetic flowmeter including Krohne/DN100 $\sim$ 300 and other type installed on the flow pipe, the measurement accuracy is better than 0.5%; three rectangular weirs for flow controlling of model, the measurement accuracy is 1.5% $\sim$ 2.5%.

4.2   Test results and error analysis

After a series of experiments and statistical analysis, the results are obtained measured with the electromagnetic flowmeter on the water supply pipes in the Yangtze River flood control model, weir and comparison test, and shown in Tab. 1.

It is noted in experimental observations that the meter readings of the three measurement devices are constantly changing (two decimal places), and flowmeter indication changes quickly than the weir, is a true reflection of the pipeline flow situation. It shows the flow in the pipeline is turbulent, rather than the stabe flow for the standard calibration device in laboratory, which is the main reason for the calibration accuracy in field is lower than in laboratory. In addition, the conditions of the pipeline at site can not be ignored, such as the measurement length of straight section of the water supply pipe is not long enough, some large diameter pipes are reel pipe (shape of pipe cross – section is uncertain) and so on.

From the measurement site analysis shows that: flow errors of weir are mainly caused by the size of the error of hydraulic construction, but its measuring stability is good that can be used to be compared with, to verify that if the two other flowmeter is working properly. The measurement results show that, the work of the two flowmeters are normal, and the measured values are closer to each other, the relative error is smaller. Actually the accuracy of two kinds of flowmeter both higher than the one of weir, it infers the reliability of ultrasonic flowmeter is good. Since the accuracy of electromagnetic flowmeter is slightly higher than the ultrasonic flowmeter, and its installation is very normative, so the results of electromagnetic flowmeter are used as a standard to compare with in the analysis.

It can be seen by results of Tab. 1 that:

(1) When pipeline of $\Phi150$ mm is measured in method – V, the measurement error by ultrasonic flowmeter is 1% to 2% compared with the electromagnetic flowmeter.

(2) When pipeline of $\Phi200$ mm is measured in method – V, the measurement errors are generally between 1% and 2%, at a few points the errors <1% or over 4%, and it increases with

the flow increases, when in method – Z measurement , the error is significantly reduced.

(3) When pipelines of $\Phi300$ mm and $\Phi400$ mm are measured in method – V and method – Z, the minimum error $<0.5\%$, the maximum error is less than 5% , the majority are from 1.5% to 3% . It means that the error in method – Z is smaller than in method – V when measur on the pipe with larger diameter.

Tab. 1 Comparison results (ultrasonic flowmeter , electromagnetic flowmeter and weir)

No	Q of weir (L/s)	Reading of Elec. flowmeter (L/s)	Reading of ultrasonic flowmeter (L/s)	Relative error A (%)	Relative error B (%)	Relative error C (%)	Notes
1	46.22	48.10	47.30	2.34	4.07	−1.66	$\Phi150$, method – V
2	72.11	75.40	74.10	2.76	4.98	−2.11	
3	82.42	85.40	83.10	0.38	3.62	−2.69	
4	19.06	19.30	18.37	−3.62	1.26	−4.82	$\Phi200$, method – V
5	19.06	19.30	18.80	−1.36	1.26	−2.59	method – Z
6	44.30	44.60	44.40	0.23	0.68	−0.45	method – V
7	75.23	75.70	76.50	1.69	0.62	1.06	method – V
8	91.50	92.30	93.80	2.51	0.87	1.63	method – V
9	51.60	54.40	54.70	6.01	5.43	0.55	$\Phi300$, method – V
10	93.21	100.10	98.60	5.78	7.39	−1.50	
11	144.80	146.70	150.90	4.21	1.31	2.86	
12	53.00	53.60	53.60	1.13	1.13	0.00	$\Phi400$, method – Z
13	53.00	53.60	55.60	4.91	1.13	3.73	method – V

Note: Relative error A is the comparison result of ultrasonic flowmeter and weir, relative error B is about electromagnetic flowmeter and weir, relative error C is about ultrasonic and electromagnetic flowmeters.

Comprehensive analysis shows that the measurement accuracy of portable ultrasonic flowmeter under the ideal conditions can reach 0.5% to 1.0% , only if the conditions in the field can meet basic measurement requiements, the accuracy can be from 1.0% to 3.0% . The test also shows that the repeatability of measurement results of the F601 ultrasonic flowmeter used is quite good, and is the appropriate equipment for test.

5 Summary and discussion

From a series of verification test by comparison method using F601 portable ultrasonic flowmeter and electromagnetic flowmeter on pipeline flow measurement, and on – line flow meter calibration and verification technology, we have achieved the following results:

(1) The accuracy and applicability of a portable ultrasonic flowmeter in practical applications have been verified. Studies have shown that it is the effective and practical way of on – line calibration for the flowmeter at large diameter pipe under the present technical conditions, and can verificate or calibrate flowmeters with the same precision or below one on line by comparison, which provides an easy way of source tracing for water pipe flowmeter metering on site.

(2) The specific tests show that the original equipment installation and working conditions must be clear before the online field calibration, the following requirements should be fulfill: highly technical and experiential requirements for professional staff, high operational requirements, to

strengthen the technical training of inspectors management and data files, further accumulation of experience in practice, the formation of the technical specifications.

(3) Due to the diversity and complexity of site conditions, it will reduce the detection accuracy of portable ultrasonic flowmeter even can not carry out online calibration. In order to guarantee and improve comparison measurement accuracy, where it is needed, improve technically the original pipeline that reserve enough section of straight pipe and space on the field water pipeline, and install the stainless steel pipe in the location with the ideal measurement conditions or other reserved means, to improve the calibration of flowmeters on - site measurement conditions is necessary.

(4) Online field calibration is the developing trend of measurement technology, it can effectively reducing the cost of measurement. Online field calibration by portable ultrasonic flowmeter is not only feasible, but easy to promote the use of, and can greatly facilitate on - site metering measurement calibration.

This study is with some limitations in the research scope and depth yet, on the next phase the research and application will be carried out at more production sites, to make online flowmeter metering is more accurate and reliable, and more targeted.

References

Cai Wuchang. Field Calibration and Verification of Flowmeters[J]. International Instrumentation & Automatio, 2006(3): 56 - 60.

Fu Xing, Hu Liang, Xie Haibo, et al. Dry Calibration Technique of Electromagnetic Flowmeters [J]. Journal of Mechanical Engineering, 2007(6): 26 - 30.

Miao Yusheng. Targe - diameter Meter Test and Calibration in the Water Supply Industry[J]. City and Town Water Supply, 2009 (2): 5 - 7.

Li Changwu, Zhang Dongfei, Yuan Ming, et al. Study of Online Calibration Method for Lliquid Flowmeter[J]. China Measurement & Testing, 2009(5): 27 - 29.

Research of Water Dispatching Model Construction for the Water Resources Regulation and Management System of the Dongjiang River

Wei Yongqiang[1,2], *Zhao Yang*[2], *Xu Zhihui*[2] and *An Dong*[2]

1. College of Water Conservancy and Environmental Engineering of Zhengzhou University, Zhengzhou, 450001, China
2. Information Center of Yellow River Conservancy Commission of Ministry of Water Resources, Zhengzhou, 450004, China

Abstract: Water Resources Regulation and Management System is a very important part for the water resources management of the Dongjiang River Basin. During the process of the system construction, through the construction of the inflow forecast model, water demand prediction model, channel evolution model and optimization allocation model of water resources, the accuracy of water dispatching scheme is improved. This paper mainly introduces the system construction goal, system composition, model function and the principle, and the relationship between them. The construction of model will help to solve the three large reservoirs, rivers and interval scheduling, and ensure each city to optimize allocation of water resources between different industries on the situation of the shortage of water and its extremely uneven seasonal distribution. Furthermore it is very important to the economic development of the Pearl River Delta, prosperity and stability of Hongkong.

Key words: inflow forecast model, water demand prediction model, channel evolution model, optimization allocation model of water resources

1 Introduction

Dongjiang River originated in Yajibo Mountain, Wu County of Jiangxi Province, from northeast to southwest into the Guangdong Province. Before into the Lion Ocean, it flows through Longchuan County, Heyuan County, Zijin County, Huiyang County, Boluo County, Dongguan city, etc. The main stream of Dongjiang River is 562 km in length and drainage area is 35,340 km^2, in which 31,840 km^2, accounting for 90% of the total area, is in Guangdong Province.

Dongjiang River is the important water sources of Hong Kong, Shenzhen, Heyuan, Huizhou, Dongguan and other city, offering life and production water for 30×10^6 people (Zhu Shulan, et al.,2009). Water resources shortage of Dongjiang River Basin, especially in the middle and lower reaches, has existed for a long time. In recent years, with the high – speed development of the economy and society in the basin, water requirement increases dramatically, and the supply and demand of water resources is increased rapidly. Water environment pollution, cutting down of the river channel, salt water intrusion and other negative factors have been a serious threat to the safety of water supply in the medium and downstream area of Dongjiang River Basin. In this case, the government of Guangdong Province promulgated the " Guangdong Province water resources allocation scheme of Dongjiang River Basin" in August 2008. Dongjiang River Basin Management Bureau (hereinafter referred to as the Dongjiang Bureau) is responsible for water resources management and scheduling, water resources allocation scheme and the annual water diversion plan, coordinate the water dispute between administrative area, to avoid ecological deterioration. With the lack of water resource and water demand increasing, only by strengthening the management measures and rapidly improving dispatching and management level, we can scientific allocate water resources, maintain healthy life and sustainable develope the ecology and economy of the basin.

2 System construction target

The construction of water resources regulation and management system for the Dongjiang River is the combination of the actual situation and Dongjiang River Basin Management Bureau functions, on the existing basis, to establish a set of "advanced, practical, reliable, and efficient" water dispatching and management system in a relatively short time. On the basis of sustainable utilization of water resources, we can realize integrated development, optimize configuration, efficient use of water resources, overall plan the relation between the production, life and ecology, ensure and provide support for population, resources, environment and economy development. With the aid of the runoff forecasting model, water demand prediction model, optimal allocation model of water resources, river evolution model, we can rapidly draw up the feasible years, months water diversion planning and regulation scheme, visual simulate on different scheduling scheme, and make decision for selection.

3 System composition

The system is designed multilayer – tier structure based on the digital Yellow River, including presentation layer, application layer and data layer (as shown in Fig. 1).

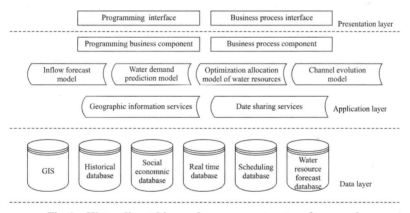

Fig. 1 Water dispatching and management system framework

The function of presentation layer is interface, used to display information, interact with decision making and decision makers through the man – machine interface.

The application layer is the system of the implementing agencies, including business component sub – layer, model sub – layer, public service sub – layer. Business component layer refers to the use of the judgment, logic and simple calculation. Model layer means developed models special for programming, including the inflow forecast model, water demand prediction model, channel evolution model and optimization allocation model of water resources. Public service layer is construction content of application service platform, which provides shared services for the business component layer and model layer.

The application layer is the system of the implementing agencies, including business component sub – layer, model sub – layer, public service sub – layer. Business component layer refers to the use of the judgment, logic and simple calculation. Model layer means developed models special for programming, including the inflow forecast model, water demand prediction model, channel evolution model and optimization allocation model of water resources. Public service layer is construction content of application service platform, which provides shared services for the business component layer and model layer.

The data layer means water dispatching database, including GIS data, historical data, social – economic data, real – time database, scheduling database, water resource forecast database. In which, scheduling database and water resource forecast database is generated from the water dispatching system, and the basis of data is by other system.

4 The function and principle of the model

The model was designed according to the practical principle, which main function has yearly, monthly inflow forecast function and water demand prediction function, scheduling optimization configuration function, as well as simulation to the configuration results, to provide basis for decision making.

4.1 The inflow forecast model

Because the basis of Dongjiang River Basin water allocation is mainly derived from the runoff forecast, the runoff forecast accuracy directly affects the accuracy of water dispatching.

The mid – long term inflow forecast model is based on multiple regression analysis method. By calling the historical hydrological data and rainfall data, through the optimization debugging, a set of model regression parameters is calibrated and three large reservoir inflow, interval inflow are predicted.

The factors for mid – long term inflow forecast are early flow and rainfall. The principle is as follows:

If the forecast object is , previous forecast factor, have been chosen, then multiple regression model is (Zhong Yonghua, 2011):

$$y = b_0 + b_1 x_1 + b_2 x_2 + b_3 x_3 + \cdots + b_m x_m$$

where, b_0, b_1, b_2, $\cdots$, b_m is the regression coefficient; y is the forecast value; x_1, x_2, x_3, $\cdots$, x_m is the forecast factor.

Only two factors is used in this model, so $m = 2$:

$$y = b_0 + b_1 x_1 + b_2 x_2$$

where, y is water quantity for dry season; x_1 is flood season water quantity; x_2 is flood season rainfall.

The estimation of b_0, b_1, b_2 is based on least square method by selecting historical data.

The generated water solution is stored in the water resources database, called by channel evolution model and optimization allocation model of water resources.

4.2 The water demand prediction model

Water demand prediction is the basis of water dispatching, and it depends on the regional economy development, society, population structure, industrial structure, and other elements. At present the methods are mainly time sequence method, grey prediction method, judgment model, quota method, et al. Non – agricultural water of basin accounts for the most, that is to say the demand of water resources is relatively stable. Water demand prediction is mainly based on the standard – year water demand of diversion project or water quantity to be determined in case of emergency, therefore the quota method is used to predict the annual water demand in Dongjiang River Basin.

Annual water demand prediction, according to the allocation indexes, water demand prediction model works by calling social and economy development data, life water quota, the second and third industry water quota of social economy and history database, it also supports water prediction of different administrative district in condition of population surge. Based on the water demand data analysis of Dongjiang River Basin, and modified by reffering to the research achievements of Zhongshan University, each industry water demand quantity and monthly water distribution ratio can be calculated under different water frequency.

The water demand data of each industry under different frequency list in Tab. 1.

Tab. 1 Annual water demand for each industry level   (unit: ×10⁴ m³)

Id	Region	F3	F4	...	F9	Second industry	Third industry	Ecology	Life
1	Heping	10,083.509	11,133.163	...	22,462.701	4,249.056,8	438.049,3	70	2,658.479,1

In the Tab. 1, from F3 to F9 is the data of agriculture demand water under different frequency, from 5% ,10% ,25% ,50% ,75% ,90% to 95%.

Monthly water demand prediction is the annual water demand prediction results allocated according to monthly allocation ratio, as shown in Tab. 2.

Tab. 2 Monthly water demand allocation for Abundant, dry and flat water years
(unit: ×10⁴ m³)

Region	Jan	Feb	...	Dec	Jan 1	Feb 1	...	Dec 1	Jan 2	Feb 2	...	Dec 2
Heping	8.8	4.1	...	4.1	11.7	3.6	...	9.3	6.5	4.5	...	5.8

In the Tab. 2, from Jan to Dec is the monthly allocation ratio of abundant water year, from Jan 1 to Des 1 is the monthly allocation ratio of flat water year, from Jan 2 to Dec 2 is the monthly allocation ratio of dry water year.

The result based on the above data is stored in the database, called by channel evolution model and optimization allocation model of water resources.

4.3 The optimization allocation model

Water allocation scheme of the Dongjiang River Basin is followed by integrated regulation according to the principle of total water quantity control, section water quantity control and water intake control, and the water dispatching objective is to ensure that daily average flow of Boluo station is not less than 320 m³/s.

Dongjiang River Basin's water allocation scheme based on the inflow forecast and water demand prediction, gives priority to life, industrial and agricultural production than in and outside the river channel, then simulates through the evolution model. when the simulation results are not consistent with scheduling boundary constraints, it is reconfigured until the recommend scheme is get, and the scheduling scheme is determined. Model relation is shown in Fig. 2.

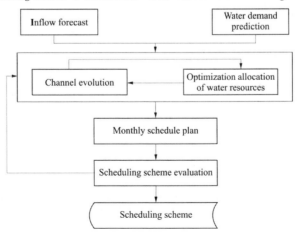

Fig. 2 Model relation

According to yearly scheduling scheme, incoming water quantity of the Boluo Station, the new monthly runoff prediction, reservoir operation scheme, information of rainfall and implementation of water regulation plan, to correct the discharges of three large reservoirs and monthly water allocation index time after time, the monthly schedule plan is worked out through optimizing configuration model.

Optimizing configuration uses multiple objective analysis method, according to the agricultural, industrial, domestic water weight rate, appropriate adjustment and balance of each user, to achieve the optimization objective (Chen Xiaohong, et al. , 2002) .

Objective 1: The ecological environment objective: to provide the necessary ecological and environment water, maintain the ecological system balance and healthy life of rivers. In order to solve the problem of water demanded for ecological system and river drying up, major control section (such as Heyuan station, Boluo station) must be ensured a certain flow to meet the ecological base flow.

Objective 2: Water resources utilization objective: the pursuit of the minimum amount of water, and the distribution is reasonable, mainly solving water allocation problems between different areas and sectors.

4.4   The channel evolution model

According to the three major reservoir flow and interval real – time water, the runoff process of control section is simulated. If minimum flow do not be required, then feedback to water resources allocation model, readjusting until required.

Channel evolution model is based on the geometry and hydrodynamic features of the main stream of Dongjiang River, using Saint – Venant equations to describe flow process, to calculate the flow, flow velocity and water level process of each section.

The calculation results compare with the measured process during the same period, debugging, with minimum error criterion to determine the parameters optimal value. Then it demonstrates the physical significance of the parameter, checking whether there is unreasonable phenomenon, and then makes further adjustments. Finally, certainty factor R of the flow velocity, the water level hydrograph fitting degree is used as target function to test the rate effect and determine optimizing configuration result.

5   Conclusions

When the water allocation model is designed, some of the parameters and coefficient can be adjusted properly, and decision makers can also adjust reservoir flow according to the experience, consider emergency water for a population surge, making the model more humane. The construction of water dispatching and management system of Dongjiang River Basin, provide technical support to the optimal allocation of water resources and alleviating the contradiction between supply and demand. It also provides a modern, intelligent decision support platform for management and water regulation, and for other river basins, it has guidance and reference value to water dispatching and the harmony between human and water under the complicated conditions of multiple sources, and multiple users.

References

Zhu Shulan, Chen Xiaohong, He Ling. Three Major Reservoirs in Dongjiang Watershed Runoff Regulation Effect Evaluation in Dry Season [J]. Guangdong Water Resources and Hydropower, 2009(8): 9 – 11.

Zhong Yonghua. The Method of Multiple Regression Analysis in Inflow Forecast Application of Miyun Reservoir[J]. Beijing Water, 2011(3): 12 – 14.

Chen Xiaohong, Chen Yongqin, Lai Guoyou. Optimal Allocation of Water Resources in Dongjiang River Basin[J]. Journal of Natural Resources, 2002(3): 366 – 372.

Realization of VR Technology – based Flood Control DSS

Ma Yidong , Wang Yuxiao and *Zhao Yusen*

Yellow River Henan Bureau, Zhengzhou, 450003, China

Abstract: A flood control DSS is established based on virtual reality technology (VR technology) which can incorporate computer – based real – time simulation of water regime in the Yellow River, flood disaster assessment, and establishment of emergency flood prevention planning in one. Virtual reality technology is a kind of high – end man – machine interface which integrates several technologies including computer graphics, graphic processing, pattern recognition, artificial intelligence, network technology and multimedia technology and can generate a vivid virtual environment and make people feel they are personally on the scene from visual sense, hearing and touch. It can overcome the shortcomings of the previous flood control decision system, such as low response, high cost, more staff involved, and high error ratio. The VR technology – based flood control DSS can incorporate the virtual reality technology, geographic information system, decision support system and expert system. The VR technology – based flood control DSS is easy to use and can make scientific decisions, have high accuracy and instantaneity, as well as strong maintainability and transportability; it is an important link in the process of transition from the prototype Yellow River to the digital Yellow River.

Key words: VR technology, flood control DSS

1 Introduction

Decision support system (DSS) is a new technology developing based on management information system and management science/ operational research. Management information system mainly focuses on the handling of mass data to finish management tasks. Management science and operational research use model for aid decision – making. The DSS can combine substantive data with several models to form a decision scheme and can support decision making through man computer interaction. DSS appeared in the 1960s; DSS theory developed sufficiently in the 1970s; in the 1990s, emergence of three technologies, namely data warehouse, on – line analytical processing and data mining promoted the further development of DSS; in the new century, owing to the internet technology, multimedia technology, artificial intelligence and improvement of computer hardware performance, decision support system further develops into a multiuser integral decision support system in the network environment.

In the new century, informatization is highly favored in all industries. The Yellow River regulation organ puts forward a modern river improvement philosophy in the century, namely Three Yellow Rivers (the prototype Yellow River, digital Yellow River and model – based Yellow River). The digital Yellow River is a counterpart of the prototype Yellow River. It gathers base data by dint of both modern means and traditional means, and builds an integrative digital integration platform and virtual environment for the natural, economic and social factors in the total catchment and relevant regions. With powerful system software and mathematical model, it can simulate, analyze and study the different programmes for the development and management of Yellow River harnessing, and provide decision support under visualized conditions to make the decision more scientific and foreseeable. Its core task is the construction of flood control decision support system. Based on the collection, calculation and processing of flood information, it finally represents the relevant information with video, images, graphs, sound and characters in a visual, simple and scientific way. Use of VR technology makes it convenient to realize man computer interaction and realistic manifestation, making people absorbed in the system completely. It can provide digitized flood control engineering information, water regime, rainfall regime, flood regime and historical hydrologic data and expert database system, and work out the flood discharge, flood elevation, flood velocity, flood arrival time, overflow area and depth on the riverbed, danger classes of hydraulic engineering, the number of workforce and materials required, as well as emergency flood control programmes based on different hydrologic data, weather conditions, flood information, hydraulic engineering information, river regime and expert strategies, so as to reproduce the flood spot virtually and organize remote dispatch and control as well as flood prevention consultations; it can greatly improve the accuracy of flood prediction and reaction under emergent conditions, and make the emergency flood control programmes more scientific. It can realize the longitudinal connection between upper and lower management layers, and make flood control commands more harmonious; meanwhile, it can also set up a lateral communication between valleys, which can realize sharing of hydrologic, weather and flood control resources and expert resources. In this way, it can effectively improve the accuracy, reliability and instantaneity of flood control command and provides a totally new means in making flood control decisions.

2   System architecture

The system is a 4 – layer remote client – based C/S system consisting of the remote client layer, access server layer, application server layer, model library and VR render layer, and resource database layer respectively. See Fig. 1.

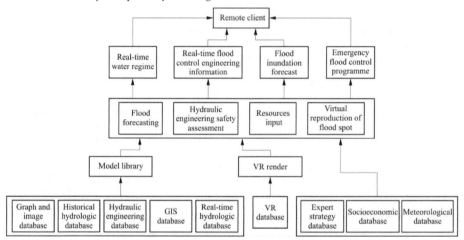

Fig. 1   General flowchart of the system

The topmost layer is the user telnet client which provides an interface between users and the system. Users can log on the remote server through the client and read the real – time water regime, real – time flood control engineering information, flood inundation forecast and emergency flood control programme and other information from server. With client access, the system's normal operation will not be affected by changes of environment and the system basically needs no intermediate support plug – in unit.

The second layer is the login server. It mainly provides logon authentication for remote users and can provide the data result of different projects and respond to the request of remote users in real time. Server performance of this layer can directly decide the speed of system access. Use of high – performance server can improve the access speed.

The third layer is processing and computing unit server. It mainly takes charge of processing of the special water conservancy model and VR render results and give real – time response to the request of the upper layer server. Complexity degree of the processing and computing unit is directly related to the complexity degree of the whole system and the stability of work, so simple, mature and stable technology shall be adopted in design.

The fourth layer is model library and VR renderer. It can set up a model for data acquired from different databases, and extract VR data of entities alongside the Yellow River from VR model database for rendering. It can give timely response to the request of the upper layer server and submit the information it processes to the server of the higher layer. VR render adopts the network composed of several workstations with powerful graphic processing ability with the assistance of professional graphic card. In this way, it can finish rendering of graphs fast to get high quality graphs.

The fifth layer is databases. In this layer, there is special water conservancy database and aid decision – making information database. Databases in this layer contain data with different functions, such as historical hydrologic data, real – time hydrologic data, hydraulic engineering data, geographic information system data, river regime data, VR resource data, socioeconomic data, expert strategy database, weather forecast data, graph data, image data and video data, documentary data, voice data, user information data etc. Among them, the weather forecast database is connected with the weather forecast server of weather bureau and the socioeconomic data is connected with the database of the ministry of land and resources. This makes it possible to acquire scientific and precise data, and in the expert strategy database, there are classical cases about flood control and disaster prevention in history; besides, techniques and experience of senior experts who are about to quit from flood fighting will be gathered to the expert strategy database, so that their techniques and experience can still play an important role even after they quit.

3 Design principle

3.1 Design of login server

Login server is a server platform which can provide support services including system information release, data inquiry, data processing, user logon, and identity authentication and so on. See Fig. 2 for its working principles:

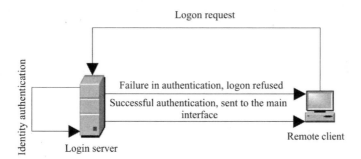

Fig. 2 Working principles of logon server

The login server mainly follows the following several procedures during working: ① The remote client sends login request to the server (including the user name, password and working place information). ② The server receives the user request and verifies it with the user list in the server; if it's correct, it will enter the next step; otherwise, it will refuse login. ③ The server sends the system main interface to the user and stop to wait for the next step of the system.

In this system, the mysql database which is based on windows 2008 sever is adopted at the login server to provide user identity authentication service. When necessary, users can read the real – time water regime, real – time flood control engineering information, flood inundation forecast and emergency flood control programme and other information from the server.

3.2 Design of processing and computing unit server

This server is the core of the system. It mainly takes charge of calculation, analysis and processing of data in the system and can generate in real – time information about flood discharge, flood elevation, flood velocity, flood arrival time, overflow area and depth on the riverbed, danger classes of hydraulic engineering, number of workforce and materials required, emergency flood control programme, remote dispatch and control, flood control consultation etc. Server performance of this layer can directly decide the speed of system access. Use of high – performance server can improve the access speed.

Kernel program of the service program of processing and computing unit server is composed with C language and the top layer program is composed with C#. This can not only improve the efficiency of source code of the kernel program to the greatest extent, but also save the workload effectively.

3.3 Design of special water conservancy model

The special water conservancy model is a business model which is mainly used for processing of information, including flood prediction, river regime information, hydraulic engineering information, weather forecast etc. This model can be further developed based on the current special water conservancy model or developed in a user – defined way according to circumstances.

3.4 Design of database

Database is an important constituent part of VR technology – based flood control DSS and is the bearer of information storage and at the same time can be used for information processing. Databases of this decision making system include: historical hydrologic database, real – time hydrologic database, hydraulic engineering database, geographic information system database, river regime database, VR resource database, socioeconomic database, expert strategy database, weather forecast database and user information database.

In order to improve the maintainability and instantaneity of database, the database is composed of the following several parts: ① Database: it's mainly composed of SQL Sever 2005 which is based on windows 2008 sever; the geographic information system database uses GEOWAY FGIS. ② Database support program: it's mainly used for insertion, deletion, search, and revision of database. ③ Collection of data definition files: it's mainly used to define the operation view.

3.5 Design of VR renderer

VR renderer is an important indispensable supporting part for VR technology – based flood control DSS to finish virtual reproduction. It's mainly used to render the 3D data of entities from VR database and send the virtual spot after rendering to the upper layer system.

VR renderer adopts the 64 – digit graphic workstation of Windows 7 and is equipped with several powerful processors, memory with the function of verification, several superior video cards and adopts CrossFire technology. System development environment is AutoCAD, 3D Max, VRP – Builder, VRML, Vega, VRPIE, VRP – SDK development library and OpenGL. Sources of VR data mainly include: Design drawings and documentary data of structures, digital map of the Yellow River, photogrammetric measurement and aerial photography data, satellite remote sensing data, available data parameters, and field survey data.

4 Performance characteristics

Compared with previous decision support systems, the VR technology – based decision support system is further improved in system stability, accuracy, instantaneity, scientific nature, maintainability and safety; besides, it also adopts expert system and virtual reality reproduction technology, which can realize the intelligent decision of electronic information of flood control and can generate virtual flood in real – time according to the flood information acquired through VR technology.

4.1   Cross – region coverage, user friendly access

This system is based on C/S structure and is characterized by high safety and convenience. This system is not limited by regions. In the Yellow River valley, remote computers of all levels of flood prevention departments can access the system and verify the identity of users automatically to assign the user rights by installing the client and input the user name, password, and working place information at the logon page. Then users can operate within the scope of their respective authority.

4.2   Fine system maintainability

The system adopts modular design. Damage of one module will not affect the normal operation of other adjacent modules, so it's only necessary to replace the damaged module. It's also very convenient for system upgrade in the future. It's only necessary to add the relevant new module to the program and define it in the master program. During system design, it's advisable to use the same operating system, database and data format. This can ensure the consistency of data format in the system, reduce the probability of system faults in the future, and improve the system maintainability and readability of source program.

4.3   More scientific, accurate and timely decision data

Because flood control task is emergent, the flood control decision making system is required to make scientific and precise response to the data and information of flood control, and send the correct conclusion to the control center, so that the control center can make correct decisions. Besides, the system also adopts the optimal man computer interface, and can make the optimal emergency flood control programmes through data comparison. Finally, it can represent the programmes in various ways and provide them to the decision makers and technologists for them to choose the best one for flood fighting.

4.4   High system safety

This system includes the weather forecast database, socioeconomic database, expert strategy database, hydraulic engineering database, and geographic information system database; many data in these databases are involved with state secrets. Therefore, the network safety of the system shall be kept in strict secret in case of data leakage. The following measures shall be taken: Install a hardware fire wall at the system entry, and equip each server with the relevant antivirus software and fire wall. The network manager shall pass the most strict identity authentication before server management. Data transmission is encrypted with asymmetric encryption, which greatly improves data safety.

The VR technology – based decision support system can be used in many fields of river improvement, which is mainly reflected in the following several aspects: ① Flood control and disaster reduction: use telemetry technology, remote sensing, satellite technology and mathematical model technology for rainfall forecast, flood forecasting, reservoir regulation and flood routing. ②Calculation of direct losses from flood disaster: use real – time flood information, hydraulic engineering information, river regime information and socioeconomic information to calculate the flooding area, water depth at the flooding area, position of danger of hydraulic engineering along the Yellow River, and direct flood damage and so on. ③ Formulation of flood control program in the expert system: read the flood information, hydraulic engineering information, river regime information, weather information, socioeconomic information and expert database system and finally formulate the flood control emergency program fast and scientifically. ④ Use in flood fighting rehearse: previous flood control rehearses usually need substantive workforce, material resources, financial resources and much time, but the final effects are usually very small; the virtual reality – based decision support system can simulate the flood peak virtually, and set information including flow, rate of flow, water level, hydrological information, silt concentration etc. , and work out the weak spots of hydraulic engineering, impacts of flood on steam channels, losses from the flood disaster, flood control emergency program and so on computer with VR technology according to the calculation and processing of flood information, hydraulic engineering information, river regime information, weather information, socioeconomic information and expert database system, which provides precious data for flood control at the spot.

5　Conclusions

With the use of VR technology and expert system in the decision support system, compared with previous decision support systems, VR technology – based decision support system can greatly improve the human – computer interaction interface and formulation of emergency disaster relief, making the establishment of strategy of the decision support system more timely, accurate and scientific, so it's a qualitative leap. This decision support system can work stably with new technology, and has more ways of man computer interaction. Users can just input the basic information to finish virtual reproduction and formulation of emergency flood control program. Therefore, it can well meet the requirements of flood control headquarters about the establishment of accurate, real – time and scientific flood control emergency programs.

References

Chen Wenwei. Textbook of Decision Support System [M]. 2nd editon. Beijing: Tsinghua University Press, 2010.

Zhuang Chunhua, Wang Pu. VR Technology and its Aspplication [M]. Beijing: Publishing House of Electronics Industry, 2010.

Zhang Jing, Zhang Tianchi. VR Technology and its Application [M]. Beijing: Tsinghua University Press, 2011.

Shen Wei, Zeng Wenqi. VR Technology [M]. Beijing: Tsinghua University Press, 2009.

Tang Guo'an. Geographic Information System [M]. 2nd edition Wuhan: Science Press, 2010.

Du Hui. Experiments for Decision Support and Expert System [M]. Beijing: Publishing House of Electronics Industry, 2007.

(U. S.) Giarratano. J. Expert Systems: Principles and Programming [M]. 4th edition Beijing: China Machine Press, 2006.

Experimental Study on Non-contact Flow Monitoring System Used at Gates of the Lower Yellow River

Yuan Zhanjun, *Zhu Zhifang* and *Zhao Liang*

Water Supply Bureau, YRCC, Zhengzhou, 450003, China

Abstract: Flow monitoring system used at gates of the lower Yellow River, with low level application of automatic system, can not satisfy the modernized and refined management in water supply. Many advanced instruments, such as electromagnetic flow-meter and Doppler current-meter, have been introduced in recent years. But each instrument has its own limitations, and can not solve all the complicated questions we meet in measuring water velocity at gates of the lower Yellow River. In the study we use non-contact flow monitoring system to suit wide and shallow channel of the Yellow River. Although it has been widely used in foreign countries, the system is used in the Yellow River for the first time. It is composed of three parts: data collection system, data analysis and calculation system, and data transmission system. The instrument of data collection, non-contact radar flow meter, is hung up in the air, without contacting the surface of water. The system has the advantages of easy installation and maintenance, with less measurement error in terms of precision of the measurement. The measurement method can affect the accuracy of the system. After the system was installed at the Liuyuankou Gate, we compared the experimental data measured by the system with that obtained from underway ADCP. It was found that precision of the data meets the flow specification requirement. Then we have compared experimental data measured in different conditions and refined the measurement method for the system. The study has paved the way for the successful application of the non-contact system in the lower Yellow River.

Key words: non-contact flow monitoring system, Liuyuankou Gate, radar flow meter, underway ADCP, the lower Yellow River

Traditional velocity apparatus has been used for many years in the gate water measurement of the lower Yellow River. Flow monitoring system used at gates of the lower Yellow River, with low level application of automatic system, can not satisfy the modernized and refined management in water supply. Many advanced instruments, such as electromagnetic flow-meter and Doppler current-meter, have been introduced in recent years. But each instrument has its own limitations, and can not solve all the complicated questions we meet in measuring water flow at gates of the lower Yellow River. In the study we use non-contact flow monitoring system to suit wide and shallow channel of the Yellow River.

1　A survey of the Liuyuankou Gate

The Liuyuankou Gate is located in Kaifeng City, Henan Province, between 33 and 34 dam of Liuyuankou vulnerable spot on the right bank of number 85 + 700 m of the Yellow River. Founded in 1966 and rebuilt in 1981, the gate has five holes whose structure is reinforced with concrete culvert. Its design discharge is 40 m^3/s and design irrigated area is 464,000 mu.

2　Design of the system

We get discharge (Q) by measuring the average velocity (v) and over-flow area (A), it is:
$$Q = vA$$
The average velocity is determined by measuring the velocity (v_i) in a point or a region, with the establishment of hydraulic model, or the relationship between calibration method and average velocity in the cross-section. The flow area is closely associated with the water level and the cross

sectional shape. And the water level and cross-sectional shape are measured to calculate the flow area.

Based on the calculation above, flow rate sensor and the water level gauge are set in the system to monitor the flow rate and flow area. And the data acquisition equipment is used to analyze, process and calculate the flow volume. The results of the experiment are conveyed to data center by transmission equipment. The controlling equipment is set in the system to control the operation of the entire system automatically. Following map is the system diagrammatic sketch (Fig. 1).

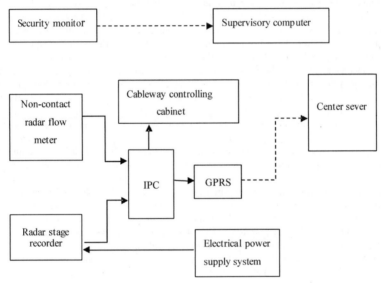

Fig. 1　Diagrammatic sketch of the measurement system

3　Introduction of flow equipment

Electric wave current-meter is a non-contact velocity measurement instruments which can measure the water flow rate automatically for a long time.

3.1　Profile of the instrument type

Electric wave current-meter measure the velocity with its radar by transmitting and receiving electromagnetic waves, so it is also called the radar velocity meter. The instrument type we choose is RQ-24 non-contact radar meter produced by Austria Sommer Engineering Company (Fig. 2). It can be used to monitor the surface flow velocity and water level, calculate and output water discharge through it's hydraulic model. The apparatus can be mounted under a bridge or a cantilever, or installed on a cable. Its installation is very simple, so it can be used easily everywhere. The accessory flow software quickly gives the integrated correlation coefficient according to the water boundary conditions and hydraulics model. At the same time it has advantages of low power consumption, maintenance-free. The system has been widely used in many foreign countries and come into China in recent years.

3.2　Working principle

Electric wave current-meter use Doppler principle to measure the speed. Radar flow meter we chosen transmit and receive electromagnetic wave which helps us get water surface velocity. Because of being microwave, electromagnetic wave attenuate slowly and well spread in the air. The wave spread to water surface with a slant, except for those absorbed or scattered, some reflected by water can be received by radar. The device calculates the frequency of the transmitted and reflected signal, and the water flow rate is obtained. The water surface is the carrier of wave, both of which have the same velocity.

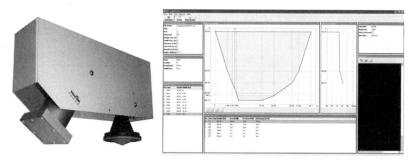

Fig. 2　RQ – 24 appearance and software interface

3.3　Structure and composition

The instrument is composed of three parts of the probe, the signal processor and battery. The probe is equipped with emitters and antennas. It can be handed or mounted on a bracket, and the signal processor can be programmed to control the probe to emit microwave, receive and precess the reflected waves of the water. The instrument has the function of self-calibration and self-discrimination. In the field measurement, the instrument can determine automatically if the emission wavelength is stable and intense enough. Satisfying all the conditions, the instrument begins to measure; if not, the machine will not work in order to avoid generating error data.

3.4　Product information

Technical data of the RQ – 24 non-contact flow meter are as follws:
Velocity measurement range: 0. 3 ~ 20 m/s; measurement angle: 20° ~ 60°; accuracy: 1 mm/s; analogue output: 4 ~ 20 mA; serial port: RS – 232; power supply: 10. 5 ~ 15 VDC; operating temperature: – 30 ℃ ~60 ℃ ;measurement freqency: 10 GHz.

3.5　Analysis of instrument's characteristics and accuracy

Radar flow meter is the one and only official domestic use as non-contact flow. It can quickly measure the water velocity from a long distance without contacting water, which makes the instrument avoid being influenced by floater on the surface or in the water, as well as water quality and flow pattern. Water velocity, floater and waves are positive factors to the machine working. It is especially suitable for radar instrument to monitor water in environmental conditions of the Liuyuankou Gate.

Ii is not much accurate when radar flow meter measures the velocity at low speed. Velocity measurement range of low speed is high. In order to get stronger echo of water waves, more apparent waves are needed to get accurate data. If the surface is smooth, even higher velocity would not make strong reflex.

With less measurement error, measurement method can affect the accuracy of radar flow meter. ①The speed of surface waves and floating material measured by instrument is not equal to the water velocity. Because of speed and direction of wind, the cause of the error caused by wind is not ignored. The waves and floater velocity is influenced not only by water flow but also by their own movement. ② The radar beam has about 10° beam angle, and forms a long and circular projection shadow. The velocity in the point with strong reflection in the shadow may be accepted as water velocity, and this uncertainty of the position will affect the accuracy of speed of water.

4 Analysis of the measured data

After the system is installed at the Liuyuankou Gate, we compare the experimental data measured by the system with underway ADCP on March 27 and 28. Tab. 1 lists the test data of radar flow meter (part) and Tab. 2 for underway ADCP test data.

Tab. 1 Test data of radar flow meter (part)

Time	Discharge (m³/s)	Time	Discharge (m³/s)	Time	Discharge (m³/s)	Time	Discharge (m³/s)
12:00	9.91	18:10	17.60	23:10	2.39	04:30	15.72
12:30	10.85	18:30	16.92	23:30	2.10	04:48	17.10
12:50	11.47	18:50	16.25	0:00	2.95	05:00	17.70
13:10	11.87	19:12	16.13	0:20	3.25	05:30	17.96
13:50	12.10	19:50	14.90	01:00	4.04	06:30	19.12
14:24	13.08	20:10	13.75	01:20	4.41	07:12	19.75
14:50	14.15	20:30	12.03	01:40	5.02	07:40	19.24
15:10	14.88	20:50	9.92	02:00	7.10	08:20	18.46
15:40	15.70	21:10	6.74	02:24	8.15	09:00	15.81
16:00	16.17	21:36	5.92	02:50	10.25	09:36	12.25
16:48	16.98	21:50	4.33	03:10	11.70	10:20	11.78
17:00	17.20	22:10	3.65	03:30	11.93	10:40	10.42
17:30	17.76	22:30	2.89	03:50	13.78		

Tab. 2 Underway ADCP test data

Time	Discharge (m³/s)	Time	Discharge (m³/s)	Time	Discharge (m³/s)	Time	Discharge (m³/s)
14:40	13.85	19:30	15.70	06:20	19.02	10:50	9.87
15:50	15.87	21:00	7.14	07:10	19.88		
16:40	16.90	22:30	2.78	09:00	15.98		

Through the comparison, the data from radar measurement fit well with data of underway ADCP (Fig. 3). After error calculation, error range of radar flow meter is between − 0.66% and 3.8%, in line with norms for flow accuracy. Precision of the data can meet the flow specification requirement.

5 Conclusions

Although radar flow meter, as a new flow measurement instrument, has been applied well in foreign countries, there are still many difficulties in the application of the system in gates of the lower Yellow River. The experiment monitoring of the water velocity at Liuyuankou Gate is successful, which extends the selection of new instruments and paves the way for the application of

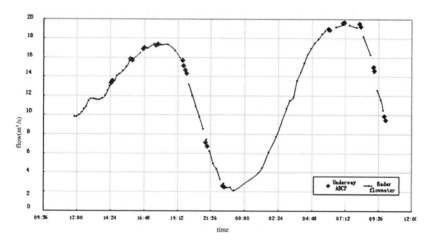

Fig. 3   Radar flow meter and underway ADCP flow process curve

the system in the Yellow River.

References

Li Guanglu, Wang Xiulian. Application of the Electric Wave Current-meter in Source Area in Qinghai [J]. Yangtze River, 2010,14.

Lin Zuoding, Zhu Chunlong, Yu Dazheng. Modernization and New Technology in Hydrology [M]. Beijing: China Water Power Press, 2008.

Response of Hydrological Simulation to the Spatial Uncertainty of Rainfall

Zhang Ang, *Li Tiejian*, *Liu Yi* and *Fu Xudong*

State Key Laboratory of Hydroscience and Engineering, Tsinghua University,
Beijing ,100084 , China

Abstract: The hydrological response of a river basin is affected by the spatial variation of rainfall. Rainfall stations with certain density and spatial distribution supply sampling records of the spatial and temporal information of rainfall. When the records are used to describe rainfall all over a whole basin, uncertainty will be introduced. In this study, the spatial uncertainty of rainfall was represented by various combinations of rainfall stations in the river basin, and its influence on the hydrological simulation was analyzed. Three rainfall events in the Qingjian River basin of the middle Yellow River in flood seasons were analyzed using clustering and correlation methods by total depth and rainfall process criteria, respectively. For each rainfall event, the stations were classified into different groups according to the differences in rainfall depth and process. Subsequently, to analyze the uncertainty of the measured rainfall data, the box plots of the basin average rainfall were drawn for each reduced number of rainfall stations according to different combinations of the stations. Furthermore, a hydrologic model, the Digital Yellow River Integrated Model, was used to quantify the uncertainty of simulated runoff that induced by the uncertainty of rainfall. The model was calibrated according to measured runoff depth and peak discharge at basin outlet, and then was used to simulate the runoff processes using different numbers and combinations of rainfall stations. The simulated results were analyzed and box plots were drawn. By comparing the uncertainties of rainfall and runoff, the response of the hydrological simulation to the spatial uncertainty of rainfall was discussed. The results showed that the uncertainty of rainfall has direct influence on the simulated runoff processes. A little difference of the rainfall input may lead to a larger change of simulated runoff. Generally, with the reduction of rainfall input information, the effectiveness of runoff simulation decreases, and the fluctuation range of the simulated runoff expands. With certain rainfall station combinations, if the storm center is not captured and applied to the hydrological model, the most part of the actual rainfall will be ignored in rainfall-runoff simulation, resulting in significant deviation in the simulated runoff.

Key words: hydrological simulation, rainfall spatial distribution, uncertainty, Digital Yellow River Integrated Model

1 Introduction

Computer simulation is a common method in current hydrological analysis. However, it is still a controversy whether the key factor influencing the accuracy of hydrological simulation is model efficiency or rainfall input (McMillan et al. , 2011). Wilson et al. (1979) changed rainfall input of Fajardo Basin with an area of 68.635 km^2 in Puerto Rico, and found out that the difference in simulated runoff is quite significant. Chaubey et al. (1999) studied the variability in estimated hydrologic/water quality model parameters solely due to the spatial variability of rainfall, and found out that a wide range of estimated parameters were resulted when the rainfall measured at each gauge location was used individually, one at a time, to estimate the model parameters. Xu et al. (2009) studied the propagation effects of rainfall uncertainty based on the fuzzy set theory, and the results showed that the propagation of rainfall uncertainty played a dominant role in influencing simulated runoff.

Studies above mainly aimed at the influence of different rainfall input on runoff and sediment simulation. In this study, rainfall information with smaller time scale is used, and the difference of the uncertainties of basin average rainfall and simulated runoff is analyzed. Therefore, the response of hydrological simulation to spatial uncertainty of rainfall is illustrated.

2　Data and methods

2.1　Study area

Qingjian River Basin, with an area of 4,078 km², is located within the longitude 109°12'E ~ 110°24'E and the latitude 36°39'N ~ 37°19'N. The length of the main stream is 167.8 km. The basin has a semi-arid warm temperate continental monsoon climate. The annual rainfall in this region is 486 mm. Short duration high intensity torrential rains in flood season, from July to September, contribute 65% of the annual rainfall. Also, the spatial distribution of rainfall is even more complex. The majority of the basin surface material is the highly erodible loess soil, covering 92% of the basin area. Consequently, runoff in this basin, mainly in the form of infiltration-excess, can cause severe soil erosion.

There are two hydrological stations in the basin, i. e. , Zichang and Yanchuan, with the contributing areas of 930 km² and 3,468 km², respectively. Both of them have measured data of runoff and sediment from the year they built (Zichang 1959 and Yanchuan 1954) to present. There are 22 rainfall stations (including the two hydrological stations above) located within and adjacent to the basin. All of their measured rainfall data from the year they built (the earliest in 1970) to 2007 are available. All the meteorological and hydrological data used in this paper were provided by the Hydrographic Bureau of the Yellow River Conservancy Commission. The drainage network and the distributions of the hydrological and rainfall stations in the Qingjian River Basin are shown in Fig. 1.

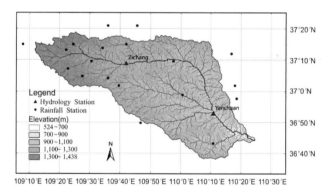

Fig. 1　Location of and stations in the Qingjian River Basin

2.2　Digital Yellow River Integrated Model

The Digital Yellow River Integrated Model developed by Tsinghua University (Wang et al. , 2007) is a distributed hydrological model based on digital drainage networks. The basic unit of the model is river reach and its corresponding hillslopes. River reaches are identified by a modified binary tree codification method (Li et al. , 2010), and this codification can be a direct expression of the connection between the river reaches and the structure of the drainage network. Meanwhile, a dynamic parallelization of the model has been achieved (Wang et al. , 2011; Li et al. , 2011) to significantly improve the simulation efficiency. The high speed rainfall-runoff simulation of the model provides the support for repeatedly simulation in this study. The rainfall-runoff model simulates the hillslope hydrologic process, including vegetation interception, evapotranspiration, infiltration-excess surface runoff, soil water discharge and etc. Confluence and flow routing process in the river network is simulated using a diffusion wave method.

2.3 Methods

Firstly, three typical rainfall events in the Qingjian River basin are selected as the research object. Then, the rainfall stations in the basin are grouped by the total depth and process of rainfall to analyze their spatial variation, for each rainfall event respectively. Subsequently, the uncertainty of basin average rainfall is quantified by using different numbers and combinations of rainfall stations. After calibrating the model according to measured runoff depth and peak discharge at basin outlet, the response of hydrologic simulation to rainfall uncertainty is analyzed by using the repeatedly simulated runoff with different rainfall inputs.

3 Spatial difference and uncertainty of rainfall

3.1 Selection of the rainfall events

As shown in Fig. 1, the density of rainfall stations downstream Zichang is much lower. If the region downstream Zichang is considered together with its upstream, the difference of rainfall station density will lead to the difficulty to characterize the real feature of the spatial distribution of rainfall. Therefore, this study only focuses on the rainfall pattern and rainfall-runoff process upstream Zichang. There are 11 rainfall stations upstream Zichang (including Zichang), and the average controlling area of the stations is 84.5 km^2. Because the recent rainfall data is more detailed, three typical rainfall-runoff processes after the year 2000 are chosen and shown in Tab. 1. Scale of each rainfall-runoff process is different from any others, with measured peak discharge in different order of magnitude, so analysis in this paper can reveal the different laws of different scale rainfall-runoff processes.

Tab.1 Characteristic parameters of the rainfall events

Number	1	2	3
Duration	2001 – 08 – 16	2002 – 07 – 03 ~ 07 – 07	2006 – 07.30 ~ 08 – 03
Basin average rainfall (mm)	34.2	144.9	27.5
Measured runoff (mm)	4.7	86.9	1.5
Simulated runoff (mm)	7.0	40.1	0.5
Measured peak discharge (m^3/s)	881	4,670	76
Simulated peak discharge (m^3/s)	1,604	2,192	28
NSE of flow discharge	0.48	0.43	0.05

3.2 Spatial difference of the measured rainfall

The rainfall stations in the basin are grouped to analyze their spatial variation, for each rainfall event respectively. Based on the consideration of total depth and the process of rainfall, two steps of grouping are conducted. Firstly, the method proposed by Park et al. (2009) is used in clustering ($P < 0.05$) the rainfall depth of different stations. Secondly, correlation statistics ($P < 0.05$) is used for the rainfall process of each station to test its difference from all the other stations. Rainfall stations, whose rainfall depths are in the same cluster and rainfall processes are all correlated with each other, are defined in a same group.

The results of grouping for each rainfall event are shown in Tab. 2 and Fig. 2. The stations within a same group are shown in a same color. There are 8 groups of stations in the 2001 rainfall event, 6 groups in the 2002 and 2006 events.

Tab. 2   Information of each station in the three rainfall events

No	Station name	Percentage of controlling area(%)	Rainfall depth (mm) and group ID of each station					
			2001-08-016		2002-07-03 ~ 2002-07-08		2006-07-30 ~ 2006-08-03	
1	Zichang	7.48	18.0	1a	283.2	2a	42.8	3a
2	Jingzeyan	16.37	75.2	1b	114.6	2b	23.8	3a
3	Lijiafen	11.17	55.4	1c	137.6	2b	24.8	3a
4	Sanshilipu	15.19	36.0	1d	130.0	2c	30.4	3a
5	Anding	12.31	19.5	1d	172.8	2d	15.4	3b
6	Zhangjiagou	10.02	27.4	1a	145.6	2b	36.2	3c
7	Siwan	12.79	19.8	1e	114.6	2e	22.4	3a
8	Xinzhuangke	10.88	0.0	1f	126.2	2b	32.6	3d
9	Yujiawan	0.13	21.6	1g	106.2	2b	46.4	3e
10	Hecaogou	0.09	16.2	1h	225.4	2f	35.6	3a
11	Yangkelangwan	3.57	31.8	1d	145.8	2b	27.4	3f
	Total	100	8 groups		6 groups		6 groups	

The 2002 rainfall is taken as an example to analyze the grouping results, which shows that the stations are mainly in group 2b. However, Zichang station (in group 2a) and Hecaogou station (in group 2f) have significant difference with the others. The total depth in Zichang station is much larger than the others, and the rainfall process in Hecaogou is different from the others (see Fig. 3). The 2002 event is a heavy rain covering the whole basin, and the measured data of 11 stations have statistically significant differences. The selected 2001 and 2006 rainfall events are in medium and small size, so their spatial distribution and process have more significant differences. Their details are not discussed here.

Rainfall stations' grouping results show that: Generally, in a rainfall event, some of the stations have the statistically same rainfall depth and process, but the rainfall records of the other stations can be significantly different. The station which records the highest rainfall is the storm center. Total depth recorded in the storm center is generally 2 ~ 5 times of the other stations. In the current density of rainfall stations, the storm center usually involves only 1 ~ 2 stations, such as Jingzeyan station in the 2001 event, Zichang station and Hecaogou station in 2002, and Zichang station in 2006. Therefore, the data recorded in storm center has important influence on basin average rainfall and then runoff simulation. Also, the rainfall process can appear significant difference, as shown in Fig. 3.

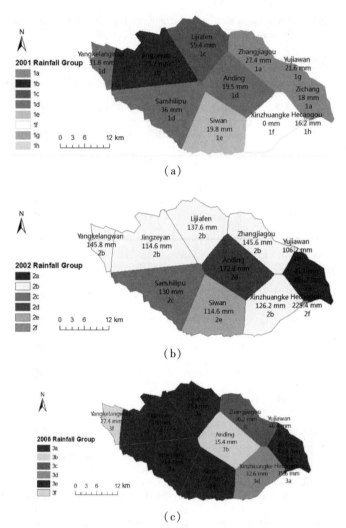

(a)

(b)

(c)

Fig. 2　Grouping results of rainfall stations

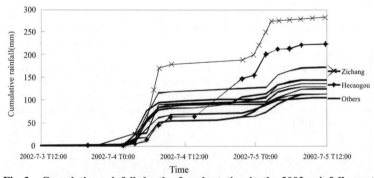

Fig. 3　Cumulative rainfall depth of each station in the 2002 rainfall event

3.3 Uncertainty of the basin average rainfall

There must be uncertainty when 11 rainfall stations are used to measure the basin rainfall with a 930 km^2 area. The uncertainty range, noted as U_0, cannot be evaluated without detailed rainfall data covering the whole basin. In order to analyze the uncertainty of basin average rainfall, the increasing uncertainty range, noted as $U_0 + U'(k)$, is introduced by reducing the rainfall stations, where k is the reduction number of stations. Then the uncertainty characteristic of basin rainfall can be concluded by analyzing the relationship between $U'(k)$ and k. For each k, basin average rainfall depths are calculated for all the combinations of stations, and a box plot is drawn. The basin average is calculated using the Thiessen polygon method. And finally, a $U'(k) - k$ plot is drawn for each rainfall event.

Box plots can reflect the range and trends of different series of data very well. A box plot is a description of data distribution which denotes the 5 percentiles of the performance of quantitative variables, i. e. , $P_{2.5}$, P_{25}, P_{50}, P_{75}, and $P_{97.5}$. The $P_{25} \sim P_{75}$ range constitutes a box of the graphics, and the $P_{2.5} \sim P_{25}$ and $P_{75} \sim P_{97.5}$ ranges constitute the two whiskers. The $U'(k) - k$ results of the three rainfall events are shown as in Fig. 4. With the reduction of rainfall input information, the range of the basin average rainfall depth expands. On the contrary, if the number of stations increases on the basis of 11 stations, the uncertainty range of rainfall is expected to decrease to less than U_0.

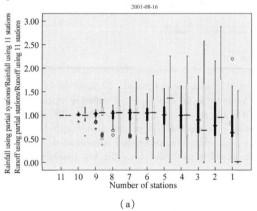

(a)

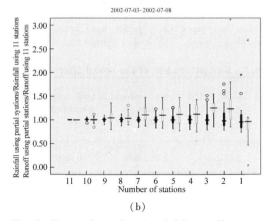

(b)

Fig. 4　Uncertainty of each rainfall-runoff process

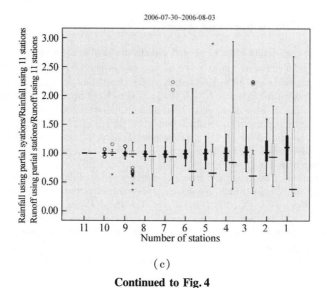

(c)

Continued to Fig. 4

4　Uncertainty of the simulated runoff

4.1　Runoff simulation

　　Firstly, measured rainfall data are pretreated. Because the time steps of rainfall records are different, varying from minutes to several hours in different stations, the time steps of all the rainfall data are adjusted to the same one, i. e. , 2 h. For rainfall records with smaller time steps, an accumulation processing is conducted. For rainfall records with larger time steps, a downscaling processing in accordance with rainfall characteristics is conducted.

　　Then the hydrologic model is calibrated, and the key parameters (Wang et al. , 2009) are shown in Tab. 3. Soil hydraulic conductivity parameters are the most sensitive to runoff simulation. They are calibrated according to the measured runoff volume and the peak discharge at the basin outlet. The initial value of soil water content in the subsoil layer is calibrated according to base flow. Besides, parts of the hydrological parameters in the model have clear physical meanings or are not sensitive to the results of simulation, so the measured values are employed without calibration or adjustment. The evaporation capacity used here are the adjusted values from evaporation pan from month to month. The spatial distribution of the above parameters is not considered.

Tab. 3　Key parameters of the model after calibration

Parameters	Topsoil vertical conductivity K_{zus}	Subsoil vertical onductivity K_{u-ds}	Topsoil horizontal conductivity K_{hu}	Subsoil horizontal conductivity K_{hd}	Topsoil initial moisture $\theta_{u,0}$	Subsoil initial moisture $\theta_{d,0}$
Value	3.7 mm/h	5.2 mm/h	5.9 mm/h	3.6 mm/h	0.15 m³/m³	0.23 m³/m³

　　Correlation of the simulated and measured runoff is expressed by the Nash-Sutcliffe coefficient of efficiency (Krause et al. , 2005), noted as *NSE*, as followed

$$NSE = 1 - \frac{\sum_{i=1}^{n}(O_i - C_i)^2}{\sum_{i=1}^{n}(O_i - \overline{O})^2} \tag{1}$$

where, C is the simulated data; O is the measured data; and the subscript i represents the sequential number of the simulated and measured data series.

An NSE equals to 1 means that the simulation results are totally consistent with the measured data. An NSE of 0 indicates that the mean of model predictions and measured values are still close. And an NSE less than 0 indicates that the simulated value cannot be a good prediction of the measured sequence.

The simulated runoff results proved the model's efficiency in representing the rainfall-runoff events, as shown in Tab. 1. The NSE is better especially for larger rainfall.

4.2 Result analysis

Different combinations of rainfall stations for differnet numbers ($1 \sim 11$) are used in runoff simulation. To reduce the repetition of simulation, the combinations that a same number of rainfall stations belongs to a same group to be permutated will be reduced into a single one, and the stations in the same group are randomly selected. Fig. 5 illustrates the NSE – number of stations and runoff volume (simulated/measured) – number of stations relationships.

As shown in Fig. 5, with the reduction of rainfall input information, the average level of the NSE decreases and its range expands. The average level of the simulated runoff volume fluctuates with no significant trend, but the expanding of the range is still notable. Meanwhile, as compared in Fig. 4, the fluctuation range of simulated runoff volume changes greater than the range of basin rainfall depth. That is to say, the hydrological simulation is sensitive to the spatial uncertainty of rainfall, and a little difference of the rainfall input may lead to a larger change of simulated runoff.

Analysis of storm center in each rainfall event is as follows: ① for the simulation of 2001 rainfall-runoff event, NSE reduces to 0.18 (using 10 stations) from 0.48 (using 11 stations), with the absence of Jingzeyan station (with the highest 75.2 mm rainfall); ② for the 2002 event, NSE reduces to 0.26 (using 10 stations) from 0.43 (using 11 stations), when Zichang station is in absence (with the highest 283.2 mm rainfall) (as shown in Fig. 3); ③ for the 2006 event, because the Yujiawan station, which recorded the highest rainfall (46.4 mm), controls a very small area (0.13%), the storm center is mainly represented by Zichang station with the second highest 42.8 mm rainfall. If it is absent, NSE reduces to -0.48 (using 10 stations) from 0.05 (using 11 stations). So the conclusion is that if the data of the storm center is not captured and applied to the hydrological model, the simulated runoff may deviate significantly. That is to say, more accurate range of storm center can be captured and then more confident simulation can be made, if higher density of rainfall stations can be achieved.

5 Conclusions

Taking the Qingjian River as the study basin, this paper analyzed the spatial uncertainty of rainfall and its influence on runoff simulation, using the Digital Yellow River Integrated Model.

(1) A little difference of the rainfall input may lead to a larger change of simulated runoff. Generally, with the reduction of rainfall input information, the effectiveness of runoff simulation decreases, and the fluctuation range of the simulated runoff expands. In a certain simulation, the NSE of the simulated runoff can appear several times or even more than ten times deviation due to the uncertainty of basin average rainfall.

(2) With certain rainfall input combinations, if the storm center is not captured and applied to the hydrological model, the most part of the actual rainfall will be ignored in rainfall-runoff simulation, resulting in significant deviation in the simulated runoff.

(3) Uncertainty of spatial distribution of the rainfall has great impact on runoff simulation in distributed hydrological modeling. Actual description of the spatial distribution of rainfall is possible

with very dense distribution of rainfall stations, or by using other techniques, e. g. , radar. Due to the limited accuracy of data, there are only 11 rainfall stations in Zichang river basin, with a drainage area of 930 km², so the practical application and parameter identification of distributed hydrological model are somewhat limited. The uncertainty of rainfall input data can be decreased and accuracy of simulation can be improved by using multi-source rainfall data or better understanding the spatial distribution law of rainfall, which should be done in further work.

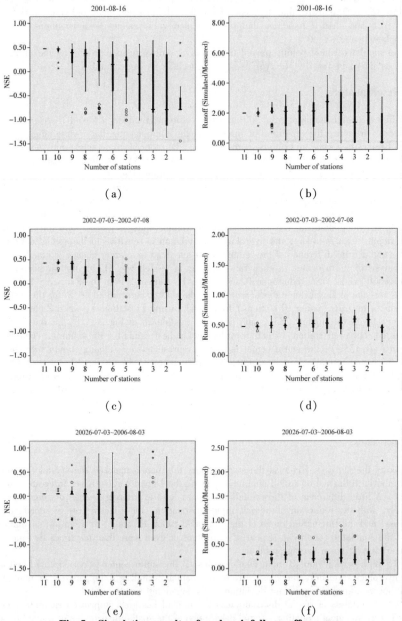

Fig. 5　Simulation results of each rainfall-runoff process

References

McMillan H, Jackson B, Clark M, et al. Rainfall Uncertainty in Hydrological Modelling: An Evaluation of Multiplicative Error Models[J]. Journal of Hydrology, 2011, 400(1 – 2): 83 – 94.

Wilson C B, Valdes J B, Rodriguez – Iturbe I. On the Influence of the Spatial Distribution of Rainfall on Storm Runoff[J]. Water Resources Research, 1979, 15(2): 321 – 328.

Chaubey I, Haan C T, Grunwald S, et al. Uncertainty in the Model Parameters Due to Spatial Variability of Rainfall[J]. Journal of Hydrology, 1999, 220(1):48 – 61.

Xu J, Ren L L, Liu X F, et al. Propagation Effect of Precipitation Uncertainty on Rainfall – runoff Modeling Based on Fuzzy Set Theory[J]. Advances in Water Science, 2009, 20(003): 422 – 427.

Wang G Q, Wu B, Li T J. Digital Yellow River Model[J]. Journal of Hydro – environment Research, 2007, 1(1): 1 – 11.

Li T J, Wang G Q, Chen J. A Modified Binary Tree Codification of Drainage Networks to Support Complex Hydrological Models[J]. Computers & Geosciences, 2010, 36(11): 1427 – 1435.

Wang H, Fu X D, Wang G Q, et al. A Common Parallel Computing Framework for Modeling Hydrological Processes of River Basins[J]. Parallel Computing, 2011.

Li T J, Wang G Q, Chen J, et al. Dynamic Parallelization of Hydrological Model Simulations[J]. Environmental Modelling & Software, 2011.

Park P J, Manjourides J, Bonetti M, et al. A Permutation Test for Determining Significance of Clusters with Applications to Spatial and Gene Expression Data[J]. Computational Statistics & Data Analysis, 2009, 53(12): 4290 – 4300.

Wang G Q, Li T J. River Basin Sediment Dynamics Model[M]. Beijing: China WaterPower Press, 2009.

Krause P, Boyle D P, Base F. Comparison of Different Efficiency Criteria for Hydrological Model Assessment[J]. Advances in Geosciences, 2005(5): 89 – 97.

Application of Remote Sensing to the Ecological Water Regulation in the Estuary Wetland of the Yellow River

Zhang Xiangjuan[1] , *Cheng Yingchun*[2] , *Feng Yun*[1] , *Liu Yongfeng*[3] and *Jiang Zhen*[2]

1. Information Center of Yellow River Conservancy Commission, Zhengzhou, 450003, China
2. Department of Personnel, Labor and Education, YRCC, Zhengzhou, 450004, China
3. Yellow River Water Resources Protection Bureau, Zhengzhou, 450003, China

Abstract: Under the background of the Yellow River water – sediment regulation, the ecological water supplement regulation for the Yellow River estuary wetland was carried out for the first time since 2008 and lasted for four years to ensure the recovery of the wetland ecosystem. During the ecological water regulation, it is necessary to acquire water and ecology information by remote sensing, in order to give effective service to the ecological water regulation in the estuary wetland of the Yellow River. We have accumulated a lot of basic data and matured monitoring methods. At the same time, we also find many problems, such as remote sensing images monitoring frequency needing to be encrypted, data acquisition processes needing further simplify in order to improve data timeliness.

Key words: estuary of the Yellow River, ecological water regulation, monitoring by remote sensing

1   Basic situation of the ecological water regulation in the estuary wetland of the Yellow River

To ensure the water supply safety and recover the ecological environment, the Yellow River Conservancy Commission (YRCC) have put into practice the unified water regulation in the Yellow River since 1999. In 2008, it is the first time to realize the ecological water regulation in the estuary wetland of the Yellow River, which stabilizes and extends the delta wetland, improves the wetland ecosystem, and reserves the wetland biodiversity. And it is also the first time to carry out the experiment of restoring water flowing in the Diaokou River in 2010. Through four years' water regulation, it is evident that the ecological environment in the estuary of the Yellow River has been greatly improved.

2   The development of remote sensing and its preliminary application to the ecological water regulation in the estuary wetland of the Yellow River

In recent years, with the development of remote sensing, the image is gradually various. And the temporal or spatial resolution of the image has greatly developed.

According to the features of those monitoring targets, good distinguishing data of remote sensing with high temporal resolution and water information, and the economic feasibility, at present, the SAR data of environmental disaster alleviation satellite are mainly selected to monitor the ecological water regulation in the estuary area of the Yellow River. TM or SPOT data are also combined with SAR data on special conditions.

In 2003, environmental disaster alleviation satellite was approved for launching the project by the State Council. The satellite is composed by two medium resolution optical small satellites (A and B) and a synthetic aperture radar small satellite C. In order to provide the scientific basis to the emergency aid, disaster relief and reconstruction work, the satellites are mainly used to dynamically monitor the ecological environment and disasters in a large scale, timely reflect the occurrence and process of the ecological environment and disasters, forecast the trends of the ecological environment and disasters, and rapidly estimate the disasters.

It is feasible for the satellites to cover the earth every two days with a flight mode of multiple satellite networks. Now, every daily transit of the satellites can basically meet the domestic demands. Due to the quality of data, such as the stability, the simplicity to be processed, the easy identification of water information, the image is mostly utilized in China.

Radar satellite is the earth monitoring satellite loaded with synthetic aperture radar (SAR). Due to the all – weather, all – time and penetrative features of the SAR image, it is superior to the optical remote sensor. One of the more important trends of SAR is to make full use of the electromagnetic features of ground body and the close relationship between the electromagnetic features of ground body and the frequency, the polarization and the incident angle of electromagnetic wave. Therefore, it is available to obtain more abundant information of ground body using monitoring the ground body by electromagnetic wave.

Meanwhile, in order to obtain the better image of target region, the radar satellite has a better capability of programming, and also can adjust the time or angle of the transit of the satellite based on the customer's demand. At present, for more accurate information of water supplement, it is mainly used to obtain spatial and temporal radar data at special time during ecological water regulation in the estuary area of the Yellow River.

3 Process and result of ecological water regulation in the estuary wetland of the Yellow River monitored by remote sensing

According to the overall arrangement of water and sediment regulation and the plan of ecological water regulation in the estuary wetlands in the Yellow River, the monitoring by remote sensing is carried out flexibly. Every year, the first monitoring of remote sensing is arranged before water supplement, so as to obtain the water body and ecological basic data before water ecological regulation. During the ecological water regulation, the image can be obtained everyday monitored by remote sensing in the estuary wetland of the Yellow River. These daily data can interpret the changes of water body and ecological habits, trace the water supplement process timely, and dynamically monitor the water body, to provide more information support and feedback of ecological water regulation in the estuary wetlands of the Yellow River.

The following data shown in Tab. 1 is based on water body information at different types of regions, which was interpreted of the monitoring by remote sensing on June 7, 2010 and July 6, 2010. It is clear to know how the different types of water body change on basis of the comparative analysis.

Tab. 1 Water body information at different types of regions

No.	Information type	Monitoring area at eastern nature reserve (Acre)			Monitoring area at northern nature reserve (Acre)				
		7th, June	6th, July	Area change	7th, June	6th, July	24th, July	16th, August	Area change
1	Water supplement area in the wetlands	6,765	55,470	48,705	0	0	11,925	34,560	34,560
2	Water area in the course of the Diaokou River	×	×	×	4,485	22,410	58,680	64,470	59,985
3	Water in the course of the Yellow River	29,335	69,134	39,799	1,343	3,546	×	×	2,203
4	Reservoir and ponds in nature reserve	10,135	21,075	10,940	92,218	95,515	×	×	3,297
5	Intertidal zone	498,043	412,862	−85,181	189,264	162,220	×	×	−27,044
6	The old course in the downstream of the Qing 8 River	44,059	42,260	−1,799	0	0	×	×	0

At the same time, the thematic maps could be drafted at different times on basis of the monitoring by remote sensing in the estuary wetlands of the Yellow River (see Fig. 1 and Fig. 2). Based on water body distribution map and the chart data, it is more visible to understand the changes and the distributions of water body during ecological water regulation of the Yellow River.

Fig. 1 The map of the monitoring by remote sensing

Fig. 2 The map of water body distribution

As to the different areas, the real – time and high – frequency monitoring can focus on the important regions. For example, the following images shown in Fig. 3 are those in the mouth of Diaokou River at three different times.

Fig. 3 Images in the mouth of Diaokou River at three different times

4 Brief conclusions

Through years of the monitoring by remote sensing, a large number of basic data and mature monitoring methods have been accumulated. These give effective service to the ecological water regulation in the estuary wetland of the Yellow River. But lots of problems still exist as follows.

Firstly, time resolution of the image needs to be further improved. Because of great changes of water supplement process, the image data require higher time resolution of remote sensing. But, due to the limit of development of remote sensing, it is unable to obtain data at any time and at any place by remote sensing. This problem is the biggest bottleneck for the high – frequency monitoring by remote sensing.

Secondly, the process from data acquisition to data interpretation needs to be more efficient, in order to interpret the data at the first time. The current data need basic correction, treatment, transmission, information extraction and other processes. These processes require a certain amount of time, which greatly reduces the limitation period of the image.

Thirdly, despite the stable and mature information extraction of water information, it is unable to acquire more information about deeper water from the monitoring by remote sensing. With the development of remote sensing, it is a trend to combine the monitoring of water supplement with the river flow in the future.

Application and Research of Digital Water Supply System (Pilot Project) in the Lower Yellow River

Zhao Liang, *Wang Hongqian* and *Yuan Zhanjun*

Water Supply Bureau, YRCC, Zhengzhou, 450003, China

Abstract: Accurate measurement of the lower Yellow River water quantity and meticulous management is important means to promote the water resources. The system is mainly composed three parts of automatic flow monitoring system, application of software system and operating environmental system. The automatic flow measurement system is the core content of "the Digital Water Supply (DWS)". Flow measurement is based on the flow velocity and water stage monitoring. At present we have chosen acoustic Doppler (H-ADCP) and electromagnetic flow meter as the measuring instrument in the pilot project. With the aid of the lower Yellow River communication engineering, the system makes full use of the resources such as water dispatching, hydrologic measurement, water quality monitoring, flood control and drought relief. The data of water supply are collected, transmitted, stored and proceeded automatically and modernization. After several years application, we have received series of experience, and improve the system more perfect. It is necessary to sum up the experience and provide scientific basis next step of the system.

Key words: DWS system, automatization, flow measurement, the lower Yellow River

1 DWS system (pilot project) construction objective

The Yellow River is China's second longest river, the lower Yellow River natural resources is meager, its population is numerous, where is one of the areas with prominent contradiction between water resources supply and demand. With the development of social economy the demand of water resources is more and more, the marketization of water resource is the inevitable trend of development. Therefore, accurate measurement for the Yellow River water supply management is the key, but the traditional method of flow measurement using artificial or cableway flow measurement, measuring two times every day, low degree of automation, high labor intensity, as well as large measurement error caused by the weather, environment and other factors, especially the test frequency can not meet the water resources management requirements. With the "digital Yellow River" construction further, Water Supply Bureau of the Yellow River Conservancy Commission (YRCC) proposed the construction of "digital Yellow River water management system" conception, from improving the water supply for accurate measurement of automatization level of the construction can realize automation, production management, water diversion precision measurement, water charging, water supply project management features such as "digital water supply" system, and the establishment of water diversion from the basic information database. Water Supply Bureau of the YRCC based on domestic and foreign advanced metering equipment selection and the different characteristics of the channels in the lower Yellow River, has selected H-ADCP, electromagnetic flow meter, automatic cableway flow measurement system, mobile flow measuring equipment as a "digital water supply" system of metering equipment. Water Supply Bureau of the YRCC installed advanced automatic measurement and transmission equipment, to achieve an online, real-time, accurate measurement, and cumulative record and display daily, monthly, annual total water supply, measurement accuracy to within plus or minus 5%.

2 Digital water supply system (pilot project) composition and function

2.1 Digital water supply system (pilot project) composition

Digital water supply system (pilot project) is mainly composed of a discharge automatic monitoring system, business application software system and the system operating environment.

2.1.1 Discharge automatic monitoring system

Discharge automatic monitoring system is the construction of the core content of "digital water supply" system, and metering equipment selection is the key to water discharge automatic monitoring system. The Yellow River differs from the general river, the high sediment content feature makes a lot of advanced equipment can not be directly applied to the Yellow River, must according to the characteristics of the Yellow River to reform or adopting engineering measure for the establishment of flow measuring device suitable for the flow conditions. Water monitoring is based on the realization of the flow monitoring, flow monitoring is based on the flow velocity and water level monitoring and implementation. The Yellow River Sluice of each monitoring section, in certain period water level amplitude is not big, the flow velocity and water level sensors are requested for higher resolution. For this project will be implemented most of the sections to the flowmeter monitoring, real-time on-line monitoring, and to the water level gauge and the hydraulic constructions using those as an alternate means to calculate water discharge. At present, acoustic Doppler (H-ADCP), electromagnetic flow meter as the measuring instrument in the pilot project has been selected for the installation, data transmission, remote monitoring and on-line monitoring using GPRS transmission main channel, the YRCC appoints microwave transmission network as the backup channel, RTU is stored as third backup plan, to ensure that the data flow continuity and integrity. According to the different equipment and transmission of different data formats, the class 1 of data type of remote transmission of the communication protocol is different, we will store different flow measuring instrument sensor test results in different database, compiled a dedicated special data conversion software, will be a variety of flow measuring sensor test data are integrated into the same database, to achieve a variety of advanced flow meter application and integration.

2.1.2 Business application software system

Business application software system is "digital water supply" pilot project important constituent, it includes water supply, water measurement, digital data collection, information services, water supply security, government affairs, digital water sites, operation maintenance and other construction elements. Among them, digital water database, water diversion measurement, water fee collection, information services and other four items (pilot project) for initial construction content.

2.1.2.1 Digital water supply professional database development

"Digital water management system of the lower reaches of the Yellow River " is in the" Digital Yellow River" engineering construction on the basis of development, in the information resources sharing principle, and digital water regulation, flood control and disaster reduction, hydrology, water quality monitoring, remote monitoring, culvert gate engineering maintenance and operation management system closely, and it will have to build the system by the useful information from the built system. Digital water supply professional database construction in addition to meet the water measurement and collection, but also meet the needs of information sharing and information service.

2.1.2.2 Water measurement system

Water measurement is core content of "the lower reaches of the Yellow River digital water supply system". Mainly include: water information receiving and processing and storing, water consumption, water diversion data management and other functions.

2.1.2.3 Water fee collection subsystem

Water fee collection subsystem includes water fee management, water supply period division management, user management, cost calculation and analysis, report generation and other functions.

2.1.2.4 Integrated information service system

According to the spirit of sharing principle, through the digital supply portal website, it is provided multi- information services that water volume, water price, water fee, water dispatch, water quality, water regime, culvert and gate, engineering and maintenance management, flood control and disaster reduction, policies and regulations etc.. Informations of water dispatch, water quality, water regime, culvert and gate, engineering, flood control and disaster reduction are derived from the business application system already built by YRCC.

2.1.2.5 Water supply security system

To ensure the goal of water supply safety, the system is provided with the function of water supply safety monitoring, forecasting and early warning, and the alarm, reporting and emergency response function when unexpected events of water supply project, water quality, sediment concentration occurs.

2.1.2.6 Government affairs system

On the basis of digital water portal website building, and the system provides all the water supply government related policies, functions, laws and regulations, as well as water supply related information, including online inquiries, telephone inquiries, information download, and information publishing et al.

2.1.2.7 Digital water portal website construction

Digital water portal website provides the network platform for government affairs. The construction contents include: water supply situation, government affairs, information services, policies and regulations, declare on the net, business training and other plates.

2.1.2.8 Operation and maintenance management system

Responsible for the "the Yellow River downstream digital water management system" for the safe and reliable operation, the system includes user management, user authentication management, system management, system function upgrading of setting management, system backup and recovery management and other functions, to take the necessary measures, ensure the data security and system security.

2.1.3 System physical environment construction

The system operation environment is "digital water supply system" headquarters, mainly including the emulated simulation screen and consultation system. According to the needs, the system has researched and developed the multi touch emulated simulation screen, and the simulation screen optimized and integrated the "digital Yellow River" project results. Simulation screen background is Yellow River water supply to the administrative regions and the irrigation area of the lower diagram, according to the relationship of administrative subordination, simulated screen embedded 14 small touch screen and 1 large touch screen and 102 supply station indicating lamp; according to the site position of the Yellow River hydrology and water quality monitoring, the 8 of river reaches are set for display unit simulation in main channel of the lower Yellow River. For an intuitive understanding of sluice engineering, the system has also established a Yellow River Sluice video database, to introduce the location, construction, project appearance of each sluice of the Yellow River by video.

2.2 DWS system (pilot project) function

Since the system running in July, 2008, the system has collected a large amount of pilot gate diversion data, via a transmission channel sent to the digital supply center server, in the flow measurement system on original data database, automatic check processing, after processing the data in real-time database of water flow monitoring, and provide basic data for water quantity statistics and water fee collection. At the same time, the data is also available through emulated simulation screen display, the central controller can be simulated screen all the data and video information freely switch, fully embodies the technology advanced, function is fully reflects the practical, simple operation characteristics.

3 "DWS system" application prospect

"DWS system" pilot project of water diversion from the established management information automatic collection system can comprehensively, accurately, rapidly access to all relevant information for management and decision making, so that the water supply management is provided with scientific and decision-making foundation and advanced technical means. "DWS system" pilot project will greatly promote the development of career of water supply in the Yellow River, and improve water management level and content of science and technology.

References

Wang Xiwen. The Lower Yellow River Digital Water Supply System (Pilot Project) Construction and Application [R]. Zhengzhou: Water Supply Bureau, YRCC, 2009.

Wang Hongqian. Study on the Water Supply Scale Variation and Its Influencing Factors in the Lower Yellow River [D]. Xi'an: Xi'an University of Technology, 2007.

An Irrigation Water Use Prediction Model for Ning – Meng Irrigation Area Based on Wavelet Neural Network

Han Jinxu, *Huang Fugui*, *Luo Yuli* and *Chen Weiwei*

Yellow River Institute of Hydraulic Research, Zhengzhou, 450003, China

Abstract: The prediction model is developed based on Wavelet Neural Network to forecast the irrigation water diverted from the Yellow River in Ning – Meng Irrigation Area. In different years, the average absolute relative error of the prediction results based on Wavelet Neural Network is less than 5%. Compared with BP Neural Network, no crops and meteorological data must be requested. The forecasting model could be used in the data – deficiency region.

Key words: irrigation water diverted from the Yellow River, wavelet neural network, prediction model

1 Introduction

Irrigation water occupies a high proportion of agricultural water consumption in irrigation district. For regional water resources allocation, water use planning and irrigation project development, it is important to predict irrigation water. Traditionally, the Penman's method (Xie Chunyan, 2004; Ma Qiang, 2011), Regression Analysis (Wang Xinrui, 2001) and Water Balance method (Luo Yi, 2008) are the common study targets, which depend on the data series and regression indicators (Huang Xiuqiao, 2004). The Grey System theory (Bai Cunyou, 2004; Shao Dongguo, 1998; Chen Zixuan, 2010; Chi Daocai, 2009) predicts irrigation water use based on some irrigation data. The calculation accuracy of gray prediction is low, mainly because of using the single exponential model to describing the time series (Huang Xiuqiao, 2004). In addition, with its powerful nonlinear mapping ability and fault tolerance property, Neural Network is gradually being used for irrigation water demand forecasting. Some authors report that the elementary irrigation water prediction model has been developed based Neural Network technology (Tang Yanfang, 2006; Zheng Yusheng, 2004; Xu Jianxin, 2005). Using BP Neural Network, Meng Chunhong (2004) calculated the irrigation water use in lower reaches of Yellow River irrigation district. BP Neural Network has good precision, however, it requires large amount of crops and meteorological data.

Combing with the advantages of the Wavelet Analysis and Neural Network, Wavelet Neural Network can effectively identify the periodic component in the data series. In recent years, Wavelet Neural Network has been used in the fields of hydrology and urban water supply systems (Luo Limin, 2005; Wan Xing, 2005). With huge water consumption and relatively single planting structure, the monthly data series of Ning – Meng Irrigation Area show inter – annual quasi – periodic variation. Such feature provides Wavelet Neural Network a chance for application. The overall objective of this study is to develop a model to forecast the irrigation water diverted from Yellow River in Ning – Meng Irrigation Area. The model, based on Wavelet Neural Network, could be used in the water resource management in the future.

2 Structure of wavelet neural network

2.1 Basic principle

Compared with Fourier analysis, Wavelet Analysis has excellent local analytical capacity (Feng Yan, 2007). Based on BP Neural Network topology structure, with the wavelet basis function as the hidden layer activation function, the Wavelet Neural Network calculates the procedure.

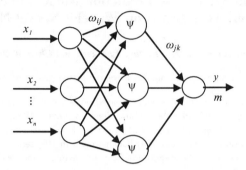

Fig. 1   The topology of Weural Network

Fig. 1 indicates the topology of Wavelet Neural Network, in which, $x_i(i = 1,2,\cdots,n)$ = input parameters; y = forecast result; ω_{ij} and ω_{jk} are network weights; ψ = wavelet basis function, where ψ is Morlet basis function, and its structure is as follow.

$$\psi = \cos(1.75x)e^{\frac{-x^2}{2}} \tag{1}$$

where, $x_i(i = 1,2,\cdots,n)$ is input , the hidden layer output formula of Wavelet Neural Network can be expressed in the following form:

$$h_j = \psi\left(\frac{\sum\limits_{i=1}^{n}\omega_{ij}x_i - b_j}{a_j}\right) \quad (i = 1,2,\cdots,n; \quad j = 1,2,\cdots,l) \tag{2}$$

where, h_j = the output of node j in hidden layer ; a_j = contraction – expansion factor of the wavelet basis function ; b_j = displacement factor of the wavelet basis function.

Eq. (3) can yield forecast result:

$$y_k = \sum\nolimits_{j=1}^{l}\omega_{jk}h_j \quad (k = 1,2,\cdots,m) \tag{3}$$

where, y_k = the output of node k in output layer.

2. 2   Prediction steps

Step 1. Network initialization. At this stage, parameter a_j,b_j,w_{ij},w_{jk} are randomly initialized, and learning efficiency is set.

Step 2. Sample classification. The data series is divided into training sample and test sample. The training sample is used to train Wavelet Neural Network, and the test sample examines network prediction accuracy.

Step 3. Output prediction. In this phase, training sample is inputted Wavelet Neural Network in order to compute network prediction output and the deviation (e_0) between the forecasted output and desired one.

Step 4. Weights value adaptation. In this step, according to the error , the weights and parameters are amended. Subsequently, the program is to be judged, if not end return to step 3.

3   Modeling

Irrigation water use is closely related with the amount and the moment of the previous water diversion. The Wavelet Neural Network is developed, based on the inter – annual quasi – periodic variation of the monthly diversion in the Ning – Meng Irrigation Area. The network is divided into input layer, hidden layer and output layer. The input parameters (n) is the irrigation times before nowadays, where $n = 4$. The output is the amount of water diversion this month. The hidden layer node is formed by the wavelet basis function, and the number of the nodes is 8. Therefore, the structure of the Wavelet Neural Network is $4 - 8 - 1$.

The monthly diversion water series, totally about 15 years (1996 ~ 2010) , are respectively divided in training sample(1996 ~ 2005) and testing sample (2006 ~ 2010) , in Ningxia and Inner Mongolia irrigation district.

Simultaneously, the BP Neural Network is built based on the meteorological and crop data of the same region, such as monthly precipitation, monthly average relative humidity, monthly mean temperature, monthly mean wind speed, monthly mean ET_0 and so on. And the data series has the same dividing without a doubt. Thus, the structure of the BP Neural Network is $7 - 7 - 1$.

4 Results and discussions

The diversion water predicted by different Neural Network in Ningxia and Inner Mongolia region is listed in Tab. 1. In which, VA_{ct} is the actual irrigation water diverted from Yellow River; VW_{av} is the prediction amount of irrigation water diverted from Yellow River by Wavelet Neural Network; V_{BP} is the prediction amount of irrigation water diverted from Yellow River by BP Neural Network; MAPE is mean absolute percentage error, and NSC is Nash – Suttcliffe coefficient. As indicated in Tab. 1, the NSCs of predictions forecasted by different neural network (threshold value of the NSC = 0.75) are more than 0.85, greater than the threshold value. On the other hand, the MAPEs are between $4\% \sim 5\%$. It is clear that the predictions of the Wavelet and BP Neural Network are precise.

The close match between VAct and VWav is shown in Fig. 2. The figure indicates that VW_{av} from the different regions show equally good agreement with VA_{ct}. Especially, for fitting the prediction to the actual value, the determinacy coefficients (R_2) are 0.94 (in Inner Mongolia Autonomous Region) and 0.92 (in Ningxia Hui Autonomous Region). It is obvious that the Wavelet Neural Network can accurately predict the changes of irrigation water use in Inner Mongolia and Ningxia Hui Autonomous Regions.

Tab. 1 Wavelet Neural Network and BP Neural Network predicted results comparison

(Units: $\times 10^9$ m^3)

Year	Inner Mongolia			Ningxia		
	VA_{ct}	VW_{av}	V_{BP}	VA_{ct}	VW_{av}	V_{BP}
2006	71.88	69.39	71.68	70.88	71.84	75.11
2007	71.15	67.82	69.81	67.27	62.29	63.93
2008	65.59	68.65	66.50	70.32	73.07	63.64
2009	74.86	70.31	66.56	66.30	70.54	66.04
2010	72.93	71.11	78.21	65.28	68.62	66.75
MAPE(%)	—	4.37	4.28	—	4.84	4.61
NSC	—	0.85	0.87	—	0.87	0.88

It is noteworthy that MAPE and NSC of the prediction based on BP Neural Network are a little greater than the wavelet's. However, in order to forecast irrigation water, too much meteorological and crop data must be inputted into BP Neural Network. Compared with BP Neural Network, Wavelet Neural Network only calls for a little data of monthly water diversion. It is surely worth emphasizing that, in the data – deficiency region, the Wavelet Neural Network can be used to predict the irrigation water.

5 Conclusions

For predicting diversion water in Ning – Meng Irrigation Area, BP and Wavelet Neural Network models are developed. As the result, some conclusion should be confirmed.

(1) The prediction of the BP and Wavelet Neural Network can be much agreement with actual

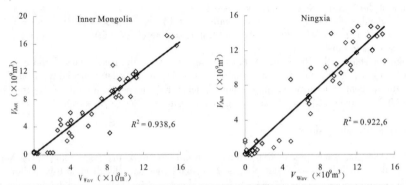

Fig. 2 Pelationship between actual water diversion and Wavelet Neural Network prediction in Inner Mongolia and Ningxia Hui Autonomous Region

water diversion in Ningxia and Inner Mongolia province.

(2) Compared with BP Neural Network, calculation accuracy of the Wavelet Neural Network is slightly lower, but no crops and meteorological data are requested. The Wavelet Neural Network forecasting model could be used in the data – deficiency region.

References

Xie Chunyan, Ni Jiupai, Wei Zhaofu. A Review on Water – saving Agriculture Crop Requirement and Irrigation Requirement [J]. Chinese Agricultural Science, 2004, 20(5) ;143 – 147.

Ma Qiang. Prediction Method of Analysis of Irrigation Water [J]. Shanxi Water Resources, 2011, 4; 7 – 8.

Wang Xinrui. Correlation Analysis of Irrigation Water and Precipitation in Baojixia Irrigation Dstrict [J]. Technique of Seepage Control, 2001(3) ;42 – 47.

Luo Yi, Lei Zhidong, Yang Shixiu. A Conceptual – stochastic Model for Predicting the Dynamic Soil Water Storage in Crop Root Zone [J]. Journal of Hydraulic Engieering, 2008, 31(8) : 80 – 83.

Huang Xiuqiao, Kang Shaozhong, Wang Jinglei. A Preliminary Study on Predicting Method for the Demand of irrigation water resource [J]. Journal of Irrigation and Drainage, 2004, 23(4) : 11 – 15.

Bai Cunyou, Feng Xu, Zhang Shengtang, et al. The Application Research on the Gray Equal – dimension and New – info Model in the Prediction of Irrigation Water Consumption [J]. Journal of Northwest Science Technology University of agriculture and forestry. (Natural Science Edition), 2004, 32(9) ; 115 – 118.

Shao Dongguo, Guo Yuanyu, Shen Peijun. Research on Long – term Forecast Model of Regional Irrigation Water Requirement [J]. Journal of Irrigation and Drainage, 1998, 17(3) ; 26 – 31.

Chen Zixuan, Li Zhanbin, Li Peng, et al. Application of Grey LS – SVM to the Forecast of Irrigation Water [J]. Journal of Water Resources and Water Engineering, 2010, 21(1) ; 75 – 78.

Chi Daocai, Tang Yanfang, Gu Tuo, et al. Prediction of Irrigation Water Use Using Parallel Gray Neural Network [J]. Transactions of the Chinese Society of Agricultural Engineering, 2009, 25(9) ; 26 – 29.

Tang Yanfang, Chi Daocai, Wang Dianwu, et al. Neural Network's Application on the Forecast of Irrigation Water Use Based on MATLAB7 [J]. Journal of Shenyang Agricultural University, 2006, 37(6) ; 867 – 871.

Zheng Yusheng, Huang Jiesheng. Forecast of Irrigation Water Use Based on Neural Network [J]. Journal of Irrigation and Drainage, 2004, 23(2) ; 59 – 61.

Xu Jianxin, Li Yanbin, Gu Hongmei. Forecasting Agricultural Irrigation Based on Neural Network [J]. Journal of Agricultural Mechanization Research, 2005, 4: 115 – 117.

Meng Chunhong, Xia Jun. The Application of Neural Network Research in the Irrigation District of Yellow River Downstream [J], Yellow River, 2004, 26(1): 37 – 38.

Luo Limin, Fang Hao, Zhong Yue, et al. The Algorithm of Wavelet Neural Network and its Application to Area Water Demand Prediction [J]. Computer Engineering and Applications, 2006, 3: 200 – 202.

Wan Xing, Ding Jing, Zhang Xiaoli. The Application of Wavelet Neural Network Research in the Runoff Forecast [J]. Yellow River, 2005, 27(10): 33 – 36.

Feng Yan. Wavelet Neural Network and its Application in Hydrology and Water Resources [D]. Northeast Agricultural University, 2007.

Study on GIS – Based Simulation of the Minor Watersheds with a Check Dam System

Wang Qingyang, *Song Jing* and *Gao Jinghui*

Upper and Middle Yellow River Bureau, YRCC, Xi'an, 710021, China

Abstract: The backbone dams of the dam system in a minor watershed were taken as the main object, while the study was focusing on the application of the "3S" – tchnology. The optimal dam sites were selected by calculating the maximum values of the storage capacity to engineering quantity ratio curves along the gully directions. The essential data acquisition of watersheds was performed using the remote sensing technique. The data was processed with the geographic information system, the computer graphics technique, the database technology and the virtual reality technology. Based on Windows platform, the object – oriented approach, the 3D visualization technology, the virtual reality technology and other advanced technologies were used to improve a conventional 2D data model to 3D space. The GIS system model, taking the digital ortho image, the digital elevation model, the digital line graph and the digital raster graphic as a comprehensive treatment object, was developed to study how to acquire the recognition of spatial relations between dam height and storage capacity, dam height and area of built land, dam height and engineering quantity, and individual dams by laying out works or cutting cross – sections on topographic maps. The dam system space – time topology model and the interval hydrological analysis model were established, and based on which, a GIS – based digital simulation system of the check dam system in a minor watershed was built, realizing the 3D solid model presentation of planned dam system layout, real – time display, tabulation and output of the engineering information (dam height, storage capacity, area of built land, etc.) of various check dams as well as top view of 3D dynamic demonstration and flight observation of dam system in different operation phases and under storm flood condition.

Key words: check dam, the minor watershed with a check dam system, simulation

A large – scale construction of check dams in the Loess Plateau region is inevitably to be launched. Traditional means and methods for planning and design can no longer meet the requirements of current working strength and depth. Firstly, there is not a practical layout optimization method available to make dam system layout more scientific and reasonable. Secondly, there is not a complete and practical planning system based on modern technological means available to meet current planning and design requirements. Thirdly, the knowledge on the development and formation of dam system remains at numerical calculation stage without a visual and intuitional digital simulation available. Traditional dam system planning on a manual basis can not conduct an analysis of the mutual effect between dams and the impact of dam system construction on the reallocation of watershed water and sediment but merely an estimate based on principles of hydrology and experience with heavy workload and low accuracy in planning and design.

1　Related theories

The check dam systems on the Loess Plateau are generally established in individual minor watersheds, where a number of large, medium and small check dams are built to form an organic system for flood control, silt retention and farmland building.

The GIS – based simulation system of the minor watersheds with a check dam system was developed using high – tech means and based on the principles of geological geomorphology, fluid dynamics, structural mechanics, soil mechanics, engineering hydrology, ecological economics, technical economics, hydraulic engineering design technique, 3S technology, CAD technology, 3D simulation technology, and software engineering technology. It can provide a scientific, practical and easy – to – use simulation environment for the planning and design of loess plateau dam systems and the technical review of dam system planning/design schemes.

Computer software technology was used to model the real environment of dam systems. Different theories and techniques were applied for various aspects of dam system planning and design.

For dam system layout (number, scale and distribution), DEM & DTM, 3D virtual simulation technique and engineering hydrological model were integrally applied. On this basis, the DEM based interval hydrological analysis model and the dam system space – time topology model were established. As a result, automatic withdrawal of tables and graphs concerning gully, minor watershed boundary, interval catchment area, gully cross – section, watershed classification and gully classification from 1: 10,000 DEM was achieved, and dynamic analysis of gully gradient, inundation loss, dam system flood and sediment reallocation, dam construction potential and yield assurance capacity can be conducted, providing a powerful technical platform for dam system layout. In addition, this technique can provide an underlying technical support for reservoir flood regulation analysis.

For dam site selection, the principles and approaches of geological geomorphology and ecological economics were integrally applied.

For typical dam design, the models of fluid dynamics, structural mechanics, soil mechanics, engineering hydrology, hydraulic engineering design technique were effectively integrated with DEM model so that various design parameters can be directly worked out from 3D terrain to generate the 2D design drawings competent for engineering construction, whose results are seamlessly integrated with CAD.

2 Technical essentials

The 3D dam system simulation systems are more difficult to establish than the 2D GIS system for their distinctive features as follows:

For 3D GIS system, exact expression of an irregular geo – scientific object, whether for vector or raster structure, always comes up against the issue of storage and processing of large quantities of data. For dam system simulation system, a dam system usually consists of dozens of soil and water conservation works together with numerous streams, roads, residential areas and other surface objects, which lead to a complex 3D scene – structure. If there is not a suitable data model and control strategy, the desired display effect for the system is difficult to achieve, much less a good interactive interface.

In 2D dam system planning, soil and water conservation works are generally represented by abstract symbols, which can not provide intuitive display of the works and their mutual association. For the 3D dam system simulation systems, simulation of real soil and water conservation works is a basic requirement of virtual reality. This makes the model more complicated, even combination construction become necessary. Therefore, selection of a suitable design pattern and organizational method is also important for establishing a dam system simulation system.

The 3D dam system simulation system is an integrated management platform that combines digital surface model, soil and water conservation engineering model and various geographic information datasets. How to combine these data together organically to provide a good interactive inquiry and maintenance function is a requirement of systematic integrity.

3 Establishment of the system model

3.1 The dam system space – time topology model

A dam system space – time topology model was developed based on the principles and methods of engineering hydrology, digital terrain models, dam system planning and check dam design. In calculation and analysis of existing dam system or planning scheme, the comprehensive influences of dam construction time – sequence and dam system operation and development were taken into full consideration. Water and sediment reallocation arising from dam system construction was automatically analyzed. Space – time topological processing was automatically conducted in light of prerequisites. With accurately calculated results, the model provides a solution for computer – assisted dam system planning.

The model has the following characteristics:

The model is not only based on space and time, but on prerequisites.

Planning scheme offers a direct drive to the model. For example, in case of any alternation (e. g. adding or deleting dams, changing dam site or type, adjusting construction time) that could lead to change of the planning scheme, the model would response to such changes immediately.

DTM also offers a direct drive to the model. The relation between any two dams needs to be worked out for topology establishment. The relation derives from hydrological analysis results. The results stem from DTM. When users correct the DTM by modifying contour lines or other ways, the system can automatically detect the "failure" of the dam system simulation system and rebuild it. Therefore, the dam system space – time topology model based on the dam system simulation system is dynamically and implicitly built under the drive of space, time, prerequisites, planning scheme and DTM. Not only does it assure the analysis and arithmetic logic of the system in line with the theories and specifications of dam system planning and check dam design, but enable users not to follow the step – by – step operating procedure when they use the system. Also, there is no need to re – input planning data when basic information changes. The model is very practical and capable.

3.2 Interval hydrological analysis model

Dam system planning work is dependent on a complicated hydrological model, including watershed demarcation, gully withdrawal, gully classification, gully gradient analysis, establishment of topological relation between any two locations, withdrawal of catchment interval (entire interval or the interval with upstream confluence sites deducted) at any confluence site, and calculation of flood data within specified range, of sediment data within specified range, and of submerged area, etc.

Although various hydrological models have already been developed based on GIS software, they usually provide a single function. Moreover, the data obtained therefrom often can not be directly used in dam system planning work. As a result, it is extremely laborious to use these conventional functions for analysis and calculations in designing the dam system construction scheme. The more critical factors leading to the inapplicability of the traditional models lie in the "special" requirements of dam system planning work itself. For example, frequent adjustments often need to be made when formulating the dam system planning scheme, including adjustments of the planning scheme and adjustments of watershed hydrological/sediment data and/or adjustments of calculation method. For the purpose of dam site survey or scheme calculation and analysis, the simulation system must have the function of hydrological analysis on the given confluence site (not only the entire watershed), and must be able to suppose a deductible upstream confluence site at discretion

The hydrologic analysis model developed based on the GIS – based simulation system of the minor watersheds with a check dam system is briefly called "interval hydrologic analysis model". The model has the following characteristics:

"Interval": The interval has two meanings. One means the entire catchment area of the given confluence site. The other means the partial catchment area of the given confluence site with one or more upstream confluence sites deducted. The former can be considered as a special case of the latter, so they can be generally called as "Interval".

"Vector operation": The GIS – based hydrological models are mostly developed based on raster. Whereas, the hydrological models of this system are all based on vector. For instance, the withdrawn catchment area and submerged area are closed polygons in three – dimensional space; gullies are polylines; and dam sites are points. It is polygons, polylines and points that are directly put into operation instead of raster data.

At the time of electronic dam site survey or analysis and calculation of planning scheme, investigation of hypothetical range or dam system topology is a drive to the model. As another example, when withdrawing gully or analyzing gully gradient, the default initial upstream point is worked out from a catchment area value which is set according to business on a flow network. Therefore, these models are subject to the direct drive of business model.

The computation models of flood/sediment data are configurable. In actual work, the methods used for flood/sediment computation are often different at different places. Besides, it is usually needed to use the data from different sources for analysis and comparison in order to pick out a suitable set. The "background data" of the model is the flow network. When users correct the DTM by modifying contour lines or other ways, the system can automatically detect the "failure" of the flow network and update it. Then, the model can also be updated accordingly through a transitive relationship.

4   Establishment of the simulation system of minor watersheds with a check dam system

Establishment of 3D physical simulation lays a foundation for establishment of entire virtual scene. Establishment of a model mainly involves modeling of 3D terrain and other 3D entities such as structures, animals and plants, vehicles, etc. Whether they are appropriately modeled or not directly affects the visual effect of virtual scene and the running speed of the system. The automation degree of modeling work is also closely related with the development cycle and workload of the whole system.

Input to the system remote sensing images and dam system planning scheme that has been formulated on digital topographic maps to automatically plot the catchment area of dams and calculate the control area of dams. In light of dam height – reservoir capacity and dam height – dammed land area relation curves and mean annual sediment transport modulus, the system can automatically calculate the sedimentation amount of dams after 1, 2, ···,20 year operation. We can see 3D dynamic presentation in different years of operation from different angles. Input to the system runoff modulus and sediment content of the flood at different frequencies to calculate the incoming flood and sediment runoffs of each dam at different frequencies of flood discharge with 3D dynamic presentation available (see Fig. 1).

The interval hydrological model and the dam system space – time topology model shall be combined for analysis of the flood control capacity and yield assurance capacity of a dam system (a planning scheme or an existing dam system) or for design of key indices of a single dam. In operational process, the data used by the interval hydrological model experiences a reverse transmission. When a request is raised to the interval hydrological model for calculation of the interval flood or sediment yield data of a dam, the interval is expressed by the dam. However, in the next step, the interval expressed by the dam is treated in combination with the topological dam system, and transformed into the interval expressed by the downstream location and deducted upstream location. After that, the expression of the interval is further transformed into the boundary of an area (polygon) by the algorithm based on flow direction and confluence cumulative data. By then, the calculation method and parameters that users prepare can be used to work out final data.

5   Applications

After investigation and research of the minor watersheds (Chenjiagou, Jiajiagou, Qingshuiping, Zhaichenggou, etc) in Shenmu County, Shanxi Province, the simulation system of minor watersheds with a check dam system was used to complete the dam system planning. In the work, the watershed dam system planning system was used to select dam sites, build dams on 3D topographic maps, query information and simulate incoming water and sediment. The results showed that the accuracy of the data provided by the watershed dam system planning system meets the requirements of feasibility study.

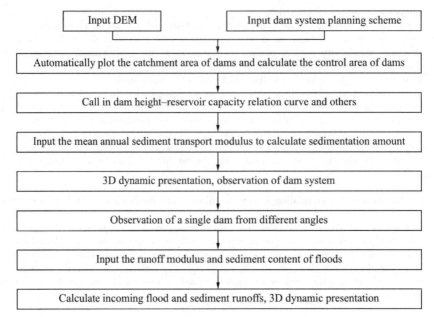

Fig. 1　Flow chart of the simulation system of minor watersheds with a check dam system

6　Conclusions

　　Establishment of the simulation system of minor watersheds with a check dam system for the study of overall layout pattern of dam systems in loess plateau minor watersheds that are subject to soil erosion is conducive to understanding and grasping of the mechanisms of how a check dam system retains, regulates and stores watershed water and sediment, and to the analysis of intercorrelation and interaction between slope erosion and gully erosion so as to make known the ecological, economic and social benefits of dam system construction, providing a powerful technical support for large – scale construction of dam system in the Loess Plateau.

References

Zheng Baoming, et al. Theory and Practice for Construction of Sam Systems in Minor Watersheds [M]. Zhengzhou：Yellow River Conservancy Press, 2004, 15 – 16.
Mao Feng, et al. GIS Library Building Technology and Application [M]. Beijing：Science Press, 1999.
Zhou Yunxuan,Wang Lei. Study on DEM – based GIS topographic Analysis and Test Methods [J]. Computer Application Research, 2002(12)：50 – 53.
He Huiming. Establishment of Digital Elevation Model (DEM) and Analysis of Related Issues [J]. China Agricultural Resources and Regional Planning, 2002, 23 (6)：55, 58.
Jenson K, Dominique J. O. Extracting Topographic Structure From Digital Elevation Data for Geographical Information System Analysis [J]. Photogrammetric Engineering and Remote Sensing, 1998, 54 (11)：1593 – 1600.

Cascaded Hydropower Station Daily Load Dispatch with Time-Shared Price

Yang Yongjian[1] , *Fu Xiang*[2] and *Xu Chenguang*[3]

1. Yellow River Engineering Consulting Co. , Ltd. , Zhengzhou, 450003, China
2. State Key Lab. of Water Resources and Hydropower,
Wuhan University, Wuhan, 430072, China
3. School of Resources & Environment, North China University of
Water Resources and Electric Power, Zhengzhou, 450011, China

Abstract: It is beneficial for improving generation benefit to optimize the daily load dispatch for cascaded hydropower station with time-shared price. It is beneficial for improving generation benefit to optimize the daily load dispatch for cascaded hydropower station in different electric price market. In accordance with the peak-flat-valley period interval electric price, the daily load dispatch models for cascaded hydropower station is built and solved by chaotic genetic algorithm (CGA). Take a cascaded hydropower station for instance; the daily optimization operation results are comparative analyzed in simple electric price and time-shared price. The results show that the dispatch models and solving method are reasonable and the economic benefit is remarkable.

Key words: water management, daily load dispatch, time-shared price, chaotic genetic algorithm (CGA), cascade hydropower station

1 Daily load dispatch of cascaded hydropower stations

As the electric system reform and marketization, it is imperative to optimize the allocation of power resources and improve the overall economic efficiency of electric power enterprise by using electricity price to adjust the economic operation of electric power. In 2006, the State Council published "The decision about strengthening the energy-saving work", says it will deepen the reform of energy prices, strengthen and improve price management, perfect the time-shared price system, guide the user to the rationality of electricity usage, saving electricity. Load distribution effect direct impact efficiency and benefit of hydropower station when hydro-power stations participate the bidding process.

Distribution of daily load of cascade hydropower stations, refers to dispatch control center implement load dispatch a few days ago on the hydropower station, or carry out optimal deployment for water which can be use in one day, it belongs to the category of short-term scheduling. Traditional short-term optimal scheduling just considers load and water, without taking into account the economic efficiency of energy and load. Under the conditions of market economy, especially for the start of power system reform in China, hydro-thermal power stations implemented "bidding". In this case, in order to maximize profits, price factor play a key role in the load distribution. How to improve power plant efficiency together ensure safety and quality which is of practical significance for the hydroelectric power plant. How to coordinate water allocate within cascade hydropower stations, which is a new challenge for water resources comprehensive utilization.

2 Research progress of optimization schedule under time-shared price

Relevant bibliographies have implement research for some problems about cascade hydropower optimization schedule. In 1997, Wan Yonghua establishes a model which consider peak-valley and wet-dry price in hydropower plant on the basis of reasonable share both capacity benefit and energy benefit. The model to be validated though example calculation and results analysis, which provide theory foundation and practical method for peak-valley and wet-dry price formulation of hydropower plant. In 1999, Zhang Yuhui and Feng Quanlong proposed hydropower plant long-term optimal operation model under peak-valley and wet-dry price, and achieved significant benefits though the Bao Zhusi hydropower station example. In 2002, Ma Guangwen establish multi-use optimization model which consider peak-valley and wet-dry price, and used genetic algorithm solution. It shows that cascade optimization model under time-shared price can improve the output of hydropower plant, reached water resources reasonable using, coordinate the conflict of energy produce and water supply through application of Zi Pingpu reservoir in SiChuan province . Also some related research, bibliography only research long-term optimization scheduling under flood and drought season price, short-term scheduling problem under peak-alley price not involved. Bibliography research level change of multi-year regulate reservoir under time-shared price, but just limited to long-term optimization problem. Bibliography although established two models of cascade hydropower stations short-term optimization under peak-valley price, that is "energy maximization under limited water" and "water minimize under certain energy", but analysis calculation results just from cascade total benefit view, the level, water head and output of each hydropower station under time-shared price not be analysis, while different regulate capacity.

The research about cascade stations daily load dispatch also comparison less, this paper established daily load dispatch optimization model of cascade hydropower station under time-shared price, while considered different regulate performance in model. Take Qingjiang cascade hydropower stations as example, analysis comparison for output and water level features of hydropower stations under single price and time-shared price through using genetic algorithm on model for has solution calculation, while for time-sharing electric price of same cascade in the different regulation performance reservoir of run way do has analysis comparison.

3 Daily load dispatch optimization model under time-shared price

3.1 Time-shared price

The time-shared price as a new price system is promoted step by step in the reform, which is established depending on the cost of produce and transport of the different periods. The so-called time-shared price is to perform different price in different periods, as a means to stimulate the power consumption, solving the problem of peak and valley difference, relatively stable load curve of the day. It consists of two classes: one is the different price in one day, which is peak-valley electricity price for short; the second is wet-dry seasonal different price or different price in summer and winter, which seasonal price for short.

For seasonal price, because it reflects the price difference of wet, flat, dry periods in one year, which could be apply in the long-term optimal operation, for short-term optimal scheduling, mainly is the application of peak-alley electricity price.

3.2 Mathematic model

According to the research needs, this paper established two models with energy maximization: ① Model which don't consider the time-shared price with single price at each interval, ② Model which consider the time-shared price. Though comparison of two models to shown that the impact of the cascade power generation, efficiency, and water level change with time-shared price.

(1)Single price:

$$F = \max \sum_{i=1}^{N} \sum_{t=1}^{T} (p_i \cdot N_{it} \cdot M_t) \tag{1}$$

(2)Time-shared price:

$$F = \max \sum_{i=1}^{N} \sum_{t=1}^{T} (p_{it} \cdot N_{it} \cdot M_t) \tag{2}$$

where, N_{it} means output of plant i at period t, use follow formula: $N_{it} = A_i \cdot Q_{it} \cdot H_{it}$; Q_{it} means turbine flow of plant i at period t ; M_t means hours of plant i at period t; A_i means output coefficient of plant i; p_i means price of plant i in calculate period; p_{it} means price of plant i at period t ,that is peak-flat-valley price; H_{it} means water head of plant i at period t.

3.3 Constraints

The constraints include: the reservoir water balance, the cascade power station water contact, reservoir storage capacity constraints, the reservoir discharge constraints and output constraints of cascade hydropower stations.

4 CGA of cascade short-term dispatch with real number code

Genetic algorithm as a intelligent of global search algorithm, shows its unique charm in numerical optimization, system control, structure optimization design and parameter identification etc. , while also exposed out many insufficient and defects, such as strong global search ability, poor local optimization and adaptation ability on search space changes. Strong randomness led to low search efficiency of algorithm while in cross and variation evolution process. To solve problem of premature convergence, local minimal etc. in genetic algorithm, and keep good performance of genetic algorithm, according to characteristics which chaos movement can traverse all state unrepeated by its itself law within a certain range, we enlarge traverse range of chaos movement to range of optimization variable, then carry out chaos optimization for some individual through genetic algorithm, last obtained optimal solutions of problem which overcome local minimal problem.

Its basic philosophy is the genetic algorithm after completing a choice, overlapping and the variation operation, then uses changes the criterion chaos optimization method to the population in adapts the value high partial individuals to carry on the chaos search, the guidance population evolution, after discovering the heredity optimization, a more outstanding chromosome, takes the reproduction next generation the male parent. The chaos optimal process is divided two stages to carry on: Firstly, in turn traversal the process in the variable value scope each spot, accepts the good spot achievement optimum point; Secondly, take the current optimum point as the center, the additional chaos small perturbation, carries on the thin search to seek for the optimum point. Chaotic motion's above nature achievement avoids falling into the confined most superior optimized search mechanism, may make up the genetic algorithm exactly easily to fall into is partially most superior, convergence rate slow flaw.

Take the cascade energy benefit maximization as the goal, the real number code short-term dispatch chaos genetic algorithm has designed. Short-term scheduler program using the Matlab language is shown in Fig. 1.

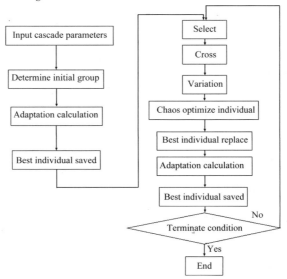

Fig. 1　Flow chart of model calculation

5 Example calculation

Take some cascade hydropower stations as example, explanation output characteristic and water level change under single electricity price and time-shared price. Cascade hydropower stations as shown in Fig. 2, the computation control conditions as follows:

The water was relatively flat season corresponding cascade power plant start, terminate the state of water level, respectively. Power Station A: 371 m ~370.6 m, Power station B: 188.56 m ~188.48 m, Power station C: 78.96 m ~78.92 m. The total calculation interval are 24 h (48 intervals), consider a single electricity price with 0.37 Yuan / kWh, with supposed peak : flat : valley price ratio of 5:2:1.

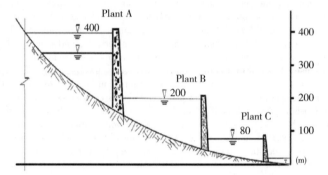

Fig. 2   The sketch diagram of cascade hydropower station

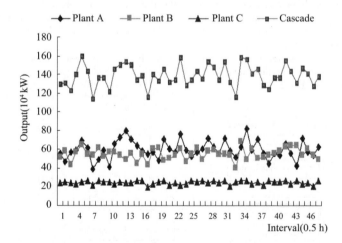

Fig. 3   Output characteristics with single price

Cascade stations daily load dispatch is calculated under single price and time-shared price. Cascade power stations output characteristics of both models are shown in Fig. 3 and Fig. 4, water level process of plant A is shown in Fig. 5 and Fig. 6, water level process of plant C is shown in Fig. 7 and Fig. 8.

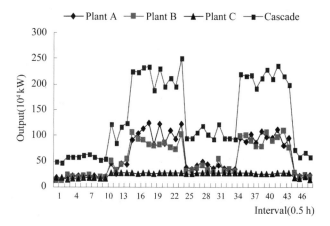

Fig. 4   Output characteristics with time-shared price

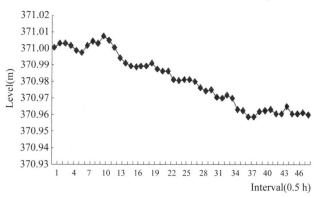

Fig. 5   Level change of Plant A with single price

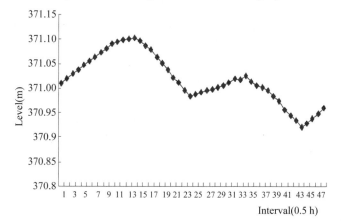

Fig. 6   Level change of Plant A with time-shared price

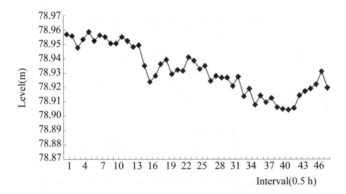

Fig. 7　Level change of Plant C with single price

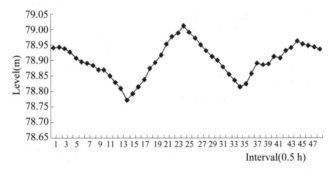

Fig. 8　Level change of Plant C with time-shared price

Output of each cascade plant and whole to be quite steady under single price, water level of each reservoir fall step by step until the goal; Under the mode of time-shared price, the flow which through turbine is big when the price is high. Therefore, the water level change process and time-shared price trend of plant A are just opposite, that is the water level raise when flow is big and the water level drop when flow is small. The vary process plant B is similar to plant A.

For Hydropower Station C, due to its small capacity, anti-regulation power plants in the upper reaches of two hydropower stations, optimized according Hydropower Station A, hydropower B, the water level will change and at peak times will produce a large number of abandoned water, does not meet the optimal conditions, therefore, be dealt with separately on Hydropower C. In order to prevent the power plants in the high - tariff period a large number of disposable water power plant in the low tariff period to increase the reference flow, so contrary to the water level in the process and hydropower stations A and Hydropower Station B power plant. From the optimization results can be seen to do so not only increases efficiency, but also to avoid a large number of abandoned water

The whole cascade energy output and total benefit under single price and time-shared price as shown in Tab. 1.

As can be seen from the table, energy productivity are similar under two models, but the total benefit of distinct. See from power generation of peak, flat and valley period view, energy is distinctly different of each period both time-shared price and single price. Under the time-shared through scheduling, assumed under the price ratio, cascade total power generation benefit of improving the 90.1%, greatly improved the power benefit, while cascade hydropower station of water level variation in a large, also put forward higher requirements for scheduling of hydropower station.

Tab. 1　Energy compare under two optimization models

Optimization mode	Plant	Peak Energy (10^4 kWh)	Flat Energy (10^4 kWh)	Valley Energy (10^4 kWh)	Whole Energy(10^4 kWh)	Benefit (10^4 Yuan)	Whole Energy(10^4 kWh)	Whole Benefity (10^4 Yuan)
Single price	Plant A	595.18	447.19	390.31	1432.68	530.09		
	Plant B	544.92	385	385.22	1315.14	486.6	3,333.07	1,233.23
	Plant C	241.49	173.25	170.51	585.25	216.54		
Time-shared price	Plant A	992.11	280.45	156.22	1428.78	1,057.59		
	Plant B	904.52	270.54	136.7	1311.76	959.35	3,307.27	2,353.03
	Plant C	262.71	178.93	125.09	566.73	336.09		

6　Concluding remarks

At present, the time-shared price system has started in most areas; practice has shown that has significantly adjusted peak effects. This paper establish daily load dispatch optimization model under single and time-shared price. Example calculation shows that we can achieve greater benefit though time-shared price under same water resources, while provider useful references in daily load dispatch for control center.

References

Wan Yonghua, Mu Haxi, Hu Tiesong. A Research on Seasonable Time-of-use Pricing of Hydropower Station[J]. Journal of Hydroelectric Engineering,1997 (1):10-15.

Zhang Yuhui, Feng Quanlong. Baozhusi Hydroelectric Station Optimizaiton Analysis [J]. Sichuan Water Power, 1999(12):82-84.

Ma Guangwen, Wang Li, Guo Xiaming. Scheduling of Hydropower Station with Multi-purpose Reservoir for Timed Power Tariff [J]. Journal of Hydroelectric Engineering, 2002 (5): 583-587.

Chen Jiankang, Wang Li, Ma Guangwen. Dispatching Optimization of Hydropower Station with Time-shared Price[J]. Journal of Sichuan University (Engineering Science Edition), 2000 (6):1-3.

Guo Xiaming,Chen Jianchun,Ma Guangwen. Research on Year-end Level of Multi-year Regulating Storage Reservoir for Timed Power Tariff [J]. Journal of Hydroelectric Engineering, 2004 (3).

Tian L F,conins c. An Effective Robot Trajectory Planning Method Using a Genetic Agorithm [J]. Mechatronics,2004,14(5):16-19.

Yu Xihui, Li Chengjun, Liu Guangyu. Analysis and Application of Short-Term Optimization operation of Cascade Reservoir Power Plant under Peak-and-valley Electricity Rates [J]. China Rural Water and Hydropower,2005(3).

Zhang Yongchuan. Principles of Economic Operation of Hydropower Station [M]. Edition 2. Beijing: China Water Power Press, 1998.

Wang Xiao'an, Li Chengjun. Research and Application of Genetic Algorithm to Cascade Hydroelectric Stations' Short-term Optimization Scheduling[J]. Journal of Yangtze River Scientific Research Institute, 2003(2).

Calculation Methods of Water Temperature
in Lakes and Reservoirs and Results Analyses

Yu Zhenzhen[1] , *Wang Lingling*[2] , *Zhang Shikun*[1] , *Zhang Jianjun*[1] ,
Yang Yuxia[1] , *Cheng Wei*[1] , *Yan Li*[1] and *Xu Xiaolin*[1]

1. Yellow River Water Resources Protection Institute, Zhengzhou, 450004, China
2. College of Water Conservancy and Hydropower, Hohai University,
Nanjing, 210098, China

Abstract: The empirical formula methods and mathematical model methods for calculating the water temperature in lakes and reservoirs were introduced in detail, and the two methods were applied respectively to the calculation of water temperature in the Xiangxi Bay of Three Gorges Reservoir, China. By contrast and analysis, it was shown that the empirical formula methods needed few data in calculation. However, the empirical formula methods were based on a lot of measured data, the general rules of water temperature changes in lakes and reservoirs could be expressed, so the error was large in Xiangxi Bay, the RMSE of temperature between calculated values and measured values was 1. 063 ℃. In this paper, the vertical temperature in Xiakou monitoring point of Xiangxi Bay was computed by the methods of mathematical model, the RMSE was just 0. 349 ℃. In addition, only vertical profile of water temperature could be obtained by empirical formula methods. A hydrodynamic – water temperature mathematical model was constructed in this paper, taking into account many factors such as the terrain, the convection, the wind force, the inflow temperature, the solar radiation, the operation of Three Gorges Reservoir, which could show the three – dimensional water temperature distributions in Xiangxi Bay.

Key words: Lakes and reservoirs temperature, empirical formula methods, mathematical model methods, results comparison, Xiangxi Bay of Three Gorges Reservoir

1   Introduction

Water temperature is one of the most important indexes for reflecting the physical characteristics in water bodies, due to its strong effects on biochemical processes of substances. Especially, the living environments of aquatic life are sensitive to the change of temperature. Many domestic and international measured data show that the Dissolved Oxygen (DO) is stratified obviously in the vertical direction during the period of water temperature stratification, meanwhile, the entire chemical stratification would happen by the effects the temperature, DO stratification and other factors (Yu et al. under press; Jason and Lars 2008).

In this study, the empirical formula methods and mathematical model methods for calculating the water temperature in lakes and reservoirs were introduced, and then the two methods were applied respectively to the calculation of water temperature in the Xiangxi Bay of Three Gorges Reservoir. Furthermore, a hydrodynamic – water temperature mathematical model was used to study the three – dimensional characteristics of water temperature distribution in the research area.

2   Empirical formula methods

The method of Northeast Survey and Design Institute is one of the empirical formula methods in China. The method was mentioned by the Dafa Zhang based on lots of measured data of reservoirs in the year of 1982, the monthly average temperature in vertical direction could be calculated as long as the top and bottom temperature got. This method is widely used in production at present, such as the hydrological criterion of water conservancy and hydropower engineering. The formula can be expressed as follows:

$$T_y = (T_a - T_b) e^{-(\frac{y}{x})^n} + T_b \tag{1}$$

where,
$$n = \frac{15}{m^2} + \frac{m^2}{35}, \quad x = \frac{40}{m} + \frac{m^2}{(1 + 0.1m) \times 2.37}$$

where, T_y is the monthly average temperature at the depth of y, ℃ ; T_a is temperature in the top layer, ℃ ; T_b is temperature in the bottom layer, ℃ ; m is month, 1 ~ 12.

3 Mathematical model methods

3.1 One – dimensional model

In the 1960s, Orlob and Selna (1970) from the water resources company in the USA and Huber and Harleman (1972) from Massachusetts Institute of Technology put forward the one – dimensional water temperature mathematical model respectively, namely the WRE and MIT model, its governing equations are all based on the convection diffusion equation. In this method, the water body would be mean divided into a series of horizontal thin layers, the thickness is dz. As far as the each layer is concerned, the heat balance equation could be established, assuming the temperature in each layer distributes uniformly:

$$\frac{\partial T}{\partial t} + \frac{\partial}{\partial z}(\frac{TQ_v}{A}) = \frac{1}{A}\frac{\partial}{\partial z}(AD_z\frac{\partial T}{\partial z}) + \frac{B}{A}(u_iT - u_0T) + \frac{1}{\rho AC_P}\frac{\partial(A\varphi_z)}{\partial z} \tag{2}$$

where, T is the unit layer temperature, ℃ ; T_i is the inflow temperature, ℃ ; A is the plane area of each unit layer, m² ; B is the average width of each unit layer, m; D_z is the vertical diffusion coefficient, m²/s ; ρ is water density, kg/m³ ; C_P is the specific heat of water, J/(kg · ℃) ; φ_z is the internal distribution of heat sources due to solar radiation absorption, W/m² ; u_i is the inflow velocity, m/s ; u_0 is the outflow velocity, m/s ; Q_v is the vertical discharge through the top boundary of layer, m³/s.

The effects of inflow, outflow, heat exchange between air and water surface on the water temperature are included in the WRE and MIT model. Its distinctions are that: the vertical diffusion coefficients change with the time and the depth in the WRE model, conversely, the vertical diffusion coefficients are constant in the MIT model; the precision of calculation on the inflow and outflow in MIT model is higher than that in the WRE model; on the other hand, the mixing function of inflow is in the MIT model, however the WRE model ignores it. Overall, the calculation procedure of the WRE and MIT model are shown in Fig. 1:

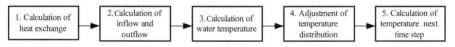

Fig. 1　Calculation procedure of WRE and MIT model

The vertical diffusion coefficient D_Z is the critical parameter in the Eq. (2), D_Z is related to some factors, such as the depth, wind, buoyancy, internal waves (Yu et al., 2012), there is no universal value at present. Often in lakes and reservoirs, surface water temperature is much higher than that at bottom, consequently, water densities become stacked vertically, forming the upper epilimnion, the bottom hypolimnion, and the metalimnion (thermocline) between them. Note that the thermocline can be regarded as a transition layer with a sharp temperature gradient and weak mixing ability. In 1983, Orlob conducted a research for determining the vertical diffusion coefficients based on four reservoirs, a practical conclusion are as follows: the diffusion coefficient range from 10 cm²/s to 100 cm²/s in the epilimnion, range from 0.1 cm²/s to 1 cm²/s in the thermocline, range from 0 cm²/s to 10 cm²/s in the hypolimnion.

The WRE and MIT model are regarded as the one – dimensional diffusion models, because of the base of the convection diffusion equation. In the middle and later period of the 1970s, another one – dimensional water temperature model was established by some researchers, namely, total energy model (Stefan and Ford 1975; Stefan et al. , 1980). The lakes and reservoirs are still regarded as a system in layers in the total energy model. In the method, the calculation for water temperature based on the energy balance, Stefan and Ford firstly mentioned this kind of model, known as Stefan – Ford model or MLTM model.

In the total energy model, the water body is mean divided into thin layers, the number is $n + m$, in which the top layers m is the uniform agitation layer, the bottom layer n is the epilimnion, the thickness of each layer is Δh (Fig. 2).

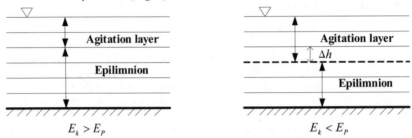

Fig. 2　Schematic of total energy model

E_k is the turbulence kinetic energy at the time step dt induced by the wind force:

$$E_k = \tau_0 W^* A_s dt \tag{3}$$

where, A_s is the area of water, m^2; τ_0 is the wind stress force, Pa; W^* is the drifting velocity on the water surface induced by the wind, m/s.

E_p is the potential energy of agitation layer induced by the blending function:

$$E_P = g \sum_{i=1}^{m} V(i,k) [P(m+1,k) - \rho(i,k)] (m+1-i) \Delta h \tag{4}$$

where, $V(i,k)$ is the volume of water body in layer of i at the time of k, m^3; $\rho(i,k)$ is the density in layer of i at the time of k, kg/m^3.

When $E_k > E_p$, the wind energy would convert to the potential energy, the agitation layer develops downward; when $E_k < E_p$, the stable water stratification would form, it is not increase in the thickness of agitation layer.

Compared with the diffusion model, the transport of the turbulence kinetic energy is in the total energy model, and the mixing characteristic could be able to describe preliminarily. However, the total energy model does not have the ability of showing the diffusion process in detail.

3.2　Two – dimensional model

The hydrodynamic and temperature have the significant two – dimensional characteristics in the river – type reservoirs, the vertical two – dimensional mathematical model could simulate the movement of the turbulence buoyancy – driven flows in the vertical section, and the formation and development process of temperature stratification (Ferrarin and Umgiesser, 2005). The two – dimensional model was applied to the predictions of water temperature structure and discharged temperature in high precision. A representative governing equation group on two – dimensional model (Liao, 2009) is:

Hydrodynamic equations:

$$\frac{\partial}{\partial x}(Bu) + \frac{\partial}{\partial z}(Bw) = 0 \tag{5}$$

$$\frac{\partial}{\partial t}(Bu) + u\frac{\partial}{\partial x}(Bu) + w\frac{\partial}{\partial z}(Bu) = \frac{\partial}{\partial x}\left(Bv_e \frac{\partial u}{\partial x}\right) + \frac{\partial}{\partial z}\left(Bv_e \frac{\partial u}{\partial z}\right) - \frac{B}{\rho_0}\frac{\partial p}{\partial x} +$$

$$\frac{\partial}{\partial x}\left(Bv_e \frac{\partial u}{\partial x}\right) + \frac{\partial}{\partial z}\left(Bv_e \frac{\partial w}{\partial x}\right) \tag{6}$$

$$\frac{\partial}{\partial t}(Bw) + u\frac{\partial}{\partial x}(Bw) + w\frac{\partial}{\partial z}(Bw) = \frac{\partial}{\partial x}\left(Bv_e\frac{\partial w}{\partial x}\right) + \frac{\partial}{\partial z}\left(Bv_e\frac{\partial w}{\partial z}\right) -$$
$$\frac{B}{\rho_0}\frac{\partial p}{\partial z} - B\beta\Delta Tg + \frac{\partial}{\partial z}\left(Bv_e\frac{\partial w}{\partial z}\right) + \frac{\partial}{\partial x}\left(Bv_e\frac{\partial u}{\partial z}\right) \tag{7}$$

Water temperature equation:

$$\frac{\partial}{\partial t}(BT) + u\frac{\partial}{\partial x}(BT) + w\frac{\partial}{\partial z}(BT) = \frac{\partial}{\partial x}\left(\frac{Bv_e}{\sigma_T}\frac{\partial T}{\partial x}\right) + \frac{\partial}{\partial z}\left(\frac{Bv_e}{\sigma_T}\frac{\partial T}{\partial z}\right) + \frac{1}{\rho C_P}\frac{\partial B_{\phi z}}{\partial z} \tag{8}$$

where, u, w are average velocity components in the x, z directions, respectively, m/s; B is the width, m; v_e is effective viscosity coefficient defined as $v_e = v_t + v$, in which v_t is the turbulent eddy viscosity coefficient and v is the molecular viscosity coefficient, m²/s; p is pressure, Pa; ρ_0 is the reference ambient density, kg/m³; β is the thermal expansion coefficient, ℃$^{-1}$; σ_T is Prandtl number.

3.3 Three – dimensional model

The flows in nature are all the three – dimensional movements, so the three – dimensional mathematical model could be able to describe the actual temperature distributions. The governing equations on the base of three – dimensional shallow water equations (continuity, momentum, temperature, state equation) in the $\xi - \eta$ coordinate system can be written as follows (Yu and Wang, 2011):

$$\frac{\partial\xi}{\partial t} + \frac{1}{g_\xi g_\eta}\frac{\partial(Hug_\eta)}{\partial\xi} + \frac{1}{g_\xi g_\eta}\frac{\partial(Hvg_\xi)}{\partial\eta} + \frac{\partial\omega}{\partial\sigma} = 0 \tag{9}$$

$$\frac{\partial u}{\partial t} + \frac{1}{g_\xi}\frac{\partial uu}{\partial\xi} + \frac{1}{g_\eta}\frac{\partial uv}{\partial\eta} + \frac{\omega}{H}\frac{\partial u}{\partial\sigma} - \frac{v^2}{g_\xi g_\eta}\frac{\partial g_\eta}{\partial\xi} + \frac{uv}{g_\xi g_\eta}\frac{\partial g_\xi}{\partial\eta} = fv - g\frac{1}{g_\xi}\frac{\partial\zeta}{\partial\xi} -$$
$$g\frac{H}{\rho_0 g_\xi}\int_\sigma^0\left(\frac{\partial\rho}{\partial\xi} + \frac{\partial\rho}{\partial\sigma}\frac{\partial\sigma}{\partial\xi}\right)d\sigma' + \frac{v_t}{g_\xi g_\xi}\frac{\partial^2 u}{\partial\xi^2} + \frac{v_1}{g_\eta g_\eta}\frac{\partial^2 u}{\partial\eta^2} + \frac{1}{H^2}\frac{\partial}{\partial\sigma}\left(v_t\frac{\partial u}{\partial\sigma}\right) \tag{10}$$

$$\frac{\partial v}{\partial t} + \frac{1}{g_\xi}\frac{\partial uv}{\partial\xi} + \frac{1}{g_\eta}\frac{\partial vv}{\partial\eta} + \frac{\omega}{H}\frac{\partial v}{\partial\sigma} + \frac{uv}{g_\xi g_\eta}\frac{\partial g_\eta}{\partial\xi} - \frac{u^2}{g_\xi g_\eta}\frac{\partial g_\xi}{\partial\eta} = -fu - g\frac{1}{g_\eta}\frac{\partial\zeta}{\partial\eta} -$$
$$g\frac{H}{\rho_0 g_\eta}\int_\sigma^0\left(\frac{\partial\rho}{\partial\eta} + \frac{\partial\rho}{\partial\sigma}\frac{\partial\sigma}{\partial\eta}\right)d\sigma' + \frac{v_t}{g_\xi g_\xi}\frac{\partial^2 v}{\partial\xi^2} + \frac{v_1}{g_\eta g_\eta}\frac{\partial^2 v}{\partial\eta^2} + \frac{1}{H^2}\frac{\partial}{\partial\sigma}\left(v_t\frac{\partial v}{\partial\sigma}\right) \tag{11}$$

$$\frac{\partial HT}{\partial t} + \frac{1}{g_\xi g_\eta}\left[\frac{\partial(g_\eta HTu)}{\partial\xi} + \frac{\partial(g_\xi HTv)}{\partial\eta}\right] + \frac{\partial\omega T}{\partial\sigma} = \frac{H}{g_\xi g_\eta}\left[\frac{\partial}{\partial\xi}\left(\frac{v_t}{\sigma_T}\frac{g_\eta}{g_\xi}\frac{\partial T}{\partial\xi}\right) +\right.$$
$$\left.\frac{\partial}{\partial\eta}\left(\frac{v_t}{\sigma_T}\frac{g_\xi}{g_\eta}\frac{\partial T}{\partial\eta}\right)\right] + \frac{1}{H}\frac{\partial}{\partial\sigma}\left(\frac{v_t}{\sigma_T}\frac{\partial T}{\partial\sigma}\right) + S \tag{12}$$

$$\rho = 999.842,594 + 6.793,952\times10^{-2}T - 9.095,290\times10^{-3}T^2 + 1.001,685\times10^{-4}T^3 -$$
$$1.120,083\times10^{-6}T^4 + 6.536,332\times10^{-9}T^5 \tag{13}$$

where, u, v, ω are the velocity components in the ξ, η, σ directions, respectively, m/s; ζ is the elevation of surface above the reference plane, m; H is the total depth, m; f is the coefficient of Coriolis force ($f = 2\bar\omega\sin\phi$); g is the acceleration of gravity, m/s²; v_t is turbulent viscosity coefficient ($v_t = c_\mu k^2/\varepsilon$); S is the heat sources term; g_ξ, g_η are the Lamé coefficients.

4 Application and results

The Xiangxi River, which is located on the north of the Xiling Gorges, is the first chief tributary of the upper reaches of the Three Gorges Dam (Fig. 3). With the impoundment of the Three Gorges Reservoir, the Xiangxi River has turned to the tributary bay, the velocity therein is the millimetresized, and the seasonal water temperature stratification has been occurred (Wang et al. , 2009). In the spring of 2007, Huang (2007) conducted an extensive field study around the Xiakou water region to estimate the trophic status of Xiangxi Bay, data on water quality parameters such as the water temperature in the vertical direction were available, The measured temperature at the Xiakou Town in Xiangxi Bay were used to evaluate the calculation methods of water temperature. As shown in Fig. 4, the values of vertical water temperature distribution in March by the empirical formula methods were not agreement well with the measured values. In the middle and bottom parts of Xiangxi Bay in April and May, the differences between the calculated and measured values were large, the RMSE was 1.063 ℃. However, the RMSE of temperature between calculated values and measured values was just 0.349 ℃ by the method of the mathematical model. It is obvious that the mathematical model could accurately capture the onset of stratification, mixed depth and water temperature. Basically, we come to the conclusion that, the results of mathematical model could be accepted as a good approximation of actual temperature status.

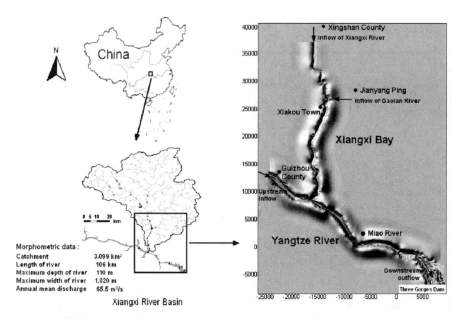

Fig. 3   Map of the basin and the modeling region for Xiangxi Bay

Meanwhile, the proposed three – dimensional model could be able to describe the water temperature distributions in the vertical and plane directions. Fig. 5 showed the vertical temperature distribution in May, 2007. From the simulation results, it is found that the water temperature stratification taken place, and the average depth of thermocline was located some 4 m beneath the water surface, with a temperature gradient of 0.35 ℃/m, the temperature difference of top layer and bottom layer reached 2.07 ℃, the horizontal surfaces of temperature were almost isothermal. In the simulation domain, the characteristics of the horizontal water temperature distributions in April, 2007 are shown in Fig. 6, in the bay the water temperature difference of surface layer was just 0.87 ℃, the overall temperature was from 16.57 ℃ to 17.44 ℃, the horizontal temperature difference appeared between the thalweg and shoal, and the micro – regions with the sharp boundary. On the other hand, the water temperature of the bottom layer ranged from 15 ℃ to 15.5 ℃, the amplitude was 0.5 ℃.

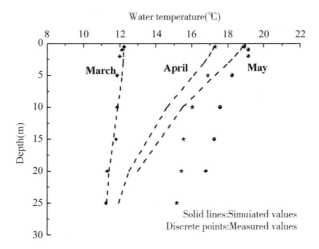

(a) Method of Northeast Survey and Design Institute

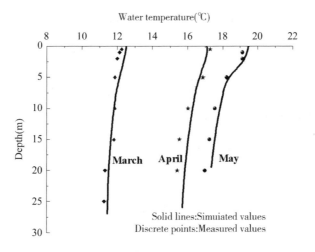

(b) Mathematical model method

Fig. 4　Calculated and measured water temperature profiles

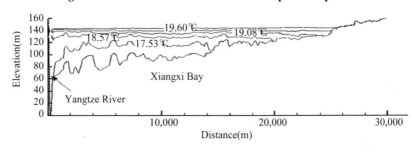

Fig. 5　Simulation result of vertical water temperature distribution in May, 2007

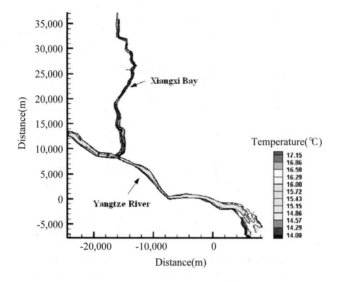

(a) Water temperature distribution in the top layer

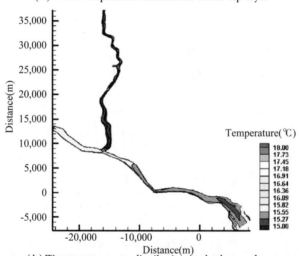

(b) Water temperature distribution in the bottom layer

Fig. 6   Horizontal distributions of water temperature in April, 2007

5   Conclusions

The empirical formula methods and mathematical model methods for calculating the water temperature in lakes and reservoirs were introduced and contrasted in this paper, the findings are as fellows:

(1) The empirical formula methods needed few data in calculation, for instance, the empirical formula method of Northeast Survey Design Institute from Dafa Zhang only got the monthly water temperature of the top layer and the bottom layer in Xingxi Bay, then the vertical temperature distribution could be calculated easily, and the temperature of the top layer and the bottom layer may also be reckoned by the correlation methods of air – water temperature or latitude – water

temperature. However, the empirical formula methods were based on a lot of measured data, the general rules of water temperature changes in lakes and reservoirs could be expressed, so the error was large.

(2) The mathematical model methods were applied to study water temperature in a specific lake or reservoir with high precision. Therefore, the mathematical model was to be established to study the water temperature in lakes and reservoirs necessarily.

(3) Only vertical profile of water temperature could be obtained by empirical formula methods. However, a hydrodynamic – water temperature mathematical model was constructed in this paper, taking into account many factors such as the terrain, the convection, the wind force, the inflow temperature, the solar radiation, the operation of Three Gorges Reservoir, which could show the three – dimensional water temperature distributions in Xiangxi Bay.

Acknowledgements

This work was supported by the National Natural Science Foundation of China (51179058) and the special funds for scientific research projects of water conservancy public welfare industry (200901023).

References

Ferrarin C., Umgiesser G. Hydrodynamic Modeling of a Coastal Lagoon: the Cabras Lagoon in Sardinia, Italy [J]. Ecological Modelling, 2005, (188): 340 – 357.

Huang Yuling. Study on the Formation and Disappearance Mechanism of Algal Bloom in the Xiangxi River Bay at Three Gorges Reservoir [D]. Ph. D. Theses. Yangling: Northwest A & F University, 2007: 57 – 60.

Huber W. C., Harleman D. R. F. Temperature Prediction in Stratified Reservoirs [J]. Journal of the Hydraulics Division, ASCE, 1972(98): 645 – 666.

Jason H., Lars U. Modelling the Tidal Mixing Fronts and Seasonal Stratification of the Northwest European Continental Shelf [J]. Continental Shelf Research, 2008, 28(7): 887 – 903.

Liao Wen – gen. Cumulative Impact of Cascade Power Station on Water Temperature and its Countermeasures in the Downstream of Jinsha River [R]. Yichang: China Three Gorges Corporation, 2009: 40 – 51.

Orlob G. T., Selna L. G. Temperature Variation in Deep Reservoirs [J]. Journal of Hydraulics Division, ASCE, 1970 (96): 391 – 410.

Stefan H. G., Ford D. E. Temperature Dynamic in Dimictic Lakes [J]. Hydraulics Division, 1975(101) (HY1): 97 – 114.

Stefan H. G., Hanson M. J., Ford D. E., et al. Stratification and Water Quality Predictions in Shallow Lakes and Reservoirs [C]. In: IAHR Second International Symposium on Stratified Flows, 1980: 1033 – 1043.

Wang Lingling, Yu Zhenzhen, Dai Huichao, et al. Eutrophication Model for River – type Reservoir Tributaries and its Application[J]. Water Science and Engineering, 2009, 2(1): 16 – 24.

Yu Zhenzhen, Wang Lingling, Mao Jingqiao, et al. Effects of Water Temperature on Chlorophyll – a Concentration Stratification in the Tributary Bay of Three Gorges Reservoir [J]. Journal of Aerospace Engineering, Under Press.

Yu Zhenzhen, Wang Lingling. Factors Influence Thermal Structure in a Tributary Bay of Three Gorges Reservoir [J]. Journal of Hydrodynamics, 2011, 23(4): 407 – 415.

Yu Zhenzhen, Wang Lingling, Zhu Hai, et al. Numerical Simulation of Internal Waves Generation by Tidal Flow in Xiangxi Bay of the Three Gorges Reservoir [J]. Journal of Sichuan University (Engineering Science Edition), 2012, 44(3): 1 – 5.

Application of Cutting – edge Technologies in Yellow River's Basic Surveying and Mapping

Zhang Yongguang, *Liu Haojie*, *Zhu Shengshi* and *Sun Dengke*

Yellow River Engineering Consulting Co., Ltd., Zhengzhou, 450003, China

Abstract: Basic surveying and mapping for the Yellow River is the preliminary work of Yellow River's integrated management and development. And its technical content is rising as the related disciplines' scientific and technological progress. Applying cutting – edge mapping technologies in the Yellow River's survey can show the survey's quick response ability for integrated management and development of the Yellow River basin. When surveying the 1:10,000 topographic map in Xiaobeiganliu of Yellow River, we integrate many new mapping technologies together, such as ADS80 digital aerial photography, ALS60 Airborne LiDAR, GPS PPP and region geoid's refinement. Compared to the traditional aerial survey, these new mapping technologies have changed the work mode of many operations, including the selection and measurement of aero – photo control point, aero – photo scanning, aerial triangulation, and survey for elevation annotation points in flood land. And they not only greatly improve the quality and operating efficiency of mapping products, but also reduce the field labor intensity. This paper elaborates the project's work process, main methods, key technologies and mapping accuracy, which will be of instructive and promotional value for the implementation of similar projects.

Key words: surveying and mapping, cutting – edge technology, basic surveying and mapping, application

1　Introduction

At present, China has entered a critical period of reform, opening – up and modernization. The development of society and economy of the Yellow River basin, especially the rapid advance of industrialization and urbanization, put new higher demands for the development and management of the Yellow River. Many tasks, such as deeply understanding of Yellow River's nature law, persistently carrying out the research on the change of the Yellow River's water and sediment, vigorously promoting the construction and application of the "Digital Yellow River", and establishing dynamic tracking mechanism for the significant water conservancy construction project, have to base on fast and real – time topographic surveying and mapping results.

Because the traditional mapping methods are time – consuming, laborious and of low technical content, they cannot meet the needs of mapping the Yellow River any more. With the introduction and application of some cutting – edge technologies, such as ADS80 digital aerial photography, ALS60 airborne Light Detection and Ranging (LiDAR), GPS PPP (precise point positioning), region geoid's refinement, and so on, a fast, efficient topographic surveying and mapping can be guaranteed. In this paper, the above cutting – edge technologies of surveying and mapping were used in surveying the 1:10,000 topographic map in Xiaobeiganliu of the Yellow River.

2　Research program

2.1　Overview of the survey area and mapping requirements

The survey area was from northern GanZe slope to southern Tongguan where turns to the east, and finally ended in Sanmenxia dam. Both sides of the area was surveyed not less than 500 m outside the top of high cliff, and it covered a total area of 3,700 km². The range is shown as

Fig. 1. 1980 Xi' an Geodetic Coordinate System is adopted as the plane coordinate system while 1985 National Height Datum is used as the elevation system. The scale of topographic maps was 1:10,000. The basic contour interval is 0.5 m in flood land, 1.0 m in plains and hills, and 5.0 m in mountains.

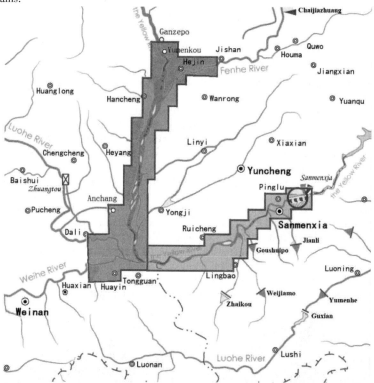

Fig. 1　The sketch map of the survey area range

2.2　Aerial photography

Before the formal aerial photography, a particular aerial photography planning should be established. It should include data preparation of flying zone, flights for camera calibration, settings of flight parameter, route planning, etc.

ADS80 digital aerial photography was used for this survey task. And it included flight data of 5 sorties and 70 air routes. These acquired data was used for mapping all land features and landscapes except contours and elevation annotation points in the area whose contour interval is 0.5 m. The ground resolution of aerial photos was 0.5 m. The ADS80 airborne digital aerial photography was the most advanced pushbroom airborne digital aerial photogrammetry system. It integrates high – precision inertial measurement unit (IMU) and global positioning system (GPS), and uses three – line array CCD camera with 12,000 pixels and professional single large – aperture telecentric lens. After only one flight, we can simultaneously obtain three – view (forward view, nadir view and backward view) panchromatic stereoscopic image, color image and color infrared image which are 100% overlapped, seamless, and with the same image resolution and good spectral characteristic.

Contours and elevation annotation points in the area whose contour interval is 0.5 m are mapped based on ALS60 airborne LiDAR scanning data whose elevation precision is no more than 0. 14 m and terrain point interval is no more than 1. 5 m. It involved flight data of 5 sorties and 70 air routes. Airborne LiDAR is a spatial surveying system integrating three modern cutting – edge technologies including laser ranging system, GPS and IMU. GPS, working based on GPS PPP, provides the location of flying platform; IMU provides the trajectory and attitude of flying platform, while laser ranging system provides the distance between the flying platform and the surveyed object. Through the integration of these three technologies, it can quickly acquire the three – dimensional information of the ground.

2.3　Setting up and surveying of the aerial photograph control survey checkpoints

The horizontal and height checkpoints were even set up in the region of ADS80 aerial photograph. And these checkpoints should be located in the places where the image is clear, or land objects have obvious turns, or the small linear features are orthometric. Before selecting the checkpoints in the field, the position of checkpoints should be firstly chosen from the digital images. If the checkpoint is at the corner of house or wall 5 m far from which the land should be flat and no shelter in the sky, the height of the house and wall should be measured. The precision of checkpoint was 0. 1 m on the spot. The example is shown as the left drawing of Fig. 2. The height checkpoint was even set up in the region of ALS60. The height checkpoint should be in a circle which centers as itself and has a radius of 5 m and the land in the circle must be flat. The position of them should be first selected according to the digital images and the old 1∶10,000 topographic maps. When on the spot, if some points don't meet the requirements, they should be adjusted. The example is shown as the right drawing of Fig. 2.

At the end, every checkpoint should be taken some pictures from not less than three different directions. And the real position of prism must be shown in the pictures.

Fig. 2　Setting up of the aerial photograph control survey checkpoint

When surveying the aerial photograph control checkpoints, the starting calculation points must be the point of the Yellow River GPS control network. And GPS observation should based on the static model with double frequency GPS receiver. The horizontal coordinates and geoid heights were worked out by the commercial software, and then the height of every point in the 1985 National Height Datum was transformed by the software of the middle Yellow River basin quasi geoid.

2.4　Coordinate transformation of aerial data

Both ADS80 images and ALS60 point cloud are based on the WGS – 84 coordinate system, but the local coordinate systems (the 1980 Xi'an Geodetic Coordinate System and the 1985 National Height Datum) are used in this project. Therefore, the coordinate transformation, including the horizontal coordinate transformation and the elevation coordinate transformation, between the WGS – 84 coordinate system and the local coordinate system is needed here. The horizontal coordinate transformation is based seven parameters: three translation parameters, three rotation parameters and one scale parameters. For the elevation transformation, the three – dimensional coordinates of WGS – 84 of land points should be first obtained by GPS. And then transform these WGS84 coordinates to geodetic heights. Finally, accordingly to high – precision and high resolution quasi geoid, geodetic heights were transformed to the heights in the 1985 National Height Datum.

The horizontal coordinate transformation is based on the result of the Yellow River GPS control network while the elevation transformation is based on the middle Yellow River basin quasi geoid. The coordinate transformation flow chart was listed as Fig. 3.

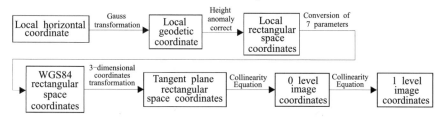

Fig. 3　Coordinates transformation flow chart

2.5　IMU/GPS solver

The IMU/GPS solver was carried out by the workstation of DELL690 while the GPS solver by WayPoint7.8 and the IMU/GPS fusion by IPAS Pro. The accuracy statistics of calibration field were shown as Tab.1 and Tab.2.

Tab. 1　The accuracy statistics table of GPS solver

Base stationtype	Position error of the positive and negative calculation(m)			Number of satellites observed
	Root meansquare error (RMSE)	Maximum error	Minimum error	
X	0.017	0.04	0	
Y	0.024	0.03	0	>5
Z	0.062	0.15	0	

Tab. 2　The accuracy statistics table of IMU solver

$X,Y(\mathrm{mm})$		$Y,H(\mathrm{mm})$		$X,H(\mathrm{mm})$		$H(\mathrm{mm})$		Roll($°$)			Pitch($°$)		
X max min		Y max min		X max min		max	min	max	min	Standard	max	min	standard
Y max min		H max min		H max min									
15	12	16	13	15	12	26	18	0	0	±3	0	0	±3
16	13	26	18	26	18								

After analyzing the resolved GPS and IMU data, we found the result is reliable and can meet the design requirements.

2.6　ADS80 aerial imager processing

2.6.1　Aerial triangulation

The aerial triangulation was executed in PixelFactory that is developed by Inforterra based on three Dell 2950 servers.

The original data loaded by Pixel Factory was divided into six partitions. Create a new project, import fused IMU/GPS data and define project coordinate system (the plane coordinate system is UTM49 while the elevation system is geodetic height). Image tie point match was automatically completed by AutomaticPointMeasure of PF. And the point match was based on L0 images. In the end, the residual distribution of tie points between encryption partitions, less than one pixel, was irregular and didn't have systematic errors. The ADS80 data processing flow chart was shown as Fig. 4.

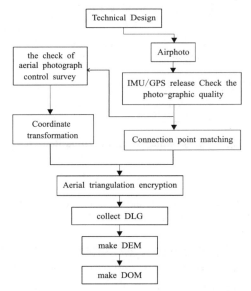

Fig. 4 ADS80 data processing flow chart

After the automatic match of tie points, tie points were optimized in Optimize module. Optimize parameters were set up based on collimation error, IMU attitude correction error and nonlinearity zoom factor correction error. The image coordinate precision statistics of the free adjustment tie points was shown as Tab. 3. The Tab. 3 showed that the residual error of tie point's image coordinate met the design requirements.

Tab. 3 the image coordinate precision table of the free adjustment tie points

(Unit: pixel)

Project Name	Xbias	Ybias	Xstd	Ystd	Xmax	Ymax
PROJECT_1Q	−0.000,5	0.004,0	0.301,7	0.272,0	0.936,0	1.043,7
PROJECT_2Q	0.003,5	0.005,7	0.295,4	0.291,7	0.998,4	0.986,2
PROJECT_3Q	0.000,1	0.005,0	0.291,8	0.263,3	0.973,7	1.115,4
PROJECT_4Q	0.000,7	−0.004,6	0.284,2	0.274,1	0.982,3	0.994,4
PROJECT_5Q	−0.000,8	−0.002,7	0.303,7	0.256,0	0.974,7	0.972,0
PROJECT_6Q	0.001,4	−0.004,2	0.259,4	0.282,1	0.982,6	0.997,9

2.6.2 Precision detection of stereo mapping

On the basis of aerial triangulation results, we rectified the L1 images and collected the coordinates of aerial photograph control survey checkpoint by the related modules of Visiontek, JX − 4CDPW, VirtuoZoADS40. The accuracy statistics was shown as Tab. 4.

Tab. 4　the precision table of stereo mapping　（Unit：m）

Point NO	X_{GPS}	Y_{GPS}	H_{GPS}	X_{JX4}	Y_{JX4}	H_{JX4}	ΔX	ΔY	ΔH
P020001	2,018.996	38,269.922	301.697			301.317			0.380
P010001	8,997.073	73,531.668	436.206			435.900			0.306
P014001	9,282.393	61,503.180	308.104			307.650			0.454
P018004	8,153.124	45,064.263	302.438			302.537			−0.099
P022001	7,806.829	30,322.484	310.999			310.900			0.099
P015001	9,081.329	28,067.794	306.726			306.815			−0.089
P015007	241.567	57,682.458	303.483			303.352			0.131
P014004	6,575.211	40,422.189	307.470			307.464			0.007
P014006	7,868.329	51,078.206	330.107			330.001			0.106
B009001	2,883.528	23,142.303	305.922			305.893			0.030
B009003	3,442.859	36,754.691	308.142			308.255			−0.113
P009006	3,193.984	49,897.625	321.107			321.255			−0.148
P001004	4,413.962	46,520.400	314.924	4,413.917	46,520.647	315.125	0.045	−0.247	−0.201
P017002	6,358.829	49,041.959	313.219	6,359.350	49,042.226	313.200	−0.521	−0.266	0.019
P015002	4,867.026	55,736.683	306.172	4,867.221	55,736.430	306.267	−0.195	0.253	−0.095
P011002	530.183	70,547.300	364.408			364.276			0.132
P014001	9,282.393	61,503.180	308.104			308.439			−0.334
P009001	3,826.244	79,200.971	314.990			314.816			0.174
P001005(1)	5,430.374	42,997.071	311.750			311.151			0.599
P001005(2)	5,862.386	42,760.462	312.729			312.851			−0.122
P018010	1,551.427	44,317.514	507.029	1,551.396	44,317.715	507.076	0.031	−0.201	−0.047
P018011	7,641.211	42,085.349	329.387	7,641.643	42,085.566	329.013	−0.433	−0.217	0.374
P022006(2)	3,217.565	27,501.719	295.964	3,217.587	27,501.463	295.843	−0.022	0.256	0.121
P023003	5,179.443	25,575.017	319.049	5,179.712	25,574.949	319.068	−0.269	0.067	−0.019
P023005	9,846.536	26,981.820	296.718	9,846.269	26,981.965	296.518	0.266	−0.145	0.200
P021009	7,379.523	31,355.403	373.942	7,379.431	31,355.574	374.206	0.091	−0.171	−0.264
P022002	6,702.078	28,868.637	413.761	6,701.970	28,868.696	413.781	0.108	−0.059	−0.020
P022003	4,805.733	30,589.849	320.210	4,805.719	30,589.825	320.293	0.014	0.025	−0.083
P022008	7,642.817	29,413.785	362.753	7,642.768	29,413.692	363.081	0.049	0.093	−0.328
P021007	3,887.669	35,383.136	475.116	3,887.741	35,383.027	475.077	−0.072	0.109	0.039
P021009	7,379.523	31,355.403	373.942	7,379.833	31,355.551	374.252	−0.310	−0.148	−0.310
P022006(1)	5,479.954	31,949.326	325.906	5,480.146	31,949.166	325.690	−0.192	0.160	0.216
P001006	5,088.659	37,959.126	360.765	5,088.898	37,959.407	360.877	−0.240	−0.281	−0.112
P020001	2,018.996	38,269.922	301.697			301.789			−0.092
P020010	6,070.230	38,052.245	327.688	6,070.158	38,051.999	328.127	0.072	0.246	−0.439
P021007	3,887.669	35,383.136	475.116	3,887.901	35,383.283	475.077	−0.231	−0.147	0.039

Continued to Tab. 4

Point NO	X_{GPS}	Y_{GPS}	H_{GPS}	X_{JX4}	Y_{JX4}	H_{JX4}	ΔX	ΔY	ΔH
P001005(1)	5,430.374	42,997.071	311.750			311.126			0.625
P001005(2)	5,862.386	42,760.462	312.729			312.701			0.029
P018011	7,641.211	42,085.349	329.387	7,641.544	42,085.813	329.126	−0.334	−0.465	0.262
P027004	5,445.232	87,548.597	542.086			541.865			0.221
P028005	9,355.840	87,552.141	319.228			319.240			−0.012
P029005	4,476.639	87,513.436	482.119			481.465			0.654
P028003	914.452	98,297.576	537.191			537.361			−0.170
P028005	9,355.840	87,552.141	319.228			319.561			−0.333
P031008	2,046.272	23,351.959	393.790			394.120			−0.330
P031010	2,312.190	11,173.453	332.640			332.833			−0.193
P029003	3,305.831	04,904.404	326.722			326.998			−0.276
P029004	5,729.885	98,216.745	359.735			360.073			−0.338
P029004	5,729.885	98,216.745	359.735			359.695			0.040
P030003	7,198.326	17,674.269	330.812			330.820			−0.008
P031007	3,979.126	34,169.692	371.489			371.902			−0.413
P033003	755.584	56,007.647	413.716			413.966			−0.250
P033007	529.111	29,164.889	342.420			342.541			−0.121
P034005(1)	3,425.447	46,310.237	357.616			357.524			0.092
P034008	1,789.635	23,914.438	337.643			337.941			−0.298
P035005	7,917.549	46,872.184	361.691			361.341			0.350
P036006(1)	1,670.882	39,831.123	393.465			393.651			−0.186
P036007(1)	1,616.123	31,722.046	446.859			447.176			−0.317
P008009(1)	2,154.842	54,886.813	391.048	2,154.721	54,886.704	390.764	0.121	0.109	0.284
P008009(2)	571.239	53,954.092	345.718	571.069	53,954.480	345.589	0.170	−0.389	0.129
P005005(1)	9,680.074	52,414.525	326.312	9,679.988	52,414.547	325.983	0.087	−0.022	0.329
P005005(2)	1,172.149	49,298.448	419.500	1,172.401	49,298.700	419.421	−0.252	−0.252	0.079
P002004	398.156	50,505.443	426.786	398.189	50,505.190	426.777	−0.033	0.253	0.009
P002005	9,515.159	4,384.679	308.760	9,515.139	44,384.813	308.464	0.020	−0.134	0.296
P001004	4,413.962	46,520.400	314.924	4,414.039	46,520.281	314.796	−0.077	0.119	0.128
P001006	5,088.659	37,959.126	360.765			360.571			0.194

From Tab. 4, we can find the X direction residual is −0.433 m in maximum, 0.014 m in minimum and the RMSE was 0.233 m. The Y direction residual is −0.456 m m in maximum, −0.022 m in minimum and the RMSE was 0.212 m. The distance residual is 0.585 m in maximum, 0.029 m in minimum and the RMSE was 0.299 m. The H direction residual is 0.654 m m in maximum, 0.007 m in minimum and the RMSE was 0.253 m. The results were far superior to the design requirements.

2.7 ALS60 point cloud data processing

The elevation RMSE of ALS60 calibration field was 0.056 m, which provided a reliable guarantee for post – processing of point cloud data. After aerial triangulation to the point cloud data of every aerial route, precision of the data between every two aerial routes is high enough to meet the design requirements. Considering point cloud data characteristics and production needs, we chose TerraSolid, ArcGIS Desktop, FME, CASS software as the data processing platform. TerraSolid was used for point cloud filter and classification, ArcGIS Desktop for contour line generation and smooth, FME for the data format conversion and the CASS for the contour line edit. The point cloud data processing flow chart was shown as Fig. 5.

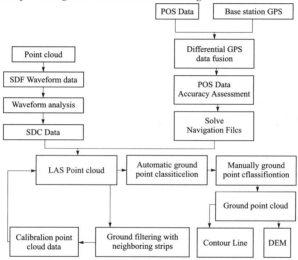

Fig. 5   Point cloud data processing flow chart

The final point cloud data was tested by the field elevation check points. The accuracy statistics was shown as Tab. 5. It was also detected by the aero – photo control points in 1: 10,000 topographic maps produced in 2003 that didn't change from then on. The results were shown as Tab. 6. From Tab. 5 and Tab. 6 we found the quality of point cloud data met the design requirements.

Tab. 5   The accuracy statistics of the point cloud data elevation checkpoints

(Unit: m)

Point No.	X_{GPS}	Y_{GPS}	H_{GPS}	LIDARH	LIDARH $- H_{GPS}$	ΔH^2
G010048	5,940.88	38,659.47	356.921	356.95	0.029	0.000,8
G011048a	5,873.99	37,186.90	354.657	354.62	−0.037	0.001,4
G011046b	7,321.84	36,252.16	350.744	350.70	−0.044	0.001,9
G036006	2,822.66	36,921.91	344.362	344.23	−0.132	0.017,4
G029004a	5,026.83	98,483.44	326.745	326.83	0.085	0.007,2
G030003	7,302.74	15,884.77	333.792	333.86	0.068	0.004,6
G029003a	3,279.05	4,840.14	330.357	330.36	0.003	0.000
G031010	2,044.80	11,148.24	332.584	332.54	−0.044	0.001,9
G035005a	5,490.56	45,173.31	355.29	355.07	−0.220	0.048,4

Continued to Tab. 5

Point No.	X_{GPS}	Y_{GPS}	H_{GPS}	LIDARH	LIDARH $-H_{GPS}$	ΔH^2
G033007a	156.51	28,156.09	344.07	344.18	0.110,0	0.012,1
G033006	1,167.52	38,277.84	353.943	353.99	0.047,0	0.002,2
G034005b	3,316.85	44,081.87	359.202	359.10	-0.102,0	0.010,4
G031009	1,587.79	19,513.35	337.076	336.97	-0.106,0	0.011,2
G016006b	4,032.28	55,886.34	305.264	305.24	-0.024,0	0.000,6
G012002b	9,369.12	65,720.72	313.547	313.47	-0.077,0	0.005,9
G010001b	3,979.97	74,114.20	315.955	315.85	-0.105,0	0.011,0
G011002b	8,223.71	71,535.00	313.277	313.06	-0.217,0	0.047,1
G015002	4,802.37	56,153.09	306.172	306.41	0.238,0	0.056,6
G016005	2,703.21	49,304.69	303.201	303.07	-0.131,0	0.017,2
G029004a	5,026.83	98,483.44	326.745	326.87	0.125,0	0.015,6
G028004a	657.96	95,605.47	326.84	326.82	-0.020,0	0.000,4
G008001	4,172.14	81,106.14	320.995	320.97	-0.025,0	0.000,6
G009003a	1,559.35	79,262.91	319.81	319.70	-0.110,0	0.012,1
G016002b	2,265.34	29,712.82	303.045	303.25	0.205,0	0.042,0
G009005	2,988.94	47,340.01	311.951	312.05	0.099,0	0.009,8
G022002c	5,578.199	30,437.72	301.972	301.95	-0.022,0	0.000,5
					REMS	0.116,5

Tab. 6 **The accuracy statistics of the point cloud data basedon the Aero – photo control points in 2003**

Serial No.	H_{GPS}	LIDARH	LIDARH $-H_{GPS}$	ΔH^2	Serial No.	H_{GPS}	LIDARH	LIDARH $-H_{GPS}$	ΔH^2
1	348.130	348.16	-0.03	0.000,9	16	332.850	332.90	-0.05	0.002,5
2	334.494	334.64	-0.15	0.021,3	17	336.880	336.82	0.06	0.003,6
3	343.486	343.49	0.00	0.000,0	18	332.900	333.03	-0.13	0.016,9
4	331.253	331.25	0.00	0.000,0	19	344.540	344.76	-0.22	0.048,4
5	333.534	333.30	0.23	0.054,8	20	347.860	347.87	-0.01	0.000,1
6	331.777	331.87	-0.09	0.008,6	21	339.890	340.06	-0.17	0.028,9
7	332.467	332.40	0.07	0.004,5	22	333.650	333.71	-0.06	0.003,6
8	333.086	333.01	0.08	0.005,8	23	348.790	348.93	-0.14	0.019,6
9	332.142	332.10	0.04	0.001,8	24	334.540	334.75	-0.21	0.044,1
10	332.358	332.19	0.17	0.028,2	25	341.350	341.19	0.16	0.025,6
11	332.750	332.70	0.05	0.002,5	26	338.520	338.49	0.03	0.000,9
12	332.703	332.67	0.03	0.001,1	27	345.540	345.66	-0.12	0.014,4
13	333.832	333.88	-0.05	0.002,3	28	347.670	347.49	0.18	0.032,4
14	337.110	336.92	0.19	0.036,1	29	350.280	350.14	0.14	0.019,6
15	341.730	341.81	-0.08	0.006,4				REMS	0.122,5

3   Conclusions

In this research the whole process was realized and tested, including the digital image / point cloud data acquisition, puncturing points and surveying for Aero – photo, coordinate conversion, the IMU / GPS solver, the image data processing, point cloud data processing and production of the 1 : 10,000 topographic maps in Xiaobeiganliu of the Yellow River. A satisfied result was acquired. Moreover, some helpful experiences were concluded: ① Based on the linear array pushbroom images of every aerial strip to collect topographic data, the workload for image mosaic and data edge matching was greatly reduced. ② GPS PPP provided the technical support for topographic surveying and mapping of hydraulic engineering in the area of high mountains and deep woods. ③Application of the refining of region quasi – geoid has successfully resolved the elevation problem that the WGS – 84 coordinates and the local coordinates cannot be transformed each other. ④It was not necessary for us to puncture and survey Aero – photo control points, and the whole aerial triangulation process was changed. ⑤ Contour lines (with an interval of 0. 5 m) and elevation annotation points were both generated by LIDAR point cloud data, which overturned the traditional work mode that we collected them only by field measurement. ⑥ we have known the key technologies, main methods and work flow for rapidly producing digital photogrammetric products with high precision as well as how to integrate so many mapping cutting – edge technologies.

References

Yang Gengyin. The Accuracy Validation and Technology of Leica ADS80 Digital Aerial Camera [J]. Bulletin of Surveying and Mapping, 2010(8):68 – 69.

Zhang Xiaohong. The Measurement Theory and Method of LIDAR [M]. Wuhan: Wuhan University Press, 2007.

Li Xiaohong. Study on ADS40 Production System [J]. Science of Surveying and Mapping, 2009, 34(6): 212 – 214.

Long Huaping, Hu Yongjian, Li Bo. Echo Detection Theory of Airborne LIDAR System and Its Precision of Determining Elevation [J]. Geospatial Information, 2007,5(3):75 – 78.

Study and Application of Hydrology Information GSM Message Releasing System in the Yellow River

Yu Hang, *Yan Yiqi*, *Luo Meng* and *Xu Zhuoshou*

The Hydrology Bureau of Yellow River Conservancy
Commission, Zhengzhou, 450000, China

Abstract:The timely releasing of hydrology information plays a very important role in embodying the value of the hydrological measured data. Message releasing technology is one of the focal points of hydrologic information service. In order to improve the efficiency of real-time information releasing and enrich the measures of early warning and forecasting means, together with the Yellow River flood control work, the Yellow River hydrology GSM message releasing system has been developed through combining the mobile communication, computer application, database and network technology. The GSM message releasing system runs automatically or semi-automatically and has the functions of statistical analysis, rapid releasing, custom querying, automatically alarming and so on. As a result, the real-time hydrology information releasing efficiency is improved. This paper presents the design and implementation of GSM message releasing system.

Key words: message releasing, database, hydrology information, the Yellow River flood control

1 Introduction

The value of real hydrology information is at its accuracy and timeliness. Hydrological departments always take "accurate in measuring, efficient in reporting" as the basic work requirements. Nowadays, with the method and technology developing, the reporting technology has a high level in terms of timeliness. It is able to report the measured data within 20 min from the observation station to the hydrology information center of YRCC. In the field of hydrology information service, with the current application of Internet and database technology, it is possible to broadcast the hydrology information at the first time. However, the information broadcasting online is passive. That means it needs audience notice it themselves. Actually, the flood control decision makers are not possible to notice the hydrology report online all the time, which makes the information delay and lowers the value of information.

The Yellow River Basin has always been a disaster-prone area. Especially in recent years, there are more extreme weathers such as storm, flood, ice, etc. which create huge damage in many areas. Now, there are over 900 stations for rainfall or water regime observation in the Yellow River basin. The density and coverage can only meet the basic monitoring requirement. In this circumstance, how to improve the timeliness of the hydrology information, to early warn the flood and drought disaster, especially the more risky ice flood and the minor watershed storm flood, is difficult and important. Because of the suddenness and randomness of the ice flood and minor watershed storm flood, their prediction is difficult. The hydrology information report is supposed to focus on real-time, intelligent and Initiative aspects.

It is a good way for improving the service of hydrology information reporting to create a high efficient and reliable Message Releasing System in which the usage of the GSM communication equipment perfectly satisfy the requirements of hydrology information reporting. The secondary development of the GSM communication equipment according to real requirements can help the information center work better on information releasing and warning.

2　System construction

2.1　System structure

The hydrology information releasing system is structured based on C/S. Supported by database server, it is installed at the client side, mainly consisted of data obtaining module, information organizing module and the message releasing module. The data obtaining module runs to fetch the data from database. The information organizing module is responsible for organizing the data according to real requirements and the message releasing module is in charge of sending and receiving messages, generating order catalogue and providing user interface (Yang Chen, 2011).

2.2　Data obtaining module and the information organizing module

The data obtaining module and information organizing module are developed based on C#, applying triple layer structure with each layer being independent, safe and portable and running on .NET platform. For different needs, each layer can fast develop relevant middleware products, which assure the extensibility of each layer.

2.2.1　Data access layer and its storage process

Connected to the data server through interactive operation and optimizing methodologies, Data obtaining module mainly provides logical conversion, standard data access interface etc. to up layers.

For historical reasons, the Hydrology Information Center of YRCC applied three types of database: Sybase, SQL server and Oracle. Oracle uses PL-SQL statement. SQL server and Sybase use T-SQL statement. Although both are based on standard SQL grammar, there are still small differences between them. Writing large quantity of SQL statements into the program enhances efficiency, but leaves potential problems in the generality. "Storage process" has been used in developing the program because only few SQL statements are referred in real-time hydrology information obtaining. "Storage process" is in the form of script. It is general and easily-modified and makes the design of this layer more concise.

2.2.2　Design pattern and component technology in the operation logic layer

In the GSM message releasing system, the operation logic layer is the core layer in which data organization module is responsible for analyzing, processing and organizing data (Liu Feng and Sun Yong, 2011). Due to the tight coupling relationship between each operation parts in the logic layer, one part change will cause a chain reaction, lead to a series of changes of related parts. To avoid this situation, data organization module has adopted "component technology" to develop the various components respectively responsible for disparate tasks: regular information generation, supplemental information generation, rainfall alarm, water regime warning, ice condition alarm, special regime and low discharge warning, custom trigger and so on (Zhang Yi, 2007). Each component has functions of data retrieval and information generation. Each component can run independently not relied on other components, and has stand-alone data port and user setup interface in the presentation layer which will be introduced in this article.

2.2.3　Presentation layer in MVC pattern

This layer mainly provides users a display interface of data obtaining and organization modules, designed with the classic mode—MVC (Model-View-Controller) that is usually used in Web application development (Zhang Yi, 2007). It is a very good reference for interactive application that the data obtaining module and data organization module are developed with WinForm way. When users change the settings, the controller will modify the properties of models (i.e. the various components of the operation logic layer). Then the event is triggered. All display objects depending on the model will automatically update, and produce a response message back to users. It is a simple and clear processing procedure.

In user interface, the information retrieval and obtaining modules usually run in the background as console applications can be opened to observe through set interface. The set interface has only start/stop button and the parameter entry of various components. This kind of design is to make the system more simple and beautiful, ease to use, and filtering out unimportant information to provide administrator a clean-cut interface.

2.3 Message releasing module

Message releasing module includes three parts: message sending and receiving, message database management and user interface.

2.3.1 The message sending and receiving part

The part is secondary developed with Visual Basic language embracing MSCOMM ActiveX which can easily and efficiently connect other devices to PC through the serial port without having to call the low-level API functions, so that the programming efficiency is greatly improved. This can reduce the risk of system instability due to improperly programming.

Messages are send out by the AT + CMGS command, using PDU mode. The program code is as following:

```
Const prex = "0891"
Const midx = "11000D91"
Const sufx = "000800"
Public Function Sendsms(csca As String, num As String, msg As String) As _Boolean
Dim pdu, psmsc, pnum, pmsg As String
Dim leng As String
Dim length As Integer
length = Len(msg)
length = 2 * length
leng = Hex(length)
If length < 16 Then leng = "0" & leng
psmsc = Trim(telc(csca))
pnum = Trim(telc(num))
pmsg = Trim(ascg(msg))
pdu = prex & psmsc & midx & pnum & sufx & leng & pmsg
sleep(1)
mobcomm. Output = "AT + CMGF = 0" + vbCr
mobcomm. Output = "AT + CMGS = " & Str(15 + length) + vbCr
mobcomm. Output = pdu & Chr $ (26)
sleep(1)
Sendsms = True
End Function
```

Messages are received by the Output property of Mscomm ActiveX and AT + CMGR command. The function code is as following:

```
Public Sub readsms(rnum As String)
mobcomm. Output = "AT + CMGF = 1" + vbCr
mobcomm. Output = "AT + CMGR = " & rnum + vbCr
End Sub
```

2.3.2 Long message sending

At present, if the length of message is more than 140 bytes, China Telecom will cut it into several parts to send out, and can not guarantee the normal sentence order on the receiving telephone. The system sent message longer than 70 characters as a whole. The long message can be displayed fully in one record. It's very convenient for reading.

At first, the system judges the message length. The long message will be send if the text is longer than 70 characters or more than 140 bytes according to certain agreement. CMPP (China Mobile Peer to Peer) has to be set as following.

TP_udhi is 0x01; Msg_Content should be 6 or 7 TP_udhi header according to agreement following with message content of USC2 code; Msg_Fmt has to be set as 0x08 UCS2 code; Pk_total and Pk_number need not be set and if set it should be consistent with MM and NN field of TP_udhi.

2.3.3 The running interface of message releasing module

It shows the frequently-used functions entrance of elements which administrator needs to focus on, including the sending progress display, sending/receiving history, automatic and manual mode switching, telephone book and so on. The interface is simple and beautiful, in line with the operating habits of the common Windows applications.

2.3.4 Message management database

Message management database is used to manage personnel information, telephone book, and subscription records, sending and receiving contents records, system logs and so on. The database provides data services only for the releasing module, not for network services. In accordance with the need of the module function, the database is designed with the telephone table, historical record table, sending rule table, sending template table, subscription information table and permissions information table. Because the form and structures are relatively simple, the message database adopts Access database. As a desktop database, with a single storage mode, the Access database is object-oriented and easy to operate. It reduces the system volume effectively, and is more convenient for management when Access database is applied in message releasing module.

3 Function features

3.1 Message sending and receiving

Message sending function includes message sending and receiving, sending and receiving history record query, automatically-sending phone list service, the management of address book, idiom and template (He Shaohua and Chen Jie, 2005). Automatically-sending phone list provide a phone list for automatical message sending. Address book management is recharge of managing phone number and other relevant information. This function is set mainly for manually sending information. In order to further improve message writing efficiency, the system also provides a series of messages generating template to make the platform automatically generate related hydrological information according to the actual needs for user to edit and send. The template and automatical sending phone list are stored in the form of text library, can be very convenient to add, delete and modify, which greatly increases the flexibility of the platform, and improve the efficiency of the messages sending.

3.2 Message query

The system also developed for the mobile phone users who want to learn about the latest situation of hydrology through sending fuzzy query messages, such as:

When the user sends P50, he will receive the station name and its rainfall values where the daily rainfall is over 50 mm.

When the user sends H1, the system will extract the flood peak data of current day in the Yellow River main stream from database and send to the user.

When the user sends H2, the system will extract flood peak data of current data in the Yellow River tributary from database and send back to the user.

3.3 Message subscriptions

Subscriptions is used for releasing the information of forecasting and bulletin, including water regime daily, ice bulletin and flood forecasting, etc. This function can sent the information to the various subscribers with message in time (Qiu Chao, 2010). The platforms set subscription items for users to optionally subscript. Ordinary users select the contents of the subscription and receiving time. In addition, privileged user can group send out this subscription information.

3.4 Message query and statistical function

Software has statistical functions through messages database operations. Administrators obtain statistical information by some keywords like date, sending phone number, receiver phone number, message content and so on (Xiao Hongxin, Lin Yuhong, 2009).

3.5 System management

The system has the management function about users through assigning users different usage and registry permissions.

4 Research direction

4.1 Combined with the web application system

Nowadays, office automation system and network query consultation system is commonly used in hydrological service. Because of B/S mode, it has certain advantages comparing with C/S mode on the convenience of information processing and management. Compared with a single client mode, hydrology message releasing system combined with the web application system can increase the releasing efficiency and conserve resources, although with some security risks. Further improving the information releasing system needs to do in the later.

4.2 Feedback mechanism.

As an important part for telecom operators, the reliability of message business is very high. But in the holidays and other peak periods, there still happen information congestion and messages loss in rare cases. Current feedback of messages is send by telecom operator other than end users. It is hard to judge whether the end users have received the message or not, which easily result in the message releasing leak, should be improved in the future.

5 Conclusions

The Hydrology message releasing system is an important part of hydrology information releasing business. It is closely connected with the real-time hydrology database, based on the fundamental needs of information services, to further explore the support ability of real-time hydrology database, and has improved the information technology level. The software has been running well and provides efficient information service for flood control personnel at all levels and the core decision-makers.

References

Yang Chen. Development of Automobile Selling and Management System Based on C / S mode [J]. The Computer Disc Software Applications, 2011, 19.

Liu Feng, Sun Yong. Application of Design Patterns and Components Technology in the Business Logic Layer[J]. Computer Systems Applications, 2011,10.

Zhang Yi. Design Essentials and Mode of Software [M]. Beijing: Electronic Industry Publishing House . 2007.

He Shaohua, Chen Jie. Research on Electronic Government Information Releasing Standard [J]. Library and Information Science, 2005, 4.

Qiu Chao. Study and Application about Real-time Hydrological Information Warning SMS Platform [J]. Yangtze River, 2010,2.

Xiao Hongxin, Lin Yuhong. Construction of Students Work SMS Response Platform [N]. Journal of Southwest Agricultural University: Social Sciences, 2009, 2.

Three – dimensional Expression of the Sediment Particles' Morphology — the Construction of "Mathematical Sediment"

Zhao Huiming[1] and *Zhang Jianmin*[2]

1. State Key Laboratory of Hydro Science and Engineering, Department of Hydraulic and Hydropower Engineering, Tsinghua University, Beijing, 100084, China
2. Department of Hydraulic and Hydropower Engineering, Institute of Geotechnical Engineering, Tsinghua University, Beijing, 100084, China

Abstract: Morphology is one of the most important individual characteristics of sediment particles. In order to overcome the deficiencies of simplicity and one – dimensional characteristics when describing the morphology of sediment particles in the traditional studies and further develop the idea of regarding the sediment particle as a "sphere", this paper presents a concept of "mathematical sediment", which uses a mathematical method to characterize the sediment shape, surface morphology or structure, thus lay a foundation for the research of the relationship between sediment particles and chemical substances or biological substances in the process of sediment transport studies. The paper introduces the principle and method of Fourier shape analysis to restore the two – dimensional projection profile of sediment particles, and provides the shape meaning of all items of Fourier descriptors. On this basis, it further constructs the "mathematical sediment" through a certain combination ways to convey the surface morphology and structure characteristics of sediment particles, and uses the Fourier descriptors to control its shape so as to generate a variety of "mathematical sediments" with various edges and shapes, which realizes the description and analysis of the three – dimensional morphology of sediment particles.

Key words: fourier shape analysis, sediment particles, "mathematical sediment", morphology

1 Introduction

The geometric characteristic is one of the most important individual characteristics of sediment particles, which is generally described in terms of roundness, sphericity, the overall shape and surface structure, and so on (Wentworth, 1919; Wadell, 1932; Sneed , Folk, 1958). In the traditional sediment transport mechanics theory, the sediment particle is usually assumed to be a sphere for the research of corresponding properties. This mathematical formulation can basically meet the needs and development of the traditional theoretical system of sediment transport mechanics, and has also been supported by a large number of laboratory flume experiments and field observation data in the theory – building process, which are consistent with the actual sediment transport (Fang et al. , 2008). However, with the understanding of the chemical and biological transport in the sediment movement, the shape and surface morphology of sediment particles has become increasingly important. Substances adsorption of the sediment particles has an important relationship with the surface properties. In the process of chemical and biological transport of sediment particles, their shape and surface morphology changes gradually with the adsorption and desorption of various substances (Zhao et al. , 2009, 2011; Fang et al. , 2008, 2011; Chen et al. , 2010), and thus the changes affect the laws of sediment transport mechanics (Shang et al. , 2012; Fang et al. , 2012). Based on the understanding of the sediment movement, the researchers hope to find a mathematical method to describe the shape, surface morphology of sediment particles. Namely they hope to further develop the method of regarding the sediment particle as a "sphere" and be able to use mathematical equations to establish a "mathematical sediment" to characterize the shape and surface morphology of sediment particles, laying a foundation for the research of the relation between sediment particles and chemical and biological substances in the process of sediment transport study.

The current methods of describing the complex morphology of sediment particles are relatively simple. The description of particle shape and surface morphology in the sediment transport mechanics is usually giving the quantitative value of overall conception through shape factors, specific surface area, etc. But the starting point of these methods is expressing the two – or three – dimensional shape information of sediment particles with a single data. This expression method of reducing the dimensions can illustrate the characteristics of particle shape, but much information of the complex particle shape, such as particle sharpness, grain size and surface structure, is also easy to be lost in the quantization process of the conceptions of shape factors, specific surface area, and so on. Virtually these quantitative values can not illustrate the morphology characteristics of sediment particles and can not research the needs of environmental sediment research.

In order to overcome the simplification of morphology description of sediment particles and the deficiency of one – dimensional characteristics, the researchers began to using mathematical analysis methods to express the geometry shape of sediment particles through some functions or many shape index. Due to the differences of sediment particles' shape, it usually calls for at least two kinds of data and their combinations to accurately describe the geometry shape of sediment particles using mathematical language. The commonly used data include the representative values of the particle size in the three – axis direction, the contour curves of the two – dimensional image projection, as well as some data related with the solid geometry, such as surface area, volume, etc. This paper describes a mathematical method basing on sediment projection shape, which uses statistical method and Fourier shape analysis to describe the planar projection shape of sediment particles and using the projection shape to restore the morphology of sediment particles.

2　The principle of Fourier shape analysis

The particle shape defined usually, which is the plane projection contour of the particle, is a closed curve in the two – dimensional plane. It can reflect the shape characteristics of the particle. The more complex the closed curve is, the more complex the edge shape of the particle is. The following is a brief introduction of the Fourier function to analyze the particle contour (Fang et al. , 2009).

The scale range of sediment particles is in the micron and millimeter level, so it first calls for the high – resolution tools such as electron microscopy to obtain the shape information. In this paper, the Scanning Electron Microscopy (SEM) is selected to conduct the morphology observation of the sediment particles and it can provide clearly and intuitively the image data of sediment particles for the researchers to make qualitative and quantitative analysis. The higher the resolution of the scanning electron microscope is, the more pixels of the SEM pictures are obtained from the same sediment particles, also the clearer the pictures are and the richer the picture information is.

On the basis of SEM observation, we chose the clearly – imaging pictures and separated the relatively independent and integrated sediment particles from the background as a basis of further study. As shown in Fig. 1(a), the picture is a gray image with a gray – value scale range of $0 \sim 255$. In graphic image study, a gray image can be described using a two – dimensional matrix $z(x, y)$ (Fig. 1(b)), and each element $z(x, y)$ corresponds to a gray value of pixel (x, y) in the gray image.

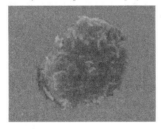

$$Z = \begin{bmatrix} z_{11} & z_{12} & \cdots & z_{1N} \\ z_{21} & z_{22} & \cdots & z_{2N} \\ \vdots & \vdots & & \vdots \\ z_{M1} & z_{M2} & \cdots & z_{MN} \end{bmatrix}$$

(a) The SEM picture of a single sediment　　　(b) The two – dimensional matrix of gray image
particle

Fig. 1

Every sediment particle extracted from the SEM pictures corresponds to a pixel matrix mentioned above, and the gray values of background pixels are to a certain value, which equals to 0 in this paper. While the gray values of particle pixels are none – zero, and the edge is the projection contour of the sediment particle, as shown in Fig. 2. The message source of the particle's shape is the location of the points on the boundary. After digitizing the pictures, we can extract the edge lines of the sediment particles' projection contour and describe them using the method of Fourier shape analysis.

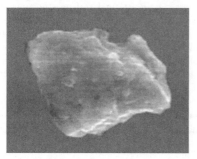 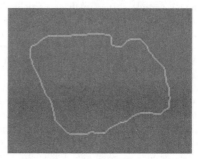

(a) The picture of an sediment particle   (b) The picture of the sediment particle's edge

Fig. 2

This method is regarding the projection boundary of a sediment particle as a closed curve constituted of a point set in the plane or space coordinate system, seeking the centroid of the closed curve as the origin to establish a coordinate system, and acquiring the (x, y) coordinate of every point on the boundary through the point set information. Transforming the cartesian coordinate system of (x, y) into polar coordinate system of (R, θ), in which, R reefers to the polar radius, and θ refers to the polar angle. Then the mathematical description of the closed curve can be expressed as a periodic equation:

$$R(\theta + 2\pi) = R(\theta) \tag{1}$$

where, 2π refers to the periodic of the curve.

The (R, θ) values of the boundary points in the sediment particle projection contain approximately all the information of its shape and size.

The periodic function of sediment particle's projection contour can be approximated to by the Fourier series (Weaver, 1983; Pete, 1997; Loncaric, 1998) as:

$$R(\theta) = A_0 + \sum_{n=1}^{\infty} [a_n \cos(n\omega\theta) + b_n \sin(n\omega\theta)] \tag{2}$$

where, n is the item number; A_0, a_n and b_n are the Fourier coefficients.

Fourier coefficients contain all the information of the sediment particle's shape and size, of which the low – order coefficients reflect the main character of particles, the high – order coefficients reflect details, and different orders coefficients reflect different information of the sediment particle's shape (Ying, 2001; Zhao, 2004). For example, a_1/b_1 reflect the information of symmetry, a_2/b_2 reflect the information of long shape degree, a_3/b_3 reflect the information of triangle degree, a_4/b_4 reflect the information of square degree, a_5/b_5 reflect the information of pentagon degree, a_6/b_6 reflect the information of hexagon degree, and so on. Fig. 3 shows the shape changing of curves described by the Fourier series, which is obtained by setting the item number n as 0 and the initial values of Fourier coefficients a_n, b_n as 0, and then adjusting respectively the first six orders of Fourier coefficients in turn. In the pictures, the black lines show the curves when the a_n, b_n are all 0, and the red lines shows the curves obtained by adjusting respectively the corresponding order of Fourier coefficients in turn, which reveals the regulation trend of the Fourier coefficients on the closed curve shape.

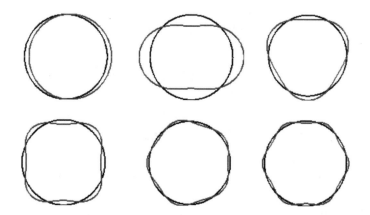

Fig. 3   The expression of shape message from the Fourier coefficients

There are different sizes of sediment particles in the sample groups. We use non – dimensional processing to normalize the polar radius? into 1, adding the shape comparability of different size of sediment particles. The normalization method is as follows:

$$R_s(\theta) = \frac{R(\theta)}{R_{mean}} \quad (3)$$

where, R_{mean} refers to the mean value of the radius sequence $R(\theta)$.

Then the Fourier can be expressed as:

$$R_s(\theta) = A_{s0} + \sum_{n=1}^{\infty} \left[a_{sn} \cos(n\omega\theta) + b_{sn} \sin(n\omega\theta) \right] \quad (4)$$

where, $A_{s0} = 1$; a_{sn} and b_{sn} are normalized Fourier coefficients.

The general features of particle's shape can be determined using the lower Fourier coefficients, yet the small differences of particle's shape couldn't be described unless the higher order Fourier coefficients are used. The more the orders are, the more clearly the shape details are described, which may also induces much calculation workload. We fit the particle boundary? with different item number of Fourier coefficients. Considering the study precision and fast calculation, the item number of Fourier coefficients n is set as 34 in this paper. Fig. 4 and Fig. 5 are respectively the fitting results of the periodic function and the contour line, which shows that the Fourier series approximation has been able to simulate the edge contour of sediment particles very well.

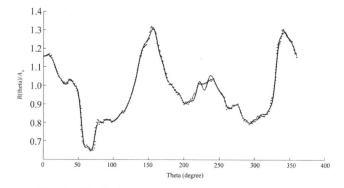

Fig. 4   The fitting test of Fourier periodic function

Fig. 5   The simulatiion comparison of particle contour

3   The construction of "mathematical sediment"

The Fourier shape analysis method has many advantages, such as high simulation accuracy, simple calculation, etc. , so it can be applied to analyze and describe the complex shape of objects. The last section introduces the methods and steps to simulate the two – dimensional projection profile of sediment particles using the Fourier shape analysis method, which could produce the outline of an sediment particle as long as make sure its Fourier coefficients. This section will continue to apply the Fourier shape analysis method to construct a "mathematical sediment" which can be used to characterize the sediment surface morphology and structural characterization, in order to achieve the description and analysis of the three – dimensional morphology and surface structure of sediment particles (Chen, 2008).

To construct the surface morphology of sediment particles in the three – dimensional spherical coordinate system through the projection profile edge of sediment particles described in the two – dimensional polar coordinate system, it needs to consider three variables of the radius vector length R, the latitude θ and the longitude φ, of which, the radius vector length R is the function of the latitude θ and $R(\theta)$ can be calculated through the known Fourier coefficients.

The construction of the "mathematical sediment" is divided into two steps: The first step is to generate a number of enclosed projection contour curves of sediment particles in the polar coordinate system; the second step is to combine reasonable these enclosed projection contour curves of sediment particles in order to restore the surface morphology of sediment particles.

The core of the construction work is to establish a reasonable description equation of the "mathematical sediment". According to the conclusions of the previous section studies, when using the Fourier series to simulate the projection profile of a sediment particle's edge, it could have a better simulation result if the item number of the Fourier coefficients n take to 34. Therefore, here the first 34 orders of Fourier series are taken, and the corresponding Fourier series equation is as follows:

$$R(\theta,\varphi) = \{A_0 + \sum_{n=1}^{34} [a_n\cos(n\omega\theta) + b_n\sin(n\omega\theta)]\} \times \cos\varphi \qquad (5)$$

The combination way of reconstruction calls for mathematical processing. As the sediment surface can be regarded as a combination of interwoven by numerous projection profile, to determine the comnination way of projection profile is to determine the surface morphology of sediment particles. In the calculation the space grid is firstly established according to $\theta = 1° \sim 360°$ and $\varphi = 1° \sim 180°$, and then the value of R is calculated. The layout of the grid is that 360 grid points are divided in the longitude direction along $\theta = 1° \sim 360°$ and 180 grid points are divided in the latitude direction along $\varphi = 1° \sim 180°$. The value of R is firstly calculated along the longitude direction to acquire $R_1(\theta,\varphi)$ and then along the latitude direction to acquire $R_2(\theta,\varphi)$. As the probabilities occurrence of $R_1(\theta,\varphi)$ and $R_2(\theta,\varphi)$ at the same grid point are equal, the following exits:

$$R = \sqrt{R_1(\theta,\varphi) \times R_2(\theta,\varphi)} \qquad (6)$$

After the values of θ, φ and R are determined, we can draw the "mathematical sediment" in the spherical coordinate space or the three – dimensional "mathematical sediment" in the Cartesian coordinate converted from the space coordinate.

The characteristic of constructing the "mathematical sediment" is establishing the space grid corresponding with θ and φ and calculating to acquire $R(\theta,\varphi)$ directly. Every (θ,φ) being given a random probability value of P, the Fourier coefficients are calculated through the Gaussian distribution established by the mean and standard deviation, and $R(\theta,\varphi)$ is calculated according to Eq. (5). The spherical coordinate system is converted to a Cartesian coordinate system as follows:

$$X = R(\theta,\varphi)\cos\varphi\cos\theta \qquad (7)$$
$$Y = R(\theta,\varphi)\cos\varphi\sin\theta \qquad (8)$$
$$Z = R(\theta,\varphi)\sin\theta \qquad (9)$$

The three – dimensional mathematical sediment can be constructed by plotting the X, Y, Z in the above three equations.

Fig. 6 shows the reconstructed three – dimensional "mathematical sediment" model, which can present directly the morphology characteristics of sediment surface. Any projection profile of the "mathematical sediment" is not the same, and the irregular morphology such as potholes and downs on the surface are clearly, performing the complex topographic characteristics of sediment particles. Compared to the sediment particle assumed in the traditional sediment movement mechanics, "mathematical sediment" overcomes the lack of particle system and smooth sphere system in the traditional sediment theory and can achieve its spatial analysis. After digitizing the surface information, the sediment surface elevation data are stored in the form of a matrix or raster data, further geomorphology methods or matrix operations are applied to make the discrimination, statistical and trend analysis of morphology, which can meet the requirement of sediment morphology research.

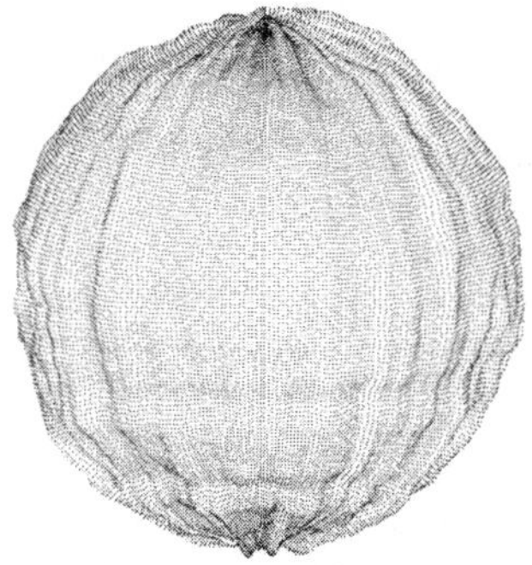

Fig. 6　"Mathematical sediment" in the Cartesian space

4　Application of the Fourier descriptors in the shape control of "mathematical sediment"

As mentioned earlier, Fourier analysis is an effective method to describe the projection contour shapes of sediment particles. The Fourier coefficients in the fitting function are important parameters to describe the particle morphology and can further control the shape change of the closed curves, which are also called the Fourier descriptors.

The "mathematical sediment" in the previous section is restored through the average values of the Fourier descriptors which are obtained by the regression of characteristic coefficient of the extracted sediment particles in the sample group. For every sample of sediment particle, a set of calculated values of a_n and b_n can be obtained after fitting the projection contour boundary line using the Fourier series. Using the 470 extracted sediment particles as a sample group and interval – estimating the corresponding parameters of its Fourier descriptors, the confidence interval of the overall sediment particles' parameters are obtained, and further the average of the Fourier descriptors are calculated for the construction of the "mathematical sediment".

The law of Fourier descriptors controlling the closed curve shape in the Section 2 is applied in the construction of "mathematical sediment" to restore the three – dimensional morphology of sediment particles of various shapes in the nature. That is on the basis of average Fourier descriptors of natural sediment samples, the Fourier coefficients of the corresponding item number are adjusted appropriately to control the curves of "mathematical sediment" in order to generate a variety of "mathematical sediments" with various edges and shapes. Meanwhile, it is known from the SEM images of sediment particles that sediment particles with shapes close to the sphere in the nature are very rare and most sediment particles are nearly – flat state with the micro – structure of the bump in one dimension. So the "mathematical sediment" is for the appropriate interception and plane reconstruction to approach the true state of sediment in the nature. Fig. 7 shows the "mathematical sediments" with more complex shapes of partial triangle, partial quadrilateral and compound shape, from which it is known that the method of Fourier descriptors controlling the shape can restore vividly the morphology characteristic of sediment particles in the nature and provide a good mathematical platform for further research on adsorption and desorption laws of various substance and its own properties changes of sediment particles in the processes of chemical and biological transportation.

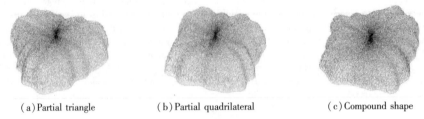

(a) Partial triangle   (b) Partial quadrilateral   (c) Compound shape

Fig. 7   "Mathematical sedimen" with different shapes

5   The effect expression of "mathematical sediment"

The mathematical software with a composite function of algorithm development, data visualization, data analysis and numerical calculation, such as Matlab, can be used in the generation and debugging of the above "mathematical sediment". Taking Fig. 7(c) as an example, its interface window of generating "mathematical sediment" is shown in Fig. 8, which can finish the work of data extraction, morphology observation, micro – structural analysis, et al. On this basis, 3Dmax is further adopted in the effect processing in order to make "mathematical sedimen" have apparent form of natural sediment particles.

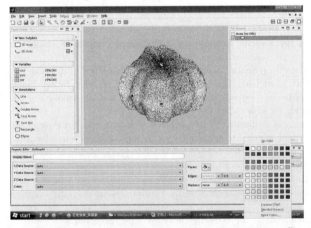

Fig. 8   Matlab expression of "mathematical sediment"

The 3Dmax expression of "mathematical sediment" is divided into the three main steps of model manufacture, scene layout of the environment and materials attachment. Firstly the digital information of "mathematical sediment" generated by Matlan is extracted, saved as wrl format and imported into 3Dmax to generate the basic model; then effect processing is finished through some corresponding operations such as rendering, hole filling, scene layout, materials attachment using maps generated by real sediment photos, and so on. The resulting effect expression of "mathematical sediment" is shown in Fig. 9, which also has the apparent visual characteristics of natural sediment particles on the basis of a combination of statistical regularity of their ever – changing shapes. It could express the morphology characteristics of the sediment particles.

Fig. 9   The effect expression of "mathematical sediment"

6   Application prospects of "mathematical sediment"

The randomly generation of the three – dimensional "mathematical sediment" provides a good research method, which can express the natural sediment particles with thousands of different forms in a numerical manner and becomes a platform of researching sediment movement and the interaction between sediment and pollutants. Whether the resulting "mathematical sediment" could reflect the true morphology and structure of sediment is the key to the universality of the "mathematical sediment". Besides, there is an important purpose of the establishment of this platform, which is to reflect the interaction process between sediment and pollutants in the "mathematical sediment" and the pollutants distribution on particle surface.

"Mathematical sediment" is a new conception and method to characterize the shape and surface of sediment particles using numerical equations. Compared to the hypothesis supposing sediment to be sphere in the past studies, it accords more with the real surface morphology of sediment particles and can also meet the research needs of environmental sediment. What the "mathematical sediment" provides is not only a research platform to reproduce the surface morphology of sediment particles in the three – dimensional space, but also an important platform to describe the interaction between sediment and pollutants in the environmental sediment research. The spatial characteristics and surface morphology features it conveys can be used as a research basis of surface adsorption of fine sediment and sediment movement mechanics. The physical and chemical properties of sediment particles surfaces can be studies on the basis of the "mathematical sediment", thus further understand the phenomenon of flocculation and adsorption, perfect the laws of suspension, settlement and resuspension of polluted sediment in rivers and reservoirs, and meanwhile complete the research of sediment adsorption kinetics and migration laws of pollutant adsorption, desorption and diffusion. Fang et al (2009) once applied the authentication method of fractal dimension to discourse the rationality of using "mathematical sediment" to simulate real sediment particles, and on this basis simulated the distribution law of copper ion on the "mathematical sediment" surface after adsorption combining with the adsorption theory of sediment particles and distribution experiments of copper ions on the sediment particles' surface. The results showed that the "mathematical sediment" can not only express completely the particle surface information, but also display the characteristics of the copper icon distribution.

7 Conclusions

The physical, chemical and biological processes on sediment surface can be described more accurately by the research of morphology of sediment from the microscopic point of view. This paper introduced the Fourier shape analysis methods which could be applied to restore and analyze the two – dimensional projection contour of sediment particles, described the three – dimensional morphology through a certain combination ways, and further control the shape changes of "mathematical sediment" using Fourier descriptors in order to restore the sediment particles of various shapes in the nature. The constructed "mathematical sediment" can characterize the complex morphology of sediment surface through mathematical equations. As "mathematical sediment" is the research platform to reproduce the surface morphology of sediment particles in the three – dimensional space, the spatial properties and surface morphology characteristics it expressed can be used as the basis of the research on surface adsorption of fine sediment particles and sediment movement mechanics, which has important practical significance for the research of the movement law of environmental sediment and its influence on hydraulic engineering.

Acknowledgements

This investigation was supported by the National Science Foundation of China No. 51109113 and No. 50909095, the Postdoctor Science Foundation of China No. 20110490413.

References

E Lestre Pete. Fourier Descriptors and Their Applications in Biology [M]. Cambridge: Cambridge University Press, 1997.

Fang Hongwei, Chen Minghong, Chen Zhihe. Surface Pore Tension and Adsorption Characteristics of Polluted Sediment [J]. Science in China Series G: Physics, Mechanics & Astronomy, 2008, 51(8):1022 – 1028.

Fang Hongwei, Chen Minghng, Chen Zhihe. The Characteristics and Model of Environmental Sediment Surface [M]. Beijing: Science Press, 2009.

Fang Hongwei, He Guojian, Liu Jinze. 3D Numerical Investigation of Distorted Scale in Hydraulic Physical Model Experiments [J]. Journal of Coastal Research, 2008(52):41 – 53.

Fang Hongwei, Shang Qianqian, Zhao Huiming, et al. Experimental Study on the Biofilm Effects on Sediment Setting I —Setting Velocity Calculation [J]. Journal of Hydraulic Engineering, 2012, 43(3):8 – 13.

Fang Hongwei, Zhao Huiming, He Guojian, et al. Experiment of Particles' Morphology Variation After Biofilm Growth on Sediments [J]. Journal of Hydraulic Engineering, 2011, 42(3): 278 – 283.

Shang Qianqian, Fang Hongwei, Zhao Huiming, et al. Experimental Study on the Biofilm Effects on Sediment Setting I —Experimental Design and Grading Variation [J]. Journal of Hydraulic Engineering, 2012, 43(2):49 – 56.

Sneed E D, Folk R L. Pebbles in the Lower Colorado River, Texas: a Study in Particle Morphogenesis [J]. J. Geol., 1958, 66:114 – 150.

Sven Loncaric. A Survey of Shape Analysis Techniques [J]. Pattern Recognition, 1998, 31(8): 983 – 1001.

Wadell H. Volume, Shape and Roundness of Rock Particles [J]. J. Geol., 1932 (40): 443 – 451.

Weaver J. Applications of Discrete and Continuous Fourier Analysis [M]. New York: John Wiley and Sons Inc. , 1983.

Wentworth C K. A Laboratory and Field Study of Cobble Abrasion[J]. The Journal of Geology, 1919(40):507 – 521.

Ying Yibin. Fourier Descriptor of Fruit Shape [J]. Journal of Biomathematics, 2001, 16(2): 234 – 240.

Zhao Huiming, Fang Hongwei, Chen Minghong. Floc Architecture of Bioflocculation Sediment by ESEM and CLSM [J]. Scanning, 2011, 33(6):437 – 445.

Zhao Huiming. Experiment on Particle's Surface Morphology after Biofilm Vegetating[R]. 33rd IAHR Congress, 2009(8):4878 – 4885.

Zhao Guangtao, Wei Zhou, Willian E Full, et al. Fourier Shape Analysis and its Application in Geology [J]. Journal of Ocean University of Qingdao, 2004, 34(3):429 – 436.

Comprehensive Drought Evaluation Model for Irrigation District Based on Variable Fuzzy Sets and Its Application

Zhang Zhenwei[1] and *Wei Rui*[2]

1. North China University of Water Conservancy and Electric Power, Zhengzhou, 450045, China
2. Yellow River Hydrological Survey and Design Institute, Zhengzhou, 450004, China

Abstract: Drought is one severe meteorological disaster in China, better comprehensive evaluation of drought can provide reliable policy – making basis for drought defending and disaster reduction. The existing methods of drought evaluation are lacking of universal, in order to solve the problem, based on the theory of Variable Fuzzy Sets, a comprehensive drought evaluation model are proposed, and the method for solution are given in the paper, the model is applied in Qucun Yellow River Irrigation Region. The result shows that the model is reasonable and feasible, it is simple and practical. The model and its method can provide academic support and policy – making reference for the comprehensive evaluation of drought in North Region of China.

Key words: drought, binary comparison, variable fuzzy sets, comprehensive evaluation, model

Drought is one severe meteorological disaster in China, and has had a serious impact on the socio – economic and living standards. During the period from 1949 to 1999, the drought – stricken areas were about $2.159,3 \times 10^6$ hm^2 annual, accounting for 60% of a variety of meteorological disasters, and about 10 billion kg food was lost every year. Drought has further increased with the economic development and the population growth. Since the year of 2008, the fewer rainfall of northern China has closed to or exceeded the historical limits, the meteorological drought of some parts of Hebei, Shandong, Henan Province has reached severe drought or even special drought, and has seriously affected crops. Northern drought has created a serious threat on food security, so the problem of drought, especially the study of agriculture drought has become a research hotspot at present.

The evaluation methods of drought have played a positive role during the drought assessment, such as single – index evaluation method, fuzzy comprehensive evaluation method, neural networks and gray system method. Based on the single – index drought evaluation, Liuwei and others introduced the model of fuzzy pattern recognition into the drought level evaluation, took Jinan as example and got a more reasonable result. But in a condition of certain space and time, the membership and the membership function which are described fuzzy concept should be relative and dynamic. As the complexity and the uncertainty of occurrence of drought, the fuzzy comprehensive evaluation method which is based on static theory of fuzzy sets has not better on the comprehensive evaluation of drought.

So, the evaluation model is set and the solutions are given based on the theory of Variable Fuzzy Sets and Engineering Fuzzy Sets in the paper, and the model is applied in Qucun Yellow River Irrigation Region. By varying the model and parameters, the methods are of the rigor theory and the simple operation, and take the boundaries ambiguity of the drought classification, so the evaluation of drought levels are given, and the credibility of the drought level evaluation is improved. The methods and the results in this paper can provide reliable policy – making basis for drought defending and disaster reduction.

1 Comprehensive Drought Evaluation Model Based on the Variable Fuzzy Sets

Assume the characteristics of drought assessment samples x are expressed by the feature vectors m, and each feature vector is divided into levels which are expressed by c.

The feature vectors of drought assessment samples:

$$x = (x_1 \quad x_2 \quad \cdots \quad x_m) \tag{1}$$

The interval matrix I_{ab} is constructed by the standard value of the drought feature vectors' levels and the actual construction:

$$I_{ab} = ([a,b]_{ih}) \tag{2}$$

where, m is index number and c is level number, $i = 1,2,\cdots,m$; $h = 1,2,\cdots,c$.

The range values matrix of the change interval I_{cd} is constructed as follows according to the interval matrix I_{ab}:

$$I_{cd} = ([c,d]_{ih}) \tag{3}$$

Determining the matrix M of the point M_{ih} when $D_A(u) = 1$ in the drought assessment attraction domain $[a,b]$ according to the actual construction of each drought index level:

$$M = (M_{ih}) \tag{4}$$

The relative difference function $D_A(u)$ is determined as follows:

(1) Assume $X_0 = [a,b]$ is the attraction domain of the fuzzy variable sets in the real axis, and $X_0 = [c,d]$ is the range interval which contains the upper and lower bounds of the point X_0, as show in Fig. 1:

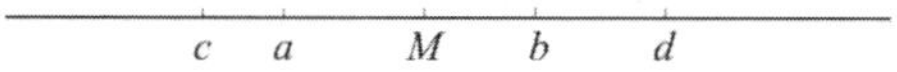

$$\underline{\qquad \quad c \quad a \quad M \quad b \quad d \qquad}$$

Fig. 1   The position relationship of point x and the interval of X_0 and X

(2) Assume M_{ih} is the point when $D_A(u) = 1$ in the attraction domain $[a,b]$, and x is any point value in the interval of X.

When the point x at the right of the point M_{ih}, the model of the relative difference function is:

$$\begin{cases} D_A(u) = \left(\dfrac{x-b}{M_{ih}-b}\right)^{\beta}, & x \in [M_{ih},b] \\ D_A(u) = -\left(\dfrac{x-b}{d-b}\right)^{\beta}, & x \in [b,d] \end{cases} \tag{5}$$

When the point x at the left of the point M_{ih}, the model of the relative difference function is:

$$\begin{cases} D_A(u) = \left(\dfrac{x-a}{M_{ih}-a}\right)^{\beta}, & x \in [a,M_{ih}] \\ D_A(u) = -\left(\dfrac{x-a}{c-a}\right)^{\beta}, & x \in [c,a] \end{cases} \tag{6}$$

where, β is non-negative index, $\beta = 1$.

(3) The relative membership degree is developed as follows:

$$\begin{cases} \mu_A(u) = [1 + D_A(u)]/2, & x \in [c,d] \\ \mu_A(u) = 0, & x \notin [c,d] \end{cases} \tag{7}$$

The target weight vector ω is determined by the method of Binary Comparison Fuzzy Decision, and the relative membership degree matrix $\mu_A(u) = (\mu_A(u)_{ih})$ is calculated according to the relative difference function model and some related data.

Calculate the non-normalized integrated relative membership degree $\mu'_h(u)$:

$$\mu'_h(u) = \cfrac{1}{1 + \left\{ \cfrac{\sum\limits_{i=1}^{m}[\omega_i(1-\mu_A(u)_{ih})]^p}{\sum\limits_{i=1}^{m}[\omega_i\mu_A(u)_{ih}]^p} \right\}^{\frac{a}{p}}} \tag{8}$$

where, p is the distance parameter, $p = 1$ representing the Hamming distance and $p = 2$ representing the Euclidean distance; a is the optimization criterion parameter, $a = 2$ representing the least squares criteria and $a = 1$ representing the weighted least involution criteria.

Evaluate the level of the sample by the level characteristic value equation as follows:

$$H = \sum_{h=1}^{c} \mu_h \cdot h \tag{9}$$

where, $\mu_h = \dfrac{\mu_h'}{\sum\limits_{h=1}^{c} \mu_h'}$.

2　Model application in the Yellow River irrigation

Qucun Yellow River Irrigation Region is selected as the typical study area in this paper. The average annual rainfall is about 581 mm, and the mean annual evaporation is 1,663 mm. Winter and spring droughts occurred frequently owning to the large annual variation and uneven seasonal distribution of rainfall. In 2001, severe drought of the land in Puyang was 0.8%, and moderate drought was 85.6%; in 2009, there was a long duration, wide scope and gravity drought since the wheat was planted in Quncun Puyang, and the rainfall was 80% less than the same period in previous years, the drought is a major natural disaster in this area. In order to evaluate the drought better, reduce the drought losses, the model and methods based on the theory of variable fuzzy sets are proposed in this paper to evaluate the drought.

2.1　Determination of drought evaluation index system

Complied with the principles of scientific, systematic, comparable and dynamic of the drought evaluation index system, according to the "Drought Assessment Criteria" which was released by the Flood Control and Drought Mitigation Engineering Research Center of the water Resources Ministry, based on the research results of the drought indexes and the grading standards in abroad and China, and combined with the actual situation of Qucun, the drought is divided into mild drought, moderate drought, serious drought and severe drought. Then meteorological drought, agricultural drought and other two were selected as the first level indicator, precipitation anomaly percentage, consecutive days without rain, dryness and other five were selected as the second level indicator, and the standard values of each level index are showed in Tab. 1.

Tab. 1　Drought scale of each index

Drought category	Drought index I	Drought intensity			
		Mild drought	Moderate drought	Serious drought	Severe drought
Meteorological drought	Precipitation anomaly percentage %	$-40 \sim -20$	$-60 \sim -40$	$-80 \sim -60$	< -80
	Consecutive days without rain d	$10 \sim 20$	$20 \sim 30$	$30 \sim 45$	$45 \sim 92$
	Dryness	$0.75 \sim 1.6$	$1.6 \sim 2.7$	$2.7 \sim 3.5$	$3.5 \sim 6$
Agricultural drought	Soil moisture content(%)	$16 \sim 20$	$14 \sim 16$	$5 \sim 14$	<5
	Disaster area percentage(%)	$10 \sim 20$	$20 \sim 40$	$40 \sim 60$	>60
	Drought area percentage%	$20 \sim 40$	$40 \sim 60$	$60 \sim 80$	>80
Hydrological drought	Anomaly percentage of river inflow%	$-30 \sim -10$	$-50 \sim -30$	$-80 \sim -50$	< -80
Socio – economic drought	Difficulty rate in drinking%	$11 \sim 20$	$20 \sim 30$	$30 \sim 40$	>40

2.2 Drought comprehensive evaluation

(1) The feature vectors of drought assessment samples are
$$x = (-43.5 \quad 85 \quad 2.5 \quad 7.6 \quad 70 \quad 68 \quad -82 \quad 23)$$

(2) According to Eq. (2) to Eq. (4), the interval matrix I_{ab}, The range values matrix I_{cd} and the matrix M are as follows:

$$I_{ab} = \begin{bmatrix} [-40,-20] & [-60,-40] & [-80,-60] & [-100,-80] \\ [10,20] & [20,30] & [30,45] & [45,92] \\ [0.75,1.6] & [1.6,2.7] & [2.7,3.5] & [3.5,6] \\ [16,20] & [14,16] & [5,14] & [0,5] \\ [10,20] & [20,40] & [40,60] & [60,100] \\ [20,40] & [40,60] & [60,80] & [80,100] \\ [-30,-10] & [-50,-30] & [-80,-50] & [-100,-80] \\ [11,20] & [20,30] & [30,40] & [40,100] \end{bmatrix}$$

$$I_{cd} = \begin{bmatrix} [-60,-20] & [-80,-20] & [-100,-40] & [-100,-60] \\ [10,30] & [10,45] & [20,92] & [30,92] \\ [0.75,2.7] & [0.75,3.5] & [1.6,6] & [2.7,6] \\ [14,20] & [5,20] & [0,16] & [0,14] \\ [10,40] & [10,60] & [20,100] & [40,100] \\ [20,60] & [20,80] & [40,100] & [60,100] \\ [-50,-10] & [-80,-10] & [-100,-30] & [-100,-50] \\ [11,30] & [11,40] & [20,100] & [30,100] \end{bmatrix}$$

$$M = \begin{bmatrix} -40 & -60 & -60 & -80 \\ 10 & 20 & 45 & 92 \\ 0.75 & 1.6 & 3.5 & 6 \\ 16 & 14 & 14 & 5 \\ 10 & 20 & 60 & 100 \\ 20 & 40 & 80 & 100 \\ -30 & -50 & -50 & -80 \\ 11 & 20 & 40 & 100 \end{bmatrix}$$

(3) According to Eq. (5) to Eq. (7), the relative membership degree matrix $\mu_{\underline{A}}(u)$ is as follows:

$$\mu_{\underline{A}}(u) = \begin{bmatrix} 0.413 & 0.587 & 0.088 & 0 \\ 0 & 0 & 0.074 & 0.926 \\ 0.091 & 0.591 & 0.409 & 0 \\ 0 & 0.144 & 0.589 & 0.356 \\ 0 & 0 & 0.375 & 0.625 \\ 0.3 & 0.7 & 0.2 & \\ 0 & 0 & 0.45 & 0.95 \\ 0.35 & 0.85 & 0.15 & 0 \end{bmatrix}$$

(4) The target weight vector of each drought Evaluation Indexes is
$$\omega' = (0.667 \quad 0.667 \quad 0.905 \quad 1 \quad 0.818 \quad 0.818 \quad 0.6 \quad 0.429)$$

The normalized weight vector is
$$\omega = (0.113 \quad 0.113 \quad 0.153 \quad 0.169 \quad 0.138 \quad 0.138 \quad 0.103 \quad 0.073)$$

(5) According to Eq. (8) and Eq. (9), the integrated relative membership degrees and the levels of the samples are showed in the Tab. 2. Where the levels for the drought is determined as follows: $H \leqslant 1$ representing mild drought, $1 < H \leqslant 2$ representing moderate drought, $2 < H \leqslant 3$ representing serious drought, $H > 3$ representing severe drought.

Tab. 2   Consolidated relative membership degree

	Mild drought	Moderate drought	Serious drought	Severe drought	Sample level
$a=1,p=1$	0. 086,0	0. 323,0	0. 386,0	0. 376,0	2. 896,0
$a=1,p=2$	0. 139,0	0. 322,6	0. 427,3	0. 391,9	2. 839,0
$a=2,p=1$	0. 008,8	0. 193,8	0. 283,4	0. 267,2	3. 073,0
$a=2,p=2$	0. 025,4	0. 184,8	0. 357,6	0. 293,5	3. 066,0

We can get that the average value is 2. 968,0 according to the Tab. 2. So this area is in a serious drought and the evaluation results are more consistent with the actual situation of the region compared with the local meteorological data.

3   Conclusions

Drought is nature disaster that has enormous impact on the economic, social and environment, and will take significant losses on the agricultural production, people lives' water and industrial production, the comprehensive evaluation of drought, especially the agricultural drought is one of the urgent problems in the northern irrigation areas. So the drought comprehensive evaluation model of northern irrigation district based on the Fuzzy Set Theory is established in this paper because the existing methods can't evaluate drought better as the complexity of drought, the fuzzy and dynamic variability of the indexes, and applied in the Qucun Yellow River Irrigation Region. At last, the result shows that the area is in the serious drought and is more consistent with the actual situation, the model and method are feasible, and it is simple and practical. The model and its method can provide academic support and policy – making reference for the comprehensive evaluation of drought in North Region of China.

Acknowledgements
The authors wish to express their thanks to the National Natural Science Funds(41071025) , the 948 project funding of Ministry of Water Resources of P. R. China(code:201047), Henan natural science research project of Education Department (2009A170004), and Henan technology funds(092102310197).

References

Yuan Wenping, Zhou Guangsheng. Theoretical Study and Research Prospect on Drought Indices [J]. Advances in Earth Science, 2004, 19(6): 982 – 991.

Luan Yulin. Serious Drought in Northern China and Severe Drought in Some Areas [EB/OL]. http://news. sina. com. cn/c/2009 – 02 – 04/021817142777. shtml

Li Xingmin, Yang Wenfeng, Gao Bei, et al. The Research and Application of The Drought Indexes in Meteorology and Agriculture [J]. Journal of Northwest A & F University(Nat. Sci. Ed.). 2007,35(7): 111 – 116.

Liu Wei, Cao Shengle, Ren Liliang. Application of Fuzzy Comprehensive Model in Drought Degree Assessment [J]. Water Resources and Power, 2008,26(2): 100 – 102.

Chen Xiaonan, Huang Qiang, Qiu Lin. Model for Estimating Agricultural Drought Based on Neural Network Optimized by Chaos Algorithm [J]. Journal of Hydraulic Engineering, 2006,37 (2): 247 – 252.

Chen Nanxiang, Yang Li, Shao Yubing. Application of Grey System Theory in Evaluation for the Arid Degree of Area [J]. Journal of Irrigation and Drainage, 2007,26(2): 26 – 29.

Chen Shouyu. The Variable Fuzzy Set Theory and Method of Water Resources and Flood Control System [M]. Dalian University of Technology Press, Dalian, 2005.

Chen Shouyu. Engineering Fuzzy Set Theory and Application [M]. National Defense Industry

Press, Beijing, 1998.

The Drought Distribution of Henan by Satellite Monitoring in 2001 [EB/OL]. http://pyx. hnnw. net/qxtd/qxtd9. htm

The Successful Implementation of Artificial Precipitation in Puyang [EB/OL] http://news. shangdu. com/category/dishi/457000/2009/02/09/2009 - 02 - 09_905663_457000. shtml

Li Xiangyun, Wang Lixin, et al. The Human Activity and Indicator Selection in Desertification Areas of Northwest Drought Region [J]. Geography Science, 2004,24(1): 68 - 75.

Lv Juan. The Improve and Revise of "Drought Assessment Criteria" [EB/OL]. http://www. iwhr. com/whr/WebNews_View. asp? WebNewsID = 222

Yao Yubi, Zhang Cunjie, Deng Zhenyong, et al. Overview of Meteorological and Agricultural Drought Indices [J]. Agriculture Research in the Arid Areas, 2007,25(1): 185 - 189.

The Determination and Classification of Drought Indexes [EB/OL]. http://nxld. gov. cn/zhengwugongkai/20090608165350. html

Shi Jianguo, Zhang Yanqing, He Wenqing, et al. Study on Spatial and Temporal Variation of Arid Index in Yellow River Basin [J]. Agriculture Research in the Arid Areas, 2009,27(1): 242 - 247.

Li Daoxi, Yang Baozhong, Lei Hongjun, et al. The Comprehensive Drought Assessment Method of Zhengzhou [J]. Yellow River, 2009,31(6): 66 - 67.

Runoff Simulation Study of the SWAT Model in Xilin River Basin[①]

Yao Suhong[1,3] , *Zhu Zhongyuan*[1] , *Li Yang*[1] , *Xiu Haifeng*[2] and *Zhang Sulu*[3]

1. Water Conservancy and Civil Engineering College of Inner Mongolia Agricultural University, Huhhot,010000,China
2. Pingyang County Water Conservancy Bureau, Pingyang,325400,China
3. Inner Mongolia Third Geological Mineral Exploration Institute,Huhhot,010000, China

Abstract: Taking the Xilin River basin as study area, the changes in runoff simulation research were studied with distributed hydrological SWAT model. Through the discussion for the applicability of the model in study area, changed trend of the runoff under change scenarios in different climates was analyzed and forecasted further. During the simulation process, data series from 1963 ~ 1975 were used to calibrate the SWAT model, and periods of 1976 ~ 1985 were employed to validate model. Results indicate that SWAT model matches well in simulating the change process of runoff. Nash-Sutcliffe coefficients during calibration period and validation period are both above 0. 60, and the correlation coefficients between recorded and simulated of monthly discharge are all above 0. 8, so it has certain applicability in the same region. It is sensitive related between the monthly discharge and the precipitation, which is slightly higher. The discharge in Xilin River Basin will decrease first and then increase in the future climate change scenarios.

Key words: Xilin River, SWAT model, discharge simulation, climate scenarios

In recent years, the meteorological factors, such as air temperature and precipitation, vary obviously under the global climate change background, especially in the Northeast, Northwest and North China. Precipitation is one of the main sources of water resources, so the climate change will directly lead to the variation of river basin water resources. Therefore, the impact of climate change on water resources assessment is regarded as a long-term task by the government and the scholars. Water resource is a basic natural resource, which is an important index of the national economic and social sustainable development. With the global climate change and effects of human activities, a series of water resource problems appears gradually with water environment pollution seriously, economic and social development demand rapid increase. Water seasonal shortage is very seriously in many areas. At present, the shortage of water resources has become the main factor that restricts social economy development. So it has become China's economic and social development of the important strategic goal that ensures the sustainable utilization of water resources.

It is the foundation of work to strengthen water resources management that simulates the water resources change and hydrological cycle. In recent years, the scholars all over the world set up the watershed hydrological model in research area to evaluate the impact of climate change on water resources. Moreover, the reasonable and effective exploitation of water resources system was put forwards by simulating and predicting the variation tendency of water resources quantity in future. The abnormal climate changes such as the scarce rainfall with uneven distribution and complicated underlying surface condition bring troubles to the hydrological model application in the Xilin River. Usually, the important factors which caused by changes in the hydrological cycle are human activities and precipitation, evaporation, surface condition changes. Distributed hydrological model SWAT on hydrological cycle changes makes positive response, which contribute to the areas with the data relatively deficient, different soil types, land use / cover style management, and the

① Supported by National Natural Science Foundation of China (51149006) & Climate Change Adaptation Project of China – UK – Swiss (ACCC/20100620)

complicated conditions. Therefore, the application of the SWAT model on basin runoff simulation process in Xilin River will contribute to analyze and predict the runoff change trend, which will provide technical support and scientific basis to basin water resources climate change and formulate corresponding measures.

1 Study area

Xilin River is an inland River, which originated in Ao Lun Knoll and Hulun Nur in Keshiketeng County in Chifeng city. Its elevation varies from 1,600 m in East to 900 m in west. The river flows from east to West in Chifeng and Xilinguole Meng, then into the Chagan Nur Lake (now dry). It covers 3,852 km² with 175 km full-length. The hill and plateau are the main topography, which undulate obviously, generally forming the Northeast - Southwest to the low hilly terrain. It is flat in river cross-strait with rich vegetation, so the loss of water and soil is not serious. The Xilin River is composed of three tributary. They are from Haolaituguoluo and Haolaiguoluo in right bank, and Husite River in left bank, which located in upper reaches of Xilin River reservoir. All the river system distributes asymmetrically.

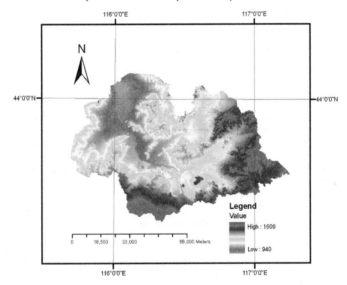

Fig. 1   DEM of Xilin River basin

2   SWAT model

The Soil and Water Assessment Tool (SWAT) model (Arnold et al. , 1998) is a continuous time model that operates on a daily time step at basin scale. The objective of such a model is to predict the long-term impacts in large basins of management and also timing of agricultural practices within a year. SWAT is a public domain model actively supported by the USDA-ARS. It is a hydrology model with the following components: weather, soil temperature and character, crop growth, land management, nutrient and pesticide loading. It has been proven to be an effective tool for assessing water resource and nonpoint-source pollution problems for a wide range of scales and environmental conditions across the globe. Other enhancements included an improved peak runoff rate method, calculation of transmission losses, and the addition of several new components: groundwater return flow, reservoir storage, the EPIC crop growth sub model, a weather generator, and sediment transport. Further modifications of SWRRB in the late 1980s included the incorporation of the GLEAMS pesticide fate component, optional USDA-SCS technology for estimating peak runoff rates, and newly developed sediment yield equations. These modifications extended the model's capability to deal with a wide variety of watershed water quality management problems.

3　Data and methods

3.1　Data information

The data adopted in the paper from the digital elevation based on Shuttle Radar Topography Mission (SRTM) with 30 m DEM in Xilin River Basin. The web site is http: // srtm. csi. cgiar. org/.

Soil type's data was the western environmental and ecological science data center (http:// westdc. westgis. ac. cn) in National Natural Science Foundation, which resolution is 1 km with the grid data format and WGS84 projection.

Land use data was the WESTDC_Land_Cover_Products1. 0 product through merger, vector and raster conversion based on the largest area processing method to 1: 10 million land resources survey results from Laboratory of Remote Sensing and Geospatial Science, Cold and Arid Regions Environmental and Engineering Research Institute, Chinese Academy of Sciences, and Environmental and Ecological Science Data Center for West China, NSFC.

3.2　Databases

The basic parameters inputted to SWAT model was composed by the spatial data and attribute data. The spatial data included the digital elevation data (DEM), land use / cover, soil spatial distribution data, the digital river information, and meteorological station and hydrological station position, etc. The attribute data is land use types, soil types and measured hydrological and meteorological data.

DEM data for Watershed runoff simulation was produced by graphics mosaic and projection transform to download images in ENVI software, and the result is DEM diagram shown in Fig. 1. Soil type source data was reclassified by FAO-90 soil classification system, which should be a standard code identified by SWAT model. The soil type and reclassification codes are shown in Fig. 2 and Tab. 1. The specific land use diagram shown in Fig. 3, and classification and code in Tab.2, which data of land use type was used by the Chinese Academy of Sciences resources environmental classification system

Tab. 1　Code conversion of soil reclassification in Xilin River Basin

Code	Soil type	SWAT code	Code	Soil type	SWAT Code
11138	Calcaric chernozems	SHXHGT	1,1413	Calcareous soil	SHXHT
11145	Luvic Kastanozems	LTLGT	1,1425	Gley soil	QYHT
11154	Transitional red sand	GDXHST	1,1540	Mollic gleysols	SRQYT
11157	Hard calcareous soil	YHGZT	1,1543	Gley soil	QYYT
11158	General use	PTLGT	1,1927	Wwater	ST
11355	Calcareous red sand	SHXHST			

Tab. 2　Code conversion of land use/cover reclassification in the Xilin River Basin

Value	SWAT code	Name	Value	SWAT code	Name
12	AGRL	Land	43	WATR	Reservoir pond
21	FRST	Forest land	46	WATR	Beach land

Continued to Tab. 2

Value	SWAT code	Name	Value	SWAT code	Name
22	FRST	Shrubbery	51	URHD	Urban land
23	FRST	Woodland	52	URML	Rural dweller dot
24	FRST	Other woodlands	53	UIDU	Other construction land
31	PAST	High grass coverage	61	SWRN	Sandy land
32	PAST	Coverage of grassland	63	SWRN	Saline land
33	PAST	Low coverage grassland	64	WETL	The swamp
41	WATR	Canals	66	SWRN	Bare rocks
42	WATR	Lake			

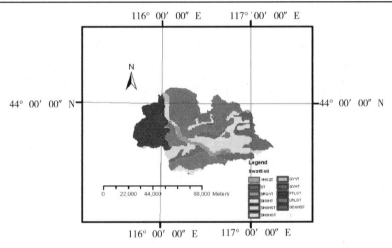

Fig. 2   Soil type of Xilin River basin

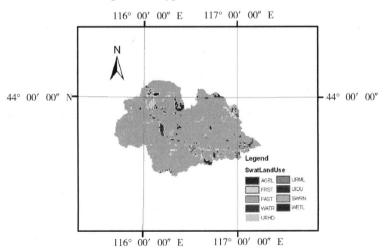

Fig. 3   Land use/cover in Xilin River basin

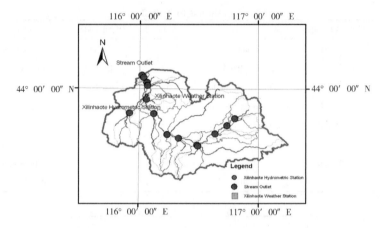

Fig. 4   The distribution of weather stations in Xilin River Basin

Meteorological data such as maximum temperature, minimum temperature, relative humidity, wind, solar radiation, precipitation are from Xilinhaote meteorological station's daily observation data from 1962 to 2008, which located in Fig. 4 Hydrological data is the runoff data of Xilinhaote hydrological station. Hydrological data are from the Inner Mongolia Autonomous Region Hydrological Bureau. Meteorological data are from Inner Mongolia meteorological scientific data sharing service center.

4   Simulation results and analysis

4.1   Parameter sensitivity analysis

The model parameters sensitivity were analyzed by the Morse classification screening method (One Factor At a Time, OAT) to understand the key influence factors in the watershed hydrological simulation. A variable in many parameters was selected to judge how the parameters variation affects on the output results by changing its' threshold value in the range.

26-input-parameters were selected for the sensitivity analysis of model parameters test in runoff simulation, which sampling interval is 10, and parameter variation is 0.05. GWQMN、ESCO, SOL _Z, SOL _AWC, CANMX and CN2 were determined as the highest sensitivity parameter on the runoff by 270 times comparison in ARCSWAT2005. The adjustment principle that was chosen in parameters determination is that upstream is prior to downstream. So surface runoff is prior to soil water, evaporation and groundwater runoff; water balance is prior to process. The results of parameter calibration is in Tab. 3 by analyzing monthly flow sequence data from 1963 to 1975 from Xilinhaote hydrological station.

Tab. 3   The definition of parameters in SWAT and their sensitivity values

Grade	Parameter	Definition	Input file	Range	Parameter value
1	Gwqmn	Shallow groundwater runoff coefficient	*.gw	0 – 5,000	2,900
2	Esco	Soil evaporation coefficient	*.hru	0 – 1	0.02
3	Sol_Z	Soil depth	*.sol	0 – 3,000	1,500
4	Sol_Awc	Available soil water	*.sol	0 – 1	0.92
5	Canmx	The maximum canopy storage capacity	*.hru	0 – 10	1
6	Cn2	SCS runoff coefficients	*.mgt	35 – 98	43

4.2 Simulation result analysis

The month runoff data is used for the coefficients calibration, the periods of 1962 ~ 1975 and 1976 ~ 1985 are the calibration and validation periods respectively. The three indexes Nash-Sutcliffe coefficients Ens, relative error of runoff R_e and related coefficient R^2 are used to evaluate the feasibility . the results are shown as Tab. 4 and Fig. 5 (a)(b), Fig. 6(a)(b).

(1) Nash-Sutcliffe coefficient Ens:

$$Ens = 1 - \frac{\sum_{i=1}^{n} (Q_0 - Q_p)^2}{\sum_{i=1}^{n} (Q_0 - \overline{Q})^2}$$

where, Q_p is simulation value; Q_0 is the measure value; $\overline{Q}$ is the mean value of measure ; n is the count of measure value.

When $Q_0 = Q_p$, $Ens = 1$; when Ens little than 0, shows the reliability of simulation mean value is lower than the measure mean value.

(2) Runoff relative error R_e:

$$R_e = \frac{Q_p - Q_0}{Q_0} \times 100\%$$

where, R_e is the relative error of simulation; Q_p is the simulation value; Q_0 is the measure value.

If R_e lager than 0, it shows the simulation result is too large; R_e is little than 0, it shows the simulation result is too small; $R_e = 0$, it shows the simulation result is good agree with the measure value.

(3) Related coefficient R^2:

The related coefficient R^2 is often used to evaluate the agree between the simulation result and measure data, when $R^2 = 1$, which shown they are agree very well each other and if $R^2 < 1$, the smaller is the value, the lower agree each other。

Tab. 4 Evaluation of the simulation results of monthly stream flow during calibration and validation periods

Simulation period		R_e	Ens	R^2
Calibration period	1963 ~ 1975	5.90%	0.815	0.862
Validation period	1976 ~ 1985	18.50%	0.722	0.826

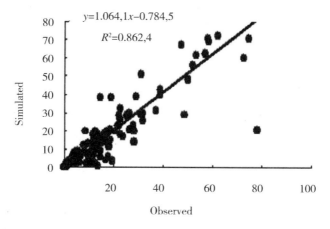

(a) Observed and simulated value fitted line in calibration period

Fig. 5 Observed and simulated value fitted line

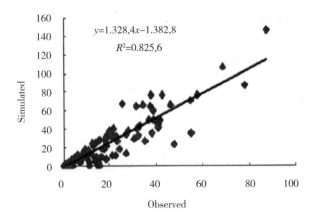

(b) Observed and simulated value fitted line in verification period

Continued to Fig. 5

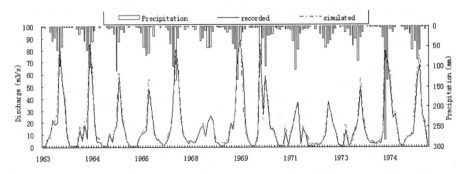

(a) Comparison between the simulated and observed monthly stream flow during calibration period

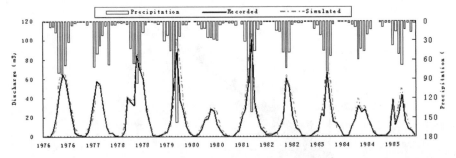

(b) Comparison between the simulated and observed monthly stream flow during validation period

Fig. 6   Flow simulation result of SWAT model

From the simulation results of Tab. 4, Fig. 5 and Fig. 6, during the calibration and validation periods the runoff related coefficient between the simulation and measure are lager than 0.8, Nash-Sutcliffe coefficient is lager than 0.60 and Years of average relative error are 5.9% and 18.5%

respectively, though validation periods result is not so good, it is still in the acceptable range, so the SWAT model can simulate the runoff process in Xilin River Basin.

5 Response analyses for the future climatic change

Based on the basin climate model HadCM3 output result of Hadley Centre for Climate Prediction and Research in England and A2, B2 and A1B daily meteorology factors data in three different climate conditions provided by Institute of Environment and Sustainable Development in Agriculture, CAAS during 1961 ~ 2050, the runoff change of Xilin River Basin is simulated in the future 40 years. The results are shown in Tab. 5 and Fig. 7.

Tab. 5 Decadal changes of annual runoff under the climate scenarios (%)

Period	The change of different decade compared with reference period		
	A1B	B2	A2
2011 ~ 2020	− 0.024	− 0.201	− 0.175
2010 ~ 2030	− 0.047	− 0.018	− 0.042
2031 ~ 2040	− 0.032	− 0.146	− 0.057
2041 ~ 2050	0.089	− 0.171	0.027

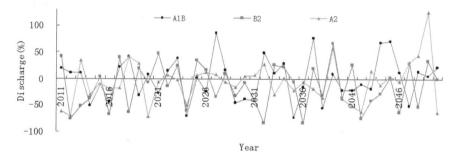

Fig. 7 Projected annual runoff variations from 2011 to 2050

Fig. 7 shows that the xilin river basin annual runoff trend to decrease compared with the reference value before the 21 century 40 s under A1B, B2 and A2 different climate conditions in future. The rate of decrease is 1.45 mm/10 a, 5.92 mm/10 a and 4.81mm/10 a respectively. But after 40 years latter, the runoffs still decrease under the B2 condition and increase under the A1B and A2 conditions.

The annual runoff reduces to 2.4%, 20.1% and 17.5% compared with the reference value under A1B, B2 and A2 condition during 2011 to 2020 period from the decadal changes of annual runoff under the climate scenarios (Tab. 5); annual runoff reduce to 4.8%, 1.8% and 4.2% compared with the reference value under those three condition during 2021 ~ 2030 period. The rate of annual runoff decrease is 14.6% under B2 condition, but it is 3.2% and 5.7% under the A1B and A2 condition during 2031 ~ 2040 period; the annual runoff still decrease compared with the reference under B2 condition, but A1B, A2 increase to 8.9% and 2.7% after the 2040s.

6 Conclusions

The results show that the Nash-Sutcliffe efficiency coefficient、related coefficient are lager than 0.60 and 0.8 respectively after calibrating the parameters many times when the SWAT model is used in xilin river basin on the monthly runoff simulation process, the years average relative error is 5.9% and 18.5% during calibration and validation period. Overall the SWAT model can simulate the runoff process quiet well in the arid and semi-arid area and similar area.

The precipitation change is highly sensitivity in the monthly runoff simulation. It further shows that precipitation is the mainly factor of runoff change. At the same time, the relative error of year's average is high because the effect of humanity activity on the runoff change is not considered.

At first the runoff total amount will increase, but decrease later under A1B, B2 and A2 future climate condition. The Xilin River Basin water resource will decrease at first and then increase in the future 40 years under the A1B, A2 condition, but runoff will continue to decrease. Because there are some uncertainties on the model prediction, so the water resource problem still should be paid more attention in order to prevent water resource crisis getting worse.

References

Zhan Yun, Cui Shubin, Hu Huifang, Zuo qiting. Discussion on the Feasibility and Key Problems of Recycled Water Utilization in Southern China [J]. South-to-North Water Transfers and Water Science & Technology, 2011, 9(1): 122 - 125.

Lu Xiaofeng. Simulation Study on Runoff in Bailian River Basin Based on SWAT Model [J]. Water Resources and Power, 2009, 27(1): 21 - 23.

Zhang Jianyun, Wang Guoqing. Impacts of Climate Change on Hydrology and Water Resources [M]. Beijing: Science Press, 2007.

Wang Guoqing, Li Jian. Grid-based Hydrological Model for Climate Change Impacting on Water Resources in the Middle Yellow River Basin [J]. Advances in Water Science, 2000, (6, supplement): 22 - 26.

Qiu Yaqin, Zhou Zuhao, Jia Yangwen, et al. A Case Study on the Evolutionary Law of Water Resources in Sanchuan River Basin [J]. Advances in Water Science, 2006, 17(6): 865 - 872.

Wang G. S., Xia, J., Chen J. Quantification of Effects of Climate Change and Human Activities on Runoff by a Monthly Water Balance Model: A Case Study of the Caobaihe River Basin in Northern China [J]. Water Resources Research, 2007, (44): 1029 - 1041.

Athanasios L., Michael C. Q. Effect of Climate Change on Hydrological Regime of Two Climatically Different Watersheds [J]. Journal of Hydrologic Engineering, 1996, 1(2): 77 - 87.

Benedikt N., Lindsay M., Daniel V., et al. Impacts of Environmental Change on Water Resources in the Mt. Kenya Region [J]. Journal of Hydrology, 2007, 343: 266 - 278.

Jin Xingping, Huang Yan, Yang Wenfa, et al. Analysis of Impact of Future Climate Change on Water Resources in Yangtze River Basin [J]. Yangtze River, 2009, 40(8): 35 - 38.

Hao Zhenchun, Li Li, Wang Jiahu, et al. Impact of Climate Change on Surface Water Resources [J]. Earth Science, 2007, 32(3): 425 - 432.

Neitsch S L, Arnold J G, Kiniry J R, et al. Soil and Water Assessment Tool User's Manual [M]. Texas: USDA, Agriculture Research Service and Grassland Soil and Water Research Laboratory, 2002. 1 - 412, 92 - 432.

Neitsch S L, Arnold J G, Kiniry J R, et al. Soil and Water Assessment Tool Theoretical Documentation Version2005 [EB/OL]. http://www. brc. tamus. edu /swat /downloads / doc/swat2005/SWAT2005 theory final. pdf, 2005.

Xiao Juncang, Zhou Wenbin, Luo Dingguiz, et al. Non - point Pollution Model SWAT User's Manual [M]. Beijing: Geology Press, 2010.

Chen Jian, Liang Chuan, Chen Liang. Parameter Sensitivity Analysis of Swat Model—A Case Study of Small Watersheds with Different Land Cover Types in Hailuogou Valley [J]. South - to - North Water Transfers and Water Science & Technology, 2011, 9(2): 41 - 45.

Xu Zongxue. Hydrological Models [M]. Beijing: Science Press, 2009.

Zhu Xinjun, Wang Zhonggen, Li Jianxin, et al. Applications of SWAT model in Zhang Wei River Basin [J]. Progress in Geography, 2006, 25(5).

Yao Yunlong, Wang Lei. Influence of Climate Change on Runoff of Typical Marsh River in Sanjiang Plain Based on SWAT—Study in Naoli River [J]. Wetland Science, 2008, 6(2): 198 - 203.

Optimal Operation Rules of Reservoir Releasing Facilities Based on Smooth Support Vector Machine[①]

Liu Xinyuan[1] , *Guo Shenglian*[2] , *Zhu Yonghui*[1] , *Qu Geng*[1] , *Guo Xiaohu*[1] and *Tang Feng*[1]

1. Changjiang River Scientific Research Institute, Wuhan, 430010, China
2. State Key Laboratory of Water Resources and Hydropower Engineering Science,
Wuhan University, Wuhan, 430072, China

Abstract: Considering the impacts of the releasing facilities on flood regulation, this study establishes a multi-objective model with the releasing facilities as constraints. The optimal operation results of the releasing facilities can be obtained by the successive optimization algorithm (POA) according to some observed historical flood hydrographs and design flood hydrographs. A support vector machine (SVM) model of operation rules of releasing facilities is established, optimized by the differential evolution algorithm (DE) using the optimal results by POA, and then used to simulate operation of releasing facilities. The results show that the optimal operation rules of releasing facilities derived from the SSVM model can efficiently increase the benefits of flood control and power generation and make the flood regulation process more smooth and steady. So it is feasible and practicable for the SSVM to achieve the optimal operation of releasing facilities.

Key words: releasing facility, support vector machine, flood control, operation rules

Reservoir operation in practice is very complex, influenced by many factors, which may result in great deviation from the theoretical results while applying optimal operation rules in practice. For example, the optimal control of releasing facilities is one of the difficult problems in engineering, and the optimal results of reservoir operation are mostly in the form of water discharge in total, which cannot be used to reservoir operation directly. There are many constraints on the use of releasing facilities, for example, the sluice gates need to be symmetrically opened, and the opening needs to be a scale in multiples, etc. Huang et al (1994) and Kang et al (2000) discussed mathematic models using the optimal square approximation function to get the optimal sluice gates operation plan in practical engineering. However, the actual operation course may deviate from the optimal operation results obtained from mathematic models because of cumulative error, and so the optimal operation results are not necessarily optimal. This article attempts to optimize the model by introducing directly constraints of the operation rules of the releasing facilities to obtain the optimal scheduling of reservoir releasing facilities. In addition, as a new machine learning algorithms developed in recent years, the support vector machine (SVM) displays a lot of advantages and has been widely applied in many fields, such as classification, regression and prediction. However, there are few studies in deriving reservoir operation rules. Zuo et al (2007) applied SVM to fitting reservoir operation function, and achieved good results. This article will use smooth support vector machine (SSVM) which is more efficient to mining the optimal operation rules of releasing facilities, and discuss its applicability and feasibility.

1 Optimal model considering releasing facility operation

There are strict service conditions for releasing facilities, such as usage, order of priority, and so on. Here a reservoir operation model for flood control is developed to consider service conditions of releasing facilities from the aspects of the model constraints.

① This study is funded by the Ministry of Water Resources (QQ0871/HL15), and basic scientific research fee of commonweal scientific research institutes at central government level(CKSF2012011, CKSF2012014).

1.1 Objective functions

Reservoirs are mostly multipurpose for flood control, power generation, navigation, and so on. Two objectives of flood control and power generation are established for the sake of simplicity as follows:
(1) The flood control objective is to gain the maximum flood safety assurance, and minimizing the highest flood-routing water level at the dam is selected as the objective function, i. e.

$$\min \quad \max\{Z_t, t = 1, \cdots, T\} \tag{1}$$

(2) The power generation objective is to achieve the maximum economic benefits for a given flood hydrograph, and maximizing the daily power generation capacity of the flood is treated as the objective function, i. e.

$$\max \quad E = \frac{1}{T} \sum_{t=1}^{T} P_t \cdot \Delta t \tag{2}$$

where, $Z_t(t = 1, 2, \cdots, T)$ means reservoir water level at the t th stage, in m. Δt is the time interval, in s. P_t is the average output of at the t th stage. T is the length of a flood series.

1.2 Subject to the following constraints

(1) Water balance equation
$$V_{t+1} = V_t + (I_t - O_t) \cdot \Delta t \quad t = 1, 2, \cdots, T \tag{3}$$
(2) Reservoir storage volumes limits
$$V_t \leqslant V_t \leqslant V_u \quad t = 1, 2, \cdots, T \tag{4}$$
(3) Power generation limits
$$P_{\min} \leqslant P_t \leqslant P_{\max} \quad t = 1, 2, \cdots, T \tag{5}$$
(4) Reservoir discharge limits

There are mainly two constraints : the maximum discharge capacity of the reservoir at water level z_t, namely $O_{\max}^{z_t}$, and the safety discharge in the downstream flood protection section, namely $O_{\max}^{down}$, of which the minimum is the maximum discharge of releasing facilities, that is,

$$Q_t \leqslant \min(O_{\max}^{z_t}, O_{\max}^{down}) \quad t = 1, 2, \cdots, T \tag{6}$$

The reservoir discharge O_t equals to the sum of the discharges of all releasing facilities in practice, i. e.

$$Q_t = \sum_{i=1}^{5} m_i \cdot q_i(z_i) \tag{7}$$

where, I_t, O_t and P_t mean reservoir inflow, outflow, and output at the time period t respectively. V_t is reservoir storage capacity at time t. V_l and V_u are the lower and upper limits of reservoir storage capacity, respectively. $P_{\min}$ and $P_{\max}$ are the minimum and maximum power limits of reservoir, respectively. q_1, q_2, q_3, q_4 and q_5 are functions of water level at the dam site, meaning releasing discharge of each power unit, bottom outlet, flotsam spill outlet, flush gallery, and crest outlet, respectively. m_1, m_2, m_3, m_4 and m_5 are the number of corresponding releasing facilities above put to use. Generally, the following constraints need to be meet, i. e.

$$\begin{cases} m_1, m_2, m_3, m_4 \text{ and } m_5 \text{ are nonnegative integers} \\ m_2 = m_3 = m_4 = m_5 = 0 \quad \text{if } m_1 < M_1 \\ m_3 = m_4 = m_5 = 0 \quad \text{if } m_2 < M_2 \end{cases} \tag{8}$$

where, M_1 and M_2 are the total number of power units and bottom outlets, respectively. In addition, these releasing facilities have strict water level conditions, and each facility has its specific requirement of water level for normal operation state. So it is necessary to make right arrangements for releasing facilities according to water level on the premise of right operation order.

1.3 Objective function and constraint handling

Constraint satisfaction and multi-objective optimization are two aspects of the same problem. Constraints can often be seen as hard objectives, which need to be satisfied before the optimization of the remaining, soft, objectives takes place. Conversely, problems characterized by a number of soft objectives are often re-formulated as constrained optimization problems in order to be solved. In this paper, the two objectives are converted to one single objective function using the weighted-sum approach. In addition, the constraint (6) is very complex, can be converted into an objective by penalty function. Firstly, constraint violation $g_t(x)$ of each solution is defined as follows:

$$g_t(x) = \min(Q_{\max}^{z_t}, Q_{\max}^{down}) - \sum_{i=1}^{5} m_i \cdot q_i(z_i), \quad t = 1,2,\cdots,T \tag{9}$$

The penalty function w_t can be calculated then, i. e.

$$w_t = \begin{cases} 0, \text{ if } g_t(x) \geqslant 0 \\ \mid g_t(x) \mid, \text{ otherwise} \end{cases} \tag{10}$$

Using the weighted-sum approach, multiple objectives can be transformed into a single objective, i. e.

$$\min c_1 \cdot Z - c_2 \cdot E + k \cdot \sum_{t=1}^{T} w_t \tag{11}$$

where, k is a large constant, usually taken from 100 to $1,000,000$. Considering reservoir flood control goals more important during the flood season, $c_1 = 0.9$ and $c_2 = 0.1$.

Dynamic programming (DP) is the classical method for solving the reservoir operation optimization problem. However, DP-based approximation algorithms are more efficient when having large numbers of variables to be optimized, and the successive optimization algorithm (POA) is used in this paper.

2　Support vector machine

Support vector machine (SVM) is a new machine study method in the field of statistical learning theory based on Vapnik – Chervonenkis theory and structural risk minimization principle, and stresses to study statistical learning rules under small samples. Being different from the artificial neural network model (ANN) based on empirical risk minimization, SVM attempts to minimize an upper bound on the generalization error by seeking a right balance between the training error and the capacity of machine. With less model parameters, SVM has overcome the drawbacks of ANN in overfitting and model structure, been successfully applied to the classification and regression, and hence become one of the focuses in the machine learning. However, SVM cannot deal efficiently with large data samples, and may be time-consuming and memory-consuming in calculation. Lee and Mangasarian developed an improved algorithm based on SVM, Smooth support vector machine (SSVM), which transforms constrained quadratic optimization of SVM into non-constrained convex quadratic optimization problem and is better able to handle the cases of classication and nonlinear regression with a larger dataset. Lee et al testified its superiority in solving effect by experiment. In this paper, operation rules of flood releasing facilities are mined by establishing a statistical relationship between reservoir water level, inflow and the number of releasing facilities put to use according to SSVM. The detail of the algorithm is described by Mason and Tippett.

There are three main parameters for the SSVM, the penalty parameter C, the radial basis kernel parameter γ and the nonsensitivity coefficient ε, among which C and γ have more significant influence on the prediction accuracy of SSVM. When C or γ is large enough, over-learning may occur and hence the worse generalization ability may be caused. The SSVM may have good learning ability, but poor extrapolation. While γ is small enough, the regression model may be difficult to achieve sufficient accuracy. The nonsensitivity coefficient ε has less influence on the prediction error, but deeply affects the number of support vectors. When ε is large enough, there are less support vectors and the model may be too simple to gain enough learning accuracy; on the contrary, when small enough, the model may be too complex to have a good generalization ability. Therefore, the complexity and generalization ability of SSVM models depends on C, γ and ε, as well as the relationship among them. It is not appropriate to optimize one single parameter each time, and also time-consuming. How to select an accurate, stable and efficient optimization algorithm with overall consideration of the three parameters is very important. As a powerful yet simple evolutionary algorithm, differential evolution algorithm (DE) has a strong rapid, parallel and global search capability, and hence is selected as the algorithm to optimize the SSVM parameters. The detail of DE is described by Price and Storn.

3　Optimization methodology

To obtain the optimal operation rules of releasing facilities, there are four steps as follows. Firstly, deterministic optimization of releasing facility operation is done for each historical inflow data series using POA, and the optimal series of releasing facility operation can be obtained. Secondly, reservoir water level series and the corresponding inflow series, obtained by POA, are selected as the input samples to establish the statistical relationships of water level, inflow and the number of releasing facilities using SSVM. Thirdly, DE algorithm is used to optimize the SSVM model and the optimal parameters can be obtained. Lastly, the simulation operation of the optimal SSVM model is done using the historical inflow series, and the benefits and risk assessment can be evaluated.

4　Results and analysis

The TGR is selected as a case study in order to demonstrate and validate the proposed methodologies, and 16 groups of 30-day flood series are used, including 11 groups of historical flood series of annual 30-day maximum flow volume with the peak flow more than 62,000 m^3/s, plus the 1998 flood, as well as 5 years of 10,000-year design flow hydrographs, i. e. 1931, 1935, 1954, 1981 and 1998, to explore the releasing facility operation rules when the extreme floods occur. POA is used to deterministic optimization, and then DE to optimize parameters of the SSVM model. The optimal results are shown in Tab. 1.

Tab. 1　Results of reservoir releasing facility operation

Flood No. (Year)	Optimal results by POA					Simulation results by SSVM			
	Maximum nflow (m^3/s)	Maximum outflow (m^3/s)	Maximum flood-routing water level (m)	Water level at the end of flood (m)	Daily electricity generation ($\times 10^4$ kWh)	Maximum outflow (m^3/s)	Maximum flood-routing water level (m)	Water level at the end of flood (m)	Daily electricity generation ($\times 10^4$ kWh)
1892	64,600	50,912	154.33	145.21	40,561	49,292	154.82	145.14	40,486
1896	71,100	53,888	156.05	145.93	39,850	53,082	160.99	149.88	43,338
1905	64,400	52,146	154.24	154.24	39,970	52,147	154.77	154.77	39,126
1921	64,800	53,610	154.38	145.06	39,837	52,972	154.38	149.29	41,327
1922	63,000	53,862	150.90	146.48	38,402	53,862	150.90	150.08	38,737
1931	64,600	53,027	155.29	145.14	40,476	51,589	156.64	149.90	41,172
1936	62,300	46,328	152.50	146.11	40,934	46,329	152.50	149.79	36,688
1945	67,500	53,914	151.79	145.51	39,300	49,555	157.86	154.32	40,471
1954	66,100	53,766	161.56	152.41	42,272	53,766	161.56	152.41	42,284
1981	69,500	51,215	155.42	145.25	40,381	51,438	159.56	153.59	39,909
1998	61,700	53,970	153.06	148.31	40,145	53,646	162.09	161.63	44,651
1931 −0.1%	88,502	53,982	173.63	173.63	42,754	61,140	174.93	174.85	42,809
1935 −0.1%	97,299	54,467	172.13	154.15	46,184	53,797	172.96	163.52	46,485
1954 −0.1%	82,560	54,610	173.33	170.36	44,181	53,950	174.03	172.41	44,191
1981 −0.1%	98,496	54,953	170.16	170.16	40,694	61,156	174.97	174.84	42,748
1998 −0.1%	97,881	54,799	174.93	172.37	45,614	59,041	174.99	174.70	45,624

4.1 Analysis of optimization results

Due to space limitations, results of the design operation scheme are not listed in this paper, but some conclusions are drawn here. The optimal results by POA increase power generation for each flood hydrograph than the design scheduling scheme, and reduce the maximum outflow greatly. For 1,000-year design flood hydrographs, the maximum outflow can be controlled at about 54,000 m³/s, and huge losses of flood diversion downstream can be avoided. The operation results of the 1954 typical flood and the corresponding 1,000-year design flood are shown in Fig. 1 and Fig. 2, respectively. It can be seen that the optimal outflow duration curves are quite stable. However, there exist minor fluctuations for the outflow curves attributed to the influence of releasing facilities. The results by POA are reasonable and can be used as samples to explore the optimal operation rules of the releasing facilities. The optimal water level series and inflow series are selected as input samples, and the statistical relationships of water level, inflow and the number of releasing facilities are established by SSVM. The simulation results of some typical flood hydrographs using the SSVM model are shown in Fig. 3, and it can be seen that the SSVM model performs well.

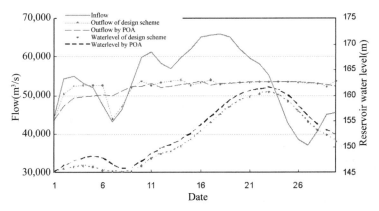

Fig. 1   Results of flood regulation for the 1954 flood hydrograph

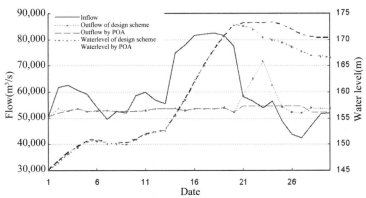

Fig. 2   Results of flood regulation for the 1,000-year design flood hydrograph in 1954

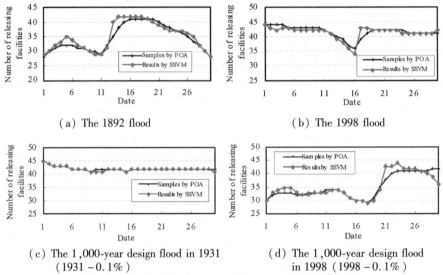

(a) The 1892 flood   (b) The 1998 flood

(c) The 1,000-year design flood in 1931   (d) The 1,000-year design flood
(1931 − 0.1%)   in 1998 (1998 − 0.1%)

Fig. 3   Results of releasing facility operation by SSVM

4.2   Analysis of the simulation results by SSVM

The SSVM model in this paper selects two input vectors, i. e. the current reservoir water level and inflow, as prediction factors to simulate reservoir operation, while the POA optimization is based on exactly known inflow hydrographs and information completeness, and hence it is reasonable for the SSVM model not to be able to fit the POA results accurately, as shown in Tab. 1. Compared with the POA optimization results, the SSVM results have slightly higher water levels and more power generation, while the peak outflows are approximate except for the 1,000-year design flood hydrographs in 1931 and 1981 of which the peak outflows are obvious larger. For the observed historical floods, the outflow discharge of can be kept below 54,000 m³/s, the peak water level at about 160 m, and hence the SSVM model can guarantee flood control safety of not only the reservoir but also the downstream; For 1,000-year design floods, the outflow discharge can be kept at about 61,000 m³/s on the premise of the peak water level not exceeding 175m, and hence it is favorable to alleviate the flood pressure downstream. Especially for the 1,000-year design flood in 1935 and 1954, the peak outflow can be kept below 54,000 m³/s. So the SSVM model in this paper is feasible and effective. The outflow hydrograph curves are also very smooth and steady as shown in Fig. 4.

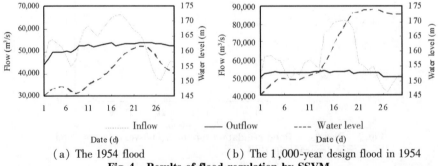

(a) The 1954 flood   (b) The 1,000-year design flood in 1954

Fig. 4   Results of flood regulation by SSVM

5　Conclusions

This study focuses on the optimal flood regulation with releasing facilities, develops flood regulation model considering the requirements of flood control and power generation based on SSVM and explores the optimal flood regulation rules for the guidance of the releasing facilities. The results show that the SSVM model can efficiently increase the benefits of flood control and power generation and make the flood regulation process more smooth and steady. So it is feasible and practicable for the SSVM to achieve the optimal operation of releasing facilities.

References

Huang D J, Liang N S, Jiang T B, et al. Optimal Control of Discharge Gates at Hydropower Station [J]. International Journal Hydroelectric Energy, 1994, 12(4): 247 – 253.

Kang L, Jiang T B. A Study on Theory of Dynamic Tracking Control of Spillway Gate [J]. Journal of Huazhong University of Science and Technology, 2000, 28(2):54 – 55.

Zuo Q C, Li C J, Fan R. Research of SVM on Optimization Function of Reservoir Regulation [J]. Yangtze River, 2007, 38(01): 8 – 9.

Vapnik V N. Statistical Learning Theory [M]. New York: Wiley, 1998.

Lee Y, Mangarasian O L. SSVM: A Smooth Support Vector Machine for Classification [J]. Computational Optimization and Applications. 1999, 20(1): 5 – 22.

Lee Y, Hsieh W, Huang C. ε – SSVR: A Smooth Sopport Vector Machine for ε – insensitive Regression [J]. IEEE Transactions on Knowledge and Data Engineering. 2005, 17(5): 678 – 685.

Mason SJ, Tippett MK. Cross – validation for Selection of Statistical Models [C]// Proceedings of 17th Conference on Probability and Statistics in Atmospheric Science, Seattle, Washington, 2004.

Price KV, Storn R: Differential Evolution—A Simple Evolution Strategy for Fast Optimization[J]. Dr. Dobb's Journal, 1997, 22: 18 – 24.

Research and Application of a Mathematical Model for Transportation and Transformation of Pollutant in Sediment – laden Flow

Mu Dewei[1] , *Yu Xuezhong*[2] , *Peng Qidong*[2] and *He Jinchao*[1]

1. Southwest Water Carriage Engineering Institute of Chongqing Jiaotong University,
Chongqing, 400016, China
2. China Institute of Water Resources and Hydropower Research, Beijing, 100038, China

Abstract: This paper analyzes the impacts of the main physicochemical processes in a water – sediment – pollutant system on pollutants in water body, and based on it is derived a basic equation describing the transportation and transformation adsorbed and dissolved pollutants. An expression for impacts of sediment erosion and deposition upon pollutants is developed from the model in regard to environment changes caused by sediment exchange in the water body and riverbed. And a model of sediment adsorption dynamics based on experiments is used for study of adsorption and desorption. The model is verified according to the measured water – sediment data and the measured water quality data from the Three Gorges Reservoir area. A mathematical model for water quality of sediment – laden flow is adopted to simulate sediment movement and its impacts on water quality in the Three Gorges Reservoir area, and analyze impacts on TP by sedimentation during the Three Gorges Reservoir impoundments. The research results show that the model not only can simulate pollutant exchanges in the sediment erosion and deposition process, but also reflect a dynamic impact of sediment adsorption on water environment and can be applied to a simulation and analysis of interaction of sediment pollutants in river as well as reservoir.
Key words: river, sediment, pollutant, mathematical model

1 Introduction

In a river water environment system, becoming a main carrier of pollutants, both sediment and water flow have the impacts upon transportation and transformation processes of pollutants in the water body, and ultimately have the impacts upon an eco – environmental condition of the water body, of which changes in the state of sediment movement and its adsorption are two main aspects affecting water quality by sediment. This effect can be called as an environmental effect given by river sediment. On one hand, sediment particles can adsorb a variety of pollutants and deposit on river bottom by a certain kinetic action, and reduce quantity of pollutants occurred in the water body in a certain period of transport; on the other hand, when there are changes in some external conditions such as the chemical processes and kinetic effects etc, pollutants adsorbed on sediment particles may change an occurrence mode and transform from the adsorbed state (solid phase) to the dissolved (liquid phase), or pollutants deposited on the bed mud make entrance into the water body again along with scouring sediment and thus it greatly changes the chemical composition in the overlying water body and even causes a secondary pollution in the water body.

In this paper, the basic equation is established for transportation and transformation of the dissolved and adsorbed pollutants, based on an analysis of the main process of the environmental impacts given by sediment, focusing on a description method of the environmental impacts of the water body by sediment erosion and deposition as well as the adsorption and desorption, verifying the established mathematical model and discussing the main problems from simulating the environmental impacts by sediment.

2 Analysis of the main process of sediment impacts upon environment

The environmental impacts by sediment in a natural river mainly consist of three processes:

①there is a dynamic change in pollutant occurrence quantity in the water body accompanied by sediment erosion and deposition and the pollutants exchanging between sediment (bed mud) and the water body; ② through the adsorption and desorption processes, the pollutants exchange between the water body and sediment (two phases), and so there is a change in pollutant occurrence state; ③ a spatial distribution of pollutants is changed by sediment transport processes. The processes mentioned above are coupled with each other and simultaneously occur in the natural state, and the combined action of several factors is reflected in changes in water environment quality and some impacts on an aquatic ecosystem.

The method to describe sediment movement and the adsorption and desorption in the form of the generalized coefficients from the equation for transportation and transformation of the dissolved pollutants is too simple in comparison with an actual process. Introducing the overall model for the sediment dynamics theory can well describe sediment movement, but a simultaneous equation is established generally according to the adsorption (desorption) dynamics or thermodynamics theory in the overall model for solving relationships between the dissolved and adsorbed pollutants. And from the point of physical analysis, it is necessary to establish the basic equation for the adsorbed pollutants because changes in the adsorbed pollutants are affected by the physical characteristics of sediment particles and its response characteristics to fluid movement and their temporal and spatial variations are intimately related to the water – sediment movement process. On the other hand, a thermodynamic equation for distribution relations of pollutants in an adsorption balance condition used as a complement of the overall equation is feasible for the water body in equilibrium, but not for environment where adsorption (desorption) processesare distinct.

Therefore, based on analysis of the main processes of the environmental impacts by sediment, the mathematical equations describing transportation and transformation of the dissolved and adsorbed pollutants are established, and a dynamics equation for adsorption (desorption) is used as a supplementary condition to determine coefficients of the equation in the paper. The basic equation describing the mathematical model for environmental effects by sediment is formed from combination of two equations above with an equation for sediment dynamics.

3   Establishment of the mathematical model

3. 1   Equations for sediment pollutants transportation and transformation

Affecting processes of the adsorbed pollutants include three parts such as transport by flow, exchanges between bed and suspended sediment and exchanges between the adsorbed and dissolved pollutants. Take a segment between two cross sections having a distance expressed by infinitesimal Δx as a control body, of which the boundary surface consists of an upstream cross section A_1, a downstream cross section A_2 and the riverbed and water surface, taking A as a cross – sectional area, A_1 as an entrance control cross – section area, and A_2 as an exit control cross – section area.

At dt time, pollutants pass through the riverbed and the interface of the sediment – laden flow include two parts, one is caused by sediment deposition, another is caused by sediment diffusion, and of which a mass is expressed as:

$$\int_B (\omega s_b c_{ks} + D_z \frac{\partial s_b}{\partial z} c_{kb}) \, db \Delta x dt \tag{1}$$

where, ω is settling velocity; s_b is sediment concentration near riverbed; D_z is a sediment diffusion coefficient near riverbed; c_{ks} is adsorption quantity of settlement sediment; c_{kb} is adsorption quantity of sediment diffusion; B is a cross – section width; d_b is a integral unit length.

According to a boundary condition of the suspended sediment diffusion in the sediment dynamics, the pollutants mass adsorbed by sediment diffusion is expressed as:

$$\int_B D_z \frac{\partial s_b}{\partial z} C_{kb} db \Delta x dt = - \int_B \omega s_{*b} c_{kb} db \Delta x dt \tag{2}$$

where, s_{*b} is sediment – carrying capacity near riverbed, taking Eq. (2) into Eq. (1), it is given by:

$$\int_B (\omega s_b c_{ks} + D_z \frac{\partial s_b}{\partial z} C_{kb}) \, db \Delta x dt = \int_B \omega (s_b c_{ks} - s_{*b} c_{kb}) \, db \Delta x dt \tag{3}$$

According to exchange characteristics of bed load and suspended sediment during sediment erosion and deposition, when riverbed scour occurs, exchanging sediment is from the riverbed scoured by flow, the adsorption quantity of sediment scoured is equal to adsorption quantity of the bed load C_{sb}, ie:

$$c_{ks} = c_{kb} = C_{sb} \tag{4}$$

When silting occurs along riverbed, the sediment exchange is from the water body, and the quantity adsorbed by bed materials is equal to the quantity adsorbed by sediment C_s in the water body, and there have been continuous exchanges between the suspended sediment and bed load when erosion and deposition is up to balance, and at the same time the quantity adsorbed by the bed load is equal to that adsorbed by the suspended sediment, which is:

$$c_{ks} = c_{kb} = C_s \tag{5}$$

Based on conditions above, respectively replacing sediment concentration and sediment – carrying capacity nearby bed surface by an average sediment concentration S and an average sediment carrying capacity S_* from cross – section, Eq. (1) can be rewritten as:

$$\int_B (\omega s_b c_{ks} + D_z \frac{\partial s_b}{\partial z} C_{kb}) \, db \Delta x dt = \int_B \omega (s_b - s_{*b}) C_k db \Delta x dt = \int_B \alpha \omega (S - S_*) C_k db \Delta x dt \tag{6}$$

where, α is a recovery saturation coefficient, B is the river width, C_k is

$$C_k = \begin{cases} C_{sb} & S_* > S \\ C_s & S_* \leqslant S \end{cases} \tag{7}$$

In the control body, the pollutants adsorbed by sediment at the period of dt is expressed as:

$$\int_A k_a s db \Delta x dt \tag{8}$$

where, k_a is a rate of change in quantity adsorbed by sediment per unit time and per unit mass.

According to the law of conservation of mass, in a period of dt the mass difference from the adsorbed pollutants in and out of the control body adding variable caused by adsorption (desorption) in the control body is equal to changes in the mass of the adsorbed pollutants in the control body in a period of dt, ie:

$$\int_{A_1} c_s s u_s \, dA dt - \int_{A_2} c_s s u_s \, dA dt - \int_B \alpha \omega (S - S_*) C_k db \Delta x dt + \int_A k_a s db \Delta x dt =$$
$$\frac{\partial}{\partial t} \int_A c_s s dA \Delta x dt \tag{9}$$

where, C_s is a mass of the adsorbed pollutants of unit – mass suspended sediment; s is the sediment concentration, u_s is sediment movement speed.

Expanding a flux integration for outlet section by the first – order Taylor series in accordance with the integral for the inlet section and taking it into Eq. (9) is given by:

$$\frac{\partial}{\partial t} \int_A c_s s dA \Delta x dt + \frac{\partial}{\partial x} \int_A c_s s u_s \, dA \Delta x dt = \int_B \alpha \omega (S_* - S) C_k db \Delta x dt + \int_A k_a s dA \Delta x dt \tag{10}$$

All items divided by Δx and dt, when $\Delta x \to 0$, a continuity equation in the integral form for the adsorbed pollutants is written as:

$$\frac{\partial}{\partial t} \int_A c_s s dA + \frac{\partial}{\partial x} \int_A c_s s u_s \, dA = \int_B \alpha \omega (S_* - S) C_k db + k_a SA \tag{11}$$

It is generally believed that sediment has good features with the flow, and sediment movement velocity u_s is a sum of muddy water speed with molecular diffusion speed, and the latter relatively to the convection can be ignored, so $u_s = u$, and by way of homogenization of the second item in Eq. (11) it is obtained as:

$$\frac{\partial}{\partial x} \overline{\int_A c_s s u_s \, dA} = \frac{\partial}{\partial x} \overline{\int_A c_s s u_s \, dA} + \frac{\partial}{\partial x} \overline{\int_A (c_s s)' u' dA} \tag{12}$$

According to the turbulent diffusion mode it is given as:

$$\frac{\partial}{\partial x}\int_A (c_s s)'u'dA = -\frac{\partial}{\partial x}\int_A D_x \frac{\partial(C_s S)}{\partial x}dA = -\frac{\partial}{\partial x}\left[AD_x \frac{\partial(C_s S)}{\partial x}\right] \tag{13}$$

where, D_x is a longitudinal dispersion coefficient. Taking Eq. (12), Eq. (13) in to Eq. (11) and omitting a homogeneous item symbol is given as:

$$\frac{\partial}{\partial t}\int_A c_s s dA + \frac{\partial}{\partial x}\int_A c_s su dA = \frac{\partial}{\partial x}\left[AD_x \frac{\partial(C_s S)}{\partial x}\right] + \int_B \alpha\omega(S_* - S)C_k db + k_a SA \tag{14}$$

Integrating all items above over the cross - section is given as:

$$\frac{\partial(ASC_s)}{\partial t} + \frac{\partial(UASC_s)}{\partial x} = \frac{\partial}{\partial x}\left[AD_x \frac{\partial(C_s S)}{\partial x}\right] + \alpha\omega B(S_* - S)C_k + k_a SA \tag{15}$$

where, U is the average velocity along the cross - section.

Based on a sediment continuity equation it can be given by:

$$C_s \frac{\partial(AS)}{\partial t} + C_s \frac{\partial(UAS)}{\partial x} = \alpha\omega B(S_* - S)C_s \tag{16}$$

Subtracting Eq. (15) and Eq. (16) is written as:

$$S\frac{\partial(AC_s)}{\partial t} + S\frac{\partial(UAC_s)}{\partial x} = \frac{\partial}{\partial x}\left[AD_x \frac{\partial(C_s S)}{\partial x}\right] + \alpha\omega B(S_* - S)(C_k - C_s) + k_a SA \tag{17}$$

Eq. (17) is an equation for the transportation and transformation of the adsorbed pollutants, and the second item of the equation right end is the quantity of pollutant exchange between the water body and the riverbed accompanied by the sediment erosion and deposition, where, $\alpha\omega B$ $(S - S_*)$ is a flux of sediment erosion and deposition, $(C_k - C_s)$ is a difference between sediment erosion and deposition and an adsorbed quantity of suspended sediment, and sediment is from the water body during siltation, thus both is equal to each other, and sediment is from the riverbed during erosion, thus the value is given by a bed material condition. This expression shows that the changes in the adsorbed quantity of sediment during sediment erosion process don't depend on the quantity of pollutants carried by wash load, but depend on a difference between the adsorbed quantity of the wash load and suspended sediment. When the adsorbed quantity of the wash load is less than that of suspended sediment, the wash load is equivalent to a self - cleaning agent, contrarily the wash load is equivalent to the pollution source. The third term of the right end of Eq. (17) represents the quantity of pollutant exchange between the two phases of water and sediment caused by action of the adsorption (desorption), $k_a > 0$ is sediment adsorption, contrarily is sediment desorption.

As for the dissolved pollutants, the following processes should be taken into account such as convection dispersion of dissolved matter, pollutants exchange in the water and solid phases, exchange and biochemical reaction processes generated by concentration gradient of pollutants between the water body and riverbed. Because of similarity to derivation of the continuity equation for the adsorbed pollutants, a differential equation can be obtained according to the law of conservation of mass, and after homogenization with integral it can be expressed as:

$$\frac{\partial(AC_w)}{\partial t} + \frac{\partial(UAC_w)}{\partial x} = \frac{\partial}{\partial x}(AD_x \frac{\partial C_w}{\partial x}) - k_a SA + A\frac{dC_w}{dt} + \sigma + r \tag{18}$$

where, U is average section velocity, m/s; C_w is average section concentration of the dissolved pollutants, g/m^3; D_x is a longitudinal dispersion coefficient, m^2/s; A is the cross section area, m^2; B is the river width, m; dC_w/dt is a biochemical reaction item, g/(m$^3 \cdot$ s); σ is a pollutant release rate unit river length, g/(m $\cdot$ s); γ is the emission source strength, g/m $\cdot$ s.

The basic equation for transportation and transformation of the dissolved and adsorbed pollutants is established above, which used for a generic description of pollutants. And as for the specific pollutants, the related items and coefficients in the equation can be determined according to their characteristics. Simultaneous usage of Eq. (17), Eq. (18) and the basic equation for describing water - sediment movement has formed into a basic equation of an one - dimensional mathematical model sediment pollutants. The specific solution procedure is as following: first a water – sediment element is obtained by an equation for the water – sediment movement, and then it is substituted into the equation to further solve concentration of the dissolved pollutants and quantity of the adsorbed pollutants, the Total Variation Diminishing (TVD) format is used for solution of equations for both water – sediment and pollutants.

3.2 Equation for adsorption and desorption kinetics

The adsorption (desorption) kinetics equation describes the quantity of pollutants exchange between the water – solid phases in the adsorption (desorption) processes, which is expressed as a coefficient ka in the basic equation, and a general expression is written as :

$$k_a = k_1 C_w (b - C_s) - k_2 C_s \tag{19}$$

where, k_1 is an adsorption rate coefficient; k_2 is a desorption rate coefficient; b is the quantity of saturated adsorption, g/kg.

Theoretical and experimental research results have shown that relation of the adsorption (desorption) kinetics processes with the adsorbent concentration is related to the adsorbate characteristics, and the adsorption (desorption) kinetics processes of the adsorbates have concentration effects, and some have not. Taking in situ water and sediment from the Three Gorges reservoir region as a sample, kinetics experiments for TP of sediment adsorption have been carries out in this study. Fig. 1 shows changes in sediment adsorbed quantity unit sediment over time when sediment concentration is respectively 0.5 kg/m^3 and 1.0 kg/m^3, under the same conditions of TP initial concentration. And experimental results show that sediment concentration has a great impact upon adsorption kinetics process of TP. And so, the adsorption and desorption coefficients in an adsorption kinetics equation must be based on experiments, which shows its relation to sediment concentration.

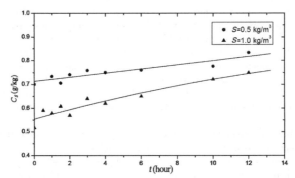

Fig. 1 Adsorption kinetics process of TP under conditions of various sediment concentration

4 Model verification

After the Three Gorges Reservoir impoundments, a synchronous monitoring of water – sediment and water quality is carried out along Qingxichang – Wanxian reach of the Yangtze River in this study. A monitoring section is placed by every 25 km from Qingxichang, and five monitoring sections are placed on a reach with a total of 100 km from Qingxichang to Zhongxian (Qingxichang section, section B, section C, section D and section E). A marine acoustic Doppler current profiler (ADCP) is used in measuring discharge along the section within the reach, and at the same time sediment concentration, concentration of TP in the dissolved state and quantity of TP in the adsorbed state are measured in water samples. The measured results can be used as a basis of the model verification.

Due to high flow velocity in the simulated region, there is no hydrodynamic conditions for significant growth of the chlorophyll, so a biochemical reaction item of TP is not taken account of calculation. TP exchange produced by bottom concentration gradient may be ignored because there is no obvious sediment pollution within the simulated region. In calculating sediment movement a formula for sediment carrying capacity and other parameters make reference to the literature, the longitudinal dispersion coefficients make reference to the literature. Because sediment in the calculated region is mostly in an adsorption or balance state, and based on the suspended sediment concentration in the calculated region, the results obtained from adsorption experiments of sediment concentration 0.5 kg/m^3 are used as the adsorption kinetics parameters, and it is expressed as:

$$k_a = 0.000,13C_w(1.538 - C_s) - 0.000,01C_s \tag{20}$$

A comparison between the calculated results such as discharge, sediment concentration, the adsorbed TP and the concentration of the dissolved TP(cross – sections B, E) from the model and the measured values is shown in Fig. 2. And from the Fig. it can be seen that a tendency of changes in the simulated results and the measured values is close to each other, and it is shown that the model is better to simulate an evolution of river natural process.

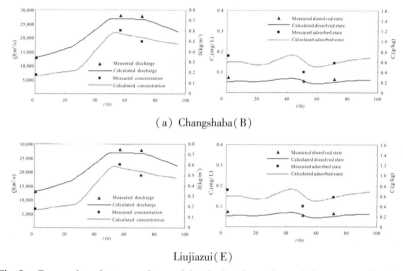

(a) Changshaba(B)

Liujiazui(E)

Fig. 2 Comparison between the model calculated results and the measured values

5   Model application and analysis

The mathematical model established by this paper is used to simulate changes in water quality before and after impoundments of the Three Gorges Reservoir and analyze an impact trend of water – sediment condition upon water quality at various periods. In order to understand the trend of changes in water quality in the reservoir area under different hydrological conditions, two typical hydrological conditions representing the low – flow year and high – flow year are taken into account in calculation. Based on an analysis of hydrological data of a long time series, the period from June 1959 to June 1960 is taken as a typical low – flow year and June 1964 to June 1965 is taken as a typical high – flow year, and reservoir inflowing pollution loading amount and reservoir border pollution loading amount in 2010 is taken as an input condition of a pollution source.

Fig. 3 and Fig. 4 respectively show comparisons of the concentration of TP in the dissolved, adsorbed and mixed state pollutants under the conditions of the low flow period in low – flow year and the higher flows season in high – flow year before and after the reservoir impoundments. In the Figure, C_w and C_s respectively represent concentration of the dissolved state, and adsorbed state pollutants.

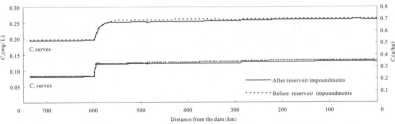

Fig. 3 Changes in TP concentration during low flow period in low – flow year before and after impoundments

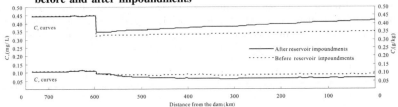

Fig. 4 Changes in TP concentration during a higher flows season in high – flow year before and after impoundments

The calculated results show that sediment impacts on water quality are not the same because there is a great difference of sediment concentration in the reservoir area corresponding to the different water storage periods. On low flow condition, there is a lower sediment concentration in the reservoir area. In spite of sediment silting within the reservoir area after impoundments, and it led to decrease in TP concentration and increase in the quantity adsorbed by sediment in the water body, but both TP concentration and the quantity adsorbed by sediment have a slight change, and there are no changes in TP concentration and the quantity adsorbed by sediment along the way in the water body, except for the impacts given by inflow of anabranches. Under conditions of the medium flow and high flow, sediment concentration in the reservoir area is significantly greater than that in the low flow period, and where which led to silting, and decreasing in amplitude of sediment variation along the way is greater than that of the low flow period. Therefore, a tendency of decrease in TP concentration and increase in the quantity adsorbed by sediment along the way in the water body is becoming more obvious. Although TP concentration in the water body is greatly reduced by sedimentation in a period of high flow, TP concentration in the water body during high flow in the reservoir area is still greater than that of the low flow period owing to impacts by upstream inflow concentration.

Different water storage periods correspond to different dam heel levels, the dam heel level in the low flow period is 175 m, and the deposition of silt began to appear at a place 670 km distance from the dam, where have changes in water quality too; when the water level at dam frontage in the periods of medium flow and high flow is respectively 156 m and 145 m, silting began to appear at a place 590 km and 550 km distance from the dam, and there have changes in water quality too.

6 Conclusions

Based on the analysis of the main physicochemical processes in the water environment system, the mathematical model suitable for the environment impacts given by river sediment is derived in this paper. The basic equations for the adsorbed and dissolved pollutants transport are respectively established, which make mathematical description of the adsorbed pollutants be more closer to practice and make a description of the environmental impacts by sediment erosion and deposition have a clear physical basis. Adoption of the kinetic mode to describe the adsorption and desorption can make the model reflect pollutant dynamic changes in process of the adsorption and desorption.

The mathematical model for river sediment pollutants can be applied to simulation and analysis of the environmental impacts by reservoir sediment. Construction of a reservoir will make pollutants and nutrients deposit in a reservoir along with sedimentation. As sediment may be scoured by flow when water level in the reservoir area should be reduced before flood season, it is necessary to consider how pollutants can be discharged as soon as possible. And it does not court any sudden risks of pollution to downstream regions, and use of the model can simulate and analyze such risks given by pollution.

Acknowledgements

This study is funded by National Science and Technology support project (No. 2011BAC09B07) of the Chinese Ministry of Science and Technology.

References

Yu Xuezhong, Zhong Deyu, Li Jinxiu. Review of Studies on Sediment in Water Environment[J]. Journal of Sediment Research ,2004,6:75 −81.

Jin Xiangcan, Wang Guilin. Sediment Effects on the Adsorption of Heavy Metals [J]. Environmental Science and Technology ,1984,2:6 −11.

Hu Guohua. Experimental Study of Influence of the Yellow River Sediment on COD_{Mn} [J]. Yellow River, 2000,22 (3) :17 −18.

Jin Xiangcan. Model Study of Heavy Metal Migration and Transformation in the Xiangjiang River [J]. China Environmental Science, 1987,7 (6) :10 −15.

Chen Junhe, Chen Xiaohong. 3D Fe, Mn Migration Model for Reservoir[J]. Advances in Water Science ,1999,3:14 −19.

Lin Yuhuan, LiQi. Model Study of Heavy Metal in River Water Body [J]. Environmental Chemistry, 1992,11 (6) :35 −42.

James R T, Martin J. , Wool T. , et al. A Sediment Resuspension and Water Quality Model of Lake Okeechobee [J]. Journal of the American Water Resources Association, 1997, 33: 661 −680.

Huang Suiliang, Wan Zhaohui, et al. Mathematical Model Study of Heavy Metal Pollutants Migration and Transformation in the Alluvial River [J], Journal of Hydraulic Engineering , 1995,1:47 −56.

Zhou Xiaode. Mathematical Model for Heavy Metal migration and Transformation in River[C]. Progress in Environmental Hydraulics. Wuhan: Wuhan University of Hydraulic and Electric Press ,1999. 279 −307.

He Yong, Li Yi, Zou Huicai, et al. Research on Sediment Pollution Water Quality Model [J]. Sichuan University (Engineering Science), 2004,36 (11) :12 −17.

Xie Jianheng. River Simulation [M]. Beijing: China Water Power Press, 1990.

Zhong Deyu, Peng Yang, Zhang Hongwu. Unsteady One − dimensional Numerical Model for Alluvial Rivers with Heavy Sediment Load and Its Applications [J]. Advances in Water Science, 2004,15 (6) :706 −710.

Pan Gang. Metastable Equilibrium Adsorption (MEA) Theory—The Challenges Faced by the Conventional Adsorption Thermodynamic Theory [J]. Journal of Environmental Science, 1998, 23 (2) :156 −173.

Pan. G. Adsorption Kinetics in Natural Waters: A Generalised Ion − exchange Model, in Adsorption and Its Application in Industry and Environmental Erotection [J]. Studies in Surface Science and Catalysis. 1999, 120:745 −761.

Changjiang Water Resources Commission. Three Gorges Project Sediment [M]. Wuhan: Hubei Science and Technology Press ,1997.

Li Jinxiu, Huang Zhengli, Lvping Yu. Study on the Flow Longitudinal Dispersion Coefficient in Three Gorges Project reservoir [J]. Journal of Hydraulic Engineering , 2000,8:84 −87.

Comprehensive Evaluation Method and Fuzzy Mathematics Method of Comparative Study

Ning Guofeng, *Zhang Yongxiang*, *Hao Jing* and *Cheng Yiping*

College of Civil Engineering and Architecture, Beijing University of Technology, Beijing,100124, China

Abstract: The serious waste and less utilization of water resource, the problem of low utilization rate in our country is always the bottleneck in the development of society in twenty – first Century. Rational water resources quality evaluation, the water quality classification, selection of reasonable, accurate method of determining the groundwater quality is precondition of improving water resources utilization. To make an accurate assessment of groundwater quality, the choice of evaluation methods is critical. Selecting total hardness, chloride, sulfate, nitrate, iron, potassium permanganate index six important evaluation factors, using the comprehensive evaluation method and fuzzy mathematics method evaluates water quality status of representative four wells in certain region respectively. The results show that: two kinds of methods can meet the precision requirement, comprehensive evaluation method boundary is very strict, not considering each sample ions composition, extremal function too obvious. But the method of fuzzy mathematics comprehensive considering wells' each ion composition, eliminating extreme qualitative effect, the evaluation results is more objective. From human health perspective, the single index exceed the standard is not conducive to health, drinking water quality evaluation must consider the extreme value problem.

Key words: water quality, comprehensive evaluation, Fuzzy mathematics

The current water quality evaluation methods varied, but all is according to the method itself request putting forward forms of expression, including the single parameter evaluation method, index method, the comprehensive method and fuzzy comprehensive evaluation method which are in the light of the physical and chemical parameters to evaluate, The latter two are more widely applied. Based on using of the fuzzy comprehensive evaluation method and comprehensive evaluation method at the same time, evaluating water quality status of representative four wells in certain region respectively to contrast and analysis advantages and disadvantages of the fuzzy comprehensive evaluation method and comprehensive evaluation method.

1 Comprehensive evaluation method

1.1 Water quality evaluation standard

Groundwater quality standards is based on China's present groundwater quality situation, human health reference value and the groundwater quality protection target, and in the light of the highest requirements of life drinking water, industrial, agricultural water quality to divide the groundwater quality into five categories. And all types of water requirements are clearly defined. Therefore the groundwater quality as a class I, II, III class can be used directly for drinking water source. After treating appropriately, groundwater quality of IV class can be used as sources of drinking water. Whether the evaluation standard of groundwater will reach the standard of drinking water depend on the class III standard of*the groundwater quality standard* GB/T 14848—93. The standards are shown in Tab. 1.

1.2 Comprehensive evaluation of groundwater quality

Groundwater quality comprehensive evaluation uses the scoring method. The specific steps are as follows:

(1) According to Tab. 1, classifies each individual component quality category.

(2) According to Tab. 2, determine evaluation score F_i of the individual component in each category.

(3) Calculating comprehensive evaluation the results of F.

According to Eq. (1) and Eq. (2) calculate the comprehensive scores F.

$$\overline{F} = \frac{1}{n}\sum_{i=1}^{n}F_i \tag{1}$$

$$F = \sqrt{\frac{\overline{F}^2 + F_{max}^2}{2}} \tag{2}$$

where, $\overline{F}$ is the mean value of each individual component score F_i; F_i is the i component evaluation value; F_{max} is the maximum value in the evaluation F_i; n is number.

According to the F value and Tab. 3, classifies groundwater quality level, such as "excellent (class I)", "good(classII)", "better (Class III)", "poor(IV)", "range (V)".

Tab. 1   Groundwater quality classification index

Project number	Category / standard / Project	Class I	Class II	Class III	Class IV	class V
1	Total hardness(mg/L)	≤150	≤300	≤450	≤550	>550
2	Sulfate(mg/L)	≤50	≤150	≤250	≤350	>350
3	Chloride(mg/L)	≤50	≤150	≤250	≤350	>350
4	Iron(Fe) (mg/L)	≤0.1	≤0.2	≤0.3	≤1.5	>1.5
5	Potassium permangana index(mg/L)	≤1.0	≤2.0	≤3.0	≤10	>10
6	Nitrate(mg/L)	≤2.0	≤5.0	≤20	≤30	>30

Tab. 2   the single component evaluation value F_i

Category	I	II	III	IV	V
F_i	0	1	3	6	10

Tab. 3   groundwater quality comprehensive evaluation

Level	Excellent	Good	Better	Poor	Very poor
F	<0.80	0.80~2.50	2.50~4.25	4.25~7.20	>7.20

1.3 The results of evaluation

Water quality monitoring data of four wells in Tab. 4. Select the total hardness, chloride, sulfate, nitrate, iron, potassium permanganate index as evaluation factors to be evaluated. The results of evaluation are in Tab. 5.

Tab. 4   The results of water quality monitoring

Serial number	Total hardness	Sulfate	Chloride	Iron	Potassium permangana index	Nitrate
The first well	302.00	65.80	68.10	0.062	0.45	9.41

Continued to Tab. 4

Serial number	Total hardness	Sulfate	Chloride	Iron	Potassium permangana index	Nitrate
Second wells	379.00	83.60	73.70	0.503	0.62	8.83
Third wells	192.00	46.10	17.00	0.119	1.13	0.82
Fourth wells	582.00	110.00	165.00	0.032	1.28	18.90

Tab. 5　Groundwater quality comprehensive evaluation statistics

Number	The first well	Second wells	Third wells	Fourth wells
Water quality	Class II	Class IV	Class I	Class V

2　Fuzzy mathematical method

2.1　Division of water quality classification boundaries with the membership degree

(1) Water level limit has vagueness. Evaluating the water quality only with a deterministic index often does not reflect the true situation. Using the theory of fuzzy mathematics for water quality evaluation is more and more extensive. The specific steps are as follows:

When the measured concentration x in the left, belong to class I degree of membership

$$y_1 = \begin{cases} 1 & x \leqslant x_1 \\ \dfrac{x_2 - x}{x_2 - x_1} & x_1 < x < x_2 \\ 0 & x \geqslant x_2 \end{cases}$$

When the measured concentration x is located in the middle, belong to class I degree of membership

$$y_1 = \begin{cases} 1 & x = x_1 \\ \dfrac{x - x_{i-1}}{x_i - x_{i-1}} & x_{i-1} < x < x_i \\ \dfrac{x_{i+1} - x}{x_{i+1} - x_i} & x_i < x < x_{i+1} \\ 0 & x \leqslant x_{i-1} 或 x \geqslant x_{i+1} \end{cases}$$

When the measured concentration x is located on the right, belong to class n degree of membership

$$y_n = \begin{cases} 1 & x \geqslant x_n \\ \dfrac{x - x_{n-1}}{x_n - x_{n-1}} & x_{n-1} < x < x_n \\ 0 & x \leqslant x_{n-1} \end{cases}$$

where, y is degree of membership of two grade water corresponding to the measured value x; x is the measured value; $x_1, x_2, \cdots, x_n$ are quality standard values for Class I, Class II, $\cdots$, Class i, $\cdots$, Class$N$.

(2) The single index evaluation. Take U for pollutant m individual index set; V is water classification collection; A is for single index. $U = \{A_1, A_2, \cdots, A_m\}$; $V = \{$ class I water, secondary water, $\cdots$, N water$\}$, Through the calculation of subordinate function, calculate membership degree of m individual index to water grade n and fuzzy matrix R.

（3）Weight calculation. Weight value is $W_i = \dfrac{C_i}{S_i}$. For fuzzy calculation, each individual weights are normalized, which is $Z_i = \dfrac{\dfrac{C_i}{S_i}}{\sum\limits_{i=1}^{m} \dfrac{C_i}{S_i}}$. Given weights to the m index of collection U, composed of a matrix of B, which is $B = (Z_1, Z_2 \cdots Z_m)$.

where, C_i is for the pollutant measured concentration i ; S_i is for the pollutants i, the arithmetic mean of special purpose of various water quality standard .

（4）The compound operation of the fuzzy matrix. Conducting compound operations to B, R matrix and determining water comprehensive grade of membership . To fuzzy matrix, the method for computing is : Multiply two numbers together with small for the "product", the numbers add up to take greater to "and", namely by ($\wedge$, $\vee$) operate matrix composite operation. Expressions is $S = B \cdot R$.

（5）According to the maximum degree of membership principle to determine quality level. From the general membership degree matrix $S = [\, Y_{11}, Y_{12}, \cdots, Y_{1n}\,]$ selects the maximum value, the number representing the level of water quality. If the matrix has two equal to the maximum value, according to the adjacent numbers and "rely on large do not rely on a small" principle to determine the degree of water quality.

2. 2　The results of evaluation

Using Tab. 4 monitoring data, selecting the six evaluation indexes to conduct fuzzy mathematics evaluation, the evaluation results are in Tab. 6 below.

Tab. 6　Groundwater quality fuzzy evaluations statistics

Number	The first well	Second wells	Third wells	Fourth wells
Water quality	Class Ⅱ	Class Ⅲ	Class Ⅰ	Class Ⅴ

3　Comparison and analysis of results

Tab. 5 and Tab. 6 show: Two evaluation results are consistent. But evaluation results still have certain differences. Limit of Comprehensive evaluation method is very strict, not considering each sample ions, extremal function too obvious. Water quality for second wells in comprehensive evaluation method is Ⅳ class, water quality for class Ⅲ in fuzzy mathematical method . In single component monitoring result , only iron is for Ⅳ, total hardness and nitrate for class Ⅲ, sulfate and chloride as class Ⅱ, potassium permanganate index for class Ⅰ. Integrated every evaluation index to be defined as class Ⅲ is reasonable. The fuzzy mathematics method can consider each ion composition, eliminating extreme qualitative effect. The evaluation result is more reasonable.

4　Conclusions

The conclusion is that the comprehensive evaluation method and fuzzy mathematics method meet the basic requirement of a certain accuracy. The fuzzy mathematical method is more objective. But the fuzzy mathematics method can not solve the evaluation information repeated problem caused by evaluation index correlation . The certainty of every factor weight have certain subjectivity. Now the general comprehensive evaluation method and fuzzy comprehensive evaluation method are combined. Contribution of information technology, neural network and other emerging discipline in the fuzzy comprehensive evaluation promote the fuzzy comprehensive evaluation to have new development.

References

Kunwar P Singh, Amrita Malik, et al. Multivariate Statistical Tech – niques for the Evaluation of Spatial and Temporal Variations inWater Quality of Gomti River (India): A Case Study[J]. Water Research, 2004, 38(18).

Peng Xiaojin, Zhang Yanhong , et al. Application of Fuzzy Comprehensive Evaluation on Underground Water Wuality Assessment [J]. Henan Water Resources & South – to – North Water Diversion, 2009(1).

You Yang. Water Quality Comprehensive Assessment Method and Its Application[D]. Xi' an: Xi' an University of Technology,2007.

Zhang Yongxiang. Chaoyang District Water Quality Security System – groundwater Quality Investigation and Water Quality Assessment Report[M]. 2009.

Application of Snowmelt-based Water Balance Model to Xilinhe River Basin in the Inner-Mongolia, China

Wang Guoqing[1,2], *Li Yang*[3], *Zhu Zongyuan*[3], *Hu Qingye*[1,2],
Dong Chong[1,2] and *Yao Suhong*[3]

1. State Key Laboratory of Hydrology-Water Resources and Hydraulic Engineering,
Nanjing Hydraulic Research Institute, Nanjing, 210029, China
2. Research Center for Climate Change, Ministry of Water Resources,
Nanjing, 210029, China
3. Inner Mongolia Agricultural University, Hydraulic and Civil Engineering Institute,
Hohhot, 010018, China

Abstract: Conceptual hydrological model are believed useful in assessing environmental change impact on hydrology. Hydrological characteristics of Xilinhe River Basin, a dry cold basin, located in the Inner Mongolia, China, were analyzed with data series from 1963 to 2008. A Snowmelt-based Water Balance Model (SWBM) was applied to the basin taking data series from 1963 to 1972 for model calibration and verification for the purpose of avoiding intensive human disturbance to flow variation. Results indicate that runoff yield in Xilinhe River catchment is dominated by both snowmelt and heavy rainfall storm. Recorded runoff at Xilinhaote station and area-average precipitation of the Xilinhe River catchment both presented a slight decline trend during 1963-2008. The SWBM performed well at simulating monthly discharge rates. NSEs in calibration and verification were 73.6% and 62.7%, respectively, while REs in both periods were quite low at less than 8%, indicating SWBM could be used in snow-covered cold catchment for hydrological modeling and climate change study.

Key words: SWBM, the Xilinhe River, discharge simulation

1   Introduction

Conceptual rainfall runoff models are widely used in hydrology. Contrary to the more complex, physically based distributed models such as SHE (Abbott et al., 1986), the required input data for conceptual models for most applications are relatively readily available; furthermore, conceptual models are usually simple and relatively easy to use (Wang G Q et al., 2008), and are believed useful in assessing impact of climate change on hydrology and water resources. However, due to complex runoff yield mechanism in cold region, hydrological modelling for cold dry areas is still a worldwide challenge. In this paper, we attempt to apply a Snowmelt-based Water Balance Model (SWBM) to Xilinhe River catchment in which temperature is low and precipitation is less.

2   Hydrological characteristics of Xilinhe River basin

Xilinhe River is the first order tributary of Chaergan river system, originating from Baoertu Mountain of Chifeng City. The river flows from east to west, running across Keshiketeng Town, Arbaga Town, turning northwestward at Beierke Pasture, and vanishing in Chaernuoer Wetland. The drainage area of Xilinhe River is 10,542 km^2 with total river length of 268.1 km. The river system consists of three big tributaries, namely, Haolaituguole River, Haolaiguole River from right bankside, Hushite River from left bankside. Xilinhaote is a key hydrometric station locating in the middle reaches of the river basin with drainage area of 3,852 km^2.

Xilinhe River Basin locates in the middle latitude Asia monsoon climate zone with climatic features of highly cold and dry. The average annual precipitation is of 266 mm with a slight decreasing rate of 0.688 mm/a (Fig. 1). The mean annual potential evaporation is approximately 1,757.3 mm, which is about 5.3 times of annual precipitation. Due to higher potential evaporation and less rainfall, water resources in Xinlinhe River Basin are very shortage. The annual runoff depth is only about 4.44 mm, with a slightly decreasing trend rate of 0.019,1 mm/a (Fig.1).

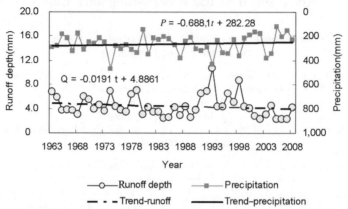

Fig. 1   Variation of area-average annual precipitation of Xilinhe River Basin and annual runoff depth at Xilinhaote hydrometric station

Rainfall is dominant factor for runoff yielding; meanwhile, potential evaporation, as well as temperature will also put great effect on runoff for arid regions (Wang, 2009). Seasonal Pattern of area-average precipitation, temperature of Xilinhe River Basin, and discharge at Xilinhaote station were shown in Fig. 2. The figure indicates that more than 80% of annual precipitation is from flood season during June to September, with the highest rainfall occurring in July. Mean annual temperature is about 3 ℃, with monthly mean temperature from April to October being higher than 0 ℃ while bellow 0 ℃ in other months. There are two peaks in seasonal runoff pattern with biggest peak occurring in April, implying snow or ice melting runoff yield mechanism. The second biggest peak occurs in August, resulting from heavy rainfall storm in flood season.

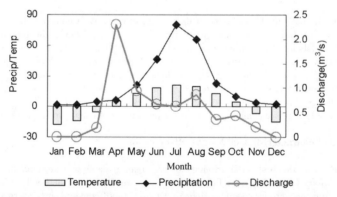

Fig. 2   Seasonal pattern of area-average precipitation (mm), temperature (℃) of Xilinhe River Basin, and discharge (m³/s) at Xilinhaote station

3   Description of SWBM

Conceptual hydrological models are useful in assessing the impacts of environmental changes on regional hydrology, and have been applied successfully on larger scales (Arnell, 1999; Yu, 2009). There are many water balance model available in literature, among which, the SWBM is a simplified one with four parameters. It could estimate monthly stream flow from monthly rainfall, temperature and potential evaporation data (Wang et al., 2000), and has been successfully applied in semi-arid and humid catchments located in China (Wang and Li, 2000; Wang et al., 2005). As comparison to other hydrological models, such as SIMHYD, TANK, SMAR, et al., the SWBM not only has the advantage of a simpler structure, including fewer parameters and more flexibility, but it also performs well for high cold areas fully or partially covering with snow. SWBM structure was shown in Fig. 3.

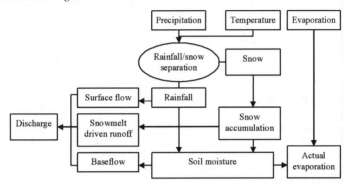

Fig. 3 Conceptual outline of the SWBM structure

Monthly precipitation is first divided into rainfall and snowfall in order to calculate surface flow and snow-melt driven runoff. The upper and lower temperature criteria, T_{H} and T_{L} are 4 ℃ and -4 ℃ respectively. Snowfall can be estimated by a linear interpolation function when temperature falls in the range between T_{H} and T_{L}, viz.

$$P_{\mathrm{SN}i} = \frac{T_{\mathrm{H}} - T_i}{T_{\mathrm{H}} - T_{\mathrm{L}}} \cdot P_i \tag{1}$$

$$P_{\mathrm{R}i} = P_i - P_{\mathrm{SN}i} \tag{2}$$

where, $P_{\mathrm{R}i}$, $P_{\mathrm{SN}i}$ and T_i, are rainfall, snowfall and air temperature of month i, respectively.

As surface flow is directly proportional to soil moisture and precipitation, it is assumed here that groundwater flow lags one time-step (Wang, 2000), and its calculation is based on a linear reservoir theory. As the Yellow River Basin is partially covered by seasonal snow, the snow-melt rate is not only an exponential function of air temperature, but it is also proportional to snow accumulation. These can be expressed mathematically as:

$$Q_{\mathrm{S}i} = k_{\mathrm{s}} \cdot \frac{S_{i-1}}{S_{\max}} \cdot P_{\mathrm{R}i} \tag{3}$$

$$Q_{gi} = k_{\mathrm{g}} \cdot S_{i-1} \tag{4}$$

$$Q_{\mathrm{SN}i} = \begin{cases} 0, \\ k_{\mathrm{sn}} \cdot \mathrm{e}^{\frac{T_i - T_L}{T_H - T_L}} \cdot SN_i \\ SN_i \end{cases} \quad \begin{cases} T_i < T_{\mathrm{L}} \\ T_{\mathrm{L}} < T_i < T_{\mathrm{H}} \\ T_i > T_{\mathrm{H}} \end{cases} \tag{5}$$

$$SN_i = SN_{i-1} + P_{\mathrm{SN}i} \tag{6}$$

where, $Q_{\mathrm{S}i}$, Q_{gi} and $Q_{\mathrm{SN}i}$ are surface flow, groundwater discharge and snow-melt driven runoff in month i, respectively; $S_{\mathrm{N}i}$ and SN_{i-1} are snow accumulation in month i and month $i-1$, respectively; S_{i-1} is soil moisture in month $i-1$, which is a basin average of the moisture holding capacity of the soil layer; $S_{\max}$ is the maximum soil moisture storage and is the upper limit for S_{i-1}; and k_{s}, k_{g} and k_{sn} are the surface flow, groundwater flow lag, and snow-melt driven runoff coefficients, respectively, which are non-dimensional parameters.

It has been shown that a one-layer soil evapotranspiration formula has enough accuracy to estimate actual evapotranspiration (Guo and Wang, 1994). The calculation formula is given as:

$$E_i = \frac{S_{i-1}}{S_{\max}} \cdot E_{601i} \tag{7}$$

where, E_i is the actual evaporation of month i; and E_{601i} is the potential evaporation measured by an evaporation pan E_{601}.

The total estimated monthly runoff in month i is the sum of the surface flow, groundwater discharge and snow-melt driven runoff. Finally, the soil moisture content at the end of month i is calculated according to the water conservation law:

$$Q_{ai} = Q_{Si} + Q_{gi} + Q_{SNi} \tag{8}$$
$$S_i = S_{i-1} + P_i - Q_{ai} - E_i \tag{9}$$

where, S_i, P_i, Q_{ai} and E_i are soil moisture, precipitation, calculated discharge and actual evaporation of the i month respectively.

4　Application of SWBM to the Xilinhe River Basin

The traditional approach to model calibration assumes that the primary objective is to obtain a "best fit" to the stream flow at each site. Usually, the Nash – Sutcliffe Efficiency(NSE) criterion (Nash and Sutcliffe, 1970) and the Relative Error (RE) of volumetric fit are employed to calibrate a model, with better simulation results giving NSE closer to 1 and RE close to 0. These two criteria are expressed mathematically as:

$$NSE = 1 - \sum_{i=1}^{n} (q_{ri} - q_{ci})^2 / \sum_{i=1}^{n} (q_{ri} - \bar{q}_{ri})^2 \tag{10}$$

$$RE = \sum_{i=1}^{n} (q_{ci} - q_{ri}) / \sum_{i=1}^{n} q_{ri} \tag{11}$$

where, q_{ri} is the observed discharge; q_{ci} is the simulated discharge; $\bar{q}_{ri}$ is the mean of the observed discharge during the calibration period; and n is the number of data samples.

As Xilinhe River Basin has been highly regulated by human being, including water intake for rapid mining development, increasing livestock demand and domestic needs, therefore, we take data series from 1963 to 1972 in which human activities are less, to calibrate and verify hydrological model. Data from 1963 to 1968 was used to calibrate model, data from 1969 to 1972 was used to verify model. The recorded and simulated discharges were given in Fig. 4.

In general, the simulated discharges correlated well with the recorded discharges, although there was a slight tendency to oversimulate peak discharges in the late period. The NSE critera in calibration (1963 ~ 1968) and verification (1969 ~ 1972) were 73.6% and 62.7% respectively. The long-term average simulated discharge was close to the recorded value, and the long-term average RE in both periods was < 8%, with a maximum RE of about 16.5%. The seasonal distribution of simulated and recorded discharges are compared in Fig. 5, which indicates a good match between the model output and observed discharges, although there was a slight undersimulation in May, June, October, and November and oversimulations in April, July and August.

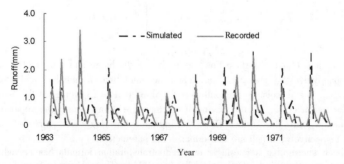

Fig. 4　The recorded and simulated discharge at Xilinhaote hydrometric station

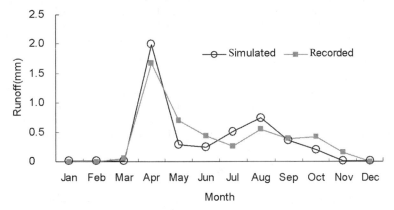

Fig. 5　Seasonal distribution of recorded and simulated discharge at Xilinhaote station

Due to much low temperature below 0 ℃ in winter, precipitation in Xilinhe River Basin was the form of snow within winter season, and water in river channel was frozen as well. With the temperature rise in April, snowmelt dominates flow generating in April and May. Data set of precipitation and discharge were divided into three groups according to three temperature ranges, lower than 0 ℃, higher then 10 ℃, ranging from 0 ℃ ~ 10 ℃. Recorded runoffs against precipitation for each group were plotted in Fig. 6.

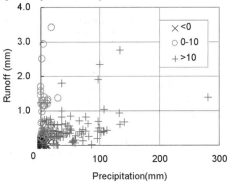

Fig. 6　Monthly recorded runoffs against precipitation for three data sets with temperature thresholds of 0 and 10 ℃

Fig. 6 indicates that discharge observations and precipitation are both very low for the first group of temperature lowing than 0 ℃. Recorded runoff varies in a high range from 0 to 3.6 mm, with low precipitation for the second group of temperature ranging from 0 ℃ to 10 ℃. For the third group set of temperature higher than 10 ℃, precipitation and runoff has a higher correlation relationship.

5　Conclusions

Xilinhe River Basin locates middle latitude cold and dry climate zone, and the basin is covered with snow during most time in dry season due to lower temperature. Runoff yield in this catchment is dominated by both snowmelt and heavy rainfall storm. Recorded runoff at Xilinhaote station and area-average precipitation of the Xilinhe River catchment both presented a slight decline trend during 1963 to 2008.

Data series from 1963 to 1972 was used to calibrate and verify SWBM. The model performed well at simulating monthly discharge rates in the Xilinhe River catchment. *NSEs* in calibration and verification were 73.6% and 62.7%, respectively, while *REs* in both periods were quite low at less than 8%, indicating SWBM could be used in snow-covered cold catchment for hydrological modeling and climate change study.

Acknowledgements
    This study has been financially supported by the National Basic Research Program of China (grant 2010CB951103), the International Science & Technology Cooperation Program of China (grant 2010DFA24330), the ACCC project funded by DFID, SDC and DECC, and National Nonprofit institute Research Program (grant Y509004). Thanks also to the anonymous reviewers and editors.

References

Abbott M B, Barthurst J C, Cunge J A, et al. An introduction to the European Hydrological System—Systeme Hydrologique Europeen, "SHE", 2: Structure of a Physically-based, Distributed Modelling System[J]. J. Hydrol. , 1986, 87:61 – 77.

Arnell N W. A Simple Water Balance Model for the Simulation of Streamflow over a Large Geographic Domain[J]. J. Hydrol. 1999, 217: 314 – 355.

Gleick P H. The Development and Testing of a Water Balance Model for Climate Impact Assessment: Modeling the Sacramento Basin [J]. Water Resour. Res. , 1987, 23: 1049 – 1061.

Guo S L, Wang G Q. Monthly Water Balance Model for the Yellow River Basin[J]. Yellow River, 1994, 112:13 – 16.

Nash J E, Sutcliffe J V. River Flow Forecasting Through Conceptual Models. Part 1: A Discussion of Principles[J]. J. Hydrol. ,1970, 10: 282 – 290.

Wang G Q, Li J. Grid-based Hydrological Model for Climate Change Impact on Water Resources in the Middle Yellow River Basin[J]. Adv. in Water Sci. ,2000, 11(6):387 – 391.

Wang G Q, Wang Y Z, Kang L L. Sensitivity of Runoff in the Middle Yellow River Basin to Climate Change[J]. J. Appl. Meteorol. Sci. , 2001, 13:117 – 121.

Wang G Q, Zhang J Y, Jing X A, et al. Study on AWBM and its Application in Semi-arid Basins [J]. Hydrol. , 2005, 25(5): 7 – 10.

Wang G Q, Zhang J Y, He R M, et al. Runoff Reduction Due to Environmental Changes in the Sanchuanhe River Basin[J]. Int. J. Sediment Res. , 2008, 23: 174 – 180.

Zhang J Y, He H. Hydrological Simulation in Ungauged Basins with GIS[J]. Adv. in Water Sci. , 1998, 9(4):345 – 351.

Entropy – based Analysis of Harmony Between Human and Water on the Digital Technology Background

Wang Jinxin[1] , *Zuo Qiting*[2] and *Wang Guoling*[3]

1. Institute of Natural Resource and Eco – environment Zhengzhou University, Zhengzhou, 450001, China
2. Center for Water Science Research Zhengzhou University, Zhengzhou, 450001, China
3. School of Marxism Zhengzhou University, Zhengzhou, 450001, China

Abstract: Technically, water conservancy informatization, whose construction is based on the very motion law of the geographical process of the water system, is the use of modern information technology to build the virtual space of the natural water system. By virtual practice, it solves a series of major problems of water system on planning, construction, management, simulation, monitoring, control, disaster prevention, disaster reduction and reconstruction by digitization, networking, automation, intelligentization, and scientization. In essence, it is to filter out the "entropy" in the virtual space to maximization, so as to achieve the goal of inputting "micro – entropy" and "negative entropy" to the natural water system, maintaining and enhancing the sustainable development of the natural environment, achieving the harmonious of people and water in objective world. From a philosophical perspective, this article cognizes and analyzes the digital technology (information technology or virtual technology), virtual space, virtual practice, and the law of entropy's operation between the virtual space and real water system, reveals important position and role of the contemporary IT in the harmonious development of man and water, provides a theoretical basis for the decision – making in sustainable development of natural water systems and human society.

Key words: harmony between human and water, entropy, digital technology, digital culture, sustainable development

1 Introduction

Harmony is a concept with small but clear connotation and epitaxial broad; it's also a familiar vocabulary for our daily life. From the general sense, harmony means the state in good order and the development mechanism in the virtuous circle of the system or between the systems; from the scientific sense, harmony is a theoretical system. Theory of harmony between human and water is the application of the Theory of Harmony on the relationship between human systems and water systems, and also is an important application field of the Theory of Harmony. As we all know, the contradictions between people and water become increasingly prominent, relations between people and water are threatened by growing crises, and harmony between human and water becomes people's dream! The formation of this situation has many reasons, the reason of irresistible natural causes, furthermore the man – made reason in the social and economic development — due to the destruction of the natural water system caused by people's ignorance and greed. After all, it's caused by the reason that people do not have a clear understanding of and compliance with the law of development of the water system. With the improvement of scientific and technological level, the rules of natural water system and the interaction between human and water system are increasingly clear to people, the ability of human to regulate the relationship between man and water continuously improves, and the worldview of harmony between human and water is basically established. This article firstly compares the harmony degree with entropy; secondly it analyzes the stages of development of human technology and their impacts on the ecological environment, especially the analysis of the characteristics of digital technology; thirdly it studies the operation law of entropy in human systems, virtual systems and water systems; finally it concludes that the contemporary information technology plays an important role in the harmonious development of man and water, and provides an effective solution for the global human water contradictions.

2 Harmony theory paradigm and the entropy criterion

2.1 The proposition of harmony theory paradigm

After years of research, the author (Professor Zuo Qiting) proposed the Harmony Theory System. The main contents include: ①harmony is the action taken to achieve the relationship of "coordination, consistence, balance, perfection and adaptation". Here "consistence" especially means the consistence of harmony goals. Harmony has five connotations: clear participations, specific considerations, hierarchical, dynamic, and extensity; ②theory of Harmony is the theory and method studying the multi – participants' behavior to realize harmony together, it's an important theory revealing the harmonious relations in nature, and it implicates the philosophy of dialectical materialism; ③it contains the five elements: the participants of harmony, harmony goals, harmony rules, factors of harmony and harmonious behavior; ④it gives the quantitative expression and grading of the degree of harmony by Harmony Degree, Harmony Degree is calculated by Harmony Degree Equation. According to the specific problem, Harmony Degree Equation is divided into three scenarios: a single factor, multi – factor and multi – level. Harmony Theory scientifically concluded Harmony as a complete theoretical system for the first time, distinctively features with the five elements and Harmony Degree Equation, formed a new theoretical paradigm.

Appling Theory of Harmony to the research of the relationship between human and water produced Theory of Harmony between Human and Water. Harmony between human and water refers to states in virtuous cycle of human systems coordinating with water systems, that is to say, with the premise of continually improving the water system's ability of self – sustaining and updating, which can make water resources provide age – long support and protection for human survival and sustainable development of economic and social. Based on Theory of Harmony Between Human And Water, the calculation of Harmony Degree of the watershed scale and regional scale, the optimal allocation model of water resources and the model of cross – river watershed Theory of Harmony were studied, and achieved fruitful results.

2.2 Entropy Theory

The concept of entropy derived from thermodynamics. In 1850, Clausius gave the initial description of the second law of thermodynamics in the study of refrigeration: It's impossible to take heat from the low – temperature objects without causing other changes. The following year, Kelvin proposed a second description completely equivalent to the second law of thermodynamics: It's impossible to remove the heat from a single heat source and change it fully into power without other effects. In 1865, Clausius officially proposed a new physical quantity "entropy", that is "hot temperature entropy" $S(dS \geqslant \frac{\delta Q}{T}, \delta Q$ is the heat absorbed by the system, T is the temperature of heat source, among it the equal sign corresponds to a reversible process, and the greater – than sign corresponds to the irreversible process). As a measure of the irreversible degree of thermodynamic process , scilicet a measure of energy degradation, it is a state parameter can reflect the disorder of the state, its increment can determine the directionality of the process. The well – known Principle of Entropy Increase can be drawn by the definition of entropy: the entropy of an isolated system or adiabatic process never reduce, it profoundly portrays the evolution direction of the isolated system, reveals the irreversibility of natural processes, or the asymmetry of natural processes for the time direction.

In 1877, based on probability theory and starting from the microscopic point of view, Boltzmann expanded the concept of entropy and proposed the so – called Boltzmann entropy ($S = k \ln \Omega, k$ is the Boltzmann constant, Ω is the number of microstate), revealed the statistical and the micOroscopic significance of entropy. Entropy is the measure of the number of microstates that the system may have, and it gives a measure of the degree of disorder in the system from the microscopic. The more the number of microstates in the system, the richer diversity the system internal motion is, the more chaotic, the greater the entropy; on the contrary, the less the microscopic state of the system, the simpler the system internal motion is, the more orderly, the smaller the entropy. Boltzmann entropy can be applied to the equilibrium and non – equilibrium, have more general significance. It contains the hot temperature entropy, as it were, Clausius entropy is the maximum value of the Boltzmann entropy.

The concept of entropy profoundly reveals the nature of the evolution of things, once it was put forth, it was quickly generalized, and new concepts continue to emerge.

In 1944, Erwin Schrodinger, the famous physicist hold that since Ω is a measure of disorder, then its reciprocal $1/\Omega$ can be used as a direct measure of order, therefore, he put forward the concept of negative entropy ($-S = k\ln(1/\Omega)$, Letters meaning as above). The introduction of negative entropy has extreme revolutionary meaning. The increasing of positive entropy means things are developing to the direction of disorder, and is a sign of degradation; the increasing of negative entropy means things are developing to the direction of order, and is a sign of evolution. Negative entropy is also a state function, is the cumulative result of the negative entropy flow. It can explain the life world, human society and the whole universe, filled the unbridgeable gap between Darwin's biological evolution and the theory of Clausius.

In 1948, the American mathematician Shannon, founder of information theory, introduced the Boltzmann entropy into information theory, he took entropy as a measure of the uncertainty of a random event, and brought out the concept of information entropy ($S = -k\sum_{i=1}^{N} P_i \ln P_i$, N is the number of signals from signal source, P is the probability for the number i kind of signal, $-\ln P_i$ is the amount of information it brings, k is the proportional coefficient, reflects the information of every signal from each signal source, and is a measure of the size of information amount). Information eliminates the uncertainty of random events, and thus the information is negative entropy. Information laid the foundation for the further generalization of the concept of entropy. In each field of natural and human society, there is a collection of a large number of random events of different levels and different types, each kind of collection of random events has the corresponding uncertainties or the degree of disorder, all these uncertainties and the degree of disorder can be described by using this unified concept of information entropy, so the information entropy is also called as the Pan – entropy or generalized entropy, used to measure the uncertainty or the disorder degree of any material movement. So far, people has brought out various concepts of entropy in many areas of natural and social sciences, the number of these concept is difficult to count, they become the joints and leads in people's scientific research.

2.3 The harmony degree and the entropy criterion

Harmony degree is the indicator expressing the degree of harmony. Corresponding to a factor F^p, the harmony degree (equation) HD_p is defined as: $HD_p = ai - bj$, a is unity degree, b is divergence degree, i is harmonious degree, j is disharmony degree. Its specific meaning and the calculation method are described(Zuo Qiting, 2012). $HD_p \in [0,1]$, The greater its value, the higher the degree of harmony. Harmony degree of the multi – factor HD is calculated out by weighting harmony degree of single factor, example as the weighted average calculation: $HD = \sum w_p HD_p$, w_p is weight, $w_p \in [0,1]$, $\sum w_p = 1$, $HD \in [0,1]$. This shows the concept of harmony degree is different from entropy or negative entropy; the calculation methods are also not the same. In the overall feeling, harmony degree is of stronger subjectivity, while entropy is totally an objective physical quantity; harmony must be orderly, but ordered not necessarily harmonious— ordered (of low entropy value)is just a necessary but not sufficient condition for harmony, and the relationship between these two concepts needs further study. However, from the physical meaning of these two, they both are the functions characterizing the things or states of the system, and are the measure of their ordering. Judging from the effect, the harmony degree is roughly equivalent to negative entropy. The greater the harmony degree, the more orderly the system is, the smaller the entropy value; on the contrary, the smaller the harmony degree, the more disordered the system is, the greater the entropy value. Entropy theory is the first rule of the whole of science; As a new physical quantity to determine the order of the system state, the harmony degree must have a close relationship with the entropy.

3   The philosophy cognition of digital technology

3.1   Analysis of the development phase of technology

The dictionary meaning of technology refers to the experience, knowledge and skills mankind accumulated in the processes of using and transforming nature and reflected in production labor (*Modern Chinese Dictionary*, The Commercial Press, 1996). Technology is a typical category of human society and a embodiment of brain intelligence. Since the birth of mankind, technology can essentially be divided into instinct, simulation and digital these three stages.

The instinct technology stage means the barbarous period accounting for 99% of the human development history, generally including the sprouting stage of the Paleolithic and Neolithic technology. The main feature of this stage is that the technical activities of the human were driven by the instinct of subconscious, mainly shows the low carbon release of individuals' or small groups' bio – energy. We should not look down on this Age of Enlightenment, that is precisely what conceived the origin of human technology and promoted human evolution. At the instinct stage, human is basically an ordinary biological factor in the natural ecosystem. Its influence on the natural world is very limited, belongs to the disturbance within the natural world, and benefits the development of the natural world on the whole.

Stage of simulation technology began from the original community to the middle of last century, roughly includes the Bronze and Iron phases. The main feature of this stage are the gist of the technical activities of the human were like simulating and enlarging the function of various organs of human body except the brain, mainly shows as high – carbon release of fossil fuels' energy. The enlarge of the function of human body by the industrial technology is represented by metal tools, is mainly driven by the high – carbon release of fossil fuels, as promoting the rapid development of human civilization, also has caused a lot of disaster to human and ecological environment. For example, a variety of advanced materials and technologies was used in the manufacture of a variety of weapons, used as the tools by the stronger of conquest, countless of living lives were destroyed by the war, resulting in a lot of tragic man – made disasters; sharp metal tools, seized numerous underground – mineral resources, cut down countless of the original forests; uncontrolled high carbon and high entropy emissions seriously polluted the ecological environment, resulting in a lot of global issues, caused a lot of shocking holocausts. The direct evil ecological consequence of industrial revolution is that has disrupted the rhythm of natural evolution, and seriously damaged the environment, thereby affected the survival of mankind itself.

In 1946, the world's first electronic computer ENIAC come out and proclaimed the birth of a new era, the era of information (or silicon utensil) entered the stage of human history. Computer technology represents the direction of the development of new technologies, for half a century, computer experienced 5 to 6 generations of development, its performance is more and more excellent, while its prices are getting and cheaper; its application is wider and wider, and it has further spread into every walk of life, each family and individual. By taking computer technology as the core, rapid development was driven in the technology fields such as the network, communications, space, 3S(GIS, GPS, PS), and precision manufacturing, automatic control and so on, formed the mainstream of modern information technology (digital technology). As described below, the development of contemporary digital technology provides new methods and ideas to integrate and resolve the contradictions of people and nature!

3. 2   The characteristics of digital technology and its filter entropy mechanism

Comparing with instinct and simulation technology, or silicon comparing with the stone and metal tools, digital technology has many unique features. Firstly, the nature of technology realized a breakthrough. The digital technology (silicon utensil) get rid of the mainly operation of the use of energy and the analog amplification of the human body's functions in the stage of instinct and simulation techniques (stone, bronze, iron), realized the direct simulation and amplification on the intelligence of the human brain, liberated the human thinking, and gave birth to a new way of digital thinking. This amplification (conversion) effect, still has no suitable mathematical law to describe, so it can only be described from philosophy at present. Secondly, the process of technical activities is different. Instinct and analog technologies are mainly corresponding to the matter and energy elements, and its technical activities are corresponding to the movement of material and the release of energy. Though the process also involves information elements, the processing of information is entirely dependent on the natural treatment method of the human brain, and continues the old model millions of years ago; But digital technology is mainly corresponding to the information elements, and then to the substance and energy elements. Its technical activities are firstly the encoding of information and computing movement, and then involving the release of material movement and energy. It can be seen as the technology established on the level of instinct and simulation techniques. Thirdly, the space where the technical activities run is different. Digital technology expands human's living space—virtual space, in the context of digital technology, the human living space is divided into three categories: real space, thinking space and virtual space. The space where matter and energy elements run can only be a real physical space, namely the natural environment. The consequences and impact of its technical activities, only the natural environment to bear, accumulate and digest; the space where information runs is a virtual space, the consequences and impact of the technology activities simulated by information movement is to the virtual environment bear, digest and absorb, but no impact on the natural environment. Virtual space can use the network technology to unite the wisdom of the world's best experts, the program through feedback and interaction of virtual practice and thinking space of human brain, repeatedly refined and optimized, is ultimately implemented in real space, namely to filter out the majority of positive entropy generated with the technical activities from a virtual space, make the impact of human activities on natural reach the minimum, even reach to the purpose of inputting the negative entropy to the nature, in order to effectively protect the environment.

The above characteristics of the digital technology essentially come down to its virtual. It abstracts all the information of human beings and their living space, including tangible and intangible, into numbers, and uses computer to code, edit, store, transmit and process these digital information, especially in the virtual space makes visualized and intelligent modeling, analysis and application of these information. It turns the colorful real world into the vivid virtual world, in order to achieve the modeling, inference and analysis of true geographic process. Here the virtual space is the technology space built by the people based on the information technology and the objective laws, the built of virtual reality model has the mathematical foundations of measurement science and a true reflection of the objective geography and physics laws of the real world (as distinct from the game space and the pure digital art space). The virtual space can be seen and "touched", but it does not have the quality, not comply with the time and space law of real world, people can modify and deal with it according to their own wills, and it can be called as a virtual testing ground. We know that the digital information comes from the objective world, the digital information contains the objective laws; the virtual space reproduces the objective world, and the virtual practice reflects the objective laws. Thus, the virtual practice with objectivity is also a criterion for truth testing. In this way, while digital technology virtualized the objective world, it also turned entropy! That is to make the possible negative effects of the realistic geographical processes virtualized into the negative effects of the virtual world, and then through the interaction of the virtual space and the human brain's thinking space, remove the negative effects from the virtual space, that is to say, filter out the entropy. Ultimately implement in the real world through optimizing the technical solution, we can achieve the purpose of inputting to the nature micro – entropy and negative entropy. In short, the virtual space has the ability to filter "positive entropy", that is to say virtual practice can make the "double – edged sword" technology "anti – blade passivation". This is the unique core value of the digital technology!

4　Analysis of entropy evolution mechanism of the cycle between human and water

Entropy and the degree of harmony between human and water are state functions, are the quality evaluation of the status or history of the man and water system, only about results, not involved in the process of motion of the system. In fact, each result has its particular specific process. This section analyses from the highly abstract point of view, the entropy operation and evolution mechanism of the entire cycle process from the natural water systems, to the thinking space, technology space, and then to the human system, and finally back to the natural water system, thus reveals the critical role of the digital technology in the harmony between people and water.

At the stages of instinct and analog technology, the cycle of the interaction between man and water is shown in Fig. 1. It is based on the understanding of human brain to the natural water system, through the technical tools, introducing water resources to human society, and then draining out wastes back to the natural water system. Negative entropy is from the natural water system to the humane system, and positive entropy is from the humane system to nature, this is a vicious cycle of one direction order flow.

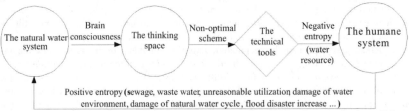

Fig. 1　Vicious cycle of man and water systems under the background of traditional technology (disharmony)

At the stage of digital technology, the cycle of the interaction between man and water systems is shown in Fig. 2. Different from Fig. 1, the circle is a ring of big and small nesting cycles. Here the small cycle is the "entropy filter" mechanism connected by virtual practice between the thinking space and the virtual space. The virtual space here is the digital virtual water space set up by using hydrological information technology (such as digital hydraulic project, digital water resource systems, digital hydraulic planning system, the digital watershed, hydraulic project simulation, digital disaster reduction and prevention systems, et al.), this space bears the objective laws such as spatial distribution law of natural water system, geological dynamic law, the laws of the nature cycle, engineering construction law, and the economic law of hydraulic project and water resource and so on. Through virtual practice "entropy filter" gives digital, networked, automated, intelligent and scientific settlement to series of major issues of watershed's planning, construction, management, simulation, monitoring, control, disaster prevention, disaster alleviation and reconstruction (currently hydraulic IT is still far from reaching this effect). In essence it is to filter out "entropy" in the virtual space to the maximization, so as to achieve the goal of inputting "micro – entropy" and "negative entropy" to the natural water system, maintain and enhance the sustainable development of the ecological environment of the water system, and realize harmony between people and water of the entire society. The people and water system in the context of digital technology is a virtuous cycle of two – way order flow and harmony between human and water.

Fig. 2　Virtuous circle of human and water systems under the background of
digital technology（harmony）

5　Conclusions

The development of the entire objective world including nature and human society is a historical process, the contradiction between human and nature is a long – term accumulating result of the of the evolution of humanities and cultural. Problems during the development of cultural must also be resolved in the cultural development. Represented by Digital Earth and based on the development of contemporary digital technology, digital culture has incomparable superiority to the traditional culture, as the nascent the Third Universe（Digital Universe）, it effectively integrates the contradiction between the First Universe（Physical Universe）and the Second Universe（the Thinking Universe or the Spirit Universe）, provides a new means of settlement to curb the deterioration of the planet's ecosystems. Thus, all mankind should vigorously develop green IT, take efforts to practice the scientific development concept, establish the cultural concept of harmony between man and nature. We have reason to believe that the integration and evolution of the ternary universe, will promote Earth's civilization to a more advanced new stage.

References

Zuo Qiting. Harmony Theory: The Theory, Method and Application [M]. Beijing: Science Press, 2012.

Zuo Qiting. Mathematical Description Method and Its Application of Harmony Theory [J]. South – to – North Water Transfer and Water Science & Technology, 2009, 7(4): 129 – 133.

Zuo Qiting. Human – water Harmony Theory: from Idea to Theory System [J]. Water Resources and Hydropower Engineering, 2009, 40(8):25 – 30.

Zuo Qiting, Zhang Yun. Harmon between Human and Water Quantitative Research Methods and Applications [M]. Beijing: China WaterPower Press, 2009.

Zuo Qiting. The Embedded System Dynamic Model Used to Human – water System Modeling [J]. Journal of Natural Resources, 2007, 22(2): 267 – 273.

Shi Rongyan. Several Researches on Entropy [J]. Journal of Jiangsu Institute of Education (Natural Sciences), 2011, 27 (2): 32 – 34, 92.

Yin Shiwei, Guo Qingwei, Li Xinru, et al. From Heat Engine to Heat Death—The Origin and Development of the Entropy [J]. University Chemistry, 2011, 26(1): 85 – 87.

Yuan Juan, Wan Yan, Chu Yixin. Theory and Application of Entropy [J]. Science and Technology of West China, 2011, 10(5): 42 – 44.

Zhang Xian. A Re – reflection over the Concept of Thermodynamic Entropy [J]. Journal of Shaoxing University, 2010, 30(8): 40 – 42, 57.

Bian Fuling. Digital Engineering Principles and Methods (second edition) [M]. Beijing: Surveying and Mapping Press, 2011.

Bian Fuling, Wang Jinxin. Material Space, Thinking Space and Cyberspace—Some Thinkings about Human Living Space [J]. Geomatics and Information Science of Wuhan University, 2003, 28(1): 4 – 8.

Wang Jinxin, Bian Fuling. On Digital Culture and Sustainable Development [J]. China

Population, Resources and Environment, 2004, 14(5): 17 – 20.

Wang Jinxin, Bian Fuling. Some Thinkings about Sustainable Development Based on the Theory of Entropy [J]. Science & technology progress and policy, 2004, (2): 52 – 53.

Wang Jinxin. Philosophy of the Imagination on Digital [J]. Science (Beijing), 2007, (5): 17 – 19.

3-D Numerical Simulation for Thermal Discharge and Residual Chlorine Distribution of Power Plant with Its Impact on Water Security and Water Environment

Wang Jingyu

The State Key Laboratory of Hydro Science and Engineering,
Department of Hydraulic Engineering, Tsinghua University, Beijing, 100084, China

Abstract: Waste heat and residual Chlorine of power plant that directly discharged into surrounding water will inevitably affect local water security and water environment. Compared with coastal power plants, inland power plants have limited water for dilution, and the activities of mankind are more frequent nearby the inland power plants, which increasingly attracting the attention of the people. This paper uses Delft3D to simulate the thermal discharge and residual Chlorine distribution of an inland power plant. The numerical results agree well with the measured data, which indicates that the results of Delft3D can well reflect the transport and diffusion laws of thermal discharge and residual Chlorine distribution under the suitable parameters. Besides, the results of this paper provide a basis for water security and water environment protection.

Key words: inland power plant, thermal discharge, residual Chlorine, water security

1 Introduction

Although considerable progress has been made in energy industry since the founding of People's Republic of China, the total energy is still unable to meet the requirements of current economic and social development due to the low-level average energy resources per capita. Under this background, considering as a clean energy, nuclear power plant has come into being (Zhang and Yan, 2006). With the development of nuclear energy, more and more people begin to concern about the problems of thermal discharge, residual chlorine distribution and radionuclide waste emissions. In general, the nuclear power plant near coastal area will not cause great harm to the nearby seawater, because the sea has a strong dilution capacity. However, compared with coastal power plants, inland power plants have limited water for dilution and have more frequent mankind activities around the river basin, which increasingly attracting the attention of the people for its influence on water security and water environment (Du et al., 2011).

Due to the inevitable distorted scale in physical models (Fang et al., 2008), the research on thermal discharge and residual chlorine distribution are still based on numerical simulation nowadays (Lin et al., 2009; Zhang et al., 2010), especially on 2-D numerical simulation (Yang et al., 2005). Nevertheless, with the continuous improvement of computer performance, rapid progress has been made to simulate hydraulic phenomenon with high accuracy (Zhou, 1995). Fang and Wang (2000) simulated suspended sediment transport using 3-D numerical model. Han et al. (2011) applied a coupled 1-D and 2-D channel network numerical model to the middle reaches of the Yangtze River. For complex boundaries, Fang et al. (2006) developed a diagonal Cartesian method for numerical simulation of flow and suspended sediment transport. At the same time, the water ecological problems brought by thermal discharge and residual chlorine distribution have become a hot issue around the world. Foreign researchers have begun to study the environmental problems caused by cooling water since mid-20th century (Harleman and Hall, 1968; Hamrick, 2000; Poornima et al., 2001) while Chinese researchers began these studied since the latter half of the 20th century (Zhou, 2006; He et al., 2008).

This paper is based on a 3-D hydrodynamic-water quality simulation system which was developed by Delft Hydraulics in Netherlands (Delft Hydraulics, 2005). By simulating an inland nuclear power plant, the 3-D temperature rise distribution and residual Chlorine distribution are obtained under different conditions, which not only provide a reference for engineering construction but also provide a basis for water security and water environment protection.

2　Numerical model

2.1　Hydrodynamics model

The 3-D hydrodynamics model consists of an orthogonal curvilinear coordinate in horizontal and a σ coordinate in vertical, which are shown as follow:

Continuity Equation:

$$\frac{\partial \zeta}{\partial t} + \frac{1}{\sqrt{G_{\zeta\zeta}}\sqrt{G_{\eta\eta}}}\frac{\partial[(d+\zeta)U\sqrt{G_{\eta\eta}}]}{\partial \zeta} + \frac{1}{\sqrt{G_{\zeta\zeta}}\sqrt{G_{\eta\eta}}}\frac{\partial[(d+\zeta)V\sqrt{G_{\eta\eta}}]}{\partial \eta} = 0 \quad (1)$$

Momentum Equations:

ζ direction:

$$\frac{\partial u}{\partial t} + \frac{u}{\sqrt{G_{\zeta\zeta}}}\frac{\partial u}{\partial \zeta} + \frac{v}{\sqrt{G_{\zeta\zeta}}}\frac{\partial u}{\partial \eta} + \frac{\omega}{d+\zeta}\frac{\partial u}{\partial \sigma} + \frac{uv}{\sqrt{G_{\zeta\zeta}}\sqrt{G_{\eta\eta}}}\frac{\sqrt{G_{\zeta\zeta}}}{\partial \eta} - \frac{v^2}{\sqrt{G_{\zeta\zeta}}\sqrt{G_{\eta\eta}}}\frac{\partial \sqrt{G_{\eta\eta}}}{\partial \zeta} =$$

$$fv - \frac{1}{\rho_0\sqrt{G_{\zeta\zeta}}}P_\zeta + F_\zeta + \frac{1}{(d+\zeta)^2}\frac{\partial}{\partial \sigma}\left[v_v\frac{\partial u}{\partial \sigma}\right] + M_\zeta \quad (2)$$

η direction:

$$\frac{\partial u}{\partial t} + \frac{u}{\sqrt{G_{\zeta\zeta}}}\frac{\partial v}{\partial \zeta} + \frac{v}{\sqrt{G_{\eta\eta}}}\frac{\partial v}{\partial \eta} + \frac{\omega}{d+\zeta}\frac{\partial v}{\partial \sigma} + \frac{uv}{\sqrt{G_{\zeta\zeta}}\sqrt{G_{\eta\eta}}}\partial\frac{\sqrt{G_{\zeta\zeta}}}{\partial \eta} - \frac{u^2}{\sqrt{G_{\zeta\zeta}}\sqrt{G_{\eta\eta}}}\frac{\partial \sqrt{G_{\eta\eta}}}{\partial \zeta} =$$

$$fu - \frac{1}{\rho_0\sqrt{G_{\zeta\zeta}}}P_\eta + F_\eta + \frac{1}{(d+\zeta)^2}\frac{\partial}{\partial \sigma}\left[v_v\frac{\partial v}{\partial \sigma}\right] + M_\eta$$

Vertical direction:

$$\frac{\partial \zeta}{\partial t} + \frac{1}{\sqrt{G_{\zeta\zeta}}\sqrt{G_{\eta\eta}}}\frac{\partial[(d+\zeta)u\sqrt{G_{\zeta\zeta}}]}{\partial \eta} + \frac{1}{\sqrt{G_{\zeta\zeta}}\sqrt{G_{\eta\eta}}}\frac{\partial[(d+\zeta)V\sqrt{G_{\zeta\zeta}}]}{\partial \eta} = 0 =$$

$$H(q_{in} - q_{out}) - \frac{\partial \omega}{\partial \sigma} \quad (3)$$

where, U and V are respectively the average velocities in ζ and η direction, and ω is the velocity in σ direction; σ is the non-dimensional coordinate in vertical direction, $\sigma = 0$ at free surface and $\sigma = -1$ at the bottom; $G_{\zeta\zeta}$ and $G_{\eta\eta}$ are coordinate conversion factors; M_ζ and M_η represent the contributions due to external sources or sinks of momentum; The forces F_ζ and F_η in the momentum equations represent the unbalance of horizontal Reynold's stresses; v_v is the vertical eddy viscosity, obtained by $k - \varepsilon$ model.

2.2　Heat flux model

Heat transport equation is as follows:

$$\frac{\partial[(d+\zeta)T]}{\partial t} + \frac{1}{\sqrt{G_{\zeta\zeta}}\sqrt{G_{\eta\eta}}}\left[\frac{\partial[(d+\zeta)U\sqrt{G_{\eta\eta}}T]}{\partial \zeta} + \frac{\partial[(d+\zeta)V\sqrt{G_{\zeta\zeta}}T]}{\partial \eta}\right] + \frac{\partial(\omega T)}{\partial \sigma} =$$

$$\frac{d+\zeta}{\sqrt{G_{\zeta\zeta}}\sqrt{G_{\eta\eta}}} \times \left[\frac{\partial}{\partial \zeta}\left(D_H\frac{\sqrt{G_{\eta\eta}}}{\sqrt{G_{\zeta\zeta}}}\frac{\partial T}{\partial \zeta}\right) + \frac{\partial}{\partial \eta}\left(D_H\frac{\sqrt{G_{\zeta\zeta}}}{\sqrt{G_{\eta\eta}}}\frac{\partial T}{\partial \zeta}\right)\right] + \frac{1}{d+\zeta}\frac{\partial}{\partial t}\left[D_v\frac{\partial T}{\partial \sigma}\right] + S \quad (4)$$

where, T, T_s and T_{back} are the water temperature, water surface temperature and background temperature, respectively; D_H and D_v are the horizontal and vertical turbulent diffusion coefficients, respectively; S is the source and sink term, $S = (d+\zeta)(q_{in}T_{in} - q_{out}T + Q_{tot})$.

2.3　Residual chlorine transport model

Transport equation of residual Chlorine is:

$$\frac{\partial[(d+\zeta)c]}{\partial t}+\frac{1}{\sqrt{G_{\zeta\zeta}}\sqrt{G_{\eta\eta}}}[\frac{\partial[\sqrt{G_{\eta\eta}}(d+\zeta)uc]}{\partial\zeta}+\frac{\partial[\sqrt{G_{\zeta\zeta}}(d+\zeta)vc]}{\partial\eta}]+K_d(d+\zeta)c=$$

$$\frac{d+\zeta}{\sqrt{G_{\zeta\zeta}}\sqrt{G_{\eta\eta}}}[\frac{\partial}{\partial\zeta}(D_H\frac{\sqrt{G_{\eta\eta}}}{\sqrt{G_{\zeta\zeta}}}\frac{\partial c}{\partial\zeta})+\frac{\partial}{\partial\eta}(D_H\frac{\sqrt{G_{\zeta\zeta}}}{\sqrt{G_{\eta\eta}}}\frac{\partial c}{\partial\zeta})]+\frac{1}{d+\zeta\partial\sigma}[D_v\frac{\partial c}{\partial\sigma}]+S-\frac{\partial(\omega c)}{\partial\sigma}\quad(5)$$

where, c is the concentration of residual Chlorine; K_d is the decay constant; S is the source and sink term, $S=(d+\zeta)(q_{in}c_{in}-q_{out}c)$.

3 Model validation and results

3.1 Validation

The water level and velocity are validated by the observed hydrology data. Fig. 1 shows the simulation range with four synchronization monitoring sections (TB2, TB3, TB4, TB5).

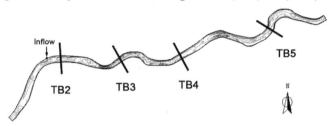

Fig. 1　Range for simulation

Fig. 2 compares the simulated and measured water level under different discharges. The water level of the downstream section TB5 maintains at about 131.8 m, regardless of the quantity of discharge. It is because there is a reservoir at downstream, which controls the water level of TB5. On the contrary, the upstream section TB2 is in the backwater zone of the reservoir, so the water level of TB2 varies a lot. When the discharge is around 200 m³/s, the water level of TB2 is about 132 m. When the discharge increases to 1,000 m³/s, the water level of TB2 rises to 133 m. When the discharge is above 2,000 m³/s, the water level of TB2 is close to 135 m.

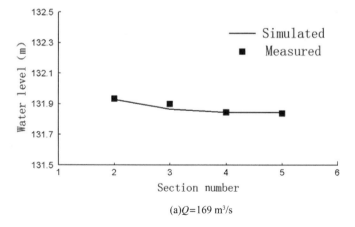

(a)Q=169 m³/s

Fig. 2　Validation of water level

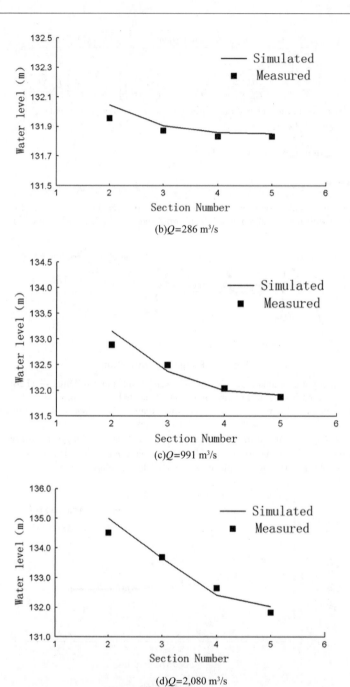

(b)Q=286 m³/s

(c)Q=991 m³/s

(d)Q=2,080 m³/s

Continued to Fig. 2

Fig. 3 shows the validation of velocity with the most unfavorable discharge $Q = 169 \ \text{m}^3/\text{s}$. Because the upstream is in the backwater zone of reservoir, the velocity distribution of section TB2 and TB3 is similar to the river section, behaved as the velocity in the central main channel is greater than the riverside. Whereas, the velocity distribution of downstream section TB4 and TB5 is close to linear distribution, showing more reservoir characteristics.

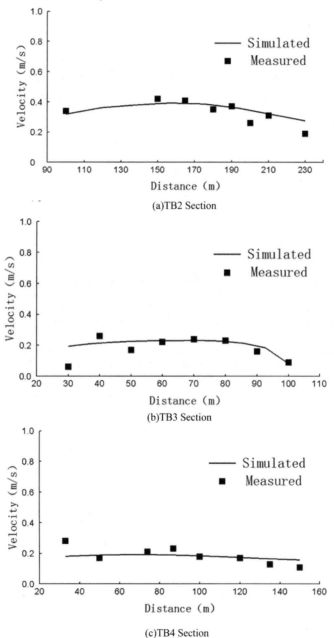

(a)TB2 Section

(b)TB3 Section

(c)TB4 Section

Fig. 3   Validation of velocity

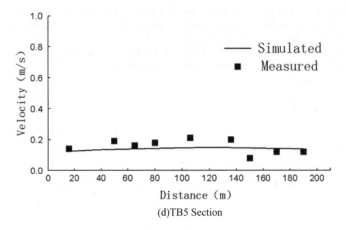

(d)TB5 Section

Continued to Fig. 3

Overall, the simulated results agree well with the observed data, which reflects that the model can simulate the water level and velocity well.

3.2 Results of thermal discharge

This paper takes the most unfavorable case (a low discharge with a return period of 33 years) as an example, considering both the first phase with 2 units and second phase with 4 units. In horizontal, as shown in Fig. 4, the thermal discharge is gradually diluted and diffuses to the downstream driven by the upstream flow. In the case of first phase, the area of range with the temperature rise greater than 1 ℃ is less than 0.01 km^2, and mainly concentrated around the outfall. However, in the case of second phase, the area with the temperature rise greater than 1 ℃ is about 0.07 km^2, and reaches the other side of the river. At the same time, in the most unfavorable case, the distance between the emission point and the depth-average 2 ℃ temperature rise contour is less than the migration distance of water in 6 h. As a result, the thermal discharge meets the environmental protection requirements.

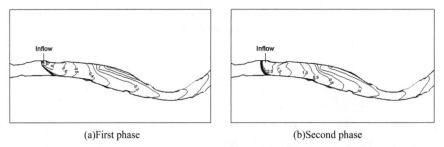

(a)First phase   (b)Second phase

Fig. 4 Distribution of thermal discharge for first phase and second phase in the upper layer of water

Fig. 5 shows the temperature rise of the section which is 0.5 km downstream the outfall. In the case of low discharge with a return period of 33 years, because the velocity is low, the cooling water gradually spread to the other side of the river, resulting in similar temperature rise of two banks. In the case of first phase, the temperature rise of the upper layer is about 0.3 ℃ higher than the lower layer of the water. Likewise, in the case of second phase, the temperature rise of the upper layer is about 0.6 ℃ higher than the lower layer of the water.

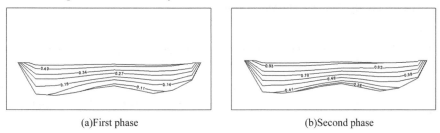

(a)First phase   (b)Second phase

Fig. 5   Distribution of thermal discharge in the section of 0.5 km downstream the outfall

3.3   Results of residual Chlorine distribution

Relative concentration is used to calculate the residual Chlorine, which means the concentration at outfall keeps 1. Similar to the thermal discharge, residual Chlorine gradually moves to the downstream driven by the upstream flow. In the process of moving to the downstream, the concentration of residual Chlorine gradually decreases, accompanied with dilution and degradation. Fig. 6 shows the residual Chlorine distribution in horizontal under the most unfavorable case.

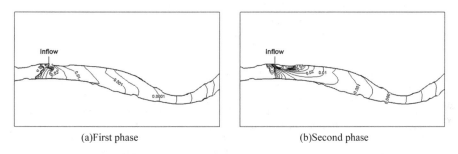

(a)First phase   (b)Second phase

Fig. 6   Residual Chlorine distribution in the upper layer of water

4   The impact on water ecology

4.1   The impact of thermal discharge on fish

Generally, fish will determine the migration time based on water temperature. Once the cooling water is discharged into its habitat, fish will misjudge the water temperature, thereby affecting the entire water ecology (Sun et al. , 2008). Domestic researchers (Jin, 1993; Sheng, 1994) have done some researches on fishery ecology. The results show that the temperature rise caused by thermal discharge has great influence on water ecology. However, it is difficult to judge the influence of thermal discharge on aquatic organisms (e. g. fish) currently. Jiang (2000) study the temperature standard of thermal discharge in fishery waters through thermal shock experiment, and it comes to the conclusion that the highest temperature of thermal discharge in summer in the Pearl River and Zhanjiang should be 36 ℃, in the Yangtze River, the Yellow River and the Qiantang River should be 35 ℃, in the Heilongjiang and the Songhuajiang should be 26 ℃, in northwest waters should be 21 ℃.

4.2 The impact of residual Chlorine on fish

The concentration of chloride ion will increase after the residual Chlorine being discharged into the waters. However, studies found that Chloride ion may damage the gills of fish and cause some lesions like epithelial tissue detachment and tissue hyperplasia, which will affect the breathing of the fish in the water (Cohen and Valenzuela, 1977; Mitz and Giesy, 1985). Certainly, fish can generate some resistance to chlorine by its own immune system. Lotts and Steward (1995) found that minnows can acclimate to residual chlorine with a concentration between 0.04 mg/L and 0.08 mg/L.

5 Conclusions

This paper applies the large-scale estuarine and coastal model Delft3D to simulate the thermal discharge and residual Chlorine distribution of one inland nuclear power plant and uses the coordinate to deal with the 3-D characteristics in the vertical direction. Model validation results show that the simulated values agree well with the observed data and the flow pattern is consistent with the actual prototype situation. In other words, the model is appropriated to simulate the thermal discharge and residual Chlorine distribution.

The cooling water and residual Chlorine discharged by the inland nuclear power plant will have great impact on water security and water environment. But after strict controls of emission standards, the impact of thermal discharge and residual Chlorine on water ecology will be reduced. At the same time, the water ecology has its own immune system to resist the external damage. Therefore, inland nuclear power plants have limited influence on the water security and water ecology unless force majeure events happen.

References

Cohen G M, Valenzuela J M. Gill Damage in the Mosquitofish, Gambusia Affinis, Caused by Chlorine in Fresh Water[J]. Science of Biology Journal, 1977, 3(4): 361 – 371.

Du X L, et al. The Impact of Sediment on Radionuclide Transportation and Deposition Outflow from Inland Nuclear Facilities [R]. Beijing: China Institute for Radiation Protection, Hangzhou: Zhejiang Institute of Hydraulics & Estuary, 2011.

Fang H W, He G J, Liu, J Z, et al. 3D Numerical Investigation of Distorted Scale in Hydraulic Physical Model Experiments[J]. Journal of Coastal Research, 2008, 52: 41 – 53.

Fang H W, Liu B, Huang B B. Diagonal Cartesian Method for Numerical Simulation of Flow and Suspended Sediment Transport over Complex Boundaries [J]. Journal of Hydraulic Engineering – ASCE, 2006, 132(11): 1195 – 1205.

Fang H W, Wang G Q. Three – dimensional Mathematical Model of Suspended Sediment Transport [J]. Journal of Hydraulic Engineering – ASCE, 2000, 126(8): 578 – 592.

Hamrick J V. Analysis of Water Temperatures in Conowingo Pond as Influenced by the Peach Bottom Atomic Power Plant Thermal Discharge[J]. Environmental Science & Policy, 2000, 3 (suppl): 197 – 209.

Han D, Fang H W, Bai J, et al. A Coupled 1 – D and 2 – D Channel Network Mathematical Model Used for Flow Calculations in the Middle Reaches of the Yangtze River[J]. Journal of Hydrodynamics, 2011, 23(4): 521 – 526.

Harleman D R F, Hall L C. Thermal Diffusion of Condenser Water in a River During Steady and Unsteady Flows with Application to the TVA Browns Ferry Nuclear Power Plant [J]. Hydrodynamics Laboratory Report, 1968(3): 98 – 115.

He G J, Zhao H M, Fang H W. 3D Numerical Simulation for Flow and Heat Transport of Power Plant Affected by Tide [J]. Journal of Hydroelectric Engineering, 2008, 27(3): 125 – 136.

Jiang L F. Effects of Thermal Shock on Fishes [J]. Journal of Fishery Sciences of China, 2000, 7 (2): 77 – 81.

Jin L. Waters thermal Introduction [M]. Beijing: Higher Education Press, 1993.

Lin Q S, Jin K, Huang L. Research and Application of Mathematical Modeling of Thermal Effluent from Power Plant [J]. Journal of Yangtze River Scientific Research Institute, 2009, 26(1): 29 – 32.

Lotts J W, Stewart A J. Minnows Can Acclimate to Total Residual Chlorine[J]. Environmental Toxicology and Chemistry, 1995, 14(8): 1365 – 1374.

Mitz S V, Giesy J P. Sewage Effluent Biomonitoring. 1. Survival. Growth, and Histopathological Effects in Channel Catfish [J]. Ecotoxicology and Environmental Safety, 1985, 10 (1): 22 – 39.

Poornima E H, Rajaduraim, Raots, et al. Impact of Thermal Discharge from a Tropical Coastal Power Plant on Phytoplankton[J]. Water Research, 2001, 1 (35): 271 – 283.

Sheng L X, Hou W L, Zhao G, et al. Entrainment Effect of Power Plant Cooling System on Young Fish and Postlarve Shrimp [J]. Acta Scientiae Circumstantiae, 1994, 14(1): 47 – 55.

Sun Y T, Wang H M, Wu X F. Impacts of Thermal Discharge on Aquatic Ecological Environment and the Countermeasures [J]. Water Resources Protection, 2008, 24(2): 70 – 72.

W | L. Delft Hydraulics, Delft3D – Flow User manual. 2005, 12.

Yang F L, Xie Z T, Zhang XF, et al. Simulation of 2 – D Cooling Water in Non – orthogonal Curvilinear Coordinate [J]. Hydro – Science and Engineering, 2005, 2: 36 – 40.

Zhang X B, Liu S H, Cui Z F. 3 – D Numerical Simulation for Residual Chlorine Distribution in Cooling Water Drainage of Heat – engine Plant [J]. Water Resources and Power, 2010, 28 (12): 61 – 64.

Zheng Y H, Yan J M. The Position, Role and Development Prospects of Nuclear Energy in China Energy System[R]. China Nuclear Science and Technology Report, 2006, S1.

Zhou L H. Numerical Simulation and Calculation of the Heated Discharge from Tiansheng Port Power Plant Based on POM Model[D]. Nanjing: Hohai University, 2006.

Zhou X Y. Computational Hydraulics [J]. Beijing: Tsinghua Publishing House, 1995.

Water Temperature Forecast in Winter Using ANFIS in the Yellow River

Wang Tao, *Yang Kailin*, *Guo Xinlei*, *Fu Hui* and *Guo Yongxin*

China Institute of Water Resources and Hydropower Research, Beijing, 100038, China

Abstract: The Inner Mongolia reach of the Yellow River lies on the northern part of the Yellow River Basin. In this reach of the river, looking like a reversed 'U', is easy for ice disaster to appear, such as flooding in the winter months. There is essential to timely and accurately forecast the ice conditions for protecting the urban and rural areas along the basin from being struck by possible ice – flooding disasters. However, in order to conduct ice condition forecasts, the freeze – up water temperature forecast in winter should be conducted at the first place. With its hybrid learning scheme, adaptive – network – based fuzzy inference system (ANFIS), constructed under the framework of the neural networks and fuzzy models, and the latter possess certain advantages over the former two, is convenient for modelling the nonlinear multivariable process; for instance, in the case of modelling the hydrological information variety. Consequently, ANFIS is applied to the freeze – up water temperature forecast in Inner Mongolia reach of the Yellow River including 4 hydrometric stations, i. e the Shizuishan Station, Bayangaole Station, Shanhuhekou Station, and Toudaoguai Station. And the forecast results of water temperature are compared with the measured records in these four hydrometric stations. Through such comparisons, it was discovered that the water temperature forecast results approximately agree with those of the field records. The ANFIS model is used for simulating the freeze – up water temperature forecast process in winter, and satisfactory results have been obtained.

Key words: ANFIS, ice condition, forecast, freeze – up water temperature, the Yellow River

1 Introduction

The Yellow River Basin is known as the cradle of the Chinese civilization, as shown in Fig. 1. However, the Yellow River, located in the north of China, is suffering from ice flooding disasters usually occurring in winter seasons. The Inner Mongolia Reach of the Yellow River, looks like a reversed 'U', suffers more frequently from ice flooding disaster in winter due to its specific geographical conditions, as shown in Fig. 1. Four hydrometric stations, including the Shizuishan Hydrometric Station, Bayangaole Hydrometric Station, Shanhuhekou Hydrometric Station, and Toudaoguai Hydrometric Station, are set up for obtaining hydrological information of the river flow timely. In this region, the averaged air temperature is usually below zero from November to March, and the lowest temperature may reach to $-35\ ^\circ\!C$. Therefore, the ice conditions of this river reach is that the freezing – up process of river flow starts from downstream, and then extends to upstream in this period. In contrast, in spring breaking – up of river ice starts on the reach upstream. The adverse effect of such freezing – up and breaking – up process may easily produce ice jams. There is essential to timely and accurately forecast the ice conditions for protecting the urban and rural areas along the river from being struck by possible ice – flooding disasters (Wang et al. , 2008). It is known that the freezing – up water temperature in winter plays a key role for ice condition. So much effort has been devoted to finding out relationship between freeze – up temperature and other filed hydrological parameters.

Several approaches of ice condition forecast have been developed, such as the empirical indexes model, the correlation analysis, and the mathematical model. Two types of empirical mathematical models have been presented for forecasting the ice conditions, one for the downstream reach of the Yellow River by Chen et al. in 1994 (Ke et al. , 2002), and the other for the upstream reach by Ke et al. in 1998 (Ke et al. , 2002). In recent years, since the Yellow River bed and weather condition have changed significantly, those models are no longer suitable for current ice forecast of the Yellow River. Therefore, Chen and Ji (2004), and Wang et al. (2008) have developed the artificial neutral network (ANN) model for predicating ice conditions of the Yellow River. The ANN approaches have been found useful and flexible in modelling of hydrological process. In the hydrological context, recent studies have reported that ANNs may offer a promising alternative for modeling hydrological variables, e. g. , sediment forecast (Cigizoglu, 2004, 2006; Alp et al. , 2007; Sarangi et al. , 2005), flood forecasting (Dawson et al. , 2006; Sahoo et al. , 2006), water level forecast (Bazartseren, 2003; Chau, 2006), flow forecast (Cigizoglu, 2003a, b, 2005), reservoir inflow prediction (Jain, 1999), rainfall – runoff model (Sudheer et al. , 2002; Riad et al. , 2004). Especially, Massie et al. (2001) applied the ANNs to the development of the elementary study for predicting ice jams.

However, ANN model are sensitive to the selected initial weight values and the networks are sometimes trapped by the local error minima on the training stage. The adaptive – network – based fuzzy inference system (ANFIS) is built under the framework of both neural networks and fuzzy models, and possess certain advantages over Artificial Neural Networks (ANNs) and fuzzy inference systems. So in this study, the adaptive – network – based fuzzy inference system (ANFIS) is tried in forecast of freezing up water flow temperature of the Yellow River.

The objective of this study is to develop an improved predictive ANFIS model for estimating freeze – up water temperature. ANFIS is constructed as a learning procedure for the fuzzy inference systems that adopts a neural network learning algorithm for designing a set of fuzzy if – then rules with appropriate membership functions from the specified input – output pairs. ANFIS was first introduced by Jang (1993), Jang and Sun (1996) and Jang et al. (1997) and later widely applied in engineering problems. In the recent years, the ANFIS model has been applied in the estimation of water resources. The evapotranspiration is successfully estimated by way of the ANFIS. Kis, i et al. (2007) by way of employing neurofuzzy models has successfully modeled the evapotranspiration process. Terzi et al. (2006) applied ANFIS to the collection of daily meteorology data from the Lake Eg irdir region in the south – western part of Turkey. Abyaneh et al. (2011) employed ANN and ANFIS to model garlic crop evapotranspiration. The literatures have reported that ANFIS may be used in flood forecasting. Chau et al. (2005) applied ANFIS models for flood forecasting in a channel reach of the Yangtze River in China and the merits and shortcomings were provided on the base of this application. Mukerji et al. (2009) employed ANFIS and ANN model to perform flood forecasting at Jamtara gauging site of the Ajay River Basin in Jharkhand. The results have shown that ANFIS model predicts better than the ANNs model in the cases. Ma and Hu (2008) employed ANFIS to predicting reservoir annual runoff. The research results show that the ANFIS has advantages such as high accuracy and strong generalization ability compared with regression analysis with the BP network and the modified Elman neural network. In addition, Azamathulla and Ghani (2011) described the use of ANFIS in estimating the scour depth at culvert outlets. Even in the hydrological ecology research, Perendeci et al. (2007) proposed ANFIS to estimate effluent chemical oxygen demand of a full – scale anaerobic wastewater treatment plant for a sugar factory operating at unsteady state. However, the application of ANFIS in ice condition modelling is limited in the literatures. Mahabir et al. (2006) once employed the ANFIS to forecast flood caused by ice jams.

The objective of this study is to develop an ANFIS model for predicating freezing – up water temperature. The inputs of the model are the hydrologic records including air temperature, river flow temperature, volumetric discharge, depth and ice conditions, which are sampled by the four stations, listed in Fig. 1 and are provided by the Hydrology Bureau of the Yellow River Conservancy Commission. In this paper, the freezing – up water temperatures in winter are forecasted with the presented approach, and then compared with the filed records.

Fig. 1   Basin of the Yellow River

2   Adaptive – network – based fuzzy inference system

Jang Roger mentions the Adaptive – Network which is based on fuzzy inference system (Jang, 1993; Jang and Sun, 2000). For convenience, it is assumed that the fuzzy inference system at present has only two input variables, x_1 and x_2, and one output variable, y. Suppose that the rule contains two fuzzy if – then rules (Jang, 1993).

Rule 1: if x_1 is A_1 and x_2 is B_1, then

$$f_1 = p_1 x_1 + q_1 x_2 + r_1$$

Rule 2: if x_1 is A_2 and x_2 is B_2, then

$$f_2 = p_2 x_2 + q_2 x_2 + r_2$$

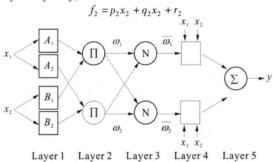

Layer 1   Layer 2   Layer 3   Layer 4   Layer 5

Fig. 2   Configuration of ANFIS model with two – input variables

Type – 3 ANFIS with two input variables is shown in Fig. 2 (Jang J – S R, 1993), which has five layers. $A_i(B_i)$ is the linguistic label (small, large, et al.) associated with this node function. In other words, the input of the first layer variable to node i is the membership function. Usually the membership function includes sigmoid membership function and bell shape function which is with maximum value equal to 1.0 and the minimum one equal to 0. The single node in output layer is a circle node labeled as Σ that computes the overall output as the summation of all incoming signals, which is expressed as follows:

$$O_{5,i} = \sum_i \overline{w}_i f_i = \frac{\sum_i w_i f_i}{\sum_i w_i} \qquad i = 1,2$$

where, $\overline{w}_i$ is expresses as the i th rule's firing strength to the sum of all rules' firing strengths.

$$\overline{w_j} = \frac{w_i}{w_1 + w_2} \qquad i = 1,2$$

$\{p_i, q_i, r_i\}$ is the parameter set. Parameters in this layer will be referred to as consequent parameters.

$$\begin{cases} f_1 = p_1 x_2 + q_1 x_2 + r_1 \\ f_2 = p_2 x_1 + q_2 x_2 + r_2 \end{cases}$$

Therefore, ANFIS provides a way to extract the appropriate information from the data sets (fuzzy rules) of learning. This learning process seems similar to that in neural network. The optimal parameters of membership functions are calculated effectively by learning process. The process can make the fuzzy inference system simulate the actual or desired relationship between input and output variable. Therefore, ANFIS may be considered as a modeling method based on the existing data.

3　Water temperature forecast based on ANFIS

3.1　Evaluation criteria of model performance

According to *Hydrographic Forecast Standards SL250 – 2000* (MWR, China, 2000), the effectiveness assessment for the water temperature is described by the coefficient of determination DC. The performance of the predictions resulting is evaluated by this coefficient. It is written as:

$$DC = 1 - \frac{\sum\limits_{i=1}^{n} [y_c(i) - y_0(i)]^2}{\sum\limits_{i=1}^{n} [y_0(i) - \overline{y}_0]^2} \tag{1}$$

where, DC is the coefficient of determination; $y_0(i)$ is the measured data; $y_c(i)$ is the orecasted data; $\overline{y}_0$ is the mean measured data; subscripts n is total number of sample sequences.

The coefficient of determination DC furnishes a quantitative indication of the model error in units of the variable. The closer the coefficient is to 1, the closer the forecast value and the measured one are.

3.2　Analysis of forecast factors

Many factors may affect the accuracy of the forecast of water temperature. Here the hydraulic factors and thermodynamic factors are considered, such as the air temperature, early water temperature, volumetric discharge, heat exchange between river bed and water, heat exchange between water surface and air, and so on. This relation is expressed by:

$$D_{wt} = f(T_w, T_a, Q, H_l) \tag{2}$$

where, D_{wt} is the forecast water temperature; T_w is water temperature; T_a is air temperature; Q is volumetric discharge; H_l is water level.

T_w, T_a, Q, and H_l include the historical data and real – time ones.

In order to diminish the influence of different dimensions of all the factors during the training and forecasting, and avoid the unit saturating, these factors are standardized as follows:

$$x_i = \frac{z_i - z_{min}}{z_{max} - z_{min}} \alpha + \beta \tag{3}$$

where, x_i and y_i are original and standardized parameters, respectively; z_{max} and z_{min} are the maximum and minimum z_i, respectively; α is a parameter in $[0, 1]$; $\beta = (1 - \alpha)/2$, so ANFIS model input units z_i is in $[0.05, 0.95]$.

Limited by the data collected, the forecast factors only include the average daily volumetric discharge, the average daily water level, and the average daily water temperature in the forecast of the water temperature. Since 1950s, hydrometric data have been collected in the different hydrometric stations of the Yellow River. Considering that the weather conditions are changed in the Yellow River because of the global warming, and the flow conditions are different from those in the past due to constructing many hydraulic projects along the Yellow River, the early hydraulic information could not be used in the current forecast. Therefore, the data collected after the Longyangxia reservoir is stored in 1987 were employed on this forecasts. In the process of the forecast at Shizuishan, Bayangaole, Shanhuhekou and Toudaoguai Hydrometric Station, the measured data for the winter period from 1987 to 1997 are the input ones. The output is the water temperature of the four years from 1998 to 2002 except 2000 in Shizuishan Hydrometric Station and except 1999 in Shizuishan Hydrometric Station because the measured data was incomplete in these years, and from 1999 to 2001 at Shanhuhekou and Toudaoguai Hydrometric Station.

3.3 Water temperature forecast in the Yellow River

For the ANFIS model, it is important to select the type and number of the membership function. First, the type of the membership function is determined. In order to fix on the values of the network output from 0 to 1, sigmoid membership function and bell – shaped membership function are employed for the selection in this study. Under the same condition of the number of the membership function, we compared the forecasted results of the sigmoid membership function with those of the bell – shaped membership function. The errors between them seem very close, but the sigmoid membership function costs much longer computing time than the bell – shaped membership function. So the bell – shaped membership function is selected in this study. The following study is to determine the number of the membership function. When the number of the membership function increases from 2 to 5, the errors decrease accordingly and the computing time increases on the contrary. The computing time for the four membership functions is longer than that for the five, but the two forecast precisions are almost the same. So the four membership functions are applied to the forecast. For the select of the forecast period, when the forecast period increases, the errors are decreasing accordingly. Taking into account the requirement of forecast precision, the longer forecast period is preferred. The four – day forecast period is selected in the paper.

The forecast of the freeze – up water temperature in winter is analyzed under the condition of the bell – shaped and four membership functions and four – day forecast period. The water temperatures are forecasted at Shizuishan, Bayangaole, Shanhuhekou and Toudaoguai four hydrometric stations. In order to diminish the influence of the different dimensions of all the factors during the mode training and forecasting, these factors are standardized according to Eq. (3). The comparison of forecast results and field measured ones is shown in Fig. 3 ~ Fig. 6. The coefficients of determination gained by Eq. (1) between the measured data and the forecast ones are shown in Tab. 1. Project effectiveness assessment for short – term forecast is in Tab. 2 from Hydrographic Forecast Standards SL 250—2000 (MWR, China, 2000). In accordance with Tab. 2, the forecast results are rated. The comments on the results are following.

Shizuishan Hydrometric Stations: With field results shown in Fig. 3, it seems acceptable that the forecast results are in good keeping with the measured values in the three years 1998, 1999, and 2001. As shown in Tab. 1, the value of the calculated coefficient of determination is more than 0.94. Else, according to Hydrographic Forecast Standard SL 250—2000 (MWR, China 2002), the grade of forecast effectiveness is labeled as grade – A, as referred in Tab. 2. The air temperature in the period of 2002 was obviously much higher than that in the same period of other years, and the higher temperature led to two freeze – up and two break – up in the winter of 2002. So it is very difficult to forecast the water temperature under such extreme weather conditions. Therefore, the value of the coefficient of determination is 0.380,2 in 2002, and the measured value and forecasted value deviate severely from November 20, 2002 to December 9, 2002.

Bayangaole Hydrometric Stations: As shown from Fig. 4, the forecast results are in good keeping with the measured values as a whole. The calculated coefficients of determination between the forecasted and measured results are more than 0.95 except 0.87 in 2000. This is because that the correlation between the water temperature and the input factors is a little poor in 2000.

Sanhuhekou Hydrometric Stations: There was no change anomaly of the air temperature in the four year. So the forecast results are in good agreement with the measured values form 1999 to 2002, as shown in Fig. 5. All of the calculated coefficients of determination between the forecasted and measured results are more than 0. 96, as shown in Tab. 3. The curve of the forecast results are in good keeping with that of the measured ones.

Toudaoguai Hydrometric Stations: The calculated coefficients of determination between the forecasted and measured results are in the range from 0. 98 to 0. 94 in 1999 ~ 2001, as shown in Tab. 1. This means that the forecast results are in good keeping with the measured values in this three years. In 2002, the water temperature increases gradually in early November, as shown in Fig. 6, which is contrary to the change trend of the multi – year average water temperature. So the coefficient of determination is 0. 55.

In summary, ANFIS can find out the change rules though studying historical data in order to forecast the future condition. When the changes in water temperature are stable, gradual, and no abnormal, the forecast results of the water temperature are in good keeping with the measured values, and all of the coefficients of determination are more than 0. 90. However, it is difficult to improve the forecast accuracy once the change anomaly of the weather happens.

Tab. 1   Coefficient of determination of the forecast water temperature

Hydrometric stations	Year				
	1998	1999	2000	2001	2002
Shizuishan	0. 951 ,9	0. 943 ,6	—	0. 966 ,8	0. 380 ,2
Bayangaole	0. 966 ,2	—	0. 871 ,6	0. 960 ,0	0. 950 ,1
Sanhuhekou	—	0. 992 ,9	0. 981 ,8	0. 982 ,3	0. 985 ,7
Toudaoguai	—	0. 984 ,7	0. 958 ,2	0. 938 ,7	0. 552 ,5

Tab. 2   Project effectiveness assessment for short – term forecast in hydrographic forecast standard

Grade of effectiveness	Grade – A	Grade – B	Grade – C
Coefficient of determination	> 0. 90	0. 70 ~ 0. 90	0. 50 ~ 0. 69

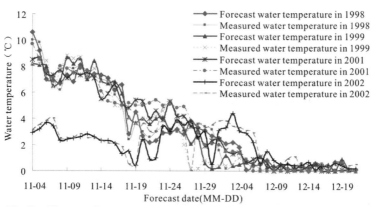

Fig. 3   Change of water temperature at Shizuishan hydrometric station from 1998 to 2002

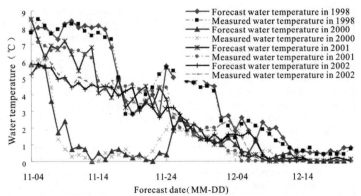

Fig. 4   Change of water temperature at Bayangaole hydrometric station
from 1998 to 2002

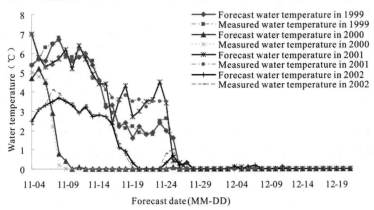

Fig. 5   Change of water temperature at Sanhuhekou hydrometric station
from 1999 to 2002

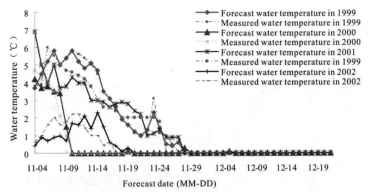

Fig. 6   Change of water temperature at Toudaoguai hydrometric station
from 1999 to 2002

4 Conclusions

The ANFIS has the advantages of the fuzzy theory and the artificial neural networks, and overcomes the shortage of both systems. In this study, ANFIS is employed to forecast the freeze – up water temperature in winter. Though the ANFIS model self – study, the change rules between the water temperature and the input factors are found out as a basis on forecast further water temperature. The forecast of freeze – up water temperature in the four hydrometric stations, Shizuishan, Bayangaole, Sanhuhekou and Toudaoguai Hydrometric Stations, is performed as an application for the ice forecast in the Yellow River in China. The evaluation of the forecast results is made in terms of the coefficient of determination between the measured and simulated data. The coefficients of determination of the 13 cases in the 16 forecast ones are more than 0.90 (seen in Tab. 1), and the rate of forecast effectiveness is grade – A, as referred in Tab. 2. The coefficient of determination for the 1 case is 0.87, which is grade – B. The coefficients of determination of the 2 cases are less than 0.70, which is grade – C. As a result, the forecast cases of A – grade, B – grad, C – grad are 81%, 6%, 13%, respectively. In conclusion, the ANFIS model can be acceptable in water temperature in the Yellow River. And the research work is as an integral part of the ice forecast of in the Yellow River. The forecast freeze – up water temperature is as a basis for the ice process forecast. Further, ANFIS models should be introduced into predicting other ice conditions based on the present study.

Acknowledgements

The authors would like to express whole – hearted thanks to the Hydrology Bureau of the Yellow River Conservancy Commission for the hydrological information used in this study and to National Nature Science Foundation of China under Grant No. 51179209 and No. 50609031 for the financial support it has provided.

References

Abyaneh H Z, Nia A M, Varkeshi M B et al. Performance Evaluation of ANN and ANFIS Models for Estimating Garlic Crop Evapotranspiration [J]. Journal of Irrigation and Drainage Engineering, 2011,137(5): 280 – 286.

Alp M, Cigizoglu H K. Suspended Sediment Load Simulation by Two Artificial Neural Network Methods Using Hydrometeorological Data[J]. Environmental Modeling & Software, 2007,22(1): 2 – 13.

Azamathulla H M, Ghani A A. ANFIS – based Approach for Predicting the Scour Depth at Culvert Outlets[J]. Journal of Pipeling Systems Engineering and Practice, 2011,2(1): 35 – 40.

Bazartseren B, Hidebrandt G, Holz K P. Short – term Water Level Prediction Using Neural Networks and Neuro – fuzzy Approach[J]. Neurocomputing, 2003,55(3 – 4):439 – 450.

Chau K W, Wu C L, Li Y S. Comparison of Several Flood Forecasting Models in Yangtze River, Journal of Hydrologic[J]. Journal of Hydraulic Engineering, 2005,10(6): 485 – 491.

Chau K W. Particle Swarm Optimization Training Algorithm for ANNs in Stage Prediction of Shing Mun River[J]. Journal of Hydrology, 2006,329(3 – 4): 363 – 367.

Chen S Y, Ji H L. Fuzzy Optimization Neural Network BP Approach for Ice Forecast (in Chinese) [J]. Journal of Hydraulic Engineering, 2004,36(6): 114 – 118.

Cigizoglu H K. Incorporation of ARMA Models into Flow Forecasting by Artificial Neural Networks [J]. Envirometrics,2003, 14(4): 417 – 427.

Cigizoglu H K. Estimation, Forecast and Extrapolation of Flow Data by Artificial Neural Networks [J]. Hydrological Science Journal, 2003,48(3): 349 – 361.

Cigizoglu H K. Estimation and Forecasting of Daily Suspended Sediment Data by Multi Layer Pperceptions[J]. Advances in Water Research, 2004,27: 185 – 195.

Cigizoglu H K. Application of the Generalized Regression Neural Networks to Intermittent Flow Forecasting and Estimation[J]. Journal of Hydrologic Engineering,2005, 10(4):336 – 341.

Cigizoglu H K, Alp M. Generalized Regression Neural Network in Modeling River Sediment Yield [J]. Advances in Engineering Software,2006, 37: 63 – 68.

Cigizoglu H K, Kisi O. Methods to Improve the Neural Network Performance in Suspended Sediment Estimation[J]. Journal of Hydrology, 2006,317(3 – 4): 221 – 238.

Dawson C W, Abrahart R J, Shamseldin A Y, et al. Flood Estimation at Ungauged Sites Using Artificial Neural Networks[J]. Journal of Hydrology, 2006,319(1 – 4): 391 – 409.

Ke S J, Wang M, Rao S Q. Yellow River Ice Conditions [M]. Zhengzhou: Yellow River Conserrancy Press,2002.

Kis, i ö, ö ztürk ö. Adaptive Neurofuzzy Computing Technique for Evapotranspiration Estimation [J]. Journal of Irrigation and Drainage Engineering, 2007,133(4): 368 – 379.

Jain S K, Das D, Srivastava D K. Application of ANN for Reservoir Inflow Predication and Operation[J]. Journal of Water Resources Planning and Management, 1999,125(5): 263 – 271.

Jang J S R. ANFIS: Adaptive – network – based Fuzzy Inference System[J]. IEEE Transactions on System, Man and Cybernetics, Man and Cybernetics, 1993,23(3): 665 – 685.

Jang J S R. Input Selection for ANFIS Learning[C]// the Fifth IEEE International Conference on Fuzzy Systems, 1996,2:1493 – 1499.

Jang J S R, Sun C T, Mizutani E. Neurofuzzy and Soft Computing: A Computational Approach to Learning and Machine Intelligence [M]. Nanjing: Nanjing Prentice – Hall, Upper Saddle River, 1997.

Ma X X, Hu T C. Reservoir Annual Runoff Forecast Based on ANFIS [J]. Journal of Hydroelectric Engineering, 2008,27(5): 33 – 37.

Mahabir C, Hicks F E, Fayek A R. Transferability of a Neuro – fuzzy River Ice Jam Flood Forecasting Model[J]. Cold Regions Science and Technology,2006, 48: 188 – 201.

Massie D D, White K D, Daly S F. Proceedings of Ice Jam with Neural Networks [C] // Proceeding of the 11th Workshop on River Ice. Ottawa: Canadian Geophysical Union, 2001, 209 – 216.

Mukerji A, Chatterjee C, Raghuwanshi N S. Flood Forecasting Using ANN, Neuro – Fuzzy[J]. Journal of Hydrologic Engineering,2009, 14(6): 649 – 652.

Perendeci A, Arslan S, Tanyolac A. Evaluation of Input Variables in Adaptive – network – bases Fuzzy Inference System Modeling for Anaerobic Wastewater Treatment Plan Under Unsteady State[J]. Journal of Environmental Engineering, 2007,133(7):765 – 771.

Riad S, Mania J, Bouchaou L, et al. Rainfall – runoff Model Using an Artificial Neural Networks Approach[J]. Mathematical and Computer Modeling, 2004,40(7 – 8): 839 – 846.

Sarangi A, Bhattacharya A K. Comparison of Artificial Neural Network and Regression Model for Sediment Loss Prediction from Banha Watershed in India [J]. Agricultural Water Management, 2005,78(3): 195 – 208.

Sahoo G B, Ray C, De Carlo E H. Use of Neural Network to Predict Flash Flood and Attendant Water Qualities of a Mountainous Stream on Oahu, Hawaii[J]. Journal of Hydrology, 2005, 327(3 – 4): 525 – 538.

Sudheer K P, Gosain A K, Ramasastri K S. A Data – driver Algorithm for Constructing Artificial Neural Network Rainfall – runoff Models [J]. Hydrological Processes, 2002, 16: 1325 – 1330.

Terzi O, Keskin M E, Taylan E D. Estimation Evaporation Using ANFIS[J]. Journal of Irrigation and Drainage Engineering, 2006,132(5): 503 – 507.

The Ministry of Water Resources of the People's Republic of China. SL 250—2000 Hydrographic Forecast Standard [S]. Beijing:China Waterpower Press, 24.

Wang T, Yang K L, Guo Y X. Application of Artificial Neural Networks to Forecasting Ice Conditions of the Yellow River in the Inner Mongolia Reach [J]. Journal of Hydrologic Engineering, 2008,13(9): 811 – 816.

Application of GR4J Rainfall-runoff Model to Typical Catchments in the Yellow River Basin

Yan Xiaolin [1,2], *Zhang Jianyun* [1,2], *Wang Guoqing* [1,2],
Bao Zhenxin [1,2], *Liu Cuishan* [1,2] and *Xuan Yunqing* [3]

1. State Key Laboratory of Hydrology-Water Resources and Hydraulic Engineering,
   Nanjing Hydraulic Research Institute, Nanjing, 210029, China
2. Research Center for Climate Change, MWR, Nanjing, 210029, China
3. UNESCO-IHE Institute for Water Education, Delft, Netherlands

Abstract: Rainfall-runoff model is an important tool for streamflow simulation. The Génie Rural à 4 paramètres Journalier (GR4J) model is a daily lumped rainfall-runoff model with four free parameters, X_1, X_2, X_3, and X_4. For the parsimonious and robust characteristics, GR4J is widely used in numerous catchments over the world. The hydrological modeling remains to be a challenge in the Yellow River basin (YRB), the mother river of China. The applicability of the GR4J model to YRB is investigated by the test of the GR4J model in four typical catchments in the YRB, and parameter sensitivity analysis using the Monte Carlo methodology. The results show that: ① the GR4J model has a good performance in the YRB, except the Loess Plateau. ② The model parameter X_1 is more sensitive in the Loess Plateau; X_2 appears to be much more sensitive at Taohe River basin located in the upper YRB; X_3 and X_4 are the most sensitive and insensitive parameters for the streamflow simulation in the YRB, respectively.

Key words: GR4J model, streamflow simulation, parameter sensitivity analysis, the Yellow River Basin

1 Introduction

Streamflow is one of the key components of water cycle and its simulation is a widely focused topic, which is always realized by Rainfall-runoff models. The GR4J rainfall-runoff model is a daily lumped watershed model with four free parameters (Edijatno, et al., 1999; Perrin, 2000). It develops from GR3J model originally proposed by Edijatno and Michel (1989) and improved by Nascimento (1995) and Edijatno et al. (1999). Both of the two models are belong to lumped soil moisture accounting (SMA) models. However, GR4J performs better in low-flow simulation compared to GR3J (Perrin, et al., 2003). In spite of simplicity of model structure, the model proves to be applicable and robust when rainfall-runoff simulating, which is tested in hundreds of catchments (Perrin, et al., 2003). What's more, Thomas Pagano et al. (2011) and the eWater Cooperative Research Centre of Australia (Dutta, et al., 2011) investigate that the performance of the GR4J model is rather good compared by AWBM (Australia Water Balance Model), SIMHYD model, Sacramento etc. Follow-up research continues. Perrin et al. (2008) introduce a so-called discrete parameterization method for the application to poorly gauged catchments. And the GR4J (CEMAGREF) model is added a snow module based on the method of the degree-days. As a simple, robust tool with a low density of parameters, it has been used to modeling more than 1, 000 catchments in France (Le Moine, et al., 2007) and hundreds of catchments form Australia, Brazil, US etc. In a word, GR4J is widely used in Europe and around the world for flood forecasting, water management planning and simulating impacts of climate change (Liersch and Volk, 2007; Harlan, et al., 2010; Simonneaux, et al., 2008).

Though the GR4J model is widely applied, there are few references and almost no applications to China. In this study, YRB is selected. The Yellow River, the second longest river in China, origins from Bayan Har Mountains, wanders east via the Loess Plateau, the

Huang-Huai-Hai Plain, and last pours into Bohai Sea. Although YRB covers 8.3% of the total area of China, the water resources of it accounts for only 2.6% of the total of China. The amount of water resources per capita in the Yellow River basin is only 1/3 of the average in China. YRB endures a severe water shortage problem. In the last decades, with the development of social development and the impact of climate change, the water crisis in YRB is becoming more critical now (Liu and Zheng, 2003).

Since it known as simple and robust, the main objectives of this study include: 1) estimating the GR4J model whether suitable for hydrological modeling in YRB or not, by applying the GR4J model to four typical catchments in the YRB; 2) with sensitivity analysis method, finding the most sensitive parameter of the GR4J for the references to model calibration and uncertainty analysis.

2　Study area and methodology

2.1　Study area and dataset

YRB, located in $96°E \sim 119°E$, $32°N \sim 42°N$, covers an area of 7.95×10^9 km^2. YRB is divided into three parts, upper reach, middle reach and lower reach. And the respective separations are Hekou town, and Taohuayu villiage. According to China Meteorological Administration, in YRB there are three climate zones, south temperate zone, mid-temperate zone and plateau climate zone. In this study, four typical catchments (Fig. 1) are selected and their hydrological control stations are Tangnaihai station, Hongqi station, Huangfu station and Baimasi station. YRB controlled by Tangnaihai station and Taohe River Basin lie in the upper Yellow River basin, and the former falls into the plateau climate zone. The Huangfuchuan River basin and Luohe River basin, both located in the middle Yellow River Basin, belong to the mid-temperate zone and south temperate zone respectively. The basic information is shown in Tab. 1.

The meteorological data in 17 stations with daily precipitation and pan evaporation from 1951 to 2008 are collected from China Meteorological Administration. Then the area precipitation and potential evapotranspiration are obtained by using Thiessen polygon method. The monthly streamflow data at the four hydrological stations are extracted from the "China's Hydrological Year Book", published by the Hydrological Bureau of the Ministry of Water Resources, China.

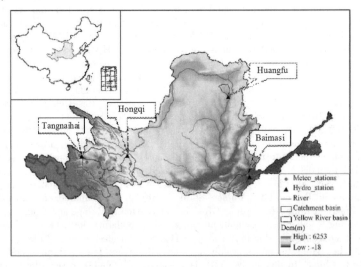

Fig. 1　Four selected typical catchments of Yellow River Basin

Tab. 1   The basic information of the four hydrological stations

Catchment	River	Lon. (°E)	Lat. (°N)	Area(km²)	P (mm)	Available data
Tangnaihai	Yellow River	100. 15	35. 50	121,972	500. 8	1956 ~ 1976
Hongqi	Taohe River	103. 57	35. 80	24,973	561. 2	1955 ~ 1975
Huangfu	Huangfuchuan River	111. 08	39. 28	3,175	394. 3	1959 ~ 1979
Baimasi	Luohe River	112. 58	34. 72	11,891	663. 5	1953 ~ 1969

2.2   Methodology

2.2.1   A brief introduction to the GR4J model

GR4J model is composed of two tanks or stores, production store and routing store, and two unit hydrographs. S and R, the two internal state variables of GR4J, mean the production store level and routing store level, respectively. As the initial value of S and R are difficult to set, usually a one-year warm-up period will be set at the beginning of each simulation and during this period, the model errors are not taken into account when calibration. Note that all water qualities' unit is mm in this model. The four parameters of GR4J model are provided in Tab. 2. The required inputs are a daily time series of precipitation (P), potential evapotranspiration (E), and observed streamflow to calibrate or verificate the model. E could be a long-term average value.

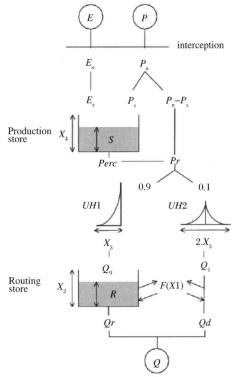

Fig. 2   Outline of GR4J rainfall-runoff model
(Perrin, et al. ,2003)

A sketch of GR4J model is illustrated in Fig. 2. Firstly, the subtraction of E from P is positive or negative determines either net rainfall (P_n) or net potential evapotranspiration (E_n). If Pn is positive but not zero, a part of it (P_s) fills the production store, otherwise a part of En (Es) evaporates from the store. A percolation leakage ($Perc$) from the production store, one of main differences with GR3J, makes contribution to low flow simulation. Pr ($Perc + P_n - P$) is routed linearly by two unit hydrographs, 90% by $UH1$ and 10% by $UH2$. F, the daily catchment water exchange, can be positive, negative or zero, respectively representing water imports, water exports and no water exchange. Change of the level in the routing store is determined by $Q9$, and F. Q_9 is flux after 90% of Pr routed by $UH1$, while Q_1 is flux after 10% of Pr routed by $UH2$. Finally, the total streamflow Q is the sum of Qr, the output of routing store, and Qd, the value of F plus Q_1.

Tab. 2 Information of the four parameters of GR4J model

Prameters	Description	80% confidence interval (Perrin, et al. ,2003)	Note
X_1 (mm)	Maximum capacity of the production store	100 ~ 1,200	>0
X_2 (mm)	Groundwater exchange coefficient	− 5 ~ 3	
X_3 (mm)	One day ahead maximum capacity of the routing store	20 ~ 300	>0
X_4 (days)	Time base of unit hydrograph	1.1 ~ 2.9	⩾0.5

2.2.2 Parameters optimization

To calibrate the model, a random sampling method and multi-objective optimization are used. Monte Carlo methods are important tools for random simulation. A Monte Carlo approach that each parameter set falls equably in its range is adopted to produce a random set of the four free parameters. And the initial ranges of sampling choose the 80% confidence intervals shown in Tab. 2. Then, the ranges will be adjusted by sensitivity analysis.

Multi-objective optimization includes Nash-Sutcliffe coefficient (Nsc) (Nash and Sutcliffe, 1970) and Relative Volume Error (Re). The objective of optimization is the highest Nsc with an absolute Re less than 20%. Nsc and Re are given as:

$$Nsc = 1 - \frac{\sum (Q_{obs} - Q_{sim})^2}{\sum (Q_{obs} - \overline{Q}_{obs})^2} \tag{1}$$

$$Re = \frac{R_{sim} - R_{obs}}{R_{obs}} \times 100\% \tag{2}$$

where, Q_{obs} and Q_{sim} are the observed and simulated daily streamflow, respectively; $\overline{Q}_{obs}$ is the average of Q_{obs} series; R_{obs} and R_{sin} refer to the observed and simulated average annual runoff, respectively.

3 Results and discussion

3.1 Hydrological simulation

Based on model calibration, the GR4J model is used for streamflow simulation in the four selected catchments. The monthly Nsc and Re values (Tab. 3) and the monthly hydrograph (Fig. 3) indicate quite good performance of GR4J. ① For Tangnaihai and Hongqi station, the Nsc values are higher than 0.85 and absolute Re are less than 8%, which denote that the streamflow simulations fit the observations well. There is a time lag and higher streamflow in valley simulation at Tangnaihai station due to the lack of snow module, which is important for the streamflow simulation in plateau climate zone. ②Although the results at Huangfu station turn not so good, both of the Nse values are greater than 0.5 in calibration and verification period. The main reasons for the low optimized *Nsc* and the missing of low peak streamflow may be that: the GR4J model is a member of the tank models family, which is suitable for saturation excess runoff theory. However, precipitation characteristics in Huangfuchuan River basin include short duration, strong intensity and small scale. What's more, located in the Loess Plateau, the Huangfuchuan River basin is dominated by infiltration excess runoff. Limited to the precipitation data and model structure, it is very hard to receive good results. ③ Baimasi station obtains a fairly satisfactory simulating result in calibration period, but the *Nsc* value is relatively low in the verification period. Note that in verification period, the Luohe River basin was experiencing a long-duration low-flow period, which is thought to be the primary reason for the little drop of Nsc value.

Tab. 3 Performance of GR4J model in the four catchments

Station	Calibration			Verification		
	Period	*Nsc*	*Re* (%)	Period	*Nsc*	*Re* (%)
Tangnaihai	1956 ~ 1969	0.85	0.52	1970 ~ 1976	0.86	−2.16
Hongqi	1955 ~ 1968	0.90	−3.77	1969 ~ 1975	0.86	7.40
Huangfu	1959 ~ 1972	0.54	−19.14	1973 ~ 1979	0.53	−4.41
Baimasi	1953 ~ 1965	0.87	−2.88	1966 ~ 1969	0.67	−9.37

Note: The first year is removed from the calculation of the Nash to account for the initialization *S* and *R*.

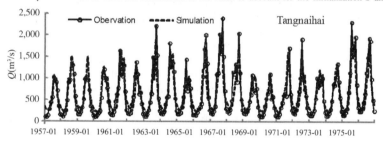

(a)

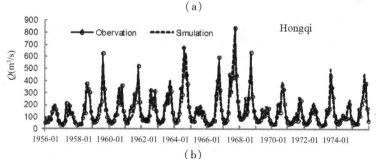

(b)

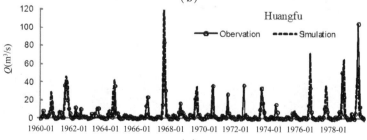

(c)

Fig. 3 Observed and simulated monthly streamflow

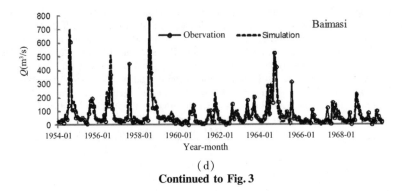

(d)

Continued to Fig. 3

3.2 Sensitivity analysis of model parameters

The model parameters sensitivity is estimated by the above mentioned Monte Carlo sampling method, with the number of 2000. The results for the four catchments (Fig. 4) show that: ①the parameter X_1 significantly influences the streamflow simulation at Tangnaihai station, Hongqi station and Huangfu station, but it is insensitive to Baimasi station; ② X_2 is insensitive to Tangnaihai and Baimasi stations but has a dramatic impact on the simulation effect at Hongqi station; ③generally, X_3 is the most insensitive parameter for the four catchment, and it appears be very sensitive at the Tangnaihai station; ④ X_4 are not sensitive to all the four catchments. X_4 determines the days of unit hydrographs. According to Perrin, et al. (2003, 2008), its 0.9 percentile were less than 3.5 days. And in this study, the maximum calibrated X_4 is 3.9 days in Tangnaihai catchment. Compared to the monthly scale of the *Nse* values in Fig. 4, the impact of X_4 on streamflow will be negligible. In a word, except X_4, the remaining parameters have their own different most sensitive catchment and for Baimasi station, there are no sensitive parameters of the GR4J model.

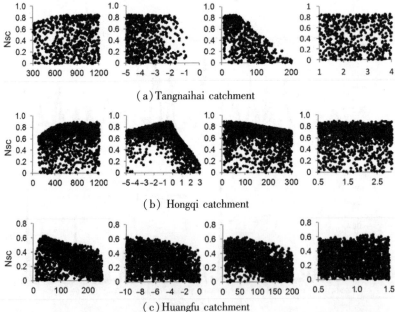

Fig. 4   Scatterplots between model parameters and *Nsc* in the four catchments

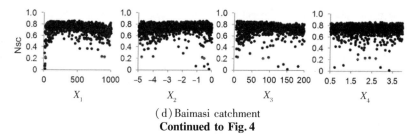

(d) Baimasi catchment

Continued to Fig. 4

4   Conclusions

Applying the GR4J model to four typical catchments of YRB and analyzing the model's parameter sensitivities, we can conclude that:

(1) The high Nse, Re values and the well fitted simulated monthly hydrograph indicate a good performance of GR4J in YRB. The Hongqi station obtains the best performance, followed by the Tangnaihai station, Baimasi station, and then the Huangfu station.

(2) Sensitivity analysis shows that the model parameter X_1, X_2, and X_3 are comparatively sensitive to the Huangfu station, Hongqi station and Tangnaihai station, and X_3 is the most sensitive parameter while X_4 is the most insensitive one.

Thought good results have been obtained, there remains some improvements can be achieved, e. g. , the simulation effect of modeling in plateau climate zone will upswing by GR4J model with snow module. The number of selected catchments in this study is small and the advantage of robustness has not been expressed adequately.

Acknowledgements

This research is jointly supported by two research programs: 1) National Basic Research Program of China (grant no. 2010CB951103); 2) International Science & Technology Cooperation Program of China (grand no. 2010DFA24330).

References

Dutta D, Welsh W, Vaze J, et al. Improvement in Short-term Streamflow Forecasting Using an Integrated Modelling Framework[J], 19th International Congress on Modelling and Simulation, Perth, Australia, December 2011, 12(12 – 16): 3405 – 3411.

Edijatno, Michel C Un modèle Pluie – débit Journalier à Troisparamètres[J]. La Houille Blanche, 1989(2): 113 – 121.

Edijatno, Nascimento N O, Yang X, et al. GR3J: A Daily Watershed Model with Three Free Parameters[J], Hydrological Sciences Journal, 1999, 44 (2): 263 – 277.

Harlan D, Wangsadipura M, Munajat C M. Rainfall – runoff Modeling of Citarum Hulu River Basin by Using GR4J[C]// Proceedings of the World Congress on Engineering 2010, London, U. K. , June 30 – July 2, 2010.

Le Moine N, Andréassian V, Perrin C, et al. How Can Rainfall – runoff Models Handle Intercatchment Groundwater Flows? Theoretical Study Based on 1040 French Catchments[J] Water Resources Research, 2007, 43.

Liersch S, Volk M. Towards Empirical Knowledge as Additional Information in Data – based Flood Forecasting Techniques, In Oxley, L. and Kulasiri, D. (eds) MODSIM 2007 International Congress on Modelling and Simulation, Modelling and Simulation Society of Australia and New Zealand, December 2007:74 – 80.

Liu C, Zheng H. Trend Analysis of Hydrological Components[J]. Journal of Natural Resources, 2003, 18(2): 129 – 135.

Nascimento N O. Appréciation à l'aide d'un modèle empirique des effets d'action anthropiques sur la relation pluie – débit à l' échelle du bassin versant, PhD Thesis, CERGRENE/ENPC, Paris, France, 1995, 550.

Nash J E, Sutcliffe J V. River Flow Forecasting Through Conceptual Models. Part I—A Discussion of Principles[J], Journal of Hydrology, 1970, 27 (3): 282 – 290.

Pagano C, Wang Q, Hapuarachchi P, et al. Hydrologic Forecasting and Uncertainty Research in Australia, European Geosciences Union General Assembly 2011, Vienna, Austria, 2011(4): 03 – 08.

Perrin C, Andréassian V, Serna C R, et al. Discrete Parameterization of Hydrological Models: Evaluating the use of parameter sets libraries over 900 catchments [J], Water Resources Research, 2008, 44.

Perrin C, Michel C, Andréassian V. Improvement of a Parsimonious Model for Streamflow Simulation[J] Journal of Hydrology, 2003, 279(1 – 4): 275 – 289.

Perrin C Vers une amélioration d' un modèle global pluie – dèbit au travers d' une approche comparative, PhD Thesis, INPG (Grenoble)/Cemagref (Antony), France, 2000, 530.

Simonneaux V, Hanich L, Boulet G Modelling Runoff in the Rheraya Catchment (High Atlas, Morocco) Using the Simple Daily Model GR4J. Trends over the last decades, In Proceedings of the 13th IWRA World Water Congress 2008, Montpellier, 1 – 4 September 2008(9):1 – 4.

Numerical Simulation of Flood Regime Caused by Dam-and Dyke-Breach with Complex Boundary

Ye Yuntao [1], *Liang Lili*[1], *Lei Xiaohui*[1], *Jiang Yunzhong*[1],
Jia Dongdong [2] and *Shang Yizhi*[1]

1. State Key Laboratory of Simulation and Regulation of Water Cycle in River Basin,
China Institute of Water Resources and Hydropower Research, Beijing, 100038,China;
2. State Key Laboratory of Hydrology-Water Resources and Hydraulic Engineering,
Nanjing Hydraulic Research Institute, Nanjing, 210029,China

Abstract: The paper presents a high resolution fully two-dimensional finite-volume numerical model based on structured grids for unsteady flow over topography with complex boundary during wet and dry movement. The model regards the fully two-dimensional shallow water equations which correctly balance the flux gradient and source terms as the governing equation, and uses the Godunov type finite volume shock-capturing schemes HLLC to compute flow flux. Its spatial and temporal accuracy are improved to two order accuracy by the slop limiter and predict-correct method called as MUSCL-Hancock and slope source terms and friction source terms are treated with central difference scheme and fully implicit method respectively. A robust procedure is adapted to efficiently and accurately simulate the movement of wet and dry boundary and a method of high simulative precision for complex boundary, simplicity of treatment and high calculation efficiency is used for dealing with the computational boundary. Finally, two tests are used to validate the solver of the presented model, and the numerical simulation results indicate that the developed model can accurately capture flow processes over irregular and shows good stability and accuracy.

Key words: hydraulics, dam breach, shallow water equations, structured grid, finite volume

1   Introduction

Frequency of the flood disaster in China has caused tremendous damage to people's lives and property, a combination measures of engineering and non- engineering is used to avoid the risk of flooding and reduce flood losses. As a part of non-engineering measures, the flow mathematical model is widely applied in the aspects of flood forecasting, dam-and dyke-breach, flood control plan, and river training, etc. , which provides a solid foundation for the study on spatial and temporal evolution laws, disaster preparedness and mitigation and risk regulation.

The conventional flow numerical model is suitable for the gradually varied flow, but the good results can not be achieved when applied in the rapidly varied flow simulation, such as dam- and dyke-breach flood, the mountain flood caused by the strong rainfall, the tidal estuary flow. However, the high performance finite volume computational scheme capturing the shock wave and contact break, along with the two-dimensional shallow water equation, provides the research idea to solve the problem.

When using the above method to simulate flow process under the condition of complex topography and boundary, the four aspects as follows merit attention: firstly, the failure to effectively balance the bottom slope and flux will cause false flow velocity ; secondly, it is difficult in the wet-and dry-boundary treatment when rapidly varied flow on the complex terrain; thirdly, complex irregular boundaries treatment will directly affect the numerical reliability, accuracy and computational time; fourthly, complex computational mesh generation is not propitious for the further application in the productive practice.

Some studies or explorations in these aspects are conducted by the scholarships at home and aboard, and good results are acquired. Zhou etc. propose the surface gradient method for treating source terms in the shallow-water equations based on accurate reconstruction of the conservative variable at cell interfaces; the numerical model which is discretized based on structured computational mesh cannot simulate the flow in the complex topography and boundary. Begnudelli etc. built the harmony numerical model based on unstructured mesh in which the modified flux is introduced to balance the slope source terms and interface flux terms which did not obey the strictly physical principle. Valiani etc. put forward a the bottom slope decomposition method with clear physical meaning to deal with the bottom slope source term, but application of the first order accuracy can not guarantee the simulation accuracy of the extreme flow changes. Sleigh, etc. and Brufau , respectively proposed the method of restrictions on water depth and bottom slope transform to treat wet and dry movement boundary in the numerical simulation. Huang etc. point out that the body-fitted grid and unstructured grid well adapted to the boundary shows more difficult in popularization and application, and especially for the rectangular grid , presents the Diagonal Cartesian Method to approximate the boundaries in two-dimensional shallow-water numerical simulation together with the wet-dry variation of the boundary points and the method is proved to study cases through the application in calculating the tidal current and achieving the good results.

On the basis of relevant research results from scholars at home and aboard, a high resolution fully two-dimensional finite-volume numerical model based on structured grids for unsteady and steady flow over topography with wetting and drying and complex boundary is presented. The model regards the fully two-dimensional shallow water equations which correctly balance the flux gradient and source terms as the governing equation, and uses the Godunov type finite volume shock-capturing schemes HLLC to compute flow flux. Its spatial and temporal accuracy are improved to second order by the slop limiter and predict-correct method called as MUSCL-Hancock and slope source terms and friction source terms are treated with central difference scheme and fully implicit method respectively. A robust procedure is adapted to efficiently and accurately simulate the movement of wet/dry boundary and a method of high simulative precision for complex boundary, simplicity of treatment and high calculation efficiency is used for dealing with the computational boundary. Finally, two tests are used to validate the solver of the presented model, and the numerical simulation results indicate that the developed model can accurately capture flow processes over irregular and shows good stability and accuracy.

2 Governing Equations

Under the assumption of the horizontal scale of water in the rivers, lakes, or flood plain being much larger than the vertical scale, hydrostatic pressure distribution, the smaller vertical acceleration, and neglecting the surface stress term, Coriolis effects term and viscous term, the two-dimensional shallow water equations derived by depth-integrating the three-dimensional Reynolds-averaged Navier-Stokes equations can be expressed as:

$$\frac{\partial \zeta}{\partial t} + \frac{\partial (uh)}{\partial x} + \frac{\partial (vh)}{\partial y} = 0 \tag{1}$$

$$\frac{\partial (uh)}{\partial t} + \frac{\partial (u^2 h)}{\partial x} + \frac{\partial (uvh)}{\partial y} = -\frac{\tau_{bx}}{\rho} - gh \frac{\partial \zeta}{\partial x} \tag{2}$$

$$\frac{\partial (vh)}{\partial t} + \frac{\partial (uvh)}{\partial x} + \frac{\partial (v^2 h)}{\partial y} = -\frac{\tau_{by}}{\rho} - gh \frac{\partial \zeta}{\partial y} \tag{3}$$

where, t denotes time; x and y are Cartesian coordinates; ζ is the free surface elevation above still water depth(hs); $h = (\zeta + h_s)$ is total water depth, h_s is still water depth; u and v are depth-averaged velocity components in the two Cartesian directions; g is the gravitational acceleration; ρ is the density of water; τ_{bx} and τ_b are bed friction stresses.

To overcome the drawback for flows over initially dry land, while retaining the mathematical balance the flux gradient and source terms, the free surface gradient term $gh\partial \zeta / \partial x$ and $gh\partial \zeta / \partial y$ in the x-and y-direction momentum equation can be split as follows:

$$gh \frac{\partial \zeta}{\partial x} = \frac{g}{2} + \frac{\partial (\eta^2 - 2z_b \eta)}{\partial x} + g\eta \frac{\partial z_b}{\partial x}, \quad gh \frac{\partial \zeta}{\partial y} = \frac{g}{2} + \frac{\partial (\eta^2 - 2z_b \eta)}{\partial y} + g\eta \frac{\partial z_b}{\partial y} \tag{4}$$

where, η is surface water level above the datum; z_b is terrain elevation above datum, and water depth is obtained by $h = \eta - z_b$.

Liang etc proved that the split mode of free surface gradient terms ensure the hyperbolic characteristics of shallow-water equations, handle the issues of still water conservation and moving wet-dry boundary, and balance the flux gradient and source terms, which avoid the condition that the numerical stability and accuracy because of not reasonable slope source terms treatment with the complex topography.

Substituting the free surface gradient terms in Eq. (2) and Eq. (3) with Eq. (4), the Eq. (1), Eq. (2), Eq. (3) may be written as in the matrix form:

$$\frac{\partial U}{\partial t} + \frac{\partial F}{\partial x} + \frac{\partial G}{\partial y} = \frac{\partial U}{\partial t} + \nabla \cdot E = S \tag{5}$$

where, U is vector representing conserved variable; E is vector representing the flux of convective flux; F and G are vectors representing fluxes in x- and y- direction; S is vector representing source term including bed slope term and friction term.

These vectors are given by

$$U = \begin{bmatrix} \eta \\ uh \\ vh \end{bmatrix}, \quad F = \begin{bmatrix} uh \\ u^2 h + \frac{1}{2}g(\eta^2 - 2\eta z_b) \\ uvh \end{bmatrix}, \quad G = \begin{bmatrix} vh \\ uvh \\ v^2 h + + \frac{1}{2}g(\eta^2 - 2\eta z_b) \end{bmatrix},$$

$$S = \begin{bmatrix} 0 \\ -\dfrac{\tau_{bx}}{\rho} - g\eta \dfrac{\partial z_b}{\partial x} \\ -\dfrac{\tau_{by}}{\rho} - g\eta \dfrac{\partial z_b}{\partial y} \end{bmatrix} \tag{6}$$

where, τ_{bx} and τ_{by} are bed stress terms representing the energy dissipation influence of bed roughness on flow and are estimate empirically from

$$\tau_{bx} = \rho g n^2 u \sqrt{u^2 + v^2}/h^{1/3}, \quad \tau_{by} = \rho g n^2 v \sqrt{u^2 + v^2}/h^{1/3}$$

The physical significance of other variables is the same as the above relevant content.

3　Finite volume model

Governing equations are distributed based on the rectangular grid cells, and Fig. 1 shows the schematic diagram of a discrete control cell. The adjacent cells of a controlling cell (i, j) are $(i, j-1)(i+1, j)(i, j+1), (i-1, j)$, and the public boundaries are $(i, j-1/2)(i+1/2, j)$ $(i, j+1/2)(i-1/2, j)$ respectively. The length of the cells at horizontal axis x is Δx, while Δy at the vertical axis y. The model basic variables of h or z, u, v are stored in the cell center and the terrain elevation is stored in the node of the cell when controlling equations are distributed, so the center elevation is get directly from the node elevation interpolation when the model calculates them.

Choosing any of the controlling cell $V_{(i, j)}$, the integral of Eq. (5) in the controlling cell is

$$\int_{V_{i,j}} \frac{\partial U}{\partial t} dV + \int_{V_{i,j}} \cdot \nabla \cdot E dV = \int_{V_{i,j}} S dV \tag{7}$$

Assuming that the average value U of the cell is stored in the cell center, converting the second term of area integral in the left of the Eq. (7) to line integral along the boundary of the controlling cell using the Gauss-Green, then Eq. (7) can be transformed into:

$$\int_{V_{i,j}} \frac{\partial U}{\partial t} dV + \int_{L_{i,j}} E \cdot n dL = \int_{V_{i,j}} S dV \tag{8}$$

where, $L_{(i,j)}$ is the boundary of the grid cell (i,j), and the Eq. (8) can be written as follows:

$$\frac{\partial U}{\partial t} \Delta A_{i,j} = -\sum_{k=1}^{4} E_{i,j}^k \cdot n_{i,j}^k \Delta L_{i,j}^k + S_{i,j} \Delta A_{ij} \tag{9}$$

where, $\Delta A_{(i,j)}$ represents the area of the cell (i,j), and $\Delta A_{(i,j)} = \Delta x \Delta y$; $E_{(i,j)}^k$ is the flux of the boundary k, $k = 1, 2, 3, 4$, corresponding to the lower, right, upper and left boundary; $n_{(i,j)}^k$ is the unit normal vector of boundary k; $\Delta L_{(i,j)}^{k}$ is the length of the boundary k, and the length of the left and right boundaries is Δy, while the upper and lower boundary is x. So you can get:

$$U_{i,j}^{n+1} = U_{(i,j)}^{n} - \frac{\Delta t}{\Delta x}(E_{(i+1/2,j)} - E_{(i-1/2,j)}) - \frac{\Delta t}{\Delta y}(E_{(i+1/2,j)} - E_{(i,j-1/2)}) + \Delta t S_{(i,j)} \tag{10}$$

where, n represents the time; (i,j) is the cell index, and $E_{(i-1/2,j)}$, $E_{(i+1/2,j)}$, $E_{(i,j+1/2)}$ and $E_{(i,j-1/2)}$ represent the fluxes of the left and right boundaries, and that of the upper and lower boundaries of the cell (i,j).

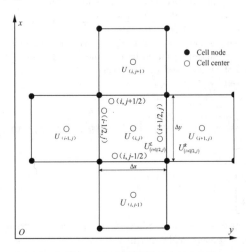

Fig. 1 Schematic diagram of a controlling cell

3.1 Numerical flux calculation

A Godunov-type scheme solves local Riemann problems at each cell interface to evaluate interface fluxes. The HLLC approximate Riemann solver is chosen to solve these local Riemann problems due to its advantages in offering automatic entropy fix and easy treatment of wetting and drying. Compared with the original HLL approach, the HLLC approximate Riemann solver also presents major benefits in modeling two-dimensional flows, especially when the solute transport is included. Ignoring the middle wave, the two-wave assumption of the HLL Riemann solver is only correct for purely one-dimensional problems and may result in excessive smearing of contact discontinuities for multidimensional simulations. Here, the two-dimensional shallow water equations are considered and hence the HLLC solver should be used. Taking $F_{(i+1/2,j)}$ as an example, the HLLC fluxes for the integrated governing equations may be defined as

$$F_{(i+1/2,j)} = \begin{cases} F_{L}, & S_{L} \geqslant 0 \\ F_{*L}, & S_{L} \leqslant 0, \ S_{M} \geqslant 0 \\ F_{*R}, & S_{M} \leqslant 0, \ S_{R} \geqslant 0 \\ F_{R}, & S_{R} \leqslant 0 \end{cases} \tag{11}$$

where, $F_{L} = F(U_{L})$ and $F_{R} = F(U_{R})$ are the fluxes in the left and right regions of the Riemann solution structure respectively, which are directly computed from the left and right Riemann states U_{L} and U_{R} defined at either side of the cell interface $(i+1/2, j)$. F_{*L} and F_{*R} are the fluxes in the middle region separated by the middle (contact) wave:

$$F_{*L} = [f_{*1} \quad f_{*2} \quad v_{L}f_{*1}]^{T} \quad \text{and} \quad F_{*R} = [f_{*1} \quad f_{*2} \quad v_{R}f_{*1}]^{T} \tag{12}$$

where, v_{L} and v_{R} are left and right tangential velocity components of Riemann states, which remain unchanged across the left and right waves, respectively; f_{*1} and f_{*2} are calculated from the HLL formula:

$$F_* = \frac{S_R F_L - S_L F_R + S_L S_R (U_R - U_L)}{S_R - S_L} \tag{13}$$

The third flux component in the middle flux vector F_* is calculated depending on which part of the middle region the flux component is evaluated. In the Eq. (13), the S_L, S_M and S_R are the left, middle and right wave speeds in the HLLC Riemann solution structure. Fraccarollo and Toro [11] recommend the following formula for estimating SL and SR to facilitate applications in wetting and drying:

$$S_L = \begin{cases} u_R - 2\sqrt{gh_R}, h_L = 0 \\ \min(u_L - 2\sqrt{gh_L}, u_* - 2\sqrt{gh_*}), h_L > 0 \end{cases},$$

$$S_R = \begin{cases} u_L - 2\sqrt{gh_L}, h_R = 0 \\ \min(u_R - 2\sqrt{gh_R}, u_* - 2\sqrt{gh_*}), h_R > 0 \end{cases}$$

$$S_M = \frac{S_L h_R (u_R - S_R) - S_R h_L (u_L - S_L)}{h_R (u_R - S_R) - h_L (u_L - S_L)} \tag{14}$$

where, u_L, u_R, h_L and h_R are the velocity and depth components of the left and right Riemann states; u_* and h_* can be evaluated from:

$$u_* = \frac{1}{2}(u_L + u_R) + \sqrt{gh_L} - \sqrt{gh_R} \quad \text{and} \quad h_* = \frac{1}{g}\left[\frac{1}{2}(\sqrt{gh_L} + \sqrt{gh_R}) + \frac{1}{4}(u_L - u_R)\right] \tag{15}$$

3.2 Time integral

MUSCL – Hancock method, where MUSCL stands for Monotone Upwind Schemes for Conservation Laws, is employed to achieve the second-order accuracy in time for the governing equations. It updates flow variables over a time interval via predictor and corrector steps in an explicit scheme.

In the predictor step, the time-marching formula for updating the cell-centered flow variables from Eq. (10) is:

$$U_{(i,j)}^{n+1/2} = U_{(i,j)}^n - \frac{\Delta t}{2}\left[\frac{(E_{(i+1/2,j)}^* - E_{(i-1/2,j)}^*)}{\Delta x} + \frac{(E_{(i,j+1/2)}^* - E_{(i,j-1/2)}^*)}{\Delta y} - S_{(i,j)}\right] \tag{16}$$

where, subscript (i, j) represents cell index, superscript n denotes the time level; Δt represents the time step; Δx and Δy are the cell size in the x – and y – direction.

The bed slope terms are approximated by central differences and interface fluxes are determined at the midpoint of each cell face by linear interpolation.

In the corrector step, the flow variables are calculated over a full time step based on flow data from the predictor step. The explicit updating formula is:

$$U_{(i,j)}^{n+1} = U_{(i,j)}^n - \Delta t\left[\frac{(E_{(i+1/2,j)}^{*n+1/2} - E_{(i-1/2,j)}^{*n+1/2})}{\Delta x} + \frac{(E_{(i,j+1/2)}^{*n+1/2} - E_{(i,j-1/2)}^{*n+1/2})}{\Delta y} - S_{(i,j)}^{n+1/2}\right] \tag{17}$$

Numerical fluxes are calculated using the HLLC approximate Riemann solver, where the face values of the flow variables are again obtained using slope – limited interpolation formula.

3.3 Discretization of Source Terms

The discretization of the bed slope term and friction term are treated respectively.

The bed slope term is calculated with the central difference scheme, compatible with the Godunov-type finite volume model, in the x- and y – direction:

$$-g\eta\frac{\partial Z_b}{\partial x} = -g\bar{\eta}\left(\frac{z_{bi+1/2,j} - z_{bi-1/2,j}}{\Delta x}\right), \quad -g\eta\frac{\partial z_b}{\partial y} = -g\bar{\eta}\left(\frac{z_{bi+1/2,j} - z_{bi-1/2,j}}{\Delta y}\right) \tag{18}$$

where, $\bar{\eta} = (\eta_{i-1/2,j}^R + \eta_{i+1/2,j}^L)/2$, $\bar{\eta} = (\eta_{i,j-1/2}^R + \eta_{i,j+1/2}^L)/2$.

The friction source terms are solved by a splitting point – implicit scheme for better stability and it is equivalent to solve the following ordinary differential equations:

$$\frac{dU}{dt} = S_f \tag{19}$$

where, $S_f = [\,0\ S_{fx}\ S_{fy}\,]^T$, $S_{fx} = -\tau_{bx}/\rho$, $S_{fy} = -\tau_{by}/\rho$. The Eq. (19) is expanded in the Taylor series to obtain the implicit expression:

$$S_f^{n+1} = S_f^n + (\frac{\partial S_f}{\partial U})^n \Delta U + o(\Delta U^2) \tag{20}$$

where n represents the time level, $\Delta U = U^{n+1} - U^n$, $(\partial S_f/\partial U)$ is the Jacobian matrix of S_f. The Eq. (20) can be transformed into:

$$[\,I - \Delta t(\frac{\partial S_f}{\partial U})^n\,]\,U = tS_f^n \tag{21}$$

In which I is identity matrix.

3.4　Treatment of boundary conditions

The boundary conditions include the two kinds of open boundary condition and wall boundary condition. The treatment of wall boundary adopts slip boundary, that is, setting the value of the normal velocity on the boundary is to be zero, as well as the normal gradient of the water depth, concentration and other variables on the boundary, but the tangential velocity not zero. On the open boundary, it is generally given the processes of water level and flow, or the stage-discharge relationship. As to the free flow boundary condition, generally setting the normal gradient of each variable is to be zero.

Wet and dry boundary treatment is a hot and difficult question in the calculation of extreme flow. The processes of wet and dry on the complex terrain are treated through amending the governing equations, and it can use a method to amend the local bed to avoid the generation of the false flow on the dry bed.

3.5　Complex boundary treatment

If the boundary is shared by two grid cells of C and G, shown in Fig. 2, the wall boundary adopts slip boundary condition, there are $\eta_G = \eta_C$, $u_G = -u_C$, $v_G = v_C$. This method of treatment will bring a false viscosity in the numerical solution when applied to the curve or irregular boundary, resulting in numerical oscillations.

Liang proposed a simple and practical boundary treatment method: the first is to calculate the flux of the midpoint O' of left boundary of the cell C in the finite volume method, and the next is to capture the nearest boundary point O from the point O', so the angle of tangent of the point O and the $X-$axis is θ. For the slip boundary, the value of the normal velocity component and the value of gradient of the tangential velocity component are zero. Therefore:

$$u_{NIO} = 0,\ \frac{\partial uT}{\partial N}\bigg|_{o} = 0$$

where, N and T are marked as normal direction and tangential direction respectively. Assumed that the boundary point O moves to point O', therefore, the slip boundary condition of O' point is shown as follows:

$$u_{NIO} = 0,\ \frac{\partial_{uT}}{\partial N}\bigg|_{O'} = 0$$

The velocity of gird cell G and cell C respectively has two components of the normal velocity and the tangential velocity, slip boundary conditions of point O' are:

Normal direction N: $-u_G\sin\theta + v_G\cos\theta = -(-u_C\sin\theta + v_C\cos\theta)$

Tangential direction T: $-u_C\cos\theta + v_C\sin\theta = -u_G\cos\theta + v_G\sin\theta$

Therefore, the slip boundary of point O' can be written as:

$\eta_G = \eta_C$, $u_G = u_C - 2(u_C\sin\theta + v_C\cos\theta)\sin\theta$, $v_G = v_C + 2(u_C\sin\theta - v_C\cos\theta)\cos\theta$

It requires an estimated value of θ in this method, it is determined usually using the following method: First of all, determining the point O of the boundary which has the nearest distance from the point O; and then determining the index of i and $i+1$ of the two boundary points which have the nearest distance from the point O, the tangent of point O uses the approximate value of point O; so finally the value of θ can be estimated by the following formula:

$$\theta = \tan^{-1} \frac{y_s(i+1) - y_s(i)}{x_s(i+1) - x_s(i)}$$

where, x_s and y_s are the coordinates of the boundary point. Obviously, the more intensive of the boundary points used, the more accurate of boundary treatment and the value of θ are gotten.

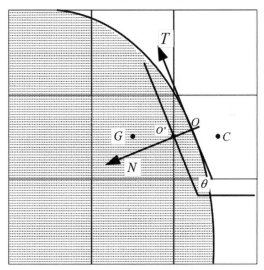

Fig. 2　Schematic diagram of computational domain boundary treatment

4　Validation using classic test examples

4.1　Breach test example of asymmetric square dam

Asymmetric square dam – breach test is one of the classic examples of dam breach testing, so this paper uses this example to validate the computational efficiency of the model adopted dealing with discontinuity and dry bed. The calculation area is a flat square area of 200 m × 200 m. There is a distance of 100 m between the edge of the calculation area and the dam on the top while 95 m at the bottom and the dam is 75 m wide. It has two scenarios in this example. The water depth of upstream and downstream is 10 m and 5 m respectively in scenario one; while in scenario two it is 10m with the upstream and dry bed with the downstream respectively. Fig. 3(a) and Fig. 3(b) show the computational results of the model with two scenarios when 6 second after.

As seen in Fig. 3 (a), the water flows to the downstream and spreads to both banks after a sudden break of the dam, meanwhile the rarefaction waves spreading upstream upspring. When the waves reach to the left bank, it generates the reflected waves because of the existence of the wall boundary, resulting in the high water level near the bank. The presence of the downstream water level to some extent hinders the advance of water, resulting in high water level at the forefront of the water body, hence two asymmetric vortex are formed on both sides of the breach, which well agree with the experimental results, and the results are also in line with the numerical calculation results of Sleigh, et al. , Wang, et al. , and Caleffi, et al. , illustrating that the model has high calculation accuracy.

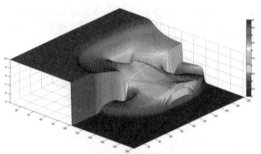

(a) Water depth of the upstream and downstream is
10 m and 5 m respectively

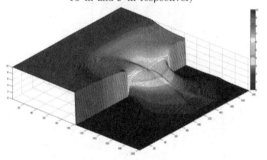

(b) Water depth of the upstream and downstream is10 m
and 0 m respectively

**Fig. 3   Computational results of asymmetric square
dam breach test at the time of 6 second**

Shown in Fig. 3 (b) , there is bigger water head between the upstream and the downstream,
the water flow rapidly to the downstream, having smaller lateral impact, which is in line with the
physical test phenomena, as well as the numerical calculation results of Sleigh, et al. , Wang,
et al. , and Caleffi, et al. , proving that the model has enough capability to handle with dry bed.

4. 2   Example of tilt hydraulic jump of supercritical flow

The purpose of this example is to test the ability of the model to deal with discontinuity and the
ability to adapt to the complex boundary. The calculation area is a flat rectangle channel of 40 m
long, the width of the entrance on the upper reach of the channel is 30 m, while it has a inward
contraction of 8. 95 degrees on the right bank of the river 10 meters away from the entrance.
Hydraulic jumps of a certain angle are formed when the water flows to the contraction of the
channel. Given a boundary of rapid flow at the entrance, that is, $u = 8.57$ m/s, $v = 0$, the initial
water depth $h = 1$ m, and then given a boundary of a free outflow at the outlet, it does not take into
account of the frictional resistance when calculating.

Analytical solution for the given boundary conditions is that: it generates a steady – state
supercritical flow, and the study area is divided into two parts by the hydraulic jump; also there is
an angle of 30 degrees between the flow direction and the hydraulic jump, and the water depth of
the hydraulic jump is 1. 5 m.

The model ability of adapt to the complex boundary is shown in Fig. 4 without treatment of the
boundary when the sharp and steep peaks appear on the water surface near the boundary, causing
numerical oscillations, and the flow direction is inconsistent with the boundary direction,
incompatible with the real physical phenomena. The model ability of adapt to the complex boundary
is shown in Fig. 5 after treatment of the boundary, and the results are to the contrary of that of

Fig. 4. There is an angle of 30 degrees between the flow direction and the hydraulic jump, and the water depth of the hydraulic jump is 1. 5 m, well capturing the shock wave in 2 ~ 3 grid cells, which shows that the model adopted can well adapt to the complex boundary and deal with discontinuous solutions with high accuracy.

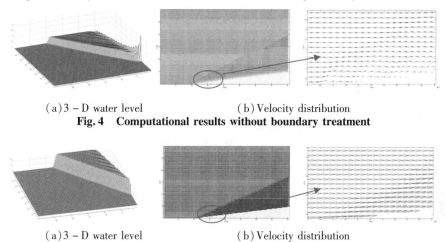

(a)3 – D water level　　　　　　　　(b) Velocity distribution

Fig. 4　Computational results without boundary treatment

(a)3 – D water level　　　　　　　　(b) Velocity distribution

Fig. 5　Computational results after boundary treatment

5　Conclusions

A finite volume mathematical model adapted to the structured grid of complex boundary shallow water flow has been built in this paper. The basic governing equations adopt the real conservative two-dimensional shallow water equations formed from the divided horizontal direction and vertical direction of the relative height on the free surface. It uses the centered discrete basic governing equations at the level of structured grid cells, and uses the approximate Riemann solver of HLLC which is able to effectively capture the shock waves to calculate the boundary fluxes. With slope limiter reconstructing variables of both sides of the boundary, the spatial accuracy to the second-order is raised; while using the integral of the MUSCL-Hancock equation of the time dimension, the time accuracy to the second order is raised. Adopting the central difference calculus to deal with the bottom slope, while implicit method to solve the resistance, which ensures the stability of the model. Further, it processes the boundary cells to improve the model ability to adapt to the complex boundary. Finally, the model performances have been proved though two classic test examples.

Acknowledgements

The paper is supported by the Open Research Fund of State Key Laboratory of Hydrology-Water Resources and Hydraulic Engineering (NO: 2011491911) and the Open Research Fund of State Key Laboratory of Simulation and Regulation of Water Cycle in River Basin, China Institute of Water Resources and Hydropower Research (IWHR- SKL-201103 ,IWHR-SKL-201117)

References

Wang Zhili, Geng Yanfen, Jin Sheng. Numerical Modeling of 2-D Shallow Water Flow with Complicated Geometry and Topography [J]. Journal of Hydraulic Engineering, 2005, 36 (4):439 – 444.

Yue Zhiyuan, Cao Zhixian, LI Youwei, et al. Unstructured grid finite volume model for two – dimensional shallow water flows[J]. Chinese Journal of Hydrodynamics, Ser. A, 2011, 26

(3):359 - 367.

Zokagoa J M, Soulaimani A. Modeling of Wetting – drying Transitions in Free Surface flows over Complex Topographies [J]. Computer Methods in Applied Mechanics and Engineering, 2010,199(33 – 36):2281 - 2304.

Huang Bingbin, Fang Hongwei, Liu Bin. Diagonal Cartesian Method for Numerical Simulation of Flow with Complex Boundary [J]. Chinese Journal of Hydrodynamics, Ser. A, 2003, 18 (6):679 - 685.

Zhou J G, Causon D M, Mingham C G. The Surface Gradient Method for The Treatment of Source Terms in the Shallow – water Equations [J]. Journal of Computational Physics, 2001, 168 (1):1 - 25.

Begnudelli L, Sanders B F. Unstructured Grid finite – volume Algorithm for Shallow – water Flow and Scalar Transport with Wetting and Drying [J]. Journal of Hydraulic Engineering, 2006, 132(4):371 - 384.

Valiani A, Begnudelli L. Divergence form for Bed Slope Source Term in Shallow Water Equations [J]. Journal of Hydraulic Engineering, 2006, 132(7):652 - 665.

Sleigh P A, Gaskell P H. An Unstructured Finite – volume Algorithm for Predicting Flow in Rivers and Estuaries [J]. Computers & Fluids, 1998, 27(4):479 - 508.

Brufau P, Vazquez – Cendon M E, Garcia – Navarro, P. A Numerical Model for The Flooding and Drying of Irregular Domains [J]. International Journal of Numerical Methods in Fluids, 2002, 39(3):247 - 275.

Liang Q H, Borthwick A G L. Adaptive Quadtree Simulation of Shallow Flows with Wet – dry Fronts over Complex Topography [J]. Computers & Fluids, 2009, 38: 221 - 234.

Fraccarollo L, Toro E F. Experimental and Numerical Assessment of The Shallow Water Model for Two – dimensional Dam – break Type Problems [J]. J. Hydraul. Res. , 1995, 33 (6): 843 - 863.

Busing T R A, Murman E M. Finite Volume Method for The Calculation of Compressible Chemically Reacting Flows [J]. AIAA Journal, 1988, 29(9):1070 - 1078.

Fiedler F R, Ramirez J A. A Numerical Method for Simulating Discontinuous Shallow Flow over An Infiltrating Surface [J]. International Journal for Numerical Methods in Fluids, 2000, 32(2): 219 - 239.

Liang Q H, Borthwick A G L. Simple Treatment of Non – aligned Boundaries in a Cartesian Grid Shallow Flow Model [J]. International Journal for Numerical Methods in Fluids, 2008, 56 (11):2091 - 2110.

Wang J S, Ni H G, He Y S. Finite Difference TVD Scheme for Computation of Dam – break Problems [J]. Journal of Hydraulic Engineering, 2000, 126(4):253 - 262 .

Caleffi V, Valiani A, Zanni A. Finite Volume Method for Simulating Extreme Flood Events in Natural Channels [J]. Journal of Hydraulic Research , 2003, 41(2):167 - 177.

Multi Index Comprehensive Drought Evaluation Based on the Entropy Method

Yi Zhizhi[1] , *Liang Zhongmin*[1] , *Zhao Weimin*[2] and *Liu Xiaowei*[2]

1. College of Hydrology and Water Resources, Hohai University, Nanjing, 210098, China
2. Hydrology Bureau of Yellow River Conservancy Commission, Zhengzhou, 450004, China

Abstract: Drought occurs more frequently, and it becomes one of the most serious natural disasters which impact on our country. Drought evaluation is the foundation of drought forecast, early warning and other works. Therefore, research on regional drought evaluation is very important. In order to solve the inconsistent problem of each index, this paper selects meteorological, hydrological and agriculture drought index, including continuous no rain days, precipitation anomaly percentage, standardized precipitation, moisture index, flow anomaly percentage, crop water shortage for comprehensive evaluation system construction. Then uses entropy method to construct a comprehensive evaluation index on which the drought assessment is conducted. Comprehensive drought assessment model based on the entropy method was applied in Xianyang, Shanxi Province. Evaluation results and the actual drought situation are basically the same. Studies show that, comprehensive drought assessment method based on the entropy is easily calculated, intuitive, with better accuracy. The method has certain credibility in drought assessment, and provides technical support for the study of regional drought.

Key words: entropy, entropy method, multi drought index, comprehensive evaluation, Xianyang region

1 Introduction

Under the background of global climate change, drought has larger influence and growth of the duration, becoming one of the most serious natural disasters which impact on our country. In order to guarantee the sustainable development of society and economy, drought relief work becomes significant. Only grasp the current drought evaluation accurately and in time, can we adopt appropriate countermeasures to deal with the effects of drought. Therefore, drought evaluation has become a hot topic. However, due to different research objects, there are different drought indices, and their evaluation results are also of great differences. There are hundreds kinds of drought evaluation indices currently. Owing to each index has corresponding temporal and spatial scales, a single index is very difficult to meet the requirements of temporal universality. In the numerous kinds of evaluation methods, comprehensive evaluation method is widely used.

Since weighting the indices is the key to drought evaluation, the paper uses entropy method. According to the main factors of drought, select meteorological, hydrological and agriculture drought index to evaluate drought of each single index. Combining entropy method, the paper constructs a comprehensive drought index to perform multi indices comprehensive evaluation. It indicates that the evaluation conclusion matches with the actual drought situation, which implies that entropy method is applicable to multi index comprehensive drought assessment.

2 Entropy method

Entropy is a physical concept of statistical thermodynamics, which was first put forward by the German physicist Clausius in 1850. Entropy is the reflection of internal disorder in material or system. The greater the entropy is, the higher the material disordered degree is. The basic theory of entropy has been widely applied to various fields, such as probability theory, physics, life science and other disciplines.

In the middle of the twentieth Century, American scientist Shannon, who is the founder of information theory and digital communication era, first introduced entropy concept into information science. In this area, entropy is used to measure the uncertainty or quantity of information. It means that the greater amount of information, then uncertainty is smaller, the entropy is smaller. Conversely, the information quantity is smaller, then the uncertainty is greater, and entropy is also bigger. In the evaluation index system, evaluation of index has greater change, which means that indicators provide more information, its entropy is smaller, and the weight should be bigger. Conversely, evaluation of index has smaller change, which means that indicators provide less information, its entropy is larger, and the weight should be smaller.

3 Comprehensive drought assessment model based on the entropy method

(1) Determine m small scale drought evaluation area, namely the m point drought evaluation of regional, $D = \{D_1, D_2, \cdots, D_m\}$.

(2) Determine n drought evaluation indices, $U = \{u_1, u_2, \cdots, u_n\}$.

(3) Determine k ratings, $B = \{B_1, B_2, \cdots, B_k\}$.

(4) Calculation drought evaluation matrix H.

$$H = (h_{ij})_{mn} = \begin{bmatrix} h_{11} & h_{12} & \cdots & h_{1n} \\ h_{21} & h_{22} & \cdots & h_{2n} \\ \vdots & \vdots & \vdots & \vdots \\ h_{m1} & h_{m2} & \cdots & h_{mn} \end{bmatrix} \quad (i = 1, 2, \cdots, m; j = 1, 2, \cdots, n) \tag{1}$$

(5) Normalize the judgment matrix.

① For index which is more quality when it is bigger:

$$l_{ij} = \frac{h_{ij} - \min\{h_{ij}\}}{\max\{h_{ij}\} - \min\{h_{ij}\}} \tag{2}$$

② For index which is more quality when it is smaller:

$$l_{ij} = \frac{\max\{h_{ij}\} - h_{ij}}{\max\{h_{ij}\} - \min\{h_{ij}\}} \tag{3}$$

Calculate the normalized matrix L:

$$L = (l_{ij})_{mn} = \begin{bmatrix} l_{11} & l_{12} & \cdots & l_{1n} \\ l_{21} & l_{22} & \cdots & l_{2n} \\ \vdots & \vdots & \vdots & \vdots \\ l_{m1} & l_{m2} & \cdots & l_{mn} \end{bmatrix} \quad (i = 1, 2, \cdots, m; j = 1, 2, \cdots, n) \tag{4}$$

(6) Calculate the entropy:

$$p_{ij} = \frac{l_{ij}}{\sum\limits_{i=1}^{m} l_{ij}} \quad (i = 1, 2, \cdots, m; j = 1, 2, \cdots, n) \tag{5}$$

$$en_j = -\frac{1}{\ln(m)} (\sum\limits_{i=1}^{m} p_{ij} \ln p_{ij})(j = 1, 2, \cdots, n) \tag{6}$$

Especially, when $p_{ij} = 0$, $p_{ij} \ln p_{ij} = 0$.

(7) Calculate the weigh:

$$W = (w_j)_{1 \times n}$$

$$w_j = \frac{1 - en_j}{\sum\limits_{j=1}^{n} (1 - en_j)} \tag{7}$$

(8) According to the calculated weight W, calculate drought multi – index comprehensive assessment level A_i.

$$A_i = \sum_{j=1}^{n} w_j \times A_{ij} , j = 1, 2, \cdots, n \tag{8}$$

4 Application examples

Xianyang city is located in the the Yellow River Central in Valley Guanzhong Basin, and drought is the main meteorological disasters. This paper selects Xianyang City, including Binxian County, Xunyi County, Yongshou County, Chunhua County, Qianxian County, Liquan County, Jingyang county, Sanyuan County, Xingping City, Wugong Town, downtown, considers rainfall, temperature, flow, soil, crops and other factors, and selects continuous no rain days, precipitation anomaly percentage, standardized precipitation, moisture index, flow anomaly percentage, crop water shortage for comprehensive evaluation system construction. See synonyms at meteorological drought grade and drought grade standard, see Tab. 1, Tab. 2.

Tab. 1   Drought classification table of single index

Index	No drought	Light drought	Moderate drought	Severe drought	Extremely drought
Continuous no rain days	< 10	10 ~ 20	21 ~ 30	31 ~ 50	> 50
Precipitation anomaly percentage	> -40%	-60% ~ -40%	-80% ~ -60%	-95% ~ -80%	< -95%
Standardized precipitation	> -0.5	-1.0 ~ -0.5	-1.5 ~ -1.0	-2.0 ~ -1.5	< -2.0
Moisture index	> -0.4	-0.65 ~ -0.4	-0.8 ~ -0.65	-0.95 ~ -0.8	< -0.95
Flow anomaly percentage	> -10%	-30% ~ -10%	-30% ~ -50%	-80% ~ -50%	< -80%
Crop water shortage	<5%	5% ~20%	20% ~35%	35% ~50%	>50%

Tab. 2   Regional drought classification table

Administrative	A			
	Light drought	Moderate drought	Severe drought	Extremely drought
County	$0.1 \leqslant A < 0.7$	$0.7 \leqslant A < 1.2$	$1.2 \leqslant A < 2.2$	$2.2 \leqslant A \leqslant 4$

The Xianyang almanac shows, there was a drought occurred in spring, 2008. So May, 2008 is selected as evaluation period. Based on the entropy method, the weigh of each index has been calculated, $W = w_j = $ (0. 164, 0. 138, 0. 185, 0. 128, 0. 153, 0. 232). Then can get the comprehensive evaluation results for each county.

Tab. 3 shows that there are of greater difference among the evaluation results of each drought index. As for the Changwu county, the results of continuous no rain days and moisture index are light drought, but they are severe drought for precipitation anomaly percentage and flow anomaly percentage, and extremely drought for crop water shortage. This result of this situation may be due to rainfall and low temperature. The evaporation is small, but the rainfall is small , leading to the river channel diameter of the flow is too small, and the rainfall does not meet the crop water requirement. The difference among the evaluation results of different indices also shows that a single index of drought assessment has some limitations and cannot reflect on the overall drought situation. According to the Xianyang almanac of 2008 data records, in 2008, Xianyang City had severe dry weather in late spring. The city's precipitation in late spring May is 6 mm to 28 mm, 50% to 89% less than normal, especially in Changwu, Xunyi, Binxian, only 6 mm to 9 mm. The city's average temperature is 17.1~21.6 ℃, 1~2 ℃ higher than the calendar year over the same period, especially southern area. Less rainfall caused some impact on wheat and milky, cut to a certain part of the plots. It claims that the assessment results consistent with the actual drought severity, meaning that the entropy method has certain credibility in drought assessment, and provides technical support for the study of regional drought.

Tab. 3   Drought evaluation results of Xianyang

County	N	Pa	SPI	M	Ir	D_W	Comprehensive evaluation
Changwu	Light	Severe	Moderate	Light	Severe	Extremely	Extremely (2.23)
Binxian	Light	Severe	Moderate	Light	Severe	Extremely	Extremely (2.52)
Xunyi	Light	Severe	Light	Light	Severe	Extremely	Extremely (2.31)
Yongshou	Light	Moderate	Light	No	Severe	Extremely	Severe (2.01)
Chunhua	Light	Moderate	Light	No	Severe	Extremely	Severe (1.99)
Qianxian	Light	Moderate	Light	Light	Severe	Extremely	Severe (2.01)
Liquan	Light	Severe	Moderate	Light	Severe	Extremely	Extremely (2.42)
Jingyang	Moderate	Severe	Light	Light	Severe	Extremely	Extremely (2.45)
Sanyuan	Moderate	Severe	Light	Light	Moderate	Extremely	Extremely (2.21)
Wugong	Light	Severe	Moderate	Light	Severe	Extremely	Extremely (2.44)
Xingping	Light	Moderate	Moderate	Light	Severe	Extremely	Severe (2.18)
Downtown	Light	Moderate	Light	No	Severe	Extremely	Severe (2.08)

5   Conclusions

Different departments have different definition on drought. A variety of factors impact on drought, which results in the evaluation of the regional drought be complicated. Considering the information of meteorology, hydrology, agriculture, and selecting continuous no rain days, precipitation anomaly percentage, standardized precipitation, moisture index, flow anomaly percentage, crop water shortage, this six drought indices aim to construct a comprehensive evaluation indices system. Then the evaluation was based on each single index and finally the multi drought index. The comprehensive drought evaluation is a good solution to the problem that the single index evaluations are of great differences in drought. The assessment, which is based on the entropy method to determine the weight coefficients, is intuitive and simple calculation, also accuracy. Studies have shown that the final evaluation results are consistent with the actual drought, and the method has a certain credibility in the evaluation of regional drought. It can provide technical support for the drought decisions.

References

Li Yuzhong, Cheng Yannian, An Shunqing. Drought in Northern Areas of Law and Drought Integrated Technology [M]. Beijing: China Agricultural Science and Technology Press, 2003.

State Flood Control and Drought Relief Headquarters Office, Ministry of Water Resources, Nanjing Institute of Hydrology and Water Resources. Floods and Drought [M]. Beijing: China WaterPower Press, 1997.

Richard R, Heim J R. A Review of Twentieth – century Drought Index Used in the United States

[J]. Bulletin of the American Meteorological Society, 2002.

Wang Qingyin. Intensional Analysis of the Uncertainty Information [J]. Journal of Northwest University for Nationalities (Natural Science), 2000,03:1 – 5.

Wang Jing, Zhang Jinsuo. Comparing Several Methods of Assuring Weight Vector in Synthetical Evaluation[J]. Journal of Hebei University of Technology, 2001,30(2):52 – 57.

China Meterological Administration. GB/T 20481—2006 Meteorological Drought Grade [S]. Beijing: Standards Press of China,2006.

PRC Minister of Water Resources. SL 424—2008 Drought Grade Standard [S]. Beijing: China WaterPower Press, 2008.

Zhang Yingmin. Xianyang Almanac [M]. Xianyang: Sanqin Press, 2009.

Research on the Measurement Techniques of Terrain and Velocity in the Model Experiment with Hyper – concentrated Flow

Zheng Jun, *Yin Yilai*, *Han Lifeng* and *Li Jie*

Beijing Sinfotek Technology Co. , Ltd. , Beijing, 100085, China

Abstract: This paper focuses on the measurement techniques of terrain and velocity in the model experiment with hyper – concentrated flow. During the measurement of the terrain, it can adjust the transmission frequency of the ultrasound terrain equipment to adapt to the variation of the sediment concentration. To reducing the interference from the electromagnetic, an electromagnetic damper has been used to measure the speed of the flow through the high – frequency sampling techniques. Meanwhile, the measurement and control system has been developed, including hardware such as the measuring bridge, surveying vehicles and other equipment equipped with the terrain meter and flow meter, and software which can achieve the terrain model experiment and automatic measurement of the flow speed through setting the parameters of the measuring section and the measurement model.

Key words: hyper – concentrated flow, model experiment, terrain, velocity, measurement

1   Background review

The common terrain measurement instrument includes photoelectric, resistance, tracking and ultrasonic. Chen Cheng(2009) summarized on the topographic ways on the domestic river model and made a comprehensive comparison between various measuring methods. Then he pointed out the direction of future research which is a wide range of non – contact measurement on the river model or topographic. The theory of the photoelectric instrument is the sludge can block the light. The disadvantage of this method is the poor sensitivity (Tang, 1995). The resistance method is easy to destroy the terrain of the bed. For ultrasonic instrument, Wang Zhengxian (2001) utilized the ultrasonic terrain instrument to achieve a fast qoasi – dynamic scanning over a large pool. The digital signal processing has been used by Lin Haili(2004) to process the ultrasonic echo. In this way, it can work under a complex environment with a quickly and accurately determine the echo which improves the reliability of the ultrasonic terrain instrument.

Velocity is also a key element for the model experiment. The method to test the speed of flow normally has propeller anemometer, hot stream anemometer and acoustic Doppler anemometer. The advantages of the propeller anemometer are easy to use, light and high sensitivity which can detect a low velocity. But it has a high standard for the water quality. When the content of the sand is high, it will influent the running of the propeller, and then need a calibration frequently. The theory of the hot stream anemometer is using a sensor to measure the loss of thermal from the resistance to decide the speed of the flow. It also has a high requirement for the quality of the water which must be clear. Otherwise, it will change the heat dissipation rate when the impurities deposited on the sensor surface (Cai, 2007). Acoustic Doppler anemometer utilizes the acoustic Doppler Effect to measure velocity which has a high – precision measurement of three – dimensional velocity. When the measuring point is very close to the border, it can measure the slow velocity (Liu, 2008). The shortage of the acoustic Doppler anemometer is it can't work in the sandy water when the sediment concentration is over 40 kg/m^3.

2   Terrain measurement

In this paper, ultrasound technology is used to measure terrain. The theory of the ultrasonic instrument calculates the position of the riverbed through the reflection of the acoustic. The propagation of sound wave is affected by temperature in the air, and mainly affected by water temperature, sediment concentration in the water.

It can use the temperature compensation to eliminate the error for the influence of water temperature. It is complex for the influence of sediment concentration. In a certain excitation frequency, it can obtain the best transmission frequency by adjusting the frequency of observation of the actual effect when the sediment concentration increases, the ultrasonic attenuation coefficient increases and a larger emission frequency.

The use of ultrasound terrain instrument to detect the riverbed surface could be connected with the water or without connection. When it works without water, there is no touch between probe and the water which can avoid adding error. When it works connecting water, it only needs to put a small part of the probe in the water which also has a small interference for the stream. When the probe works in the water, there is a penetrating difference between different frequencies. The low frequency wave has a higher penetration than the high frequency wave. It can change the frequency of the wave adapt to the content of the sand by using the software. In this way, it can make sure to get an accuracy data under different concentration of the sand. For example, when the sediment concentration increases, it can reduce the acoustic frequency, and then enhance the ability of sound waves penetrate which could get a more accurate data of the riverbed.

As shown in Fig. 1, the measuring result is corresponding with the actual terrain of the riverbed. The sediment concentration is about 200 kg/m^3.

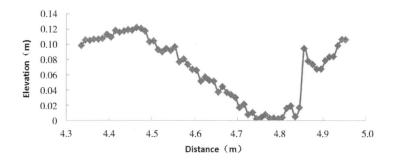

Fig. 1   The measuring data of the terrain

3   Flow velocity measurement

In this paper, it discussed the use of an electromagnetic flow meter for velocity measurement of model experiment in the sandy environment.

The theory of the electromagnetic flow meter is based on Faraday's law of electromagnetic induction, and using the water body as a conductor to measure the flow velocity. It can measure the flow velocity under hyper – concentrated flow, because the electromagnetic flow meter is a fixed instrument which can avoid the influence from the concentration of the sand. High – frequency sampling technique has a less interference from the electromagnetic. Compare to the other velocity measuring equipment, the electromagnetic velocity has a high accuracy, wide measurement range (including the low velocity measurement), a small flow disturbance, and can be adapted to the hyper – concentrated flow stream.

Fig. 2 indicates the data of velocity measurement. The sediment concentration is about 400 kg/m^3. With a continuous collection of about 2,000 data acquisition at some point in the state of constant flow and 1 second interval, the data accuracy is about ± 2.5 mm / s.

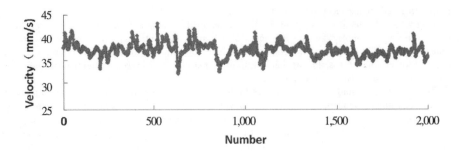

 is the velocity chart.

Fig. 2   Data for flow velocity measuring

4   Measurement control system

To achieve the automatic measurement of the terrain and flow velocity in the model experiment, it have to design a hardware control system which controlled by the software system through the whole process of operation.

The hardware system is mainly consisted of measuring bridge and vehicle. Crossing the track over the model, the measuring bridge can be moved on track by the motor to the appropriate position where measurement need. There is a measuring car, which also has a motor inside, on the bridge. To ensure the accuracy and efficiency of the movement, there is a rack and gear belt on the bridge which could effectively prevent the slipping. Terrain instrument, velocity meter and other equipment can be equipped in the measuring car. Along with the movement of the car, it can be operated on the section of the model experiment. As shown in Fig. 3, there are the measuring vehicle and the bridge.

Fig. 3   Hardware system

The software part provides the parameter settings of the measuring section and the measurement mode. After the setup, the system can achieve an automatic measurement on a particular section under the automatic control mode. The system has two modes of horizontal scanning: a move & stop measurement and constant speed measurement which could adapt to different control tasks and acquisition speed requirements.

5 Conclusions

The measurement technique of the model experiment is an important method to ensure an accurate data obtaining from the experience. In this paper, a new measuring control system has been developed during the research on the measurement techniques of terrain velocity in the model experiment of hyper – concentrated flow.

For an ultrasound terrain instrument, to measuring the surface of the riverbed, it can adapt the frequency according to the content of the sand, through which can maintain the penetration of the acoustic wave in the stream with hyper – concentrated flow. For an electromagnetic damper, it utilized high – frequency sampling techniques which can improve the ability to reduce the interference from the electromagnetic field, and also can avoid the influence from the content of sand. Then, this electromagnetic damper can measure the speed of the stream with hyper – concentrated flow.

To achieve the auto – control measurement, a measuring system has been developed which contains two parts: the hardware including measuring bridge, surveying vehicle and other equipment; the software which can change the parameter setting of the target section, and choose the model of the measurement.

References

Chen Cheng, Tang Hongwu, Chen Hong. Review of the Research on Topographic Survey Methods for Physical River Models in China[J]. Advances in Science and Technology of Water Resources,2009,29(2):76 – 79.

Tang Hongwu, Li Jing, Zhou Haoxiang. Development and Application of Photoelectric Reflecting Topographic Apparatus[J]. Journal of Hohai University,1995,23(1):80 – 84.

Wang Zhenxian, Jin Huanyang. Development of Ultra – acoustic Topographic Surveying Meter for Use in Laboratory[J]. The Ocean Engineering,2001,19(1):94 – 98.

Lin Haili, Qu Zhaosong, Wang Xingkui. Application of DSP to Ultrasonic Topographic Surveying meter[J]. Water Power,2004,30(11):78 – 80.

Cai Shouyun, Yang Daming, Zhu Qijun. Study of Velocity Instruments by the Model Test[J]. Journal of Water Resources and Water Engineering,2007,18(3):36 – 38.

Liu Guoting. Measuring Method for Current Velocity in Hydraulic Model Test and Data Processing Program[J]. Port & Waterway Engineering,2008(12):1 – 4.

Application of BP Artificial Neural Networks Model to Forecast Sediment Concentration of Longmen Station of the Yellow River

Cao Yanxu[1], *Liang Zhongmin*[1], *Huo Shiqing*[2] and *Xu Keyan*[2]

1. College of Hydrology and Water Resources, Hohai University, Nanjing, 210098, China
2. Hydrology Bureau, YRCC, Ministry of Water Resources, Zhengzhou, 450004, China

Abstract: Artificial Neural Networks (ANN) is applied more and more in various fields as its powerful nonlinear processing capacity and the ability of self-learning and self-organizing. This article built a BP network using the rate of sediment transportation and precipitation of areas of the middle and lower reaches of YR from Wubao Station (WBS) to Longmen Station (LMS). It is showed that the forecast precision of BP network is quite high and can be used for reference in engineering practice.

Key words: Artificial Neural Networks, sediment concentration process forecast, BP network, genetic algorithm, self-adjusted momentum-learning rate algorithm

1　Preface

Artificial Neural Networks (ANN) is one branch of artificial intelligence. It is an information processing system which numerous simple artificial neurons of working in parallel are connected to each other and aims to imitate the structure and function of human brain. It has powerful nonlinear processing capacity and the ability of self-learning and self-organizing, and it has good effects on suppressing noise and the impact of information losses of optimal solution.

ANN has been applied in hydrologic forecast since the 1990s. Karuuanithi made the research of CASCADE network on calculation of riverbank conflux first, then Tokar, Yuan Ximing and others also studied on application of ANN to hydrologic forecast. This article, based on VB assembly language, built one BP network which took synthetic sediment concentrations and rainfall of each subinterval of the middle and lower reaches of the Yellow River (YR) from Wubao station (WBS) to Longmen station (LMS) as the objects of this study, and made a preliminary try on the application to sediment concentration forecast.

2　Artificial Neural Networks (ANN)

There are tens of ANN and can be divided into several kinds of networks according to their different topological structures, transfer modes of information flow and training methods, including back propagation network, perception, self-organizing feature mapping, Hopfield network, boltzmann machine, adaptive resonance theory and so on. This paper used one BP network which is most widely used and have good effects by practice now.

2.1　The working principle of BP neural network

BP(Back Propagation) neural network, suggested by a group of scientists led by Rumelhart and McCelland, is a kind of multilayer feedforward neural networks, and it has been one of the networks which are most widely used in the world at present. BP network is divided into input layers, hidden layers and output layers. Input layers receive information from outside, then hidden layers restore and express information, finally output layers make discriminations and decisions for input information.

The basic idea of BP network algorithm is: the learning process is made up of signal forward-transmission process and error back-propagation process. When signals transmit forward, input samples are input from input layers and then processed step by step by hidden layers, finally they are transmitted to output layers. If the output values do not match with the expected values, then the errors propagate back from output layers to input layers by hidden layers step by step. All the neurons of layers share the error, thus they will get their own error signals, and according these signals, weights of input of each neuron of the network can be corrected. The process which corrects weights of each neuron of each layer through signal forward-transmission and error back-propagation is called training process of network, and it needs numerous iterations until the error is limited in an allowed range.

2.2　The structure of BP network

This article built a BP network of 3 layers shown in Fig. 1, including one input layer, one hidden layer and one output layer. P represents input information, T represents output information, w and b are the weight value and the threshold value of each layer, and f represents the transfer function of neurons of each layer.

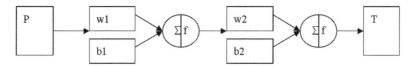

Fig. 1　The structure of BP network

2.3　Several methods to improve forecast precision

2.3.1　Genetic algorithm

It can optimize initial weights of network effectively and avoid network training falling into local minimum point, so as to improve the fitting precision of the network.

2.3.2　Self-adjusted momentum- learning rate algorithm

The standard BP algorithm often makes the learning process oscillate, and the momentum theory is able to reduce the sensitivity of the network to local details of the error curve. The improved algorithm is proposed as follows:

$$W(k+1) = W(k) + \alpha[(1 - \eta)D(k) + \eta D(k-1)] \tag{1}$$

$$D(K) = \frac{-\partial e}{\partial w(k)} \tag{2}$$

Self-adjusted learning rate contributes to shorten learning time. Choosing the improper learning rate is the most important reason to slow convergence speed of standard BP algorithm, so there is an improved algorithm of self-adjusted learning rate.

$$W(k+1) = W(k) + \alpha(k)D(k) \tag{3}$$

$$\alpha(k) = 2^\lambda \alpha(k-1) \tag{4}$$

$$\lambda = \text{sgn}[D(k)D(k-1)] \tag{5}$$

where, $D(k)$ is the negative gradient at k time point; α is the learning rate, $\alpha > 0$; η is the momentum factor, $0 \leqslant \eta < 1$.

3　Practical application

3.1　Basic datum of the basin

YR is the second longest river in China with a 5,464 km main stream and a 7.95×10^5 km^2 area. Its average annual nature runoff volume is about 58 billion m^3, and its average annual precipitation is about 476 mm. The geo-graphical conditions over YR basin are very different, and the distribution of the source regions of water and sediment is very uneven. The upper stream of YR is the main source region of water, while the middle stream of YR is the main source region of sediment. The research object of this article is the stream segment from WBS to LMS which belong to the middle stream of YR, and there are 5 tributaries in the left bank and 4 tributaries in the right bank.

For convenience the research, the research region is divided into three sub-basins. A sub-basin consists of Qingjian River (QR), Wuding River (WR), Sanchuan River (SR) and Quchan River (QCR). B sub-basin comprises of the area which is the lower reaches of QR in the left bank of the main stream, and C sub-basin is the area in the corresponding right bank. The specific situation of division is shown in Fig. 2 as follows.

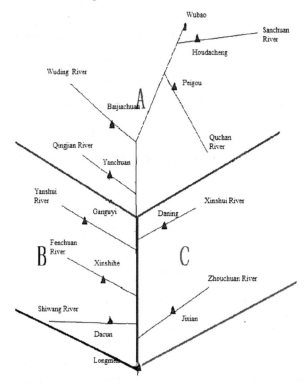

Fig. 2　Division of the researched basin

3.2　The choice of neurons of input layer

Water and sediment coming from the upper stream have an effect upon the sediment concentration process of LMS, thus the impact factors have to be flow, sediment transportation rate and rainfall in the upper stream. Since sediment concentration is coursed by sediment transportation and flow, the impact factors can be determined two: sediment concentration and rainfall in the upper stream.

This article aims to forecast the sediment concentration process of LMS according to the conditions of water and sediment concentration from the upper stream. The Fig. 2 shows that there are several control stations in each sub-basin. Refer to the concept of synthetic flow, here make the sum of instantaneous sediment concentrations of all control stations in each sub-basin synthetic as the synthetic sediment concentration of each sub-basin at this time. Strictly speaking, the synthetic sediment concentration should consider about the time of sediment transportation from each control station to the confluence cross-sections. Because of that travel time of sand is difficult to determine, and most sand peaks lag behind the corresponding flood peaks from conditions of water and sediment concentration of floods selected, using travel time of floods instead of travel time of sand is unreasonable. So temporarily the sum of instantaneous sediment concentrations of all control stations in each sub-basin synthetic is made as the synthetic sediment concentration of each sub-basin at this time. The formula of the synthetic sediment concentration of each sub-basin is shown as Eq. (6).

$$Q_{s1} = Q_{sW} + Q_{sH} + Q_{sP} + Q_{sB} + Q_{sY}, Q_{s2} = Q_{sG} + Q_{sX} + Q_{sC}, Q_{s3} = Q_{sN} + Q_{sJ} \qquad (6)$$

where, Q_{s1}, Q_{s2}, Q_{s3} are the synthetic sediment concentration of A, B and C sub-basins respectively; Q_{sW}, Q_{sH}, Q_{sP}, Q_{sB}, Q_{sY}, Q_{sN}, Q_{sG}, Q_{sX}, Q_{sC}, Q_{sJ} are the instantaneous sediment concentrations of each control station.

The synthetic sediment concentrations at different time may all have effects upon the one of LMS. Considering factors including the simplification of model structure, the leading time of model and so on, this paper defines that only choose the synthetic sediment concentrations at time of $t - \tau_1$ in each sub-basin as its sediment concentration input factor (t is the current forecast time of LMS), which is recorded as $Q_{si}(t - \tau_1)$.

As for the rainfall, the formula of accumulation of no-point average rainfall of each sub-basin is shown as follows:

$$\overline{P}_{m_i,i} = \sum_{j=1}^{n_i} P_{m_i,j}/n_i, \ i = 1 \sim 3, \ j = 1 - n_i \qquad (7)$$

where, $\overline{P}_{m_i,i}$ is the accumulation of no-point average rainfall of m_i hours in the i – th sub – basin; $P_{m_i,i}$ is the accumulation of rainfall of m_i hours of each precipitation station in the i – th sub – basin; n_i is the number of precipitation stations in the i – th sub-basin.

As the input factor of sediment concentration which is referred to in the preceding part of the article, each sub-basin also only chooses accumulation of no-point average rainfall of m_i hours which before the $t - v_i$ time. After several experiments, it finds that when $m_i = 48$ h there is a highest sensitivity in the impact of the input factor of no-point average rainfall to the sediment concentration of LMS. So the input factor of rainfall of the three sub-basins is recorded as $\overline{P}_{48,i}(t - v_i)$.

By above knowable, there are six input factors in total, and among them, $t - \tau_1$ represents the travel time of the sediment of each sub-basin transporting to LMS, and it can be calculated by the lag time of the sand peak, $t - v_1$ represents the impact time of no-point average rainfall of each sub-basin impacting in the sediment concentration of LMS, that is, in each sub-basin, the accumulation of no-point average rainfall of 48 hours which before the $t - v_1$ has a biggest impact in the sediment concentration of LMS, which can be got from correlation analysis. The lag time of the sand peak of A, B and C sub-basins are respectively 20 h, 21 h and 17 h. according to the flow-sediment data of twenty-five floods of hydrologic stations from WBS to LMS between 1980 and 2007.

3.3 The determination of the number of neurons of the hidden layer

There is no unified method of calculation to choose the number of neurons of the hidden layer, if it is too few, the convergence speed of the network will be too slow, and on the contrary, the calculated quantity will be increased and it will cause over-fitting, and the generalization ability of the network will be reduced. Thus, the number of neurons of the hidden layer should be determined by experiments.

Generally, the number of neurons of the hidden layer should meet this requirement:

$$\min(n_i, n_0) \leqslant n \leqslant 2n_i + 1 \qquad (8)$$

According to several experiments, when the number of neurons of the hidden layer is ten, the training effect of the network is the best.

The transfer function of hidden layer and output layer in BP network can have several options: (s type) logarithm function, (s type) tangent function, pure linear function and so on. Tangent function (s type) which is differentiable function can help the input range of neurons map from $(-\infty, +\infty)$ to $(-1, +1)$ and it is appropriate for the training of BP network. The expression of (s type) tangent function is:

$$f(x) = \frac{1}{1 + e^{-x}} \tag{9}$$

According to extensive experiments and experience, choosing tangent function (s type) as the transfer function of hidden layer and output layer could get a good effect.

3.4 The assessment indexes of the error of the process of sediment concentration forecasting

It should be considered of the inspection of the forecast model from two aspects: one is that if the model can be available, the other is the precision of the results of forecast. The standard to assess the precision of the forecast model is considered about similarity between the forecast results and the actual values. As the uncertainty factors of sediment concentration changes are more than the ones of flood changes, and the sediment concentration process forecast is still in the stage of exploration, there is no reasonable standard to assess a forecast model of sediment concentration, so here only comparing the shape and sediment transportation of forecast process with the ones of actual measurement process. The assessment index of shape uses the coefficient of determination and the ones of sediment transportation during a flood use absolute error and relative error. The coefficient of certainty is calculated by the following expressions:

$$DC = 1 - \frac{S_{\mathrm{C}}^2}{\sigma_y^2} \tag{10}$$

$$S_{\mathrm{C}} = \sqrt{\frac{\sum\limits_{j=1}^{n_i} (y - y_i)^2}{n}} \tag{11}$$

$$\sigma_y = \sqrt{\frac{\sum\limits_{j=1}^{n_i} (y_i - \bar{y})^2}{n}} \tag{12}$$

where, S_{C} is the mean square deviation of forecast errors; σ_y is the mean square deviation of forecast elements; y_i and y respectively represent the measured values of sediment concentration and forecast results of sediment concentration; $\bar{y}$ is the average of the measured values of sediment concentration.

3.5 Input data preprocessing

Because of the difference between orders of magnitudes of the input factor of sediment concentration and rainfall, the input data should be normalized in order to speed up the network convergence speed,

This article applies the following function to normalize data, that is maximize-minimize method.

$$f(x_i) = \frac{x_i - x_{\min}}{x_{\max} - x_{\min}} \quad (i = 1, 2, \cdots, N) \tag{13}$$

where, $x_{\min}$ is the minimum of input data; $x_{\max}$ is the maximum of input data; N represents the number of training samples.

3.6 Forecast modeling

There are flow-sediment data of twenty-five floods of hydrologic stations from WBS to LMS between 1980 and 2007 to be used as network training samples in this article, and floods of 19820729, 19990719 and 20010721 are used as calibration sessions.

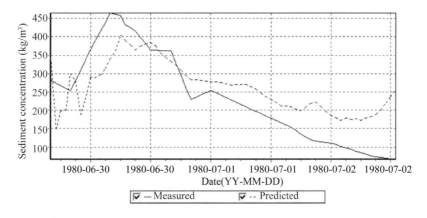

Fig. 3   19800629

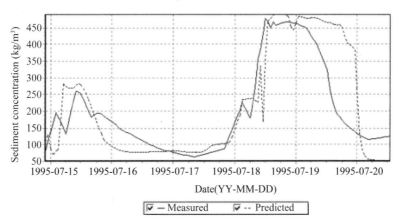

Fig. 4   19950715

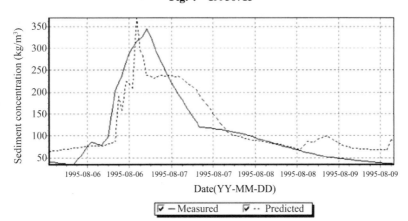

Fig. 5   19950804

By above knowable, using the synthetic sediment concentration of A sub-basin at (t – 20 h) time, B sub – basin at (t – 21 h) time, C sub-basin at (t – 17 h) time, and the accumulation of no-point average rainfall of 48 hours before the (t – 23 h) time in A sub-basin, before (t – 21 h) time in B sub-basin, before the(t – 18 h)time in C sub-basin as input neurons to forecast the sediment concentration of LMS at time. The training times is 10,000, the learning rate is chosen as 0.05, the momentum factor is chosen as 0.9, and the selective probability, the crossover probability and the mutation probability in genetic algorithm are all 0.1. And using (s type) tangent function as the transfer function of hidden layer and output layer to train 981 sets of data. The calibration results are shown as the figures from Fig. 3 to Fig. 5. The coefficient of certainty, the forecast values of sand peak and the lag time of sand peak are shown in Tab. 1.

Tab. 1　Forecast Results of the Model

Flood number	Measured sand peak	Forecast sand peak	Relative error(%)	Lag time	The coefficient of certainty
19800629	465.0	404.9	12.9	+2	0.567
19950715	468.3	490.0	– 4.63	+2	0.512
19950804	344.3	369.8	– 7.4¹	– 3	0.786

4　Conclusions

This article first analyzed the theory and mechanism of sediment forecast based on neural network technology, and further discussed how to optimize the network and improve the forecast precision, and then tested the feasibility of this neural network model combining flow-sediment data of hydrologic stations from WBS to LMS in the middle and lower reaches of YR. The inspection results show that the sediment forecast model based on ANN has worthiness in practical application.

References

Zhang Guiqing. The Introduction Theory of ANN [M]. Beijing: China Water Conservancy and Electricity Press, 2004.

Zhao Qi, Wu Sufen, Xue Yan, et al. Application of Neural Network Model in The Medium and Long Term Forecasting [J]. Heilongjiang Science and Technology of Water Conservancy, 2010, 38(3).

Wang Guangsheng, Su Jialin, Shen Biwei, et al. Application of Neural Network in Channel Flow Forecasting[J]. Hydrology, 2009, 29(5).

Zhang Defeng. Designation of neural network in MATLAB [M]. Beijing: China Machine Press, 2009.

Jin Shuangyan. Study on The Model of Sediment Forecast in The Middle and Lower Yellow River [D]. Nanjing: Hohai University, 2007.

Study on Soil Moisture Retrieval Based on Evapotranspiration Fraction Using Remote Sensing in the Yellow River Irrigation Area

Chen Liang, Shen Yuan, He Houjun, Han Lin,
Song Shuchun and *Zhang Xiangjuan*

Information Center of Yellow River Conservancy Commission, Zhengzhou, 450004, China

Abstract: Soil moisture in field scale of the Yellow River irrigation area is critical to decision making for drought resistance and disaster relief. Evapotranspiration fraction (EF) can reflect the capacity of evapotranspiration and has a high relationship with soil moisture, which is one of index in common use for soil moisture retrieval based on remote sensing. In this paper, People's Victory Canal (PVC) of the lower Yellow River was selected as study area to study soil moisture retrieval model based on EF using remote sensing. On the basis of surface energy balance algorithm for land, EF was calculated by SEBAL model using TM images and meteorological data in 2010. Field observation of soil moisture was applied to validate the soil moisture models. The model was applied to map the distribution of soil moisture in PVC. The result indicated that the empirical retrieval model was suitable for accessing soil moisture of the Yellow River irrigation area in field scale.

Key words: soil moisture, evapotranspiration fraction, the Yellow River irrigation area, SEBAL, remote sensing

1 Introduction

As global warming, severe drought continuing, development of society and economic and population increasing, water shortage is more serious in the Yellow River Basin, and it is becoming one of the restricted factors to implement sustainable development of the basin. In the recent years, the lower Yellow River irrigation area, one of main food production area in China, suffered from drought every year and it threatened safe food production. Soil moisture in field scale of the Yellow River irrigation area is critical to decision making for drought resistance and disaster relief. Acquiring soil moisture of the Yellow River irrigation area, it provides information for drought assessment and prediction, which supports decision making of drought resistance.

Soil moisture are multifaceted functions that depends upon soil properties, climatic conditions, land use, vegetation, and topography which cause these parameters to vary in both space and time (Parlange et al., 1995). Traditional field-based methods are cumbersome, expensive, and provide data for limited spatial and temporal coverage. Due to the advantage of short-revisit and wide cover, remote sensing is a powerful tool for estimating soil moisture over wide spatial scales and has been widely used for soil moisture retrieval. However, many soil moisture retrieval models are limited by coverage of vegetation and retrieval depth of soil moisture. The thermal inertia based retrieval models are suitable for bare soil or lower coverage of vegetation, while the vegetation index based retrieval models are suitable for higher coverage of vegetation. Meanwhile, most optical and microwave techniques can only estimate the near-surface soil moisture content, i. e., the upper few centimeters of the soil profile (Xiao, 1994; Engman and Chauhan, 1995; Jackson, 1997). Many water resource applications, such as drought resistance and water regulation, require information on the root zone soil moisture content, which coincides with approximately the first meter of soil depth at a resolution of one hectare or finer and over wide spatial scales.

Evapotranspiration Fraction (EF) can reflects the capacity of evapotranspiration and has a high relationship with soil moisture. The study of Bastiaanssen et al. (Bastiaanssen et al., 1997) has built an empirical retrieval model of root zone soil moisture based on EF and the model has been tested and applied in many countries, such as Pakistan, Mexico and Australia (Christopher et al.,

2003; Hafeez, 2007). However, few studies focus on applying the model to retrieve soil moisture in China.

In this paper, EF was estimated by SEBAL (Surface Energy Balance Algorithm for Land) using Landsat TM data in the People's Victory Canal (PVC) of the lower Yellow River. The soil moisture retrieval model based on EF proposed by Bastiaanssen et al. was validated using ground observation data of study area.

2　Study area and data

2.1　Study area

PVC is the first large-scale water diversion and irrigation project constructed in the lower and middle reaches of the Yellow River since the founding of P. R. China in 1949. The canal flows from southwest to northeast from Wuzhi-Qinchang Dike at the northern bank of the Yellow River. The canal supports 7 counties and 1 suburban of Xinxiang, Jiaozuo and Anyang City, Henan Province with the population of 2.658 million. The irrigation area locates at alluvial plain of the Yellow River and Qin River. The annual rainfall is 617 mm, but the seasonal distribution of precipitation is uneven and 83% of that is in summer. This area suffers from drought, flood and salinity disaster. The main crops are wheat, corn and rice in PVC.

2.2　Study data

Remote sensing data, DEM, meteorological data and ground observed soil moisture data were used in this study.

Landsat 5 TM is a multi-spectral scanning radiometer that was carried on board Landsat 5, which has been widely applied in many fields. TM data cover six visible and near-infrared spectral bands with the resolution of 30 m and one thermal infrared spectral band with the resolution of 120 m. In this paper, Landsat 5 TM data was acquired on May 11, 2010 and June 12, 2010. Wheat was in filling stage with higher vegetation cover on May 11 and corn was in sowing stage on June 12.

Meteorological data was acquired by local meteorological department, including daily maximum air temperature, daily minimum temperature, wind speed and relative humidity.

DEM was obtained by Shuttle Radar Topography Mission (SRTM) of NASA. Resolution of the data is 90 m and it can be downloaded from website (http://srtm. csi. cgiar. org/).

The soil moisture observed by ground observation stations in study area was used to validate the accuracy of the model. Depth of the data is 10 cm, 20 cm, 30 cm, 40 cm, 50 cm, 60 cm, 80 cm, 100 cm.

3　Method

3.1　SEBAL

According to surface energy balance, land surface absorbs energy from sun and transforms to other energy in several ways. One part of net radiation is used to heat near surface air named as sensible heat flux, one part is used to heat surface soil named as soil heat flux and another part is used to vapor water named as latent heat flux. SEBAL (Surface Energy Balance Algorithm for Land, SEBAL) is a remote sensing flux algorithm that solves the surface energy balance on an instantaneous time scale and for every pixel of a satellite image. The method is based on the computation of surface albedo, surface temperature and vegetation index from multi-spectral satellite data. Net radiation, soil heat flux and sensible heat flux are estimated combining with meteorological data, such as air temperature, wind speed, and relative humidity. SEBAL is a well-tested and widely used method from many study cases (Tenalem et al. , 2003; Allen et al. , 2003; Li et al. ,2005; Bastiaanssen et al. , 2003).

The SEBAL procedure consists of a suite of equations that solve the complete energy balance:

$$R_n = G + H + \lambda E \qquad (1)$$

where, λ_E is latent heat flux; R_n is net radiation at the surface; G is soil heat flux; and H is sensible heat flux to the air.

R_n is computed for each pixel using albedo and transmittances computed from short wave bands and using long wave emission computed from the thermal band. Soil heat flux is predicted using vegetation indices computed from spectral data and net radiation. Sensible heat is calculated from several factors: surface temperature and a single wind speed measurement at the ground, and estimated surface roughness and surface-to-air temperature differences predicted from thermal infrared radiances. To compute sensible heat flux, "hot" pixel and "cold" pixel are used for estimating surface-to-air temperature differences. And iterative predictions of sensible heat are improved using atmospheric stability corrections based on Monin-Obukhov similarity. Iterations of sensible heat flux were conducted at least five times. The detailed process of the model can be found in the literature (Du, 2010).

The EF is defined as the ratio between the latent heat flux and the difference between net radiation and ground flux:

$$\Lambda = \frac{\lambda E}{R_n - G} \qquad (2)$$

where, Λ is the value of EF. EF expresses the ratio of the actual to the vegetation evaporative demand when the atmospheric moisture conditions are in equilibrium with the soil moisture conditions. The value of EF usually ranges from 0 to 1, which represents zero to maximum evapotranspiration.

3.2 Soil moisture retrieval model

Since EF can be calculated over large areas using satellite imagery, it is a suitable indicator for the description of soil moisture conditions on a regional scale, while the traditional methods of study of soil moisture storage in the unsaturated zone may present many problems (F. Sini, 2005). An empirical relationship between evaporative fraction and volumetric soil moisture content (θ) was developed by Bastiaanssen (Bastiaanssen et al., 2000).

$$\Lambda = 0.421\ln\theta + 1.284 \qquad (3)$$

4 Results and discussion

The SEBAL procedure consists of a suite of algorithms, the procedure was implemented in the ModelMaker module of the ERDAS software in this case. After the TM data was rectified, EF was calculated based on SEBAL on May 11, 2010 and June 12, 2010. And soil moisture was estimated using Eq. (3), the results were shown in Fig. 1 and Fig. 2.

From Fig. 1, the soil moisture of wheat is higher and the values were bigger than 0.25 cm^3/cm^3 in most area. It was benefit for wheat growth since wheat was in the filling stage during this period. Bare soil and village had lower value of soil moisture. Comparing Fig. 2 with Fig. 1, soil moisture in June was smaller than that in May because it was time for sowing corn and most of the irrigation area was bare. The higher values on the northwest and northeast were caused by cloud. Therefore, the estimated results were consistent with the actual situation both of spatial and temporal distribution.

The soil moisture observed by ground observation stations was used to validate the accuracy of the model. The depths of root zone soil moisture were 0 ~ 80 cm on May 11 and 0 ~ 10 cm on June 12. Difference between root zone soil moisture content estimated from empirical model and field observation over a range of evaporative fractions was shown in Fig. 3.

From Fig. 3, the mean value of difference between soil moisture content estimated from empirical model and field observation was 0.042 cm^3/cm^3 and most of the difference was controlled under ± 0.05 cm^3/cm^3. It showed that estimated value could maximally differ from actual measurements by 0.115 cm^3/cm^3, with 0.87 of EF. The nonlinear relationship of the evaporative

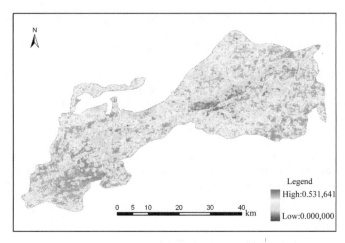

Fig. 1 Estimated soil moisture map on May 11, 2010

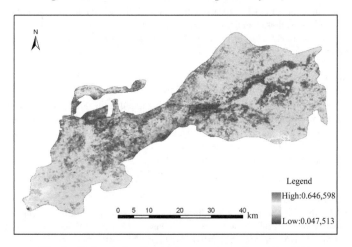

Fig. 2 Estimated soil moisture map on June 12, 2010

fraction and soil moisture demonstrates that little crop water stress exists if the soil is near saturation. The value of EF > 0.8 is generally associated with favorable crop growing conditions without crop water stress (Christopher et al., 2003). Moreover, there is a systematic trend in the error: a lower error in is found that when EF ranges from 0.65 to 0.85, being the usual range of Λ for irrigated agriculture, and the error increases at high and low values of EF(Wang et al., 1998; Bink, 1996; AHMAD et al., 2003). Therefore, the results of this study were consistent with the previous studies and the empirical model proposed by Bastiaanssen can be applied to PVC.

5 Conclusions

In this paper, EF was calculated by using SEBAL and the empirical retrieval model of soil moisture based on EF applied in the Lower Yellow River irrigation area. Through comparing the estimated soil moisture with field observation data, the mean estimated error of the model was 0.042 cm^3/cm^3 and the results were consistent with the previous studies. The results indicated that the empirical retrieval model of soil moisture based on EF was suitable to estimate soil moisture in

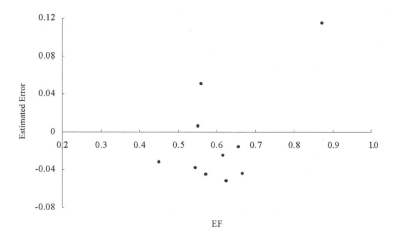

Fig. 3 Difference between soil moisture content estimated and field measurement

the Yellow River irrigation area in field scale and could provide supports for drought resistance and disaster relief.

Acknowledgements

The study was sponsored by governmental public industry research special funds for projects of MWR (fund No. 200901021) and the Excellent Program of Technology Foundation for Selected Overseas Chinese Scholar, Ministry of Personnel of China (No. 03) in year 2010.

References

Parlange M B, et al. Regional Scale Evaporation and The Atmosphere Boundary Layer [J]. Reviews of Geophysics, 1995, 33: 99 – 124.

Xiao G, Chen W, Sheng Y. A Study on Soil Moisture Monitoring Using NOAA Satellite[J]. Journal of Applied Meteorological Science, 1994, 5(3): 312 – 318.

Engman E T, Chauhan N. Status of Microwave Soil Moisture Measurements with Remote Sensing [J]. Remote Sens. Environ. , 1995, 51,189 – 198.

Jackson T J. Soil Moisture Estimation Using Spectral Satellite Microwave/imager Satellite Data over a Grassland Region [J]. Water Resour. Res. , 1997, 33(6): 1475 – 1484.

Bastiaanssen W G M, Pelgrum H, Droogers P, et al. Area – average Estimates of Evaporation, Wetness Indicators and Top Soil Moisture During Two Golden Days in EFEDA [J]. Agric. Meterol. , 1997, 87, 119 – 137.

Christopher A S, Bastiaanssen W G M, Ahmada M. Mapping Root Zone Soil Moisture Using Remotely Sensed Optical Imagery [J]. Journal of Irrigation and Drainage Engineering, 2003, 129(5): 326 – 335.

Mohsin Hafeez, Shahbaz Khan, Kaishan Song, et al. Spatial Mapping of Actual Evapotranspiration and Soil Moisture in the Murrumbidgee Catchment: Examples from National Airborne Field Experimentation [C]// International Congress on Modelling and Simulation, 2007, 2611 – 2617.

Tenalem A. Evapotranspiration Estimation Using The Matic Mapper Spectral Satellite Data in the Ethiopian Rift and Adjacent Highlands [J]. Journal of Hydrology, 2003, 279, 83 – 93.

Allen R, Morse A, Tasumi M. Application of SEBAL for Western US Water Rights Regulation and Planning [R]. Montpellier, France: ICID Workshop on Remote Sensing of ET for Large Regions, 2003.

Li H, Lei Y, Zheng L, et al. . SEBAL Model and Its Application in the Study of Regional

Evapotranspiration [J]. Remote Sensing Technology and Application, 2005, 20 (3): 321 – 325.

Du J. Evapotranspiration Estimation Based on Multi – Remote Sensing Data over Sanjiang Plain [J]. Beijing: Chinese Academy of Sciences, 2010.

Sini F. Satellite Data Assimilation for the Estimation of Surface Energy Fluxes [D]. Potenza: University of Basilicata.

Wang J, Bastiaanssen W G M , Ma Y, et al. Aggregation of Land Surface Parameters in the Oasis – desert Systems of Northwest China [J]. Hydrological Processes, 1998, 12, 2133 – 2147.

Bink N J. The Structure of the Atmospheric Surface Layer Subject to Local Advection [D]. Wageningen: Wageningen University, 1996.

Ahmad M, Bastiaanssen W G M. Retrieving Soil Moisture Storage in the Unsaturated Zone Using Satellite Imagery and Bi – annual Phreatic Surface Fluctuations [J]. Irrigation and Drainage Systems, 2003, 17, 141 – 161.

An Application Case Study of the Yellow River's Flood Control and Drought Relief Network Platform

Ding Bin, *Yao Baoshun*, *Du Wen*, *Li Yong* and *An Dong*

Information Center of Yellow River Conservancy Commission, Zhengzhou, 450003, China

Abstract: The report introduces Internet/Intranet based "Flood Control and Drought Relief Network Platform on the Yellow River", which is developed on the basis of current leading technologies such as java technology, internet technology, report server and middleware technology. Basic frame of the system, logic structure and function configuration are provided, in addition, key technologies used in this system are investigated and discussed. According to the demands of the Yellow River Conservancy Commission and several military departments, the project accomplished constructing Virtual Private Network (VPN) connections between those users. It connects the flood control information between the Yellow River Flood Control and Drought Relief Headquarter and related military area command. Meanwhile, it meets the demands of related military area command on flood control and ice jam flood control information, which can provide sufficient evidences on flood control and drought relief's dispatch. The result from practices suggested that the application could contribute to the "Digital Flood Control Project" by providing users with strong and reliable information support.

Key words: flood control and drought relief, digital flood control, network platform, middleware

1 Introduction

A flood control and drought relief network platform has been recently developed, and provided efficient functions to report rainfall state, water regimen, freeze regimen and emergent risk information for flood control projects. The application has also been proved that it could enhance the communication efficiency between the Yellow River Flood Control and Drought Relief Headquarter and several military commands (ie. Beijing, Lanzhou and Jinan).

According to the demands of Yellow River Conservancy Commission (YRCC) and several military departments, the project accomplished constructing VPN connections between those users by using existed network and other IT recourses. Various technologies have been applied in this application, such as java, internet, report server, middleware, et al. Displayed as the style of B/S (Brower/Server), the application could provide users with rich information of water (ie. river and reservoir) regimen and rainfall state. Distinctive from other similar applications, the project supplies abundant information demonstration solutions, such as tables, texts, and graphs. As follows, the paper introduces our application by elaborating its basic framework, logical organization and functional structure. Moreover, it also includes detailed investigation and discussion on key technologies we applied (Hui, Wu and Boshun, Yao, 2007).

2 Framework of system

"Yellow River's Flood Control and Drought Relief Network Platform" focuses on processing water regimen information of the Yellow River Basin. It involves the work of information integration, network connection, database construction and program development. According to the proposal of "Digital Yellow River", the platform took the design of three layers' framework. It includes infrastructure, application support platform, and application system. As follows, the framework's detail is displayed in Fig. 1.

The main task of infrastructure is to transport the data of water regimen and system management and to manage software and hardware that were related to data storage. It is the

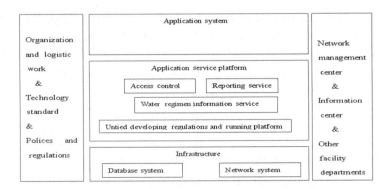

Fig. 1　Framework of Flood Control and Drought Relief Network Platform

foundation of the "Flood Control and Drought Relief Network Platform". The information center of YRCC provides the data needed to be process, and a specific computer network is used to transmit those data to application systems.

Application support platform is the layer providing software's support for the "Flood Control and Drought Relief Network Platform". For instance, it supplies a standard environment for developing software, which includes standard application servers and developing tools. Furthermore, it also furnishes some essential services like data accessing, report forms, et al.

"Flood Control and Drought Relief Network Platform" aims to serve the members of the Yellow River Flood Control and Drought Relief Headquarter. Thus, the application layer is the key part of this project. Various kinds of vivid functions have been developed, such as statistical analysis report forms, hydrographs, et al. .

Moreover, regulated and well-designed technical standards, organization work and policies regulations are the most fundamental parts for the "Flood Control and Drought Relief Network Platform". In the beginning of current project, several standard regulations and policies for operation and management are set up. And it does not only guarantee the financial investment, but also build up a good system in which has a completed management framework for technology and human resources (YRCC, 2002).

3　Functions' structure

The foundation of "Yellow River's Flood Control and Drought Relief Network Platform" is the "Digital Yellow River" project, as they share lots of common resources between different systems and between different departments.

The application system can be divided into these parts: watercourse regimen, reservoir regimen, newest regimen, essential regimen, daily forecasting report, rainfall per station, rainfall in multi-stations, rainstorm retrieval, rainfall report, ice flood state, daily ice flood report, real-time ice state, analysis of ice flood. The detailed structure is shown in Fig. 2.

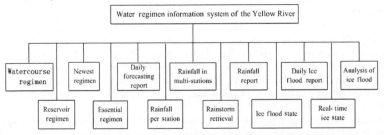

Fig. 2　Functions' structure

4 Analysis of key technologies

4.1 Systems' running and developing environment

The water regimen information system is built up on the basis of system's software and hardware support. The hardware and software facilities use existed resources from flood control office. It includes hardware servers, database software and server's software (Bin, Ding, 2008).

Oracle database is used in data management, as this solution is easily to extend and had good performance considering its cost. The system of platform is Windows 2000 Server, and BEA Weblogic Server 9.1 is installed for server platform.

After considering the performance, usability and cost of developing tools, the tools, such as Myeclipse, Flex, are used in this project. The result suggests that those tools have ample templates and controls, and the utilization of them significantly decreases developing time, cost and promotes project's quality.

The information system is a database application in which has strong interaction property with users. By applying the well-known MVC framework and JSP technology, it has sufficient ability to deal with different kinds of information, and supplies users with efficient interaction functions. The technology of JSP realizes the function of accessing database, while JAVA was used to process data. With the cooperation of these different technologies, the system can provide users with completed and handy functions.

4.2 J2EE framework

Previous work of the "Digital Yellow River" project has obtained abundant experience and successful cases in applications of J2EE. J2EE is well known for its openness and transplantable ability. For current project, J2EE is used as the technology framework. Considering the complexity of system, the design also considers the integration problem with net platform (Bin, Ding and Shaofei, Ren, 2008).

According to the work of "Yellow River's reservoirs dispatching information platform", application support platform was constructed. It can be divided into two parts. The first part is called as software framework platform, as it is a basic framework platform that based on basic control function models. On the other hand, the other part is built up according to the Yellow River's professional models, and it is named as professional servers. Application support platform is set up on hardware and operation systems. It is the collection for different services. Those services are efficient and professional, and they serve the flood control work of the Yellow River. Moreover, the application support platform services are well encapsulated. In other words, relevant top applications can use the function of service without knowing the service's processing detail.

As for application system, it is import to notice its problem of integration, expansion and security. Result from the use of "Yellow River's Flood Control and Drought Relief Network Platform" reveals significant advantages of J2EE technology concerned on these problems. It has powerful ability of transporting, processing and displaying information.

4.3 Middleware technology

Middleware technology is also widely used in development. The technology encapsulates the detail of processing. Hence, programmer can use or adapt the function without known the detail. The utilization of middleware can greatly reduce the difficulty of software development, shorten the development cycle, and improve the quality of software (Zhang Yunyong and Zhang Zhijiang, 2005).

Middleware is deployed in the application service platform, and it includes access control middleware, report service middleware and data access middleware. Middleware can assist system to realize most of functions in the system.

4.3.1　Access control middleware

The use of access control middleware is to set up users' information and arrange the limits of authority of users. By using this type of middleware, users can use the system safely.

4.3.2　Report service middleware

The goal of report service middleware is to build the general report forms and provide the environment for generating report forms. This type of middleware can construct report forms faster, and hence improve work efficiency.

4.3.3　Data access middleware

Data access middleware constructs access channels from application system to database, and provided fundamental applications for application service platform. The use of this type middleware can optimize the access to database, improve the processing performance of database and therefore promote the efficiency of systems' reaction time.

5　Conclusions

The use of application support platform can eliminate the diversity of hardware and operation system. The professional servers can satisfy users' demands and make the "Yellow River's Flood Control and Drought Relief Network Platform" accessible cross platforms. At the meantime, the setup of application support platform makes the whole system more stable. The professional servers satisfies users' demands when information is changed. The "Yellow River's Flood Control and Drought Relief Network Platform" uses J2EE as the platform framework, and builds up profession work platform by applying MVC model. It did not only have the property of stability, but could also satisfy users' future demands. Moreover, the platform could also be used as an independent information system. The result suggests that the "Yellow River's Flood Control and Drought Relief Network Platform" is not only compatible with different platforms and databases, but it also convenient to extend its functions.

The application has successfully applied to connect the communication work between military commands (i. e. Beijing, Lanzhou and Jinan) and the Yellow River's Flood Control and Drought Relief Headquarter. It significantly improves the efficiency of searching and reviewing flood control and drought relief information. The result from the practice suggested that the application can contribute to the "Digital Flood Control Project" by providing users with strong and reliable information support.

References

Wu Hui, Yao Baoshun. Applications of GIS in Flood Control Work of Yellow River [J]. Geomatics World, 2007(1).

Yellow River Conservancy Commission. The Scheme of "Digital Yellow River" [R]. Zhengzhou: Yellow River Conservancy Commission, 2002.

Ding Bin. WebGIS-based Consultation System in Flood Control Engineering and Risk Affairs for Yellow River's Lower Reaches[J]. Journal of North China Institute of Water Conservancy and Hydroelectric Power, 2008(05).

Ding Bin, Ren Shaofei. A Case Study of Water Dispatching and Integrated Supervisory Control System in Black River[J]. Gansu Water Conservancy and Hydropower Technology, 2008(07).

Zhang Yunyong, Zhang Zhijiang. The Middleware Technology and its Applications[M]. Beijing: Tsinghua University Press, 2005.

Application of Fingerprint-based Method in the Attribution Study of Observed Precipitation and Runoff Changes

Ding Xiangyi[1], *Jia Hongtao*[2] and *Jia Yangwen*[1]

1. China Institute of Water Resources and Hydropower Research, Beijing, 100038, China
2. Yellow River Conservancy Technical Institute, Kaifeng, 475004, China

Abstract: Many observational facts and studies have shown that the climatic and environmental conditions in the Hai River Basin, which is the political and cultural centre of China, have changed significantly over last half of the 20th century. Based on dualistic water cycle simulation technology and the fingerprint-based attribution method, this study attempts to attribute the observed precipitation and runoff changes in the basin over a 40-year period (1961 ~ 2000) to different factors, including natural climate variability, climate change induced by anthropogenic forcing of greenhouse gas emissions (referred to as anthropogenic forcing hereafter) and local human activity. The results indicate that, during the past 40 years in the basin: ① natural climate variability may be the factors responsible for the observed precipitation changes; ② natural climate variability and local human activity may be two factors responsible for the observed runoff changes, with local human activity being the main factor and accounting for about 60% of the changes. This study may provide a new method for the attribution study of hydrological changes under changing environment.

Key words: fingerprint-based attribution method, dualistic water cycle, precipitation change, runoff change, the Hai River Basin

1 Introduction

The characteristic of water cycle processes has been gradually transformed from the original natural mode to natural-artificial dualistic mode due to the continuous enhancement of human activities including water resources development and utilization as well as water conservancy construction (Wang et al., 2005). Climate change with the main feature of global warming in recent years has exacerbated the complexity of water cycle system, resulting that the simulation and prediction of water cycle is becoming increasingly difficult. Due to the lack of awareness of the climate change effects on water cycle, current models and methods cannot discriminate the contribution of natural climate variability and local human activity in hydrological changes, which may increase the difficulty and uncertainty in the prediction and management of water resources under changing environment in the future.

The results of some existing researches show that up to 60% of the climate-related trends of river flow, winter air temperature, and snow pack between 1950 and 1999 are human-induced in western United States (Barnett et al., 2008). Thus, the impacts of climate change on water cycle are increasingly intensive, and its contribution in hydrological changes may exceed those of natural factors and human activities in some basins. Current attribution methods are mostly based on the fingerprint-based method (Hasselmann et al., 1997). Many studies researched the attribution of the changes for atmospheric and oceanic climatic variables at a global scale (Barnett et al., 2001, 2005; Thorne et al., 2003; Lambert et al., 2004; Gillett et al., 2005). A review of previous attribution studies is given by the International Ad Hoc Detection and Attribution Group (2004). Additionally, in China, the methods used to discriminate the impacts of natural factors and human activities on water cycle can be divided into two categories: investigation methods and hydrological model methods (Wang et al., 2006; Li et al., 2007; He et al., 2007; Chen et al., 2007). However, these methods are incapable of discriminating among the impacts of natural climate variability, anthropogenic forcing and local human activity in hydrological changes. Therefore, it has been a key scientific problem in modern hydrology and water resources researches to quantitatively distinguish the impacts of climate change, natural factors and human activity in hydrological changes and thus to provide practical guidance for integrated water resources management.

As the political and cultural center of China, climatic and environmental conditions in the Hai River Basin (320,000 km^2) in recent decades have significantly changed. Many observations and studies have shown that water resources amount in the basin significantly decreased over the last half of twentieth century, and the reduction magnitude of runoff has been the greatest among China's major rivers over the past 40 years (Zhang and Wang, 2007). This study attempts to attribute the observed precipitation and runoff changes in the basin during 1961 ~ 2000 to different factors, including natural climate variability, anthropogenic forcing and local human activity using the fingerprint-based attribution method.

2 Methodology and data sources

The general outline of the methodology is combining the dualistic water cycle model with the fingerprint-based attribution method. Specifically, this study uses a global climate model to obtain the precipitation and temperature data under the scenarios of natural climate variability, anthropogenic forcing and solar radiation, a statistical downscaling model to downscale the climate model output, the dualistic water cycle model to obtain runoff series under different scenarios, and the fingerprint-based attribution method to investigate the possible contribution of different factors to the observed variable changes, with the assistance of observed meteorological data made throughout the basin. The methods and data mentioned above are each described in the following sections.

2.1 Dualistic water cycle model

The dualistic water cycle model (Jia et al., 2010), which consists of two sub-models, i. e. distributed hydrological model WEP-L and water resources allocation model ROWAS, is used to obtain runoff series under different scenarios.

WEP-L model could simulate natural water cycle processes under historical and different scenarios in the basin through the simulation of water movement in land surface, soil, underground and river. ROWAS model could simulate the allocation of water resources from water sources to different water users considering both natural water cycle and artificial water cycle based on generalization of water resources systems. The coupling way of the two sub-models is: WEP-L provides surface water resources amount as well as groundwater recharge and discharge amount at each node and calculation unit for ROWAS, while the spatial and temporal outputs of ROWAS provide a basis for reservoir operation and water supply in WEP-L.

2.2 General Circulation Model

The factors affecting the evolution of hydrological elements considered in this study consist of natural climate variability, anthropogenic forcing, solar radiation and local human activity. The precipitation and temperature series under the scenarios of natural climate variability, anthropogenic forcing and solar radiation can be deduced from the outputs of General Circulation Model (GCMs).

The GCM selected in this study is the PCM version 2.1 at a resolution of T42L26 (Washington et al., 2000), which has been widely used in hydrological studies and realistically reflects main characteristics of actual climate and the amplitude of natural climate variability (Barnett et al., 2008). Comparing with other coupled climate models such as CGCMA3 and MPI-ECHAM5, PCM has a higher resolution in the ocean and sea-ice components and the physical processes are simulated more realistically. The precipitation and temperature under the scenarios of natural climate variability, anthropogenic forcing and solar radiation are characterized by run B07.20, B06.22 and B06.69 respectively.

2.3 Statistical Downscaling Model

Generally, the spatial resolution of GCMs is too coarse, and there are two kinds of methods for downscaling the GCM outputs, i. e. dynamical downscaling method and statistical downscaling method. Statistical downscaling methodologies have several practical advantages over dynamical downscaling approaches. In situations where low-cost, rapid assessments of localized climate change impacts are required, statistical downscaling represents the more promising option currently (Chu, 2009). Therefore, in this study, we select the statistical downscaling methodology, concretely the Statistical Down-Scaling Model (SDSM) (Wilby et al. , 2002), to downscale the GCMs output.

The SDSM, which is a widely used statistical downscaling model in the world, allows the construction of climate change scenarios for individual sites at daily time scales using the gridded output of the GCM. Based on the statistical relationship between large-scale climatic factors and local variables, simulate local climate conditions and future climate scenarios could be obtained.

2.4 Fingerprint-based attribution method

Fingerprint-based attribution method is a kind of technology for the attribution analysis of variable changes in meteorology (Hegerl et al. , 1996; Hasselmann et al. , 1997), which uses fingerprint and signal strength as the quantitative evaluation indexes of variable changes. Fingerprint refers to the leading Empirical Orthogonal Function (EOF) of the dataset (model or observation). The EOF analysis is the method of choice for analyzing the variability of variables and finding the spatial patterns of variability that are referred to as the EOFs. Given the fingerprint, the signal strength is calculated as the least-squares linear trend of the projection of a dataset onto the fingerprint.

The general idea for attributing variable changes is to reduce the problem of multiple dimensions to a low dimensional problem. This operation can also be understood as applying a filter to the observations or simulations. This has advantages compared with the two extremes of trying to assess a significant variable change in the full variable space or simply using a mean value. In the low dimensional space, observed variable changes can be attributed to different factors by comparing the signal strengths under different scenarios with the signal strength calculated from observations.

2.5 Data sources

The source of Digital Elevation Model (DEM) in this study is GTOPO30 data developed by the U. S. Geological Survey's EROS Data Center, which can be downloaded from the website: http://eros. usgs/gov/#/Find_Data/Products_and_Data_Available/gtopo30_info.

Land use data for three periods 1980s, 1990s and 2000s are downloaded from the Data Sharing Infrastructure of Earth System Science, http://www. geodata. cn/Portal/? isCookieChecked = true.

Soil data including map of soil types and correspondent characteristic parameters are from National Second Soil Survey Data and Soil Types of China.

Vegetation information (Leaf Area Index and area fraction of vegetation) from 1982 to 2000 is obtained using the NOAA AVHRR data.

Time series (1961 ~ 2000) of meteorological observations including those of precipitation and temperature at 26 stations in the basin are provided by the China Meteorological Administration (CMA).

Time series (1961 ~ 2000) of observed river runoff at main hydrological stations used for validation of the dualistic water cycle model are provided by the Hai River Water Conservancy Commission (HRWCC), Ministry of Water Resources.

Reservoirs data are from map collections of the Hai River Basin compiled by HRWCC. Irrigation districts data are from Introduction to Irrigation Districts in the Hai River Basin compiled by HRWCC.

The SDSM is obtained from https://co-public.lboro.ac.uk/cocwd/SDSM/. SDSM calibration data are taken from NCEP reanalysis data, which are obtained from http://www.cdc.noaa.gov/.

The climate model outputs of three runs are obtained from http://www.ipcc-data.org.

3　Results and analysis

3.1　Application of dualistic water cycle model

The dualistic water cycle model is applied to the Hai River Basin to simulate the dualistic water cycle processes based on different kinds of data such as hydrological and meteorological information, DEM, land use, soil, vegetation, population, gross domestic product, water usage, and so on. The model is validated using observed daily runoff data at main hydrological stations and groundwater monitoring data in the basin. Details of the model application can be referred in Jia et al. (2010).

Verification results of the dualistic water cycle model at 12 main gauging stations in the basin indicate that the average errors of annual runoff are less than 15%, the average Nash-Sutcliffe efficiency coefficients of monthly runoff at the main gauge stations exceed 0.6, and the coefficients of correlation between simulated and observed monthly runoff series exceed 80%. Thus the dualistic water cycle model performs well in simulating runoff processes. A validation example for four gauging stations is shown in Fig. 1.

A comparison between simulated and monitored groundwater table contour in 2000 in the basin is shown in Fig. 2, which indicates that, the simulation results reproduce the monitored conditions well, and the dualistic water cycle model can simulate groundwater flow with high accuracy.

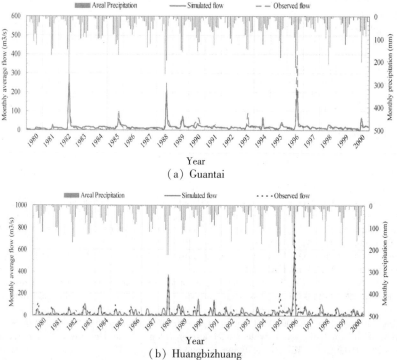

(a) Guantai

(b) Huangbizhuang

Fig. 1　Comparison between simulated and observed monthly runoff at main gauging stations

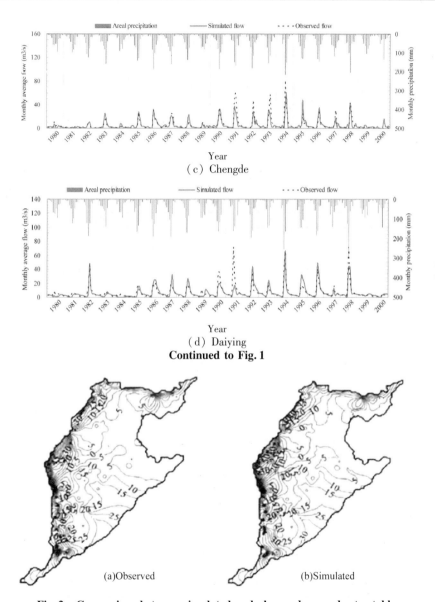

(a)Observed   (b)Simulated

Fig. 2   Comparison between simulated and observed groundwater table contour in 2000

3.2   Application of SDSM

The applicability of the SDSM and GCM to the Hai River Basin is affirmed by comparing the annual statistical characteristics of downscaled historical climate variables including precipitation and temperature simulated by PCM with those of observations (1961 ~ 2000) at 26 selected meteorological stations. Limited to the paper length, a validation example for Beijing station is shown in Fig. 3.

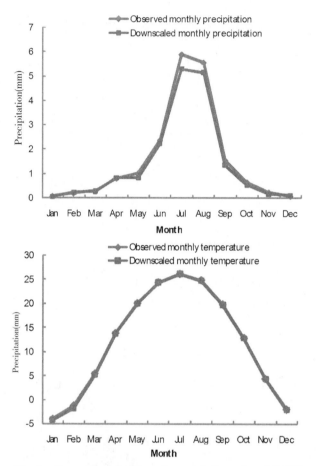

Fig. 3　Comparison between downscaled results of SDSM and
observation at Beijing station

3. 3　Attribution analysis

For the attribution of observed precipitation changes, three scenarios are set: natural climate variability, anthropogenic forcing and solar radiation. For the attribution of observed runoff changes, five scenarios are set: natural climate variability, anthropogenic forcing, artificial water use, land use change, and local human activity (combining artificial water use with land use change).

Under the three scenarios of natural climate variability, anthropogenic forcing and solar radiation, the precipitation and temperature outputs from the three control runs of PCM are firstly downscaled using the SDSM and then interpolated to the calculation units of the dualistic water cycle model. It should be noted that, to eliminate the effects of human activity on water cycle processes, we use land use data for the 1980s in the basin and do not consider artificial water use in the dualistic water cycle model. Runoff series under different scenarios can be obtained through dualistic water cycle model simulations.

Under the three scenarios of artificial water use, land use change and local human activity, precipitation and temperature data required in the dualistic water cycle model simulations are the same as the data for natural climate variability scenario, and runoff series can be directly obtained through changing the simulation conditions including land use data and artificial water use data.

Using the fingerprint-based attribution method and the 1961 ~ 2000 time series of precipitation and runoff for the whole basin, the fingerprints of precipitation and runoff changes under different scenarios could be obtained. We then further calculate the signal strengths of precipitation and runoff changes under different scenarios as shown in Fig. 4, thus attribute the observed changes of precipitation and runoff to different factors by comparing the signal strengths.

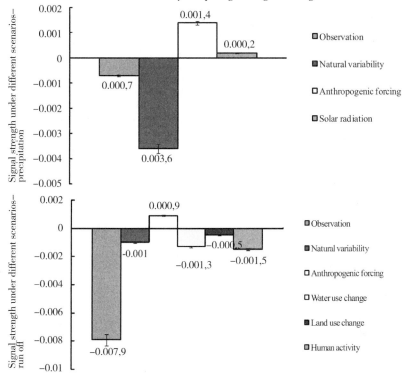

Fig. 4   Signal strengths of precipitation and runoff changes under different scenarios

As can be seen from Fig. 4 that, the signal strengths of precipitation change under anthropogenic forcing and solar radiation scenario are inconsistent with the observed, and thus we consider that these two factors may be not the factors responsible for the observed precipitation change during the past 40 years in the basin. While the signal strength of precipitation change under natural variability scenario is consistent with the observed and is larger than the observed, thus we consider that natural variability may be the factors responsible for the observed precipitation change during the past 40 years in the basin.

Additionally, Fig. 4 shows that the signal strengths of observed runoff changes is $-0.007,9$, and the signal strengths of runoff changes under the scenarios of natural variability, anthropogenic forcing, water use, land use change and local human activity are -0.001, $0.000,9$, $-0.001,3$, $-0.000,5$ and $-0.001,5$ respectively. The signal strength of runoff change under anthropogenic forcing scenario is inconsistent with the observed, and thus we consider that anthropogenic forcing may be not the factor responsible for the observed runoff change during the past 40 years in the basin. Among the factors that have signal strengths consistent with the observed, natural variability accounts for 36% of the observed runoff changes, artificial water use accounts for 46%, land use change accounts for 18%, and local human activity accounts for 60%. Thus, we consider that natural variability and local human activity may be two factors responsible for the runoff evolution during the past 40 years in the basin, with local human activity being the main factor.

3.4 Discussions

The attribution results are generally consistent with the actual conditions in the basin. Actually, the average water vapor content over the basin has undergone a significant change since 1950s. The water vapor was abundant from 1950s to the mid-1960s, and then began to decline, resulting in the decrease of precipitation in the basin. On the other hand, artificial water use amount has greatly increased during the past 40 years in the basin, from about 5 billion m^3 in the 1950s to about 4.0×10^{10} m^3 in the 1990s. Accompanying the increase in artificial water use, evapotranspiration in the basin may simultaneously increase, resulting in a decrease in runoff. Additionally, land use condition has also changed during the period. For example, comparing with the conditions in the 1980s, the area of sparse forest had decreased by 14% and the areas of dense forest, farmland and grassland had increased by 3%, 8% and 3%, respectively by 2000. The land use changes may be responsible for the decrease of runoff in the basin.

The attribution results also show that the contributions of artificial water use, land use change and human activity to the observed runoff changes are 48%, 16% and 60% respectively. Although the human activity scenario refers to the combination of artificial water use and land use change, the contribution of human activity does not equal to the sum of the contributions of artificial water use and land use change. As described in this study, the signal strengths are calculated based on the dualistic water cycle model simulations which are affected by many physical processes, and the simulation results of runoff under human activity scenario are apparently not equal to the simple sum of those under artificial water use and land use change scenario. Therefore, the contribution of human activity should be greater than that of artificial water use or land use change singly, and may not simply equal to the sum of the contribution of artificial water use plus that of land use change. The relation between them needs to be further studied in the following studies.

4 Conclusions

This study demonstrates a multidisciplinary study by integrating the fingerprint-based attribution method and the dualistic water cycle model to help discriminate the impacts of different factors on precipitation and runoff evolution in the Hai River Basin. The research results indicate that natural climate variability may be the factor responsible for the observed precipitation changes, and natural climate variability and local human activity may be two factors responsible for the observed runoff changes during the past 40 years in the basin with local human activity being the main factor and accounting for 60% of the changes. This endeavor has led to expand the application of fingerprint-based method from climatic and meteorological fields to hydrological fields and provides a new method for discriminating and quantitatively estimating the effects of climate change and human activity on hydrological changes in large river basins that have been greatly affected by climate change and intense human activity.

It should be noted that many uncertainties arising from the models such as those of climate model and hydrological model may affect the results. These aforementioned issues will be listed as our major focal areas of future researches.

Acknowledgements

This work is financially supported by the National 973 Program of China (2006CB403404), the National Natural Science Foundation for Young Scholars of China (51109223), and the National Scientific Foundation of China (50939006).

References

Barnett T P, Pierce D W, Schnur R. Detection of Anthropogenic Climate Change in the World's

Oceans [J]. Science, 2001(292): 270 – 274.

Barnett T P, et al. Penetration of A Warming Signal in the World's Oceans: Human Impacts [J]. Science, 2005(309): 284 – 28.

Barnett T P, et al.. Human – Induced Changes in the Hydrology of the Western United States [J]. Science, 2005(319): 1080 – 1083.

Chen L Q, Liu C M. Influence of Climate and Land Cover Change on Runoff of the Source Regions of the Yellow River [J]. China Environmental Science, 2007(4): 559 – 565.

Chu J T. Statistical Downscaling Methods and Their Application in Haihe River Basin of China [O]. Dissertation of Graduate University of Chinese Academy of Sciences, 2009.

Gillett N P, et al. Detection of External Influence on Sea Level Pressure with a Multi – model Ensemble [J]. Geophysical Research Letters, 2005(32), L19714: 4.

Hasselmann K, Cubasch U, Mitchell J F B, et al. Multi – fingerprint Detection and Attribution of Greenhouse – gas and Aerosol – forced Climate Change [J]. Climate Dyn. , 1997(13):613 – 634.

Hegerl, Storch, et al. Detecting Greenhouse – gas – induced Climate Change with an Optimal Fingerprint Method [J]. Journal of Climate, 1996(9):2281 – 2306.

He R M, Wang G Q, Zhang J Y. Impacts of Environmental Change on Runoff in the Yiluohe River Basin of the Middle Yellow River [J]. Research of Soil and Water Conservation, 2007(2): 297 – 301.

International AD Hoc Detection and Attribution Group. Detecting and Attributing External Influences on the Climate System: a review of recent advances [J]. Journal of climate, 2004 (18): 1291 – 1314.

Jia Y W, Wang H, Zhou Z H, et al. Development and Application of Dualistic Water Cycle Model in Haihe River Basin: I. Model Development and Validation [J]. Advances in Water Science, 2010, 21(1): 1 – 8.

Jia Y W, Wang H, Zhou Z H, et al. Development and Application of Dualistic Water Cycle Model in Haihe River Basin: II. Strategic Research and Application for Water Resources Management [J]. Advances in Water Science, 2010, 21(1):9 – 15.

Lambert, et al. Detection and Attribution of Changes in 20th Century Land Precipitation [J]. Geophysical Research Letters, 2004,31, L10203: 4.

Li X H, Zhao X, Ren L L . Quantitative Analysis of Water Resources Variation for Shiyang River Basin for Recent 50 years [J]. Journal of Hohai University (Natural Sciences), 2007(2): 164 – 167.

Thorne P W, et al. Probable Causes of Late Twentieth Century Tropospheric Temperature Trends [J]. Climate Dynamics, 2003(21): 573 – 591.

Washington W M, et al. Parallel Climate Model (PCM) Control and Transient Simulations [J]. Climate Dyn. , 2000(16): 755 – 774.

Wang G Q, Zhang J Y, He R M . Impacts of Environmental Change on runoff in Fenhe River Basin of the Middle Yellow River [J]. Advances in Water Science, 2006(6): 853 – 858.

Wang H, Jia Y W, Wang J H, et al. Evolutionary Laws of the Yellow River Basins Water Resources under the Impact of Human Activities [J]. Journal of Natural Resources, 2005 (20)(2): 157 – 162.

Wilby R L, Dawson C W, Barrow E M. SDSM — A Decision Support Tool for the Assessment of Regional Climate Change Impacts [J]. Environmental Modelling & Software, 2002 (17): 147 – 159.

Zhang J Y, Wang G Q. Study of Climate Change Impacts on Hydro – cycle and Water Resources [M]. Beijing: Science Press, 2007.

River Networks Fluidity Classification Analysis Based on Two Dimensional Numerical Model and Hydraulic Characteristic

Yang Fuxiang[1], *Pan Jie*[2], *Li Dongfeng*[3] and *Zhang Hongwu*[4]

1. Henan Vocational College of Water Conservancy and Environment, Zhengzhou, 450008, China
2. Yuyao Municipal Bureau of Water Resources, Yuyao, 315400, China
3. Zhejiang University of Water Resources and Electric Power, Hangzhou, 310018, China
4. State Key Laboratory of Hydro-science and Engineering, Department of Hydraulic Engineering, Tsinghua University, Beijing, 100084, China

Abstract: Water Diversion from outside of an urban is the effective method of solving city water pollution. Because the boundary condition is different, quantity of flow distribution is different for the every river water. The velocity is also different. The velocity of some river channels is even much smaller. Only the fluidity of every river is known well, can the measures of improving river water environment be found out. The two-dimensional finite element method mathematical models of the river network is established and verified. Bases on the established and verified planar 2-D unsteady river networks flow finite element model, the velocity size and orientation of the every river and water levels are calculated. The numerical simulation research and analysis shows that the quantity and velocity of flow distribution are both different for every river water. The velocity of some river channels is even much smaller. According to the hydraulic characteristic such as velocity size and fluidity, five classifications are made, and some measures of improving and harness of water fluidity are provided and analyzed

Key words: river network, flow hydraulic characteristic, classification, two dimensional finite element method

1 Introduction

Nowadays, urban river networks is criss-crossing, and water environment pollution is critical and complex problem. Water diversion from outside of an urban is the effective and actual method of solving city water pollution. Otherwise, there is an obvious question that the quantity of flow and velocity distribution varies with different river. Because the boundary condition is different, quantity of flow distribution is different for the every river water. The velocity is also different. The velocity of some river channels is even much smaller. Therefore, only the fluidity of every river is known well, can the measures of improving river water environment be found out. In order to investigate quantity, depth-averaged planar 2-D shallow water finite element method -Numerical Model is a good method.

2 Numerical simulation basic theory methodology

2.1 The governing equations and deterministic conditions (Li Dongfeng, 1999)

Using depth-averaged planar 2-D shallow water equations as the governing equations for computation.

The equations of continuity:

$$\frac{\partial z}{\partial t} + \frac{\partial hu}{\partial x} + \frac{\partial hv}{\partial y} = 0$$

The equations of motion:

$$\frac{\partial u}{\partial t} + u\frac{\partial u}{\partial x} + v\frac{\partial u}{\partial y} + \frac{\partial z}{\partial x}\frac{gn^2 u \sqrt{u^2 + v^2}}{H^{4/3}} - fv - \varepsilon(\frac{\partial^2 u}{\partial x^2} + \frac{\partial^2 u}{\partial y^2}) = 0$$

$$\frac{\partial v}{\partial t} + u\frac{\partial v}{\partial x} + v\frac{\partial v}{\partial y} + g\frac{\partial z}{\partial y} + \frac{gn^2 v \sqrt{u^2 + v^2}}{H^{4/3}} + fu - \varepsilon(\frac{\partial^2 v}{\partial x^2} + \frac{\partial^2 v}{\partial y^2}) = 0$$

where, u, v is x, y direction components of depth averaged velocity; z, h is water level (or tidal level) and depth; g is acceleration due to gravity; f is Coriolis force coefficient, $f = 2\omega\sin\Phi$, ω is rotation angular velocity of earth, Φ is the latitude of computed reach; ε is turbulent viscosity coefficient.

C is Checy's coefficient, C is calculated by Checy's formulation:

$$C = \frac{1}{n}R^{1/6}$$

where, n is Manning roughness coefficient.

2.2 The deterministic conditions

The deterministic conditions involve boundary conditions and initial conditions.

2.2.1 Boundary conditions

Boundary conditions include opening boundary and closing boundary. The former opening boundary is inlet and outlet water boundary, which is governed by inlet flow quantity process and outlet water levels process for model. The latter closing boundary is land boundary and the normal velocity is treated as zero for model.

2.2.2 Initial conditions

The initial water level and the initial velocity are given by measured tidal level or given value zero, initial conditions does not affect the precision of computed result. At initial time, tidal level, velocity and other varibles are given.

3 The finite element solution of the equations and verification of model

3.1 Numerical procedure

According to the above formulation, finite element analysis is carried out first, then composing the each element and the overall finite element equations can be obtained.

3.2 Establation of finite element model

The finite element model is employed to solve the equations. In this model, quadrangle grids are selected to disperse calculation zones. Applying Galerkin's weighted residual method is used in every element where weighted function is the function of interpolation, the weak formulation of the above equations are:

$$\int_\Omega (\frac{\partial h}{\partial t} + \frac{\partial hu}{\partial x} + \frac{\partial hv}{\partial y})\delta h \mathrm{d}\Omega = 0$$

$$\int_\Omega (\frac{\partial v}{\partial t} + u\frac{\partial u}{\partial x} + v\frac{\partial u}{\partial y} + g\frac{\partial(z_0 + h)}{\partial x} - fv + \frac{gu \sqrt{u^2 + v^2}}{C^2 h}) - \varepsilon(\frac{\partial^2 u}{\partial x^2} + \frac{\partial^2 v}{\partial y^2}) \cdot \delta h \mathrm{d}\Omega = 0$$

$$\int_\Omega (\frac{\partial u}{\partial t} + u\frac{\partial v}{\partial x} + v\frac{\partial v}{\partial y} + g\frac{\partial(z_0 + h)}{\partial y} - fu + \frac{gv \sqrt{u^2 + v^2}}{C^2 h}) - \varepsilon(\frac{\partial^2 v}{\partial x^2} + \frac{\partial^2 u}{\partial y^2}) \cdot \delta h \mathrm{d}\Omega = 0$$

where, the δh, δu, δv are the variation of variables; Ω is calculation zones.

Employing the Green's formulation to the above two order differential items, supposing weighted function and integration function of the mean velocities and water levels for some finite subspace element:

$$u = \sum_{i=1}^{3} \varphi_i u_i \qquad\qquad \delta u = \sum_{i=1}^{3} \varphi_i u_i^*$$

$$v = \sum_{i=1}^{3} \varphi_i v_i \qquad\qquad \delta v = \sum_{i=1}^{3} \varphi_i v_i^*$$

$$h = \sum_{i=1}^{3} \varphi_i h_i \qquad\qquad \delta h = \sum_{i=1}^{3} \varphi_i h_i^*$$

Then their finite element equations can be given.

$$M_{ij}\frac{\mathrm{d}z_{si}}{\mathrm{d}t} = -P_{ij}(hu)_j - P_{zij}(hv)_j$$

$$M_{ij}\frac{\mathrm{d}u_j}{\mathrm{d}t} = -N_{ij}u_j - gP_{1ij}z_{sj} - M_{ij}/\rho - \varepsilon(Q_{ij} + R_{ij})u_j$$

$$M_{ij}\frac{\mathrm{d}v_j}{\mathrm{d}t} = -N_{ij}v_j - gP_{2ij}z_{sj} - M_{ij}\tau_{yj}/\rho - \varepsilon(Q_{ij} + R_{ij})v_j$$

$$i, j = 1, 2, 3, \cdots, I$$

In which

$$M_{ij} = \iint_{\Omega} \Phi_i \Phi_j \mathrm{d}x\mathrm{d}y$$

$$N_{ij} = N_{1ijk}u_k + N_{ijk}v_k$$

$$N_{1ijk} = \iint_{\Omega} \Phi_k \Phi_i \frac{\partial \Phi_j}{\partial x}\mathrm{d}x\mathrm{d}y$$

$$N_{2ijk} = \iint_{\Omega} \Phi_k \Phi_i \frac{\partial \Phi_j}{\partial y}\mathrm{d}x\mathrm{d}y$$

$$Q_{ij} = \iint_{\Omega} \left(\frac{\partial \Phi_i}{\partial x}\frac{\partial \Phi_j}{\partial x} + \frac{\partial \Phi_i}{\partial y} + \frac{\partial \Phi_j}{\partial y}\right)\mathrm{d}x\mathrm{d}y$$

$$D_{1ij} = \iint_{\Omega} \frac{\partial \Phi_j}{\partial x}\Phi_i \mathrm{d}x\mathrm{d}y$$

$$P_{2ij} = \iint_{\Omega} \frac{\partial \Phi_j}{\partial y}\Phi_i \mathrm{d}x\mathrm{d}y$$

$$R_{ij} = \oint_{r} \left(\frac{\partial \Phi_j}{\partial y}\Phi_i \mathrm{d}x - \frac{\partial \Phi_j}{\partial x}\Phi_i \mathrm{d}y\right)$$

$$i, j, k = 1, 2, \cdots, I$$

A forward time stepping method is used to approximate the u, v and h derivatives, i. e:

$$\frac{\partial u}{\partial t} = \frac{u_{n+1} - u_n}{\Delta t} \qquad \frac{\partial v}{\partial t} = \frac{v_{n+} - v_n}{\Delta t} \qquad \frac{\partial z}{\partial t} = \frac{z_{n+1} - z_n}{\Delta t}$$

3.3 Solution of finite element model

The predictor – corrector method is used to estimate the iteration equations.

3.4 Verification of the Model

The comparison of calculated data and filed show that all these numerical simulation results, the water level and velocity process, are well agreement with the field ones. The detailed verification of the model is shown in reference (Li Dongfeng, 2009)

4 The analysis of calcualting results

4.1 Calcualting river channel networks

As a example, the Cixi city of Zhejiang province, its river channel networks is shown in Fig. 1.

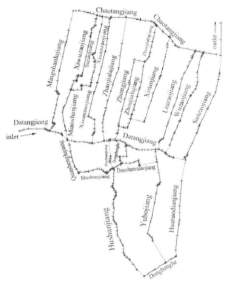

Fig. 1 Cixi urban river channel networks and boundary

4. 2 Calcualting river channel networks initial grid meshes

River channel networks initial grid meshes is shown in Fig. 2.

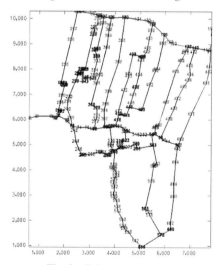

Fig. 2 Initial grid meshes

4. 3 Calculating boundary conditions

According to the adives the calculating boundary conditions is that at the inlet flow quantity is 18 m³/s and at outlet water levels is 0. 90 m . The closing boundary is land boundary and the normal velocity is treated as zero for model.

4.4   Velocity size distribution of every river

According to the above condition, the finite element software is used to the computation, putting the average velocity size data into their river channels. Velocity size distribution of every river is shown in Fig. 3.

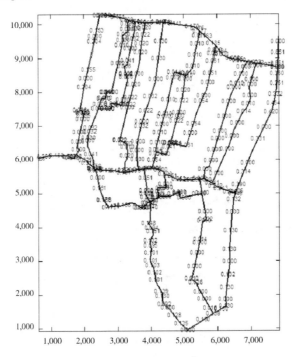

Fig. 3   Velocity size distribution of every river

4.5   River network fluidity hydraulic characteristic classification analysis

Regarding of the computing rusults data, six classification is given by the velocity size and hydraulic characteristic. They are shown in Tab. 1 to Tab. 5.

4.5.1   Hydraulic characteristic classification I —fluidity very weak and slow
These rivers flow very slowly and these are in urgent need of harnessing, and the measures can be taken such as using sluce gate to control water quantity and improve water fuildity.

4.5.2   Hydraulic characteristic classification II — fluidity weak and slow
These rivers flow slowly and these are in need of harnessing

4.5.3   Hydraulic characteristic classification III— a little bit weak and slow
These rivers flow a little bit weakly and slowly.

4.5.4   Hydraulic characteristic classification IV —a little bit quick
These rivers flow a little bit quickly

Tab. 1 Hydraulic characteristic classification I — fluidity very weak and slow

River No.	Fluidity classification	Name of river	Beginning location	Endding location	Velocity (m/s)	Flow orientation
1		Liuzaojiang	Datangjiang	Chaotangjiang	0.005	South to North
2		Middle of Zhongjiang	South of Zhongjiang	North of Zhongjiang	0.007	South to North
3		Yubojiang	Datangjiang	Donghenghe	0.008	North to South
4		Danshanmiao jiang	West of Dongchenghe	West of Yubojiang	0.009	West to East
5		Wuzaojiang	Datangjiang	Chaotanghengjiang	0.011	South to North
6		North of Zhongjiang	Middle of Zhongjiang	Chaotanghengjiang	0.013	South to North
7	Fluidity very slow and these need harnessing	West of Xiwuzaojiang	Datangjiang	Chaotanghengjiang	0.019	South to North
8		West of Xisanzaojiang	Datangjiang	Chaotanghengjiang	0.031	South to North
9		Zhoujialujiang	South of Zhongjiang	North of Zhongjiang	0.035	South to North
10		Miaoshanjiang	Datangjiang	West of Xiwuzaojiang	0.039	South to North
11		Westof Xisizaojiang	Datangjiang	Chaotanghengjiang	0.041	South to North
12		South of Zhongjiang	Datangjiang	Middle of Zhongjiang	0.046	South to North
13		Zhaojialujiang	Datangjiang	Chaotanghengjiang	0.049	South to North
The measures		Such as using sluce gates which located in the inlet of some river to control water quantity and improving their water fuildity and mending water enviroment				

Tab. 2 Hydraulic characteristic classification II -fluidity weak and slow

River No.	Fluidity classification	Name of river	Beginning location	Endding location	Velocity (m/s)	Flow orientation
14		North of Hushanjiang	Datangjiang	Shishanjiang	0.061	North to South
15	Fluidity weak and Slow	Xinanjiang	South of Zhongjiang	Chaotanghengjiang	0.071	South to North
16		Jinshanhou hengjiang	East of Hushanjiang	West of Dongchenghe	0.089	West to East
17		Dongchenghe	Datangjiang	Danshanmiaojiang	0.091	North to South
The measures		Such as using sluce gates which located in the inlet of some river to control water quantity and improving their water fuildity and mending water enviroment				

Tab. 3　Hydraulic characteristic classification III — a little bit weak and slow

River No.	Fluidity classification	Name of river	Beginning location	Endding location	velocity (m/s)	Flow orientation
18		Dongheng he	West of Yubojiang	Huatuodian jiang	0.131	West to East
19		Huatuodian jiang	Donghenghe	Datangjiang	0.141	South to North
20		Donghenghe	Hushanjiang	West of Yubojiang	0.144	West to East
21	Fluidity a little bit weak and slow	Datangjiang	Zhongjiang	Dongchenghe	0.153	West to East
22		Chaotanghengjiang	Zhongjiang	Xinanjiang	0.177	West to East
23		South of Hushanjiang	Datangjiang	Donghenghe	0.179	North to South
24		Chaotanghengjiang	Zhaojialujiang	Zhongjiang	0.192	West to East
25		Shishanjiang	Xianshenglujiang	Hushanjiang	0.196	West to East
26		Chaotanghengjiang	West of Xisizaojiang	West of Xiwuzaojiang	0.199	West to East
The measures		Such as using sluce gates which located in the inlet of some river to control water quantity and improving their water fuildity and mending water enviroment				

Tab. 4　Hydraulic characteristic classification IV-a little bit quick

River No.	Fluidity classification	Name of river	Beginning location	Endding location	velocity (m/s)	Flow orientation
27		Datangjiang	Miaoshanjiang	Fangwanglujiang	0.202	West to East
28		Mingshanlujiang	Zongxinglu	Chaotanghengjiang	0.206	South to North
29		Chaotanghengjiang	Mingshanlujiang	West of Xiwuzaojiang	0.209	West to East
30		Datangjiang	Hushanjiang	Zhongjiang	0.211	West to East
31		Xianshenglujiang	Datangjiang	Shishanjiang	0.212	North to South
32		Datangjiang	Fangwanglujiang	Zhaojialujiang	0.225	West to East
33		Mingshanlujiang	Datangjiang	Zongxinglu	0.227	South to North
34		Datangjiang	Dongchenghe	Yubojiang	0.229	West to East
35	Fluidity with a little bit quickly	Datangjiang	Wuzaojiang	Sanzaojiang	0.239	West to East
36		Datangjiang	Liuzaojiang	Wuzaojiang	0.243	West to East
37		Chaotanghengjiang	West of Xiwuzaojiang	West of anzaojiang	0.251	West to East
38		Chaotanghengjiang	West of Sanzaojiang	Zhaojialujiang	0.253	West to East
39		Datangjiang	Yubojiang	Liuzaojiang	0.255	West to East
40		Chaotanghengjiang	Xinanjiang	Liuzaojiang	0.258	West to East
41		Chaotanghengjiang	Wuzaojiang	Sanzaojiang	0.261	West to East
42		Datangjiang	Zhaojialujiang	Hushanjiang	0.263	West to East
43		Chaotanghengjiang	Liuzaojiang	Wuzaojiang	0.267	West to East
The measures		Fluidity is a little bit quickly, and no other measures are needed				

Tab. 5　Hydraulic Characteristic Classification IV - flow quick

River No.	Fluidity classification	Name of river	Beginning location	Endding location	velocity (m/s)	Flow orientation
44		Datangjiang	Mingshanlujiang	Miaoshanjiang	0.458	West to East
45	Fluidity quickly	Datangjiang	Datangjiang (Inlet)	Mingshan lujiang	0.532	West to East
46		Sanzaojiang	Datangjiang	Outlet	0.539	South to North
The measures		Fluidity is quikly, and no other measures are needed				

4.5.5　Hydraulic characteristic classification V—flow quick

These rivers flow quickly.

6　Conclusions

Bases on the established and verified planar 2-D unsteady river networks flow finite element model, the velocity size and orientation of the every river and water levels are calculated. The numerical simulation research and analysis shows that the quantity and velocity of flow distribution are both different for the every river water. The velocity of some river channels is even much smaller. According to the hydraulic characteristic such as velocity size and fluidity, five classifications are made; some measures of improving and harness of water fluidity are provided and analyzed

Acknowledgments

This research was supported by the National Natural Science Foundation of China (No. 51039003) and the 2011 projects the Water Resources bureau of Zhejiang Province (RC11092011), the 2009 Zhejiang provincial Education bureau key projects (No. Z200909405), the Zhejiang provincial Education Science Plan Office Project[(2009)12, No. SCG220] and the "325"Talent Training Program of Water Resources Department of Zhejiang Province.

＊All authors contributed equally to this work.

References

Li Dongfeng, Zhang Hongwu, Zhang Junhua. Finite Element Method of Yellow River and Sediment Movement [J]. Journal of Sediment Research, 1999 (4): 59 – 63.

Li Dongfeng, Zhang Hongwu, Zhong Deyu. 2 – D Mathematical Model for Flow and Sediment Transport in the Estuary of Yellow River [J]. Journal of Hydraulic Engineering, 2004(6): 1 –7.

Li Dongfeng, Zhang Hongwu, Zhong Deyu. Numerical Simulation and Analysis on Tidal Current and Sediment Silting Process in Yellow River Estuary [J]. Journal of Hydraulic Engineering, 2004(11): 74 – 80.

Zou Bing, Li Dongfeng, Zhong Deyu. Effect of Training Levees on Movement of Water Flow and Sediment in Yellow River Estuary [J]. Journal of Hydraulic Engineering, 2006, 37(7): 880 – 885.

Li Dongfeng, Zhang Hongwu, Chen Bin. Numerical Simulation Analysis of Bridge Construction Flood Flow Effect Based on Two Dimensional Finite Element Model. In: Zhou Baozhi [C]// Proceedings of 2010 International Conference on Modern Hydraulic Engineering. Xi'an: London Science Publishing, 2010.

Li Dongfeng, Zhang Hongwu, Chen Bin. Dammed Water Level Comparison of Widening Bridge Piers Based on 2D FEM Numerical Simulation and Empirical Formula. In: Zhou Baozhi[C]// Proceedings of 2010 International Conference on Modern Hydraulic Engineering. Xi'an:

London Science Publishing, 2010.

Li Dongfeng, et al. Numerical Simulation and Analysis of Hydraulic Characteristic of Cixi rural river networks N Technical Report [R]. Report of ZJWCHC, 2009.

Li Dongfeng, Mao Lianming, Chen Dongyun, Zhang Hongwu. 2D Numerical Calculation Analysis on River Water Flow Control [J], Journal of Zhejiang Water Conservancy and Hydropower College, 2011, 23(1):9 – 12.

The Application of GIS in the Ice Prevention of the Yellow River

He Gaofan, *Yao Baoshun* and *Li Yong*

Information Center of Yellow River Conservancy Commission, Zhengzhou, 450004, China

Abstract: The technique of the integration of 2D and 3D can change the means and modes of the application in the ice prevention of the Yellow River. SuperMap GIS is applied to information management of the Yellow River. The Yellow River ice prevention integrated database and geographic space frame is established. Each function module of the ice prevention information management of the Yellow River is developed. The improved issue about the Yellow River ice prevention is put forward.

Key words: ice prevention of the Yellow River, SuperMap GIS , GIS, database

1 Introduction about the Yellow River ice

The Yellow River basin is across the 23 longitude from east to west, and 10 latitude from north to south, the terrain differs and the temperature varies greatly. The ice floods may generate from December to March in the next year in the Yellow River each year.

Du to the effect of cold air from the Siberia and Mongolia area in winter and spring, there are more northerly wind, dry and cold climate, scarce rain and snow. The average temperature in January is below 0 ℃. Annual minimum temperature on the upper reaches is $-53 \sim -25$ ℃, on middle reaches $-40 \sim -20$ ℃, and on downstream reaches $-23 \sim -15$ ℃. The shape of the river from upstream to downstream likes a " 几 " of Chinese characters.

The river is from low latitude flow to circumpolar latitude from Ningxia to Inner Mongolia, Henan to Shandong. . The lower is prior to the upper when river is freezing up and the upper is prior to the lower at the time of thawing and the upper ice thickness is thin and the lower ice is thick. Some ice flood disasters such as backwater by ice dam are easily occurred when thawing.

Just after 1949, since the poor quality of lower river of the Yellow River and inexperienced ice prevention, two ice flood crevasses had been happened which caused 27,500 people affected and the affected area is 1.29×10^6 mu. The ice flood damage of lower river has slow down since the Sanmenxia Reservoir has been put into the application of ice prevention. ice flood disasters have often happened in the Ningmeng, Hequ and Longmen reaches in the Yellow River. Since the 1970s, ice flood disaster has happened once in every three or four years. Especially in 1993 ~ 1994, a dam burst has happened in Dengkou channel segments below the Sanshenggong Sluice which caused 13,000 people affected and the affected area is 120,000 mu.

With the development of society, national economy has shown a rapid and stable development trend and Yearly accumulation of wealth in the provinces of the Yellow River Basin. However, the casualty loss of ice flood is heavier and heavier and the influence on policy, economy and society is larger. So the ice prevention of the Yellow River has been the task of primary importance in national flood prevention in winter and spring.

2 The workflow in the ice prevention of the Yellow River

2.1 Information acquisition

We collect all kind of information including weather, water and rain information, work situation, disaster situation and ice condition. Only ice prevention information is collected timely and accurately and organized effectively, then we can have clear recognition and judgment for the ice prevention situation and make scientific decision.

2.2　Forecast and prediction

We make timely and accurate prediction about the trend of flood, atmosphere temperature, project dangerous case and ice flood according to collected information. The ice prevention decision is decision – making in advance, namely decision – making before the disaster, and forecast and prediction is significant.

2.3　Project design

The ice prevention consists of water, rain, work and disaster situation and the prediction and forecast for development tendency. The content and target of the decision can be obtained through analysis and induction. The solution is designed based on the decision objective and the ice prevention method, which is used to realize the decision objective. Then the solution is simulating displayed and evaluated.

2.4　Consultation and decision – making

On the basis of getting a clear understanding of the ice prevention situation, the decision – making level should conform to the principle of "the part obey the overall, and the least disaster is the objective" and select the best solution via consultation and analysis.

3　The system logic structure and function

By making use of 3D and 2D geographic information system and database technology, the ice prevention synthesis database including the space database and attribute database is established, the application system framework is built, and the ice prevention information management function module is developed. .

3.1　The ice prevention consultation synthesis database

By utilizing the SuperMap desktop tools, remote sensing image processing tools and Oracle database management system, the electronic map is built with the scale of 1:50,000 in the Yellow River ice prevention area and important area map is with the sale of 1:10,000. We should obtain DEM data with the accuracy of 30 m and remote sensing image data with the resolution of 30 m. The high resolution remote sensing image data is adopted in the important flood and ice prevention channel segment. The ice prevention consultation space database is established. The ice prevention attribute database is established according to the flood prevention project database standard issued by the Ministry of Water Resources.

The geographic information data, which are needed to be processed, mainly include Digital Line Graphic (DLG) in the ice prevention area, Digital Orthophoto Map (DOM) and Digital Elevation Model (DEM).

The maps in the ice prevention mainly include watercourse topographic map with the scale of 1:50,000 and the ice prevention terrain map of the Yellow River and so on. There are 17 types of vector coverage data needed to be processed with some emphasis. The data is imported to map database after processed by SuperMap desktop software, which is mainly applied into 2D geographic information system.

The basic work, namely vector data, is a time – consuming effort with a lot of manpower job which lasted a long time. The coverage management table is as follows (Tab. 1).

Tab. 1 The coverage management table

No.	Coverage name	The content of construction
1	Administrative boundary	Administrative border at or above the county units
2	Resident place	The position and note of county, villages and towns
3	Mainstream and tributaries	The rivers have appellatives with the scale of 1 : 50,000 in the ice prevention area. The expansion code is obtained based on coding rule from 'Code for China River Name', SL 249—1999. The unknown rivers are reserved, and the rivers are classified and filled into attribute table
4	Reservoir in the ice prevention zone	The large and medium reservoirs concerned with the ice prevention control, including Longyangxia, Liujiaxia, Wanjiazhai, Sanmenxia, Xiaolangdi and so on, is coded based on coding rule from 'Code for China Reservoir Name', SL 259—2000
5	Hydrological control station	All hydrometric stations and gauging stations in the ice prevention areas
6	Video monitoring point	Bayangaole, Sanhuhekou, Zhaojunfen, Huajiangyingzi, Deshengtai bridge, Toudaoguai, Lamawan and other points of the 3G video
7	Dike segment	Important embankment sections in the rivers of ice prevention zone
8	River Regulation Project	Major in engineering and controlling hydraulic engineering in ice prevention zone
9	Sluice	The important water and flood control sluice gate in ice prevention zone
10	The past records of breach place	9 important breach place since 1949
11	River Crossing Engineering	Bridges and pontoons over the river course in the ice prevention zone
12	Dike — crossing structures	Important pipeline through the Yellow River in the ice prevention zone
13	The use of electrical irrigation and drainage pumping station	The large and medium – sized Electromechanical irrigation and drainage pumping stations in the ice prevention zone
14	The ice flood storage and detention basin	Including Xiaobaihe, Zhaojunfen, Pugebu, Hangjinnaoer, Wuliangbuhe and WuLiangsuhai
15	Easy to grip the ice dam	Mainly in the history of Ningxia and Inner Mongolia often stuck ice dams on the river. According to the historical data collection
16	Flood storage and detention basin	Dongping Lake
17	Irrigation area	Four large irrigation area in Ningxia and Inner Mongolia reach

3.2 Application system framework

The 2D /3D visual operating environment is established in order to meet the ice prevention application by using SuperMap middleware. The data query and space analysis function based on the map visually is developed in order to meet the requirements such as rapid information query of the ice prevention, practical function and novel display form.

3.3　The development of the ice prevention information management function module

The functions of the ice prevention information management development task mainly include the development of the ice prevention trends bulletin, the information collection of the freeze – up and break – up data, the drawing of trends thematic map of the freeze – up and break – up, the statistics and query of the reach of freeze – up and break – up, the surveillance of the ice prevention, the special element position service of the ice prevention, the monitoring and prediction of the ice prevention, the map plotting, the historical disasters and the data query of the ice prevention.

4　Technology roadmap

The system is developed based on J2EE architecture; it adopts the complete B / S mode and service – oriented architecture design and MVC model for the design idea of layer.

The 3D objective world can be well described by 3D GIS. However, there are some disadvantages in the aspects of query, analysis and macroscopic description. At present, 2D GIS and 3D GIS should be operated alternately for different applying objectives.

Realspace GIS repertoire can guarantee the seamless fusion between 2D and 3D GIS, including the integration of data model, data storage scheme, data management, visual and analysis function between 2D and 3D GIS, It provides massive two – dimensional data directly in the 3D scene in high performance visualization, the direct operation of 2D analysis function in 3D scene and 3D analysis function which is more and more abundant. Realspace GIS has made a breakthrough that the check function of 3D GIS in the past and promoted the profound application of 3D GIS.

5　Conclusions

The ice prevention synthesis database and geographic space framework of the Yellow River is established by using SuperMap GIS, each consultation function module of the ice prevention in the Yellow River is developed and the improved issue about the Yellow River ice prevention is put forward.

References

Cui Jiajun. The Yellow River Flood Control Decision Support System Research and Development [M]. Zhengzhou: Yellow River Conservancy Press, 1998.

Wei Xiangyang. The Yellow River Downstream Factor and Control Countermeasures of the Yellow River[J]. Yellow River, 1997(12).

Cai Lin. The Yellow River Flood of 50 Years[J]. Yellow River ,1996(12).

Zhai Jiarui. The Yellow River Ice Flood Control and Dispatch[J]. China Water Conservancy, 2007(3).

Applications of Remote Sensing in Obstacle Clearance within River Channels of the Lower Yellow River

Shen Yuan, *Ma Xiaobing* and *Li Meng*

Yellow River Conservancy Commission, Zhengzhou, 450003, China

Abstract: Large number of obstacles against flood passing occupies large flood plain areas, which affects the safety of draining flood seriously. Aimed at implementing the most stringent management system of river channel, it must strengthen supervision, strict monitor and control the changes of obstacles along the lower river to minimize adverse effects. The paper presents the situation of flood plain areas and the businesses of obstacle clearance between river channels of the lower Yellow River. Then it analyses problems on the flood passing impacts against obstacles. According to the actual requirements of law enforcement of obstacle clearance, this paper utilizes remote sensing for monitoring the distributions of obstacles including patch forest, production dykes and brick – kiln fields, and comparison between images in different phase. It also illustrates the application of the information system of obstacle clearance monitored by remote sensing built on three – dimension landforms of Yellow River Basin. At the end, it shows the effect upon actual obstacle clearance works.

Key words: remote sensing monitoring, obstacles against flood passing, the lower Yellow River, the three – dimension landforms of the Yellow River Basin

1 Introduction

With the rapid development of society and economy, the phenomenon of contending for the ground and water with the river has become increasingly prominent, which is a threat to river channels. More and more water actives happened within the limit of river channels management, such as the rapid spread of illegal plants, illegal sand extraction and illegal mining. Many projects occupying river channels are built rapidly, some of which are illegal and unauthorized buildings against the management requirement of flood controlling.

Recent years, the idea that the implementation of the most stringent river basin management system providing supports for keeping healthy life of rivers in management, is put forward by Yellow River Conservancy Commission (YRCC), one of which is that the implementation of the most stringent channel management system (Li Guoying, 2010). Thus, the paper aims to be well aware of the grim situations and the outstanding problems in management brought by obstacles against flood passing of the Lower Yellow River firstly, and then monitors and manages the changes of obstacles by Remote Sensing and other information technologies, which provides technical support for the management and decision of obstacles in flood plain areas.

2 Analysis on obstacles within river channels of the lower Yellow River

2.1 The general situation in floodplains area of the lower Yellow River

The lower Yellow River is 878 km long, and the floodplain area between levees along river is up to 4,416 km², now where there are a total of 189 million people living in Henan province and Shandong province with distribution in 15 cities and 43 countries covering 2,036 villages and 375 million cultivated lands. The floodplain area is the way of flood passing and detention and also the place for people living.

As a result of long – term sedimentation, elevation of the lower river bed is generally 4 m to 6 m higher than that of the ground surface outside the levees, and it becomes worse and forms a secondary suspended river. The floodplain area between levees is known as high water level of flood area, where flood passing withstand by the dikes to ensure flood control safety.

The places of flood passing areas are changed with flood, so some areas without flood passing for several years are occupied by illegal buildings, those are called as obstacles preventing from flood passing.

2.2 Analysis on obstacles

A thorough checking finds that there are four major obstacles including patch forests, production dykes, brick – kiln fields and floating – bridges.

Lots of production dykes were built by villagers living in flood plain area to avoid flood submergence, which made Secondary Suspended River worse. And some beyond controlling hydraulic engineering influenced the engineering layout for controlling river regime. When a largebig flood happens, the sharp drop of the volume of sedimentation makes the water level slowing down which can not reduces impact on levees. The situation of production dykes reinforced constantly raises height difference between channel and floodplain, which causes erosion ditches and rolling river formed once production dykes were broken. It is dangerous to channel. will greatly affect the safety of flood control, and increase the difficulty of flood control.

Recently, a forest is worth considerable. With disregarding the policy that it is prohibited to plant trees or long – stalk crops impeding flood discharge, villages planted so many trees that they cut flood control capacity down and lower the flood control standard.

To gain economic interest, villages sold land to the owner of brick – kiln fields who built many brick – kilns. These fields occupy a large area of land impeding flood discharge, and mining soil on spot where brick – kiln is made ecological environment worse, which is a disadvantage to change channel route. And the vehicles with bricks damaged roads where are the flood – control passageway and caused serious negative effects on the river engineering and embankment.

A floating – bridge makes cross travel convenient and helps boost the local economy. The piers of them were wrapped around which made inside section too narrow to flood discharge, the narrowest part is only 190 m wide. And some bridges beyond controlling hydraulic engineering occupy main channel. The human behavior increased the wandering risk, and endangered the safety of flood – control.

Except for four obstacles above – mentioned, there are other obstacles, for example, big places for recreation and leisure were rebuilt and extended without authorized. Illegal mining excavated so much river – sand for long time that made river – bed uneven, which aggravated the danger to embankments collapsed and broken. These bad behaviors damage the passageway and the basic function of river, which lower its capability of soil and water and self purification ability.

3 Obstacles clearance monitored by remote sensing

With features of wider scope management, longer river, wider floodplains and uneven distribution of obstacles within the lower Yellow River Basin, the traditional enforcement methods used by grassroots staffs for a long time, can not get the distribution of obstacles clearly. To know of all obstacles well and clear them up, it takes advantages of remote sensing to get ground scene information which were not got before.

3.1 The Remote Sensing survey for background data of floodplains

The vegetation vigorous season of 2009 was chosen as the best time for background data including basic information of floodplains, the distribution, property and scale of obstacles along the lower Yellow River from dam of Xi Xiayuan project in Mengjin County of Henan Province in the west to section of Tao Chengpu in Liangshan County of Shandong Province in the east, with 500 km long, transverse average width of 12 km and total area of about 350,000 m^2.

Taking the latest orthophotoes by aerial Remote Sensing as the underlying data source and referring to interpretation signs to describe features of pitch forests, production dykes, buildings, high crops and so on. Then with the topographic map for floodplains and expert knowledge, it uses visual interpretation to trace and draw out borders of obstacles saved as vector data. Then checking in field survey is the last step to confirm the result from images.

3. 2 The dynamic monitoring for obstacles in floodplains

Before flood season of every year, taking aerial Remote Sensing to get data about key area of the lower Yellow River is done. Compared with background data to find out changes of obstacles is used for knowing of the clearance progress and monitoring implementation effect well.

Combining comprehensive monitoring with local monitoring makes dynamic monitoring. Taking low – resolution images to find out the changes area macroscopically and then taking high – resolution images to find out the detail changes of key area to make statistics. In order to enhance comparability between dynamic data and background data, the two kinds of data are needed to be with similar resolution. The time of contrastive monitoring data should be crops non – growing season, which are best for interpreting information about scales and locations of obstacles and improving analysis accuracy.

4 The information system of obstacle clearance monitored by Remote Sensing

In order to make business informationization based on actual requirement of clearance business, "The information system of obstacle clearance monitored by Remote Sensing" is built on database technology and network technology, with functions of information release, browse, query, statistics and comparative analysis.

4. 1 Business process

Business process of obstacles clearance of the lower Yellow River is sa Fig. 1.

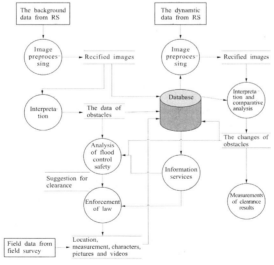

Fig. 1   Business process of obstacles clearance of the lower Yellow River

4. 2 The system structure

The system is of three – tier structure. The uppermost tier is for user interface, the middle tier is for business logic processing, and the bottom tier is for data access.

In the system, the middle tier is the only way for user interface accessing to data, which ensures data secrecy. Spatial data are visual displayed on graphics, which makes staffs find out geographic locations of engineering and obstacles on map through map services, and then read business information.

After getting request from users, business logic layer is with function of getting the data required from data access layer, and then send data back to user interface.

Data access layer defines the operations of original data, and realize data consistency in multiple servers.

4.3　Data and data management

There are several spatial data including Remote Sensing images from different phases with multi - resolutions, vector layers of river channel, vector layers of pitch forests, vector layers of production dykes, vector layers of brick - kiln fields, vector layers of villages in floodplains, and non - spatial data such as statistic data and user information.

The system releases geographic data based on WebGIS. The data managements are mainly includs:

(1) Organization of Web resources. There are many resources with spatial distribution features on internal network of YRCC. So it is useful to make good use of these resources together.

(2) Spatial data release. To release images, spatial data of obstacles and their attribute data through map services.

(3) Spatial query and retrieve. Taking graphs and characters combined method to query the attributions of graphs and natures.

4.4　The system implementation

4.4.1　Integrated information services of obstacles

It is a three - dimensional landform platform based on Skyline technology (Skyline Company, 2005) for main obstacles above - mentioned of the lower Yellow River (Liu Zhanping etc., 2002), with functions of integrated information query, layers choices, and statistics of all obstacles of each unit, area measurement, and screenshots.

All obstacles layers are shown by default when system starts, and users can select layers they need. The display window shows vector graphs that are related with selected layers. Operating map by panning, zooming and rotating on three - dimensional platform, users can know of the spatial relationship between obstacles and flood control projects very well, and get attributions when clicking obstacles. Summary statistics data on left tells users the scales and quantities of each unit, and users can go to the jurisdiction of unit when users click the unit name in Fig. 2.

Fig. 2　Results of integrated information services

4.4.2 Monitoring for obstacles

It is a platform based on ArcGIS Server and Ajax technologies for monitoring pitch forests, production dykes, brick – kiln fields and floating bridges. Taking pitch forests as an example, users can query what they want, make statistics and comparative analysis of random years.

This part only involves layer of pitch forests with thematic map to show the latest situation, and put out statistics data as excel document. And map is located on the jurisdiction of unit when users click the unit name in Fig. 3.

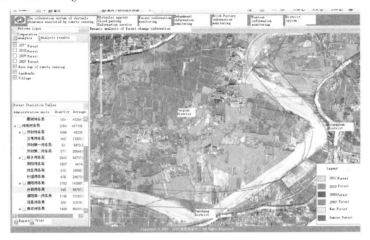

Fig. 3　Results of pitch forests within the Taiqian County

Based on the comparison of background and dynamic data, across processing of different layers of years, it demonstrates the increasing and decreasing trends of pitch forests between two random years. Analysis results are shown in fresh layers, chopped layers forms. In Fig. 4, the chopped layers are highlight with red, where you can find the area and department in charge of it.

Fig. 4　Analysis results of comparative analysis for pitch forests

5　Conclusions

Through comparative analysis between background data and monitoring data, the scales and quantities of obstacles are known very well. The respective statistics of obstacles of Henan Province and Shandong Province are good scientific basis for scheme of pitch forests. Obstacles such as factories, recreation places, other kinds of bridges found in actual work, will be monitored by remote sensing in future. And it will provide informationzation method of working on field and submitting information for grassroots staff.

References

Li Guoying. Report on the Work of Yellow River of 2010. http://www. yellowriver. gov. cn. Jan. 2010.

Skyline Company. Terra Explorer User Manual [M], 2005.

Liu Zhanping, et al. Research on Key Issues for Digital Earth Oriented Virtual Reality System [J]. Journal of Image Graphics, 2002, 7 (2):160 – 164.

Water Resources Vulnerability Evaluation of the Yellow River Delta Based on GIS and Variable Fuzzy Set Theory

Shi Yuzhi, *Liu Haijiao*, *Fan Mingyuan* and *Li Fulin*

The Water Research Institute of Shandong Province, Jinan, 250013, China

Abstract: The Yellow River delta is the Lord area of national high efficient ecological-economic region and Shandong peninsula blue economic region, and it has an important strategic status. This paper integrates variable fuzzy set theory with geographic information system (GIS) to construct the vulnerability evaluation model of the Yellow River delta water resources. Firstly, The study area is partitioned into different evaluation zones (sub-area) based on the spatial recognition technology of GIS; Secondly, the evaluation index system is formulated in terms of the two aspects of water resources vulnerability, natural and human factors, Finally, city of Dongying is selected as study area, which accounts for 93% of the Yellow River delta, to verify the proposed model. The results indicate that the water resources vulnerability of the Yellow River delta greatly changes in space, the region of coastal, Xiaoqing river and Zhimai river shows high vulnerability, while the region along the Yellow River has low vulnerability. In conclusion, the proposed model can effectively identify water resource vulnerability in space.

Key words: variable fuzzy set evaluation model, GIS, the Yellow River delta, water resources vulnerability

1 Introduction

The Yellow River delta is the national high efficient ecological-economic as well as Shandong peninsula blue economic development area, while the United Nations Development Program makes "the Yellow River delta resource development and environmental protection" as a support "China agenda 21" privilege aid project. Due to its abundant natural resources, huge development potential and fast development speed, it is the important regional strategic position with the focus of world attention. At the same time, the Yellow River delta not only is the youngest, fastest growth rates, and the most dramatic landscape changes zone in China, but also is the birds migration, habitat and breeding inland in winter from northeast Asia and western Pacific. Therefore, it is the national ecological environment protection area. Water resources are the source of life, the essential of industrial production, the foundation of ecology, with the development of economy and society, the local ecological environment and water resources have been affected, especially because of shortage of water resources and unreasonable utilization, its vulnerability is severely changing. Therefore, it is of important significance to evaluate the Yellow River delta water resources vulnerability and to realize the local sustainable utilization of the water resources.

The study of water resource vulnerability originated from the concept of groundwater vulnerability in 1968 put forward by French Margat, then, hydrological geologists at home and abroad gradually began to study groundwater vulnerability, including groundwater vulnerability concept, evaluation index system and evaluation methods. Nowadays, scholars widely agree that groundwater vulnerability is the capability of recovering groundwater resource system and element compositions damaged to original, its assessment methods, such as iterative index method, the fuzzy mathematics method, hierarchical analysis method were applied to evaluate the water resources vulnerability, SunCai et al represent the groundwater vulnerability evaluation factors, methods and research prospect based on the overview of the foreign achievement. However, the existing evaluation merely involves the groundwater resource vulnerability, its comprehensive evaluation of groundwater and surface water resources is less focused on, in addition to not represent the spatially characteristics of the vulnerability. With the development of modern information technology, the Geographical Information System (GIS) effectively deals with spatial data analysis and process. Water resources vulnerability has large number of impact factors or indexes as well as their different distributions in space, so GIS technology is obviously used to process spatial water resources vulnerability data, meanwhile, the evaluation model chosen is the key to reasonably evaluate vulnerability. Therefore, two aspects of the natural and artificial factors for constructing the vulnerability index system of surface water and groundwater resources is comprehensively considered. GIS technology is applied to partition evaluation sub-area, and variable fuzzy set model is adopted to represent water resources vulnerability of the Yellow River delta for supporting its economic-social development.

2 Variable fuzzy set evaluation model

This paper integrates GIS with variable fuzzy set theory to construct water resources vulnerability evaluation model of the Yellow River delta, the proposed model is mainly composed of four parts: first, introduce the concepts and definitions of variable fuzzy set theory; Second, set up evaluation index system and partition evaluation sub-area by the GIS spatial analysis technology; third, construct water resources vulnerability evaluation model based on sub-area set and variable fuzzy set theory; Forth, determine the index weight of variable fuzzy evaluation model with the improved entropy weight method. Each part in details is as follows.

2.1 Overview of variable fuzzy set theory

2.1.1 Concepts and definitions of relative difference degree, relative difference function and fuzzy variable set

Definition 1: Set domain U as a fuzzy concept (things and phenomena) $\underset{\sim}{A}$, to any element u of U or $u \in U$, in relative membership functions of the continuum system axes, the relative membership degree of u to $\underset{\sim}{A}$ attract nature is described as $\mu_A(u)$, and on the contrary, to the relative membership degree of its to repel nature $\underset{\sim}{A^c}$ is described as $\mu_{Ac}(u)$, $\mu_A(u) \in [0,1]$, $\mu_{Ac}(u) \in [0,1]$.

Set
$$D_A(u) = \mu_A(u) - \mu_{Ac} \tag{1}$$
where, $D_A(u)$ is called as the relative difference degree of u to $\underset{\sim}{A}$.

Mapping
$$\begin{cases} D_A: & D \to [-1,1] \\ u \mapsto D_A(u) \in [-1,1] \end{cases} \tag{2}$$
is indicated as the relative difference function of to. The schematic diagram of relative difference function, as shown in Fig. 1.

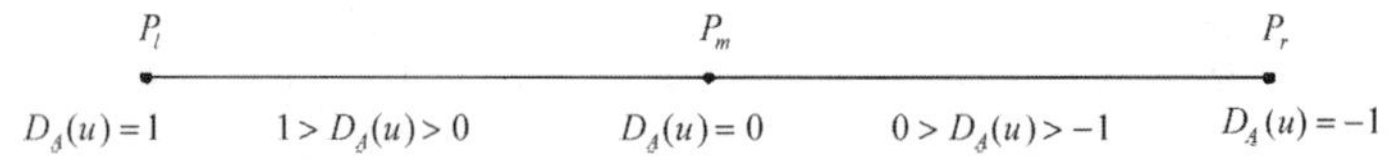

Fig. 1　Schematic diagram of relative difference function

Definition 2:
Set
$$V = \{(u,D) \mid u \in U, D_A(u) = \mu_A(u) - _{Ac}(u), D \in [-1,1]\} \tag{3}$$
$$A_+ = \{u \mid u \in U, 0 < D_A(u) \leqslant 1\} \tag{4}$$
$$A_- = \{u \mid u \in U, -1 < D_A(u) \leqslant 0\} \tag{5}$$
$$A_0 = \{u \mid u \in U, 0 < D_A(u) = 0\} \tag{6}$$
where, $\underset{\sim}{V}$ is called as fuzzy variable set; A_+ is the attract region; A_- is the repel region; A_0 is the boundary point.

2.1.2 The relative difference function model

Set $X_0 = [a,b]$ as the attract region of fuzzy variable set $\underset{\sim}{V}$, that is the interval $0 < D_A(u) \leqslant$

1, $X = [c,d]$ is Upper and Lower boundary including $X_0(X_0 \subset X)$, the position relation is shown as Fig. 2.

$$c \qquad x \quad a \quad M \qquad b \qquad d$$

Fig. 2　Position relation between points x, M and intervals X_0, X

In terms of the definition of fuzzy variable set V, $[c,a]$ and $[b,d]$ are called as repel regions, that is the interval $-1 \leqslant D_A(u) < 0$. Set M as the point of attract region $[a,b]$ under the condition $D_A(u) = 1$, x is the arbitrary point value of interval X, and when x drops into the left side of M, then the relative difference function model $D_A(u)$ is written as

$$\begin{cases} D_A(u) = (\dfrac{x-a}{M-a})^\beta, x \in [a,M] \\ D_A(u) = -(\dfrac{x-a}{c-a})^\beta, x \in [c,a] \end{cases} \tag{7}$$

When x drops into the right side of M, then the relative difference function model is presented as

$$\begin{cases} D_A(u) = (\dfrac{x-b}{M-b})^\beta, x \in [M,b] \\ D_A(u) = -(\dfrac{x-b}{d-b})^\beta, x \in [b,d] \end{cases} \tag{8}$$

Otherwise, x drops into the outside of $[c,d]$, then the relative difference function is shown as

$$D_A(u) = -1; \ x \notin [c,d] \tag{9}$$

where, β is the nonnegative index, generally $\beta = 1$, that is linear function relation.

Eq. (7) and Eq. (8) always meet the relations: ①when $x = a$, $x = b$, then $D_A(u) = 0$; ②when $x = M$, then $D_A(u) = 1$; ③when $x = c$, $x = d$, then $D_A(u) = -1$.

2.2　Construct variable fuzzy evaluation model

2.2.1　Evaluation index system

The vulnerability evaluation index system of water resources refers to many indexes, such as natural factors and human factors. In order to facilitate evaluation analysis, this paper underlines the principle of scientific, completeness, leading of the index system, and applies method in reference to construct evaluation index system from two aspects of nature and human in study area, as shown in Fig. 3, it contains hydrological geology, topography, climate and other natural conditions and social economic status.

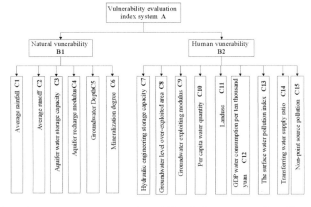

Fig. 3　The water resources vulnerability index system of study area

2.2.2 GIS sub-area partition

To release the spatial vulnerability of water resources, the study area is partitioned into sub-area in terms of the basic data set, such as "Shandong province environment geology atlas", "Water resources bulletin of Dongying city (2010)", and other relevant data materials. The groundwater salinity index distribution is shown in Fig. 4, all index spatial layers are analyzed by MAPGIS software to get the evaluation sub-areas, and the study area is partitioned into 310 sub-areas, as shown in Fig. 5.

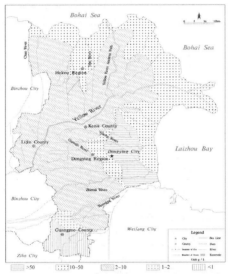

Fig. 4   Spatial distribution map of degree of mineralization of ground water

Fig. 5   The GIS evaluation map of study area

2.2.3 Variable fuzzy set model

Set n sample (sub – areas), $X = (x_1, x_2, \cdots, x_n)$, the ith sub-area mth index vector $x_i = (x_{i1}, x_{i2}, \cdots, x_{im}$, thenindex matrix is written as $X = (x_{ij})$, $i = 1, 2, \cdots, n$; $j = 1, 2, \cdots, m$.

The mth index of sample is divided into l grades, and then the index standard interval matrix $m \times l$ is presented as

$$Y = (y_{jh}) = \begin{pmatrix} [a_{11}, b_{11}] & [a_{12}, b_{12}] & \cdots & [a_{1(l-1)}, b_{1(l-1)}] & [a_{1l}, b_{1l}] \\ [a_{21}, b_{21}] & [a_{22}, b_{22}] & \cdots & [a_{2(l-1)}, b_{2(l-1)}] & [a_{2l}, b_{2l}] \\ \vdots & \vdots & & \vdots & \vdots \\ [a_{m1}, b_{m1}] & [a_{m2}, b_{m2}] & \cdots & [a_{m(l-1)}, b_{m(l-1)}] & [a_{ml}, b_{ml}] \end{pmatrix} \tag{10}$$

where, Y_{jh} is the ith index hth grade standard interval matrix, a_{jh}, b_{jh} are the jth *grade* hth upper and lower point respectively.

In terms of the index standard interval matrix and sub-areas, each index attractive region and boundary matrix are shown as

$$I_{ab} = ([a_{jh}, b_{jh}]) \tag{11}$$
$$I_{cd} = ([c_{jh}, d_{jh}]) \tag{12}$$

The Eq. (11) and Eq. (12) are suitable for two type's index, which is the larger the better as well as the smaller the better index.

On the bases of variable fuzzy set definition and the relative difference degree concept, the point matrix M is determined under the condition of the relative difference degree $D_A(\mu) = 1$, the matrix M is shown as

$$M = (m_{jh}) \tag{13}$$

where

$$\left. \begin{aligned} & m_{j1} = a_{j1}, \ h = 1, \ j = 1, 2, \cdots, m; \\ & m_{jh} = \frac{c-h}{c-1}a_{jh} + \frac{h-1}{c-1}b_{jh}, \ h = 2, 3, \cdots, c-1; \\ & m_{jc} = b_{jc}, \ h = c \end{aligned} \right\} \tag{14}$$

Using the Eq. (7) or Eq. (8) to obtain $D_A(u)$, then the relative membership degree is calculated by

$$\mu_A(u) = [1 + D_A(u)]/2 \tag{15}$$

Integrating the index relative membership $\mu_A(\mathring{u})$ and index weight ω_i with the variable fuzzy model, shown as Eq. (16), as well as the index weight is written as Eq. (19), and then water resource vulnerability of each GIS sub-area is calculated by

$$v_A(u) = \frac{1}{1 + (\frac{d_g}{d_b})^\alpha} = \frac{1}{1 + \left\{ \dfrac{\sum\limits_{i=1}^{m}[w_I(1 - \mu_A(u)_t)]^p}{\sum\limits_{i=1}^{m}w_i\mu_A(u)_t)^p} \right\}^{\alpha/p}} \tag{16}$$

where, α is the optimal rule parameter, generally equal to 1 or 2.

Due to the unsuitability of fuzzy concept graded under the maximum membership principle, this paper applies the relative grade to evaluate the vulnerability. Set the relative membership u to A as $h \sim v_A(u) (h = 1, 2, \cdots, l)$, where, h is the evaluation grade, the final evaluation result of each sub-area is indicated as

$$H = \sum_{h=1}^{l} u_h h \tag{17}$$

where, $u_h = v_A(u) / \sum\limits_{h=1}^{l} v_A(u)$.

In order to improve the evaluation accuracy, the judgment rule is formulated by

$$\begin{cases} 1.0 \leqslant H \leqslant 1.5, \text{ belong to } 1 \\ h - 0.5 \leqslant H \leqslant h, \text{ belong to } h, \text{swing to } (h-1)(h=2,3,\cdots,l-1) \\ h \leqslant H \leqslant h - 0.5, \text{belong to } h, \text{ swing to } (h+1)(h=2,3,\cdots,l-1) \\ c - 0.5 \leqslant H \leqslant c, \text{ belong to } l \end{cases} \tag{18}$$

2.2.4 The improved entropy weight method

The kernel idea of entropy weight method is that the difference between indexes is larger and its weight is more important and larger. Many studies indicate that entropy weight method efficiently recede the human influence on calculating index weight, and get more reasonable evaluation results, however, when the difference of index entropy is small, then its weight is almost difficult to distinguish. Therefore, the new method is proposed to calculate index weight W, it is shown as

$$W = (\omega_i)_{1 \times m} \tag{19}$$

where, $\omega_i = \begin{cases} \dfrac{\sum\limits_{i=1}^{m} H_i + 1 - 2H_i}{\sum\limits_{i=1}^{m} (\sum\limits_{i=1}^{m} H_i + 1 - 2H_i)} & \text{when } H_i \neq 1, \ i = 1, 2, \cdots, m \\[4mm] \dfrac{1 - H_i}{m - \sum\limits_{i=1}^{m} H_i} & \text{when } H_i \neq 1, \ i = 1, 2, \cdots, m \end{cases} \tag{20}$

and ω_i meets, $\sum\limits_{i=1}^{m} \omega_i = 1$, $H_i = -\dfrac{1}{\ln n}(\sum\limits_{j=1}^{n} f_{ij} \ln f_{ij})$ $(i = 1, 2, \cdots, m; j = 1, 2, \cdots, n)$.

In order to ensure $\ln f_{ij}$ to have geometrical significance, $f_{ij} = \dfrac{1 + r_{ij}}{\sum\limits_{j=1}^{n} (1 + r_{ij})}$ is presented.

3 Study case

3.1 Overview of study area

The Yellow River delta covers an area of $13,230.8 \text{ km}^2$, including all five counties of Dongying city (account for 93%), Zhanhua, Wudi of Binzhou city and Shouguang (account for 7%). Due to limitation of basic materials, Dongying city (account for 93%) is selected as study area in this paper, to evaluate regional water resource vulnerability.

Study area is located in the northeast of Shandong province, belongs to the mouth of the Yellow River delta region, the latitude is from $36°55'$ to $38°10'$, longitude is $118°07'$ to $119°10'$, and the whole land area is $7,923 \text{ km}^2$, covers three counties of Guangrao, Lijin, Kenli and two districts of Dongying, Estuary. North of it near the Bohai Sea, east of it next to Laizhou Bay, west of it adjacent to Binzhou city, south of it adjacent to Zibo and Weifang city. It belongs to the continental climate of the warm temperate monsoon type. The average temperature of Dongying city is 12.8 ℃, the average annual sunshine hours is $2,728.5 \text{ h}$, the average rainfall is 554.5 mm. Rainfall spatial distribution is not uniform, the overall trends is decline from 575 mm in southern to 533 mm in the north. On the one hand, the annual precipitation greatly change during years, the multiple of the maximum 991.2 mm (1964) by the minimum 317.8 mm (1992) is 3.02 times; on the other hand, rainfall is mainly concentrated in flood season, accounts for about 70% of the annual precipitation.

3.2 Index evaluation criterion

Generally, water resources vulnerability evaluation criterion is established with large subjectivity and without spatial recognition. Therefore, the cluster analysis method, maximum variance, is applied to construct 3 – level evaluation criterion. Especially for qualitative indexes, like as landuse, its evaluation criterion is artificially set according to the influence on water resources. The index evaluation criterion at study area is shown in Tab.1.

3.3 Index weight determination

According to the proposed water resource vulnerability evaluation model, each index weight is calculated by Eq. (19) and Eq. (20), as shown in Tab. 2.

Tab. 1 The index evaluation criterion at study area

Number	Index name	Unit	Lower vulnerability degree(I)	Mid vulnerability degree(II)	High vulnerability degree(III)	Type
1	Average rainfall	mm	[600,552]	[552,526]	[526,450]	↑
2	Average runoff	mm	[60,57]	[57,55]	[55,40]	↑
3	Hydraulic engineering storage capacity	10^4 m^3/km^2	[30,18]	[18,6]	[6,0]	↑
4	Per capita water quantity	m^3/p	[690,504]	[504,243]	[243,50]	↑
5	GDP water consumption per ten thousand yuan	m^3/10^4 yuan	[50,64]	[64,107]	[107,150]	↓
6	The surface water pollution index	—	[1,5]	[5,9]	[9,15]	↓
7	Transferring water supply ratio	%	[10,60]	[60,98]	[98,100]	↓
8	Non – point source pollution	kg/km^2	[20,102]	[102,361]	[361,600]	↓
9	Landuse	—	[1,4]	[4,7]	[7,10]	↓
10	Aquifer water storage capacity	m^3/d	[4100,1413]	[1413,391]	[391,0]	↑
11	Mineralization degree	g/L	[0,2]	[2,45]	[45,60]	↓
12	Groundwater Depth	m	[30,9]	[9,4]	[4,0]	↑
13	Groundwater exploiting modulus	10^4 m^3/km^2	[5,22]	[22,37]	[37,50]	↓
14	Aquifer recharge modulus	10^4 m^3/(a · km^2)	[40,27]	[27,14]	[14,0]	↑
15	Groundwater level over-exploited area	m	[0,9]	[9,24]	[24,40]	↓

Note: The symbol " ↑ " represents that the larger the better, " ↓ " represents that the smaller the better.

Tab. 2 The weight of evaluation indexes

Number	Index name	Index code	Weight
1	Average rainfall	C1	0.039
2	Average runoff	C2	0.043
3	Aquifer water storage capacity	C3	0.038
4	Aquifer recharge modulus	C4	0.078
5	Groundwater Depth	C5	0.008

Continued to Tab. 2

Number	Index name	Index code	Weight
6	Mineralization degree	C6	0.103
7	Hydraulic engineering storage capacity	C7	0.078
8	Groundwater level over-exploited area	C8	0.031
9	Groundwater exploiting modulus	C9	0.068
10	Per capita water quantity	C10	0.086
11	Landuse	C11	0.061
12	GDP water consumption per ten thousand yuan	C12	0.062
13	The surface water pollution index	C13	0.120
14	Transferring water supply ratio	C14	0.085
15	Non-point source pollution	C15	0.101

Tab. 2 indicates that weights of human vulnerability indexes are more significant than that of natural vulnerability index. The top 8 list are the surface water pollution index, mineralization degree, non-point source pollution, per capita water quantity, transferring water supply ratio, aquifer recharge modulus, hydraulic engineering storage capacity, and groundwater exploiting modulus respectively.

3.4 Result analysis

According to index evaluation criterion and actual situation of study area, three parts of the Yellow River delta vulnerability index, such as the attractive region matrix I_{ab}, range domain matrix I_{cd}, point matrix M, are determined by Eq. (11), Eq. (12) and Eq. (14) respectively, so the relative membership degree and relative difference degree are solved. The weight vector and relative membership degree matrix are put into the variable fuzzy recognition model to evaluate vulnerability of each sub-area, as shown Tab. 3, the result spatial distribution map is illustrated in Fig. 6. It is especially note that the optimal parameters of Eq. (17) and Eq. (18) are selected with $d = 2$, $p = 1$ in this article.

Tab. 3 The evaluation results of water resources vulnerability of study area

Sub – area	H	Sub-area	H	Sub-area	H	Sub-area	H	Sub-area	H
1	2.22	63	2.44	125	2.09	187	2.15	249	2.40
2	2.23	64	2.16	126	1.96	188	2.08	250	2.31
3	2.38	65	2.24	127	2.03	189	2.23	251	2.57
...	...	...	...	...	...	...	...	...	...
60	2.29	122	2.00	184	2.30	246	2.25	308	2.04
61	2.10	123	2.47	185	2.08	247	2.69	309	1.72
62	2.33	124	2.19	186	2.15	248	2.15	310	2.06

The results show that the water resources vulnerability of the whole study area is greater or equal to mid degree, there is no lower degrees vulnerable area. The spatial distribution of local water resources vulnerability is illustrated that the coastal area, Xiaoqing River, as well as Zhimai River are highly vulnerable; the zone along the Yellow River and parts of Guangrao are mid vulnerability.

The reasons are noted that the most region of Dongying city, where it depends on inter-basin transferring water, groundwater salinity is high, especially because of seawater intrusion, that of the coastal areas are higher, so the water resources vulnerability is higher; the Xiaoqing River and

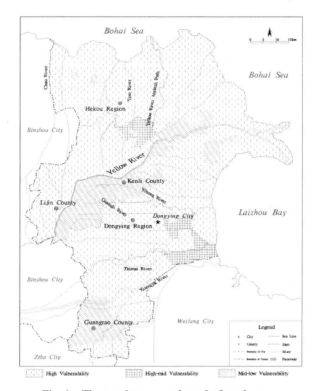

Fig. 6   The result map evaluated of study area

Zhimai River are affected by groundwater over-exploitation, surface water pollution and agricultural non-point source pollution, so water resources vulnerability is high, especially the downstream of Zhimai River, where is the Dongying industrial zone, water resources vulnerability is higher; along the Yellow River, the water resources are relatively rich, perfect hydraulic project, the water resources vulnerability is relatively low, belongs to medium-low degree.

4   Conclusions

The water resources vulnerability evaluation establishes the foundation for water resources optimization allocation and the rational development and utilization. This paper applies GIS technology to realize the space data process of water resources vulnerability evaluation index, and builds a perfectly integrated water resources evaluation index system. The proposed evaluation model, variable fuzzy assessment model, is obviously used to recognize the water resources vulnerability of Dongying city in Yellow River delta, the evaluation results show as: on the one hand, the proposed model efficiently identifies regional water resources vulnerability characteristic; On the other hand, the whole water resources vulnerability degree of the Yellow River delta is higher, and the spatial variation is obvious.

Acknowledgements

The work reported in this article was fund by MWRPRC, the Ministry of Water Resources of the People's Republic of China (Project number: 200801026; Project number: 201201114).

References

Doerfliger N J, Eannin P Y, Zwahlen F. Water Vulnerability Assessment in Karst Environments A New Method of Defining Protection Areas Using A Multi-attribute Approach and GIS Tools [J]. Environmental Geology, 1999,7 (2): 165 – 176.

Aller L et al. DRASTIC: A Standardized System for Evaluating Groundwater Pollution Potential Using Hydrogeologic Settings [M]. Ada, OK: U. S. Environmental Protection Agency, 1985.

Zhang Baoxiang. Groundwater Vulnerability Assessment and Wellhead Protection Area Delineation in Huangshuihe River Basin [D]. Beijing: China University of Geosciences, 2006.

Sun Caizhi, Pan Jun. Concept and Assessment of Groundwater Vulnerability and Its Future Prospect [J]. Advances in Water Science, 1999, 10 (4): 444 –449.

Chen Shouyu, Fu Guangtao, Zhou Huicheng. Fuzzy Analysis Model and Methodology for Aquifer Vulnerability Evaluation [J]. Journal of Hydraulic Engineering, 2002(7): 23 –30.

Chen Shouyu. Theory and Model of Variable Fuzzy Sets and Its Application [M]. Dalian: Dalian University of Technology Press, 2009.

Liu Luliu. Concept and Quantitive Assessment of Vulnerability of Water Resources [J]. Bulletin of Soil and Water Conservation, 2002, 22 (2):41 –44.

Zhou Huicheng, Zhang Gaihong, Wang Guoli. Multi – objective Decision Making Approach Based on Entropy Weights for Reservoir Flood Control Operation [J]. Journal of Hydraulic Engineering, 2007, 38 (1): 100 –106.

A Pilot Study of the Internet of Things in the "Digital Yellow River" Project

Song Shuchun

Information Center of Yellow River Conservancy Commission,
Zhengzhou, 450003, China

Abstract: The "Digital Yellow River" project started in 2001, and it has gained significant achievement that improved the work of the Yellow River conservancy. Previous works of the "Digital Yellow River" project are lack of checking methods for flood control works, and the solutions of water resource management only covers the very fundamental applications. Currently, the Ministry of Water Resources of China (MWR) schedules future informatization work in the field of water conservancy. It is suggested that there should have much more to do with the technology of the internet of things. Therefore, it is possible to build up "Intelligent water conservancy project". so that it would be much easier to obtain real-time water yield data and water quality data. The paper introduces the background knowledge of the "Digital Yellow River" project. It highlights the explanations of the framework and principle for the internet of things' application in water conservancy work. It is suggested that, combined with the existed resources from the "Digital Yellow River" project, applications of the internet of things can construct more intelligent flood dispatching system and water resources management system, which are key parts for the proposal of "Intelligent Yellow River".

Key words: "Digital Yellow River" project, the things of internet

1   Introduction

The "Digital Yellow River" project has been building up since 2001. After years of construction and improvement, six fundamental applications and their infrastructure were almost finished. It is the first step to achieve the goal of the "Digital Yellow River" project. Recently, the things of internet are attracted more and more attentions with the development of information technology. In 2009, the Chinese Primer Wen Jiabao put forward the concepts of "perception of China and intelligence of China". Therefore, the MWR suggests that there should be more real applications of the internet of things in water conservancy projects so that ample information, such as the amount of water and the quality of water, can be obtained. The paper shed light on introducing how the internet of things can be applied in the "Digital Yellow River" project. Relevant technology detail and applications' structures are proposed and discussed.

2   Background knowledge of the "Digital Yellow River" project

The MWR set up the goals for the work of water conservancy. It suggests that there should be more information technology used in the work of Yellow River management. In July, 2001, the Yellow River Conservancy Commission (YRCC) proposed the "Digital Yellow River" project. It aims to construct a more modern and sustainable water conservancy project.

The project is huge and complex, which needs different resources and technologies in its construction. For instance, the technology from water controlling, remote sensing, GIS, GPS, broadband internet, high-capacity data processing and storing, virtual reality and computer simulation has been widely applied. Results suggest that it has significantly improved the efficiency of water conservancy work for the Yellow River.

2.1 Framework of the "Digital Yellow River" project

The "Digital Yellow River" project covers almost all main operations in management work on the Yellow River, such as information collection, information transmission, information storage, and information processing. The framework is organized by three layers: the infrastructure layer, the application service platform and the application system layer (Fig. 1).

Fig. 1 Framework of the "Digital Yellow River" project

The "Digital Yellow River" project is organized as a three tier architecture, and the three layers are infrastructure, application service platform and application system. Infrastructure includes the functions or resources like data collection, data transmission, computer network, data storage and processing system.. Application service platform is not only the place to manage sharable resources, but also provides middle-wares or services which assist to finish functions of applications. It is the information resources' manager in "Digital Yellow River" project. Application system is composed of applications which assist to accomplish the tasks in the work of the Yellow River control and exploit, such as flood control and disaster relief, water dispatching, water resource conservancy, soil and water conservation and engineering management.

2.2 The achievement of the "Digital Yellow River" project

The "Digital Yellow River" project was approved by the MWR in 2003. Since then, lots of new infrastructure has been built up. The remote sensing center, the computer network system, the data management centers were constructed. The remote sensing center realizes the functions like data collections. The data types cover different affairs, such as hydrology, flood-hazard, and water quality. It has finished the data collection work such as the remote sensing image data for the whole lower reaches of the Yellow River starting from Sanmenxia. Remote sensing data has significantly promoted the work of flood and ice flood control, water and sediment diversion, engineering management and polices management. Besides, SDH ultra-short wave network from Sanmenxia (in Henan Province) to Dongying (in Shandong Province) has been built up and put into use. In the lower reaches of the Yellow River, wireless network has also been designed and constructed. Until now, the computer network with strong security and stability has almost covered the whole YRCC and its partner's departments. In addition, the setup of data management center further provids essential resources for the application service platform. Moreover, the hierarchy of flood forecasting, flood dispatching and flood control command is built up, and various application systems are developed and come into use as well, such as the water resources dispatching and automation monitoring system and water resources events processing system. It is proved to significantly help to solve the water drying up problem for the Yellow River. Furthermore, for the regions where exist serious water and soil loss and water pollution problems, relevant real time water quality reporting system and water-soil loss detection system have been constructed and obtained good work efficiency. In order to improve the work of engineering management, a system aimed to manage flood control projects is developed. Last but not least, those years, several well-designed e-government websites, office automation systems are put into use. With the help of them, it is much easier and convenient to share and transmit policies and management information, which were relevant with the Yellow River control affairs, between different departments. Besides, an online video conference system is also developed, and it has dramatically improved the efficiency of decision consultation meeting for the Yellow River.

The "Digital Yellow River" project enhances the efficiency of the Yellow River control and management work. It do not only provide abundant experience from relevant technologies' applications, but also obtain huge improvement and development for social values.

3　Framework and content of the river management system that used the internet of things

The Internet of things is the new stage of informatization development, and it is the road that must take to realize information intelligence. After the technology of computer, internet and communication network, internet of things is the third wave for information revolution. To this end, the MWR has proposed the "Twelfth Five-Year Plan", and it should devote more efforts to put for exploring of the internet of things in the water conservancy application. The goal is to manage the real-time water amount and quality by constructing a huge intelligent water management system. Along with the Internet of things deeply applied in water control work, in Tai-lake, an example project using the internet of things has been successfully constructed. For the "Digital Yellow River" project, there is also an emergent need to combine the technology of internet of things with those existed collection systems, communication networks, and applications. By using the technology of internet of things, we expect this update work that can further improve the management work of the Yellow River.

3.1　Framework of the river management system that used the internet of things

Framework of the internet of things in river basin management are composed of three layers : perception layer , network layer and application layer (see Fig. 2).

3.2　Content of the river management system that applied the internet of things

The perception layer is the foundation for building and developing the application of internet of things, and it consists three layers: data acquisition sub-layer, short distance communication technology and collaborative information processing sub-layer. The data acquisition sub-layer is composed of sensors, 2-dimensional bar code, radio frequency identification (RFID), global positioning systems, remote sensing, monitoring probe and other equipments. The short distance communication technology and collaborative information processing layer could collaboratively process the gathered data in a local range, so that they could improve the precision of information and reduce the redundancy of information. Therefore, they have self-organizing ability to connect the short distance sensor network to a wide area network. The internet of things is the network which connected objects. The key of the technology depends on sensing technology. Through devices, such as RFID and sensors on various objects, after using its interface connecting with wireless network, the technology realizes the function of perceiving objects. Hence, sensing technology is the key technology to construct the applications like intelligent flood control, intelligent water resources management, water and soil conservation monitoring, and water control project security monitoring.

The network layer is the nerve center of the internet of things. On the basis of existed communication network and Internet, it transfers all kinds of information from the perception layer to the application layer through FPBN. The FPBN includes the Yellow River private communication network and public communication network. In the early stages of building up the applications of internet of things, those resources could be enough to satisfy the needs of data transmission. Network layer focuses mainly on the data transmission problems, and the technologies, such as intelligent router, heterogeneous network, and self-organizing communication network, should be well studied and designed.

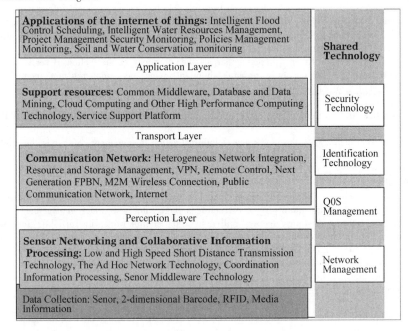

Fig. 2　Framework of the internet of things in river basin management

The application layer is realization procedure of internet of things, and it contains the service support layer and the application layer. The software development and intelligent technology in this layer could provide scientific and technological support for water conservancy work. The essential function of the internet of things is to collect, develop and use information resources. Sensory data management and processing technology is one of these core technologies. It includes relevant theories or technologies in perception data storage, data query, data analysis, data mining, result understanding and decision making. As a key part of the technology internet of things, cloud computing platform is the platform to store and analyze data. Therefore, it is the foundation of application layer.

The application layer involves lots of intelligent computing work on those data from perception layer. Thus, the results from computing could promote the efficiency of works for managing the Yellow River, such as flood control, water resources management, water and soil conservation monitoring, water control project security monitoring and so on. Furthermore, those information from the application layer may also provide strong support for intelligent decision making.

4 Applications of the internet of things in the "Digital Yellow River" project

4.1 The internet of things in flood control and flood dispatching affairs

The overall structure of intelligent flood control system is shown in Fig. 3. In flood control affairs, the collected information includes water and rainfall data, flood-hazard data, meteorological data, goods and materials data, and personnel arrangement data. The perception layer of the internet of things in flood control affairs consists of RFID, water level sensor and water quantity sensor. Through M2M (Machine-to-Machine) intelligent terminals, it could really realize the function of connecting object to object, people to machine and machine to people. RFID could be used to collect data in the goods and material management. The remote sensing data could be applied in modeling 3D visual model in the affairs like flood routing, river regime change, flood hazard and flood submergence. The transport layer is composed of communication network and computer network. The communication network could use resources from public internet or VPN, and it depends on the performance and security needs to select suitable resources. The goal of the application layer intends to process the perception data by using different technologies, such as data mining, cloud computing, and fuzzy diagnosis. Hence, after processing those huge information, it could really supply the users with the function of intelligent computing, intelligent diagnosis, intelligent forecasting, intelligent dispatching and intelligent processing. The final results could be published as different formats, such as Webpage, document in PDA, and mobile phone's messages.

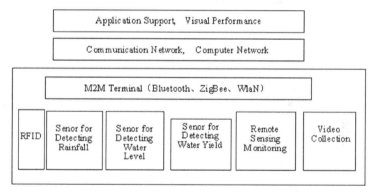

Fig. 3 Framework of intelligent flood control system

4.2 The applications of the internet of things in water resources management

As shown in Fig. 4, in water resources management, there is needed to combine with the water inflow forecast information and the water demand to make more scientific and reasonable water resources allocation scheme and water quality monitoring evaluation. Information collection work includes hydrologic data, drought soil moisture data, groundwater dynamic data and dispatching water data. The perception layer includes sensor equipment like RFID, water quantity sensor, water level sensor, water quality sensor, and underground water level sensor. Through M2M (Machine-to-Machine) intelligent terminals, it could connect object to object, people to machine and machine to people. The transport layer includes communication network and computer network. The communication network could use resources from public internet or VPN. In application layer, after analyzing and mining perception data by using cloud computing, fuzzy diagnosis or other technologies, system could automatically publish the result information in different terminals, such as web browser, PDA or mobile phone.

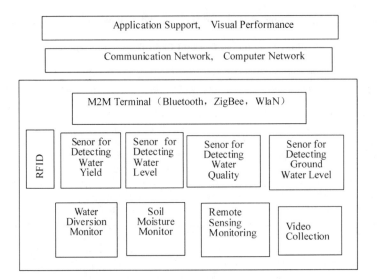

Fig. 4　Framework of intelligent water resource dispatching management system

5　Conclusions

In China, the technology of the internet of things has been widely applied in the fields of intelligent traffic, power dispatching, logistics distribution, environment monitor and disaster relief. In the industry of water conservancy, Jiangsu province has already set up successful applications of the internet of things for the Tai Lake. The Yellow River was recognized as the most complex and difficult sediment-laden river in the world. "Digital Yellow River" project has already achieved huge improvement for the work of water conservancy. However, until now, the gap between the application of the internet of things and the work of the Yellow River conservancy is still there. Therefore, there are needed to enhance the development of the internet of things' applications in the framework of "Digital Yellow River" project. Current research would be usefull for the Yellow River management by putting forward the idea and direction of constructing the applications of internet of things for the Yellow River. Future work could be done by extending and realizing our proposal in real applications. We expect that the applications of the internet of things would be good supplements for the "Digital Yellow River" project, and hence, it must significantly promote the performance of exploiting and managing the Yellow River.

References

Yellow River Conservancy Commission. Planning of "Digital Yellow River" Project [M]. Zhengzhou: Yellow River Conservancy Press, 2003.

Yang Gang. The Theory and Technology of Internet of Things [M]. Beijing: Science & Technology Press, 2010.

Li Jingzong, Kou Huaizhong. The Current Situation and Future Development of "Digital Yellow River" Project [J]. Yellow River, 2011, 33(11): 1 - 6.

Jiang Yunzhong, Ye Yuntao, Wang Hao. The Internet of Things Based Intelligent Dispatching Technology in Water Conservancy [J]. Informatization Work of Water Conservancy, 2010 (4):1 - 5.

Calculation of Pressurized-free Flow
in Long Distance Water Conveyance Canal

Luo Qiushi[1,2] , *Li Bin*[2] and *Geng Bo*[2]

1. Postdoctoral Research Institute of Yellow River Engineering Consulting Co. , Ltd. ,
Zhengzhou, 450003, China
2. Yellow River Engineering Consulting Co. , Ltd. , Zhengzhou, 450003, China

Abstract: The 1D unsteady open channel flow mathematical model was modified for pressurized-free flow in this paper with the "slot method" brought forward by Preissmann and the validation results coincide with the exiting studies. The design data on Zhengzhou Section 1 of South-to-North Water Diversion Middle Line Project are utilized for model application and the result shows that the shorter the gate closure time, the smaller the water level fluctuation range and the water level will be obviously reduced if the water release gate open synchronously with the regulation gate.

Key word: South – to – North Water Diversion Project, water conveyance canal , pre ssurized – free flow

1   Introduction

Long distance water conveyance canal is one of the engineering measures to mitigate the uneven distribution of water resources in space, and now a series of water conveyance project have been successfully built, such as Luan River to Tianjin water division project, Yellow River to Shanxi Province water division project and South to North Water Division Project. Hydraulic factors in long distance water conveyance canal may change suddenly while there's a sudden regulation of the discharge in it. The hydraulic factors of the unsteady flow caused by sudden regulation of discharge may out of the design specification for the safety of the project, so it is one of the focused questions during the engineering design and operation stage.

The 1D mathematical model of unsteady flow is one of the efficient tools to study the unsteady hydraulic process in the water conveyance canal. The 1D mathematical model of unsteady flow usually is solved by the Characteristic Method or the Preissmann Method. It is excellent in keeping the balance of the quantity of water by using the Characteristic Method, but the computational efficiency need to be improved in nowadays. Though the Preissmann Method has high efficiency, it can't keep the balance of the quantity of water, besides the discrete equation usually too complex. Long distance water conveyance canal usually delivers water as pressurized-free flow. The simplify approach and simulation technology of pressure flow and free flow is the important and difficult problem of this question. Now, different models usually built for them, which usually bring in a lot of difficulties to determine the boundary conditions for different models.

A numerical method based on Finite Volume Method (FVM) has been built for 1-D mathematical model of unsteady flow, which can not only keep the balance of the quantity of water, but also show a clear physical relationship in the discrete equations. The 1D unsteady flow mathematical model of open channel was modified for pressurized-free flow in this paper with the "slot method" brought forward by Preissmann. The design data on Zhengzhou Section 1 of South-to-North Water Diversion Middle Line Project are utilized for model application.

2   Mathematical model

The governing equations for 1D mathematical model of unsteady open channel flow can be written as follows:

Continuum equation of water:

$$\frac{\partial z}{\partial t} + \frac{w^2}{gA}\frac{\partial Q}{\partial x} = q_l \tag{1}$$

Momentum equation of water:

$$\frac{\partial Q}{\partial t} + \frac{\partial}{\partial x}\left(\frac{Q^2}{A}\right) = -gA\frac{\partial z}{\partial x} - J_f \tag{2}$$

where, x represents the coordinate along the flow direction; t represents the time; Q represents the discharge; z represents the water level; A represents the discharge area; q_l represents the change in the discharge per unit river length along the route; g is the gravity acceleration; J_f represents the water surface slope, and w represents the water wave.

For open channel flow:

$$w = \sqrt{\frac{gA}{B}}$$

$$J_f = \frac{gn^2\,|Q|\,Q}{A(A/B)^{\frac{4}{3}}}$$

where, n represents the roughness coefficient and B represents the river width.

The boundary conditions of one-dimensional unsteady flow model include the entrance boundary and the exit boundary. The entrance boundary generally gives the discharge hydrograph and the exit boundary generally gives the water level-discharge curve hydrograph.

3 Numerical method

3.1 Discretisation of governing equations

A FVM was employed to discrete the governing equations of one-dimensional unsteady flow model, and an implicit Semi-Implicit Method for Pressure Linked Equations (SIMPLE) scheme based upon staggered grids is proposed to settle the coupling relation ship between the pressure (or water level) and the discharge.

Shallow part in Fig. 1 is selected as the control volume, the cross section P is the i th nodes for calculation, E and W are the neighbor point of it, e is the interface between control volume i and control volume $i+1$, and w is the interface between control volume i and control volume $i-1$. The interface usually set at the center between W and P or P and E.

Fig. 1 Sketch of control volume

3.1.1 Discretization of the momentum equation

Integrate the momentum equation along the control volume, and the convection term is discretized with the first order upwind scheme. The final discretized momentum equation can be written as follows:

$$A_p Q_p = A_W Q_W + A_E Q_E + b_0 \tag{3}$$

where:

$$A_W = \max(U_w, 0)$$
$$A_E = \max(-U_e, 0)$$
$$A_P = A_W + A_E + \frac{\Delta x}{\Delta t} + g\frac{n^2\,|Q|}{A(A/B)^{4/3}}\Delta x$$
$$b_0 = \left(\frac{BQ^2}{A^2} - gA\right)^0 (z_e - z_w) + \frac{Q^0}{\Delta t}\Delta x + \left(\frac{Q^2}{A^2}\right)^0 (A_e - A_w) + (U_e - U_w)Q^0$$

where, U_w and U_e are the interface mass flow; Δx is the control volume length; Δt is the calculation time step length, the superscript 0 represents that the variable uses the calculation result of the previous time level.

The final discrete form of the equation of motion of water flow can be obtained by substituting the velocity under-relaxation factor α_0 into the Equation.

$$\frac{A_p}{\alpha_0} Q_p = A_W Q_W + A_E Q_E + b_0 + (1 - \alpha_0) \frac{A_P}{\alpha_0} Q^0 \tag{4}$$

3.1.2 Correction equation for water level

According to the idea of momentum interpolation, the following interface discharge calculation equation and discharge correction calculation equation are introduced:

$$Q_e = \frac{1}{2}(Q_P + Q_E) - \frac{1}{2} g \left[\left(\frac{A}{A_p} \right)_P + \left(\frac{A}{A_p} \right)_E \right] (z_E - z_P) \tag{5}$$

$$Q'_e = -\frac{1}{2} g \left[\left(\frac{A}{A_p} \right)_P + \left(\frac{A}{A_p} \right)_E \right] (z'_E - z'_P) \tag{6}$$

where, A_P is the principal diagonal element coefficient of the equation of motion.

The interface discharge Q_e^* can be obtained by substituting the initial value of previous water surface level and the initial value of flow velocity which is derived by solving the momentum equation into the above equation. The correction equation for water level can be obtained by substituting $Q_e^* + Q'_e$ into the continuity equation:

$$A_P^P z'_P = A_W^P z'_w + A_E^P z'_e + b_0^P \tag{7}$$

where, the superscript P represents the coefficient of the correction equation for water level:

$$A_W^P = \frac{1}{2} g \left(\left(\frac{A}{A_p} \right)_P + \left(\frac{A}{A_p} \right)_W \right)$$

$$A_E^P = \frac{1}{2} g \left(\left(\frac{A}{A_p} \right)_P + \left(\frac{A}{A_p} \right)_W \right)$$

$$A_P^P = A_W^P + A_E^P + B \frac{\Delta x}{\Delta t}$$

$$b_b^P = q_l \Delta x + Q_w^* - Q_e^*$$

After the corrected value of water level is derived, the water level and the velocity are corrected according to the following equations respectively:

$$z_P = z'_P + \alpha_1 z'_P \tag{8}$$

$$Q'_P = -gA(z'_e - z'_w)$$

The discretized algebraic equations composed of the water momentum equation and the water level correction equation, and the algebraic equations are solved by the Gauss-Seidel iteration method in this article, the iteration steps are as follows: ① giving the guessed value for Q and Z; ② calculating the coefficients in the momentum equation, and solving water momentum equation; ③ calculating the coefficients in the correction equation for the water level, solving the correction term of the water level, and then correcting the water level and the discharge; ④ checking whether the results satisfy the precision criterion according to the residual fluxes of each control volume and whole flow field. If the answer is yes then resume next, otherwise go to Step 2. The converging criterion in this article is that the ratio between the error discharge term and the input discharge should be less than 0.5%.

3.2 Treatment of related problems

(1) The Intersection of main canal and its branch. The discharge of branch canal was submit into the Continuum equation in form of ql, which taken a positive value if discharge flow into the main channel, and taken a negative value if discharge drained out of the main channel.

(2) Treatment of pressured section. In the calculation process, the flow in the open section of the canal and the flow under the top of the closed section of it were both treated as free flow. Assumes that on top of the closed section there is a slot with a width of $B = \frac{gA}{w^2}$ (w is the water

hammer wave velocity) which does not increase the discharge area A and the hydraulic radius R when the water flow level reaches the top of closed section. In this way, the pressurized flow is converted into a special open channel flow, which can be solved according to the method for unsteady open channel flow.

4 Verification

4.1 Jihe ~ Yanhe Section of South to North Water Diversion Middle Line Project

Further research has been carried out on the unsteady flow caused by the discharge regulation in water conveyance canal by Zhang Cheng and the Jihe (Section: 501.398) ~ Yanhe (Section: 529.866) section of south to north water diversion middle line project was one of his examples. The length of calculation canal is 28.468 km, the averaged bottom width is 15.02 m, the average stream wise slope is 1/280,88 and the averaged side slop is 3.077: 1. The designed discharge of the canal is 265 m^3/s and the checked discharge is 320 m^3/s.

Tab. 1 gives verification results of Maximum values of the water surface increase at Yanhe sluice during the hydraulic process caused by the sluice regulation of Yanhe. It can be known from the result that the calculated maximum values of water surface increase here are relatively close to the research findings of Zhang Cheng et al. and that the maximum errors between the two is not over 10 cm in general.

Tab. 1 Verification result of the maximum water surface increase at the regulation sluice

Operating Time (s)	Designed condition(m)		Checked condition(m)	
	Literature	The artical	Literature	The artical
10	1.653	1.609	1.718	1.779
15	1.644	1.590	1.701	1.761
20	1.625	1.573	1.687	1.742
30	1.590	1.537	1.643	1.702

4.2 Zhengzhou 1st Section of South-to-North Water Diversion Middle Line Project

Zhengzhou 1st section of south-to-north water diversion middle line project lies in Zhengzhou City, which is 9.773 km long with the start stake number is SH201 + 000. The canal in research pass cross the rivers, which is Jialu River, Jiayu River and Xushui River, and several hydraulic structures are set along it, such as Jialu River inverted siphon project, Jiayu River inverted siphon project, Jiayu River water release canal project, East Zhongyuan road diversion gate, and Xushui River inverted siphon project. The water in canal was conveyed by open channel combined with inverted siphon project. The designed discharge of the canal is 285 m^3/s at the start, and 265 m^3/s at the end. The designed discharge of Jiayu river water release canal is 142 m^3/s and the designed discharge of East Zhongyuan road diversion gate is 12 m^3/s.

The flow at the designed condition of the canal was simulated by the mathematical model built here and the calculation continues until the numerical flow reaches a steady status. Refer to Fig. 2 for the calculation results. It can be seen from the Figure that the water surface profile obtained from mathematical model calculation is basically consistent with the canal design value.

5 Engineering application

The unsteady flow in the canal of Zhengzhou section 1 of South-to-North Water Diversion Middle Line Project was simulated by the 1D numerical mathematical model built here, and the variation of the hydraulic characters, such as the water pressure (for pressed flow) and the water level (for free flow), has been analyzed according to the result.

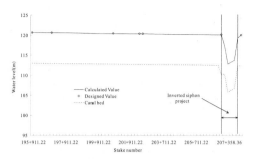

Fig. 2　Comparison of calculation and designed water level in canal

5.1　Calculation condition

Five different conditions were prepared for calculation. Under Case1 ~ Case3, the regulation gates at the upstream and downstream of the canal was closed synchronously in 5 min, 12 min and 18 min respectively without considering the water release gate. Under Case4, the regulation gates at the upstream and downstream of the canal was closed synchronously in 18 min respectively with the water release gate was closed in 3.5 min.

Tab. 2　Calculation condition and the increased values of the water surface

Calculation condition		Case1	Case2	Case3	Case4
Gate regulation time(min)	Gate inlet(closed)	5	12	18	18
	Xushui River Regulation gate(closed)	5	12	18	18
	Jiayu River Water release gate (opened)	—	—	—	3.5
Jialu River regulation gate (201 +000.00)	Maximum values(m)	0.39	0.3	0.27	0
	Time(min)	32	37	43	0
Jiayu River Water release gate (202 +506.95)	Maximum values(m)	0.46	0.43	0.33	0
	Time(min)	13	20	26	0
East Zhongyuan road diversion gate (202 +614.00)	Maximum values(m)	0.54	0.46	0.34	0
	Time(min)	13	20	26	0
Xushuihe inverted siphon Overhaul gate (207 +322.37)	Maximum values(m)	1.31	1.26	1.04	0.42
	Time(min)	12	16	19	17
Xushui River Regulation gate (207 +322.38)	Maximum values(m)	1.32	1.30	1.13	0.46
	Time(min)	12	14	19	17

5.2　Results

5.2.1　Results of water surface increase

The increased values of the water surface for different calculation cases are lists in Tab. 2, form which we can see that:

(1) Under the designed condition, the maximum increased values value of water surface appeared before the Xushui river Regulation gate, the maximum value was 1.32 m,1.30 m,1.13

m,0. 46 m under Case1 ~ Case4 respectively.

(2)The shorter the gate closure time, the smaller the maximum increased values of water surface. Under the designed condition, the maximum value was 1. 32 m,1. 30 m,1. 13 m under Case1 ~ Case3 respectively.

(3)The water level will be obviously reduced if the water release gate opens synchronously with the regulation gate. The calculation results show that the maximum increased values of water level is 1. 13 m (Case 3) if the regulation gate closed in 18 min, which will be reduced to 0. 46 m if the water release gate was opened at the same time (Case4).

5. 2. 2   Analyzing of water level fluctuation process

Form the water level fluctuation process at typical sections, we can see that the water level fluctuation process under Case1 ~ Case3 are qualitatively consistent but quantitatively different. Fig. 3 shows that the water level fluctuation process in the channels under case1. From Fig. 3, we can see that the unsteady hydraulic process will be appeared if the regulation gate is being closed, the water level at the down stream of the canal will be increased and the water level at the upstream of the canal will be decreased. With the march of time, if the water level at the downstream of the canal reaches the peak it will decrease down and if the water level at the upstream of the canal stream it will raise up. The flow in the canal oscillated back and forward, and the oscillation amplitude decreased due to the existence of the flow resistance.

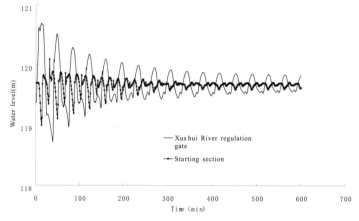

Fig. 3   Water level-time curve of typical sections(Case1)

Fig. 4 shows the water level fluctuation process in the canal under Case4. From Fig. 4, we can see that the water level at the down stream of the canal will be increased and the water level at the upstream of the canal will be decreased while the regulation gate is being closed, which is qualitatively consistent with the process under Case1 ~ Case3. But the water level was decreased quickly and the canal was empty in 2 h because the water release gate was opened while the regulation gate is being closed.

6   Conclusions

(1)The 1D mathematical model of unsteady open channel flow was modified for pressurized-free flow in this paper with the "slot method" brought forward by Preissmann and the validation results coincide with the exiting studies.

(2)The design data on Zhengzhou Section 1 of South-to-North Water Diversion Middle Line Project are utilized for model application and the result shows that the shorter the gate closure time, the smaller the water level fluctuation range and the water level will be obviously reduced if the water release gate open synchronously with the regulation gate. The maximum value was 1. 32 m,

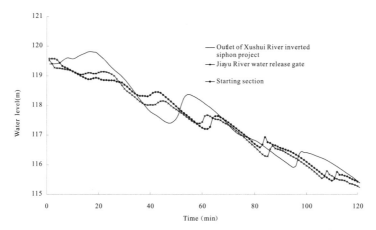

Fig. 4　Water level -time curve of typical sections(Case4)

1. 30 m,1. 13 m,0. 46 m for Case1 to Case4 respectively.

References

Wan Wuyi. Study on Unsteady Flow in Long-Distance Water Diversion Projects [D]. TianJi: Tianjin University, 2004.

Fan Jie, Wang Changde, Guan Guanghua. Study on the Hydraulic Reaction of Unsteady Flows in Open Channel[J]. Advances in Water Science, 2006, 17(1): 55 – 59.

Xie Jianheng. River Simulation [M]. Beijing: China Water Conservancy and Electric Power Press, 1998.

Yang Guolu. River Mathematical Model [M]. Beijing: Ocean Press, 1993.

Shi Yong, Luan Zhenyu, Hu Siyi. Numerical Modeling of Flow-sediment Rransport in the Middle-Lower Reach of the Yangtze River [J]. Advances in Water Science, 2005, 16 (6): 840-848.

Feng Pulin, Chen Nailian, Ma Xueyan. Study on 1D Mathematical Flood Routine Model in Lower Wei River[C]//7th National Symposium on Sediment Basic Theoretical Research: 836-844.

Zhou Xiaolan, Liu Jiang, Luo Qiushi. The Further Research of 1-D Numerical Simulation of Unsteady Channel Flow [J]. Journal of Wuhan University (Engineering), 2011, 43(4): 443-445.

Xu Jingxian. Experimental Study of Free and Pressed Flow Transition Process in Ertan Hydropower Station Tailrace Tunnel [J]. Journal of Hydroelectric Engineering, 1990(4): 58-70.

Zhang Cheng. Research on Response of Unsteady Water Transport and Operation Control in the Middle Route of the South-to-North Water Diversion Project [D]. Beijing: Tsinghua University, 2008.

K. Others

Topography Generation Method of Main Channel on Lower Yellow River and Its Application

Liang Guoting , *Lai Ruixun* and *Zhang Xiaoli*

Yellow River Institute of Hydraulic Research, YRCC, Zhengzhou, 450003, China

Abstract: With change of inflow and sediment, the river course of the lower Yellow River is scoured or silted unceasingly, in particular, radical variation in the main channel and quick change in the river pattern take place. For the sake of observing those changes, large cross sections are usually monitored both before and after flood season annually. However, it is very difficult in acquiring main channel's relief because of a large spacing of those cross sections. This paper describes the method of generating main channel topography based on such observed data as river cross sections, river pattern maps and major flow line, with an example of Huayuankou ~ Lijin section prior to the flood season of 2004.

Key words: downstream of the Yellow River, main river channel, topography, TIN

1 Introduction

The main channel relief of the lower Yellow River as initial riverbed condition is essential for 2D mathematical model of water and sediment, directly relating to the calculatation results and precision of the model, and also, is an important basic data for visualization of 2D flow field, flood advance, river course souring and siltation change, and river regime change. With change of inflow and sediment, the river course of the lower Yellow River is scoured or silted unceasingly, in particular, radical variation in the main channel and quick change in the river pattern can be witnessed. For the sake of observing the change in the river course, large cross sections are usually monitored before and after flood season annually. However, it is very difficult in acquiring main channel's relief because of the large spacing of those cross sections laid out. The methods for obtaining the relief include: making use of topographic map and satellite image or remote sensing image, or manual interpolating of river sections with reference to recorded cross section data and water level ~ discharge relationships. The first is hard to satisfy the requirements of the water and sediment model due to longer duration of relief measurement, big cost and inaccessible to the underwater part for measurement. The second one as time and labor consuming one, with relatively big density, can not meet the requirements of production and living in time effectiveness of river channel relief and of the model neither.

This paper mainly outlines the method of generating main channel topography based on the observed data of river cross sections, river pattern maps and major flow line, with an example of Huayuankou—Lijin section prior to the flood season of 2004.

2 Data preparation for landform generalization

To generalize the relief of river channel needs such data as river pattern map, major flow line of main channel, level ~ discharge relationships, observed parameters of big sections. The coordinates locating training works, vulnerable spots and so on shall also be provided if their impact on the main channel relief is taken into consideration.

2.1 River pattern map and major flow line of main channel

Both water edge and the channel line are interpreted from remote sensing photos. Since the photos taken at one time are unable to provide the data for the whole downstream of the Yellow River, the interpretation has to be obtained from the ones taken at different time periods. For Huayuankou—Lijin river reach, the data for Huayuankou is acquired from TM of June 27, 2004, for Dongbatou from radarsat data on July 6, 2004, for Dongpinghu from radarasat data on June 26, 2004 and 2000's data are used for Shandong because of little change of the river channel.

To acquire the water edge and the channel line, two procedures are needed. Firstly, to collect linear elements, i. e. interpretation of remote sensing photos. Some elements, for instance, water edge, the Y R dikes, etc. can be read directly from the photos, and automatic identification by computer can also be utilized. As for other linear elements, visual interpretation has to be performed, as the channel line are hard to be interpreted under the conditions of present precision of photos and heavy sediment laden water.

2. 2   Observed cross section data and level ~ discharge relationships

2. 2. 1   Observed cross section data

The data of coordinate points of both banks for laying out monitoring cross sections and of observed sections shall be offered, as well as the coordinates of wetted sub-section of main channel for the river section being of compound and the suspended river.

2. 2. 2   Water level ~ discharge relationships

Provide the actual level ~ discharge relationship of monitoring points within the river course. The level ~ discharge relationships are classified according to discharge, including the minimum and overbanking discharge, and that classification shall be identical at every monitoring point. The closer the monitoring point locates, the higher the precision of the river channel relief created is.

3   Topography generalization

The generalization shall be accomplished in two steps. ① By making use of prepared five sorts of data including water edge, the channel line, observed cross sections, level ~ discharge relationships, sub-sections), the interpolating cross sections, coinciding with actual relief, can be generated by self-developed RGTOOLS programe (Yellow River Observed Cross Section Database Management and Analysis System). Such sections shall be perpendicular to the major flow line of the river. ② By using recorded cross section data for both upstream and downstream and level ~ discharge relationships to calculate the leveling, area and width curve of the interpolated sections, those, with elevations, will create irregular TIN. The interpolation of sections in the two steps is critical in relief generalization and also a hot potato in technique.

3. 1   Interpolating cross section

Between the two given neighboring observed sections, the level area and water surface width of the given cross section at a certain level are calculated on the basis of level ~ discharge relationships respectively, thereby, the location and elevation of the interpolated sections can be determined. The interpolated sections are integrated by several points with elevations.

The recorded sections 1 and 2, and the channel line, as shown in Fig. 1, are known, it is required to interpolate cross sections between them. First of all, based on level ~ discharge relationships, the two sections' levels at discharge Z_1 are calculated and then the level at any locations between them linearly interpolated as shown in Plan DEFG. After deciding the location of a section to be interpolated (Plan AACC), intersecting line CC between Plan AACC and Plan DEFG is computer and Projective point C' of two ends of CC in 2 D plan can be acquired. The elevation values at discharge Z_1 are finally sampled.

The elevation discrete points B' and A' of interpolated sections at discharges Z_2 and Z_3 can be found out by the same way, thus, points A', B' and C' are the elevations of the interpolated sections.

There are 3 points need to be clarified. Firstly, selection of interpolating section location. The river channel between two given sections is normally meandering and straight one is seldom seen, therefore, in order to better display the change of river channel topography and pick up good samples, the location of interpolated section is usually rotated or its spacing is increased. Secondly, within the limit of water level ~ discharge relationships, the level and its elevation discrete points, corresponding to any discharge, are calculated as required. Thirdly, any number of sections can, between two adjacent given observed sections, be interpolated.

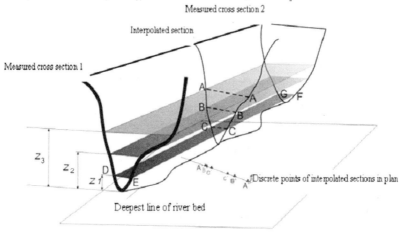

Fig. 1   Sketch of interpolated cross sections

3.2   Generating river bed topography by means of interpolating cross section

After interpolation of sections, irregular TIN needs to be created by using Editor module in ArcInfo, for the preparation of picking up elevations at any point of river channel(see Fig.2).

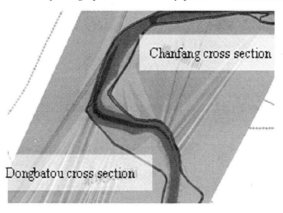

Fig. 2   TIN generated from discrete point

4   Comparison between generalized and actual topography and critical technique

4.1   Comparison between generalized and actual topography

Cite Gaocun—Luokou section is an example for comparison between generalized and real landform, in which the thick line stands for the generalized cross section. Fig. 3 indicates the actual and generalized fit well in the elevation, pattern and area that can satisfy the requirements of the model calculation.

4.2 Critical technique

(1) Bending section treatment. In relief generalization, both curved pattern of river channel and the location of interpolating cross section shall be taken into consideration. To solve this problem, an evenly divided method for segmental arc is adopted as shown in Fig. 4.

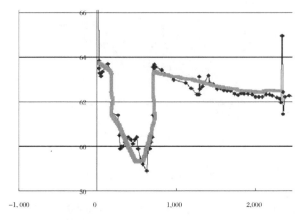

Fig. 3　Comparison of generalized and actual topography

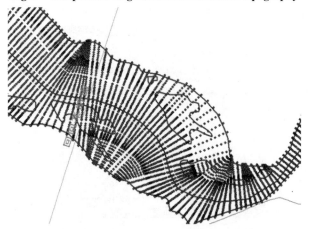

Fig. 4　Bending section treatment of interpolated cross section

(2) Creation of irregular TIN. Ideal method for its creating shall be the one that every discrete elevation point at the same cross section is first elevation interpolated, then elevation interpolation follows, whereas, the traditional method conducts interpolation between every two cross sections, producing the elevation somewhat higher than the actual. The solution to this problem is to add some discrete elevation points accordingly.

(3) Time effectiveness of interpolated sections. Limited by technical and cost conditions, the

data of water edge, the channel line and the water level ~ discharge relationships of Baihe—Huayuankou interpolated cross sections, can not be obtained at one time, but "make up" from several time periods. The method available at present is to shorten time interval as far as possible.

5   Find out elevation values from generalized topography

The TIN of relief can, after its development, be used to generate triangle or quadrangle nets and to get elevation value for any points.

5.1   Create TIN for river channel

The TIN for Baihe—Lijin will be generated by means of Delaunay triangulation theorem and self developed program of Triangular Mesh Generators.

5.2   Take elevation from generalized relief

The 2D triangle net after its creation will be superimposed on relief TIN for obtaining the elevation of any nodes. At last, 2D net with elevation values will break away from the original relief TIN.

6   Conclusions

River channel cross section interpolating technique and the channel relief generated have, during the course of 2D dimensional mathematical water and sediment model study based on GIS for the lower Yellow River, been utilized to calculate the elevations of 2D grid, which provides effective and reliable data source to mathematical model computation and the flow field of water and sediment visualization.

Acknowledgements
    This paper is supported by the National Foundation of Natural Science Commission of China and Yellow River Research Foundation (No. 50439020)

References

Detailed Design Report for 2D Mathematical Water and Sediment Model Study Based on GIS for Lower Yellow River, YRCC, October, 2004.

Wu Xiaobo, Wang Shixin, Xiao Chunsheng. Study on Calculation Method for Delaunay Triangle Net Generation[J]. Surveying & Mapping Newspaper, 1999,28(1):28 – 35.

Mei Anxin, Peng Wanglu, Qin Qiming. On Remote Sensing [M]. Beijing: Higher Education Press, 2003.

The Incentive Mechanism Design of Water Project Ex-Post Evaluation under Incomplete Information[①]

Xiao Yi[1] , *Zhang Huaxing*[2] , *Li Xiaowei*[1] and *Yong Chaoqun*[1]

1. State Key Laboratory of Water Resource and Hydropower Engineering Science of Wuhan University, Wuhan, 430072, China
2. Yellow River Conservancy Commission, Zhengzhou, 450003, China

Abstract: The water project ex – post evaluation is an important part of the project management. With the reform of construction project management system and the model of sustainable development proposed in China, the theory and method system of water project ex – post evaluation is increasingly paid attention to. The current mechanism of construction market can not fully play its role, and there are a large number of incomplete information situations in water conservancy construction, In this article, the multilayer agency relationship and structure of water conservancy project is analyzed, being based on ideas of incentive design, and making use of the incentive model of absolute performance and relative performance, the selection ideas of comprehensive index and the necessity of horizontal contrast evaluation of project are proposed in the ex – post evaluation, and related incentives are researched. The project agents are promoted to take the initiative to provide true information and the consistent behavior and water administrative department are ensured in charge of the social welfare maximization objectives by making the use of an effective incentive mechanism.

Key words: incomplete information, water conservancy project, ex – post evaluation, incentive mechanism CLC: N949

1 Introduction

The water project ex – post evaluation is an important part of the project management process, it is also an important way for the feedback of project information. The techniques and methods of project ex – post evaluation have a development with the change of the concept of project management. In the concept of sustainable development strategy, faced with the amount of investment, capital sources of diversity and management complexity are greatly exceeds the previous case that the large and medium – sized domestic water conservancy project have,? combining with the actual situation in China, it should establish innovative ex – post evaluation incentive mechanism, improve investment supervision and operation supervision, identify the problem of incomplete information in the management of water conservancy construction, and design the ex – post evaluation mechanisms which suitable for China's national conditions and water conservancy industries.

2 The research of incomplete information of the ex – post evaluation

2.1 The definition of incomplete information

The founder of the information theory is Shannon who defines information as: "The information is used to eliminate the uncertainty", but Wiener who is the founder of Cybernetics, defines information as: " Information is the content and name of the mutual exchange between the human and the outside world during the process that people adapt to the outside world , and make this adaptation process give an reaction to the external world".

Incomplete information is commonly exited, professor Julong Deng holds the view that the

① Supported by "the Fundamental Research Funds for the Central Universities(No. 2010004)" system reflects the incomplete information and the principle of the least information, which is to say the system is complicated in general, and it is in the moving and changing conditions. In the absolute sense, incomplete information comes from two aspects, one is that the interaction between system and environment as well as the not entirely clear system factor causing by uncertainty, the not entirely clear relationship of factor, the not entirely clear system structure and the not entirely clear principle of the system interaction. The other is the limited informed capacity of human, cognitive bias as well as in the condition with little information (Since time constraints & information costs, one can not collect all information), the system will encounter vague or uncertain or even no information.

2.2 The research of ex – post evaluation under incomplete information

The incomplete information system herein can be defined as: if the formula $S = (U, AT)$ represents an information system, where U is a finite nonempty set of plant, called a main or object space, AT is a finite nonempty attribute set; when an attribute $a \in AT$, and an arbitrary object $x \in U$, then there is a mapping $a : x \rightarrow a(x)$, in other words, it is the value taken of object x from the attribute a. If $a(x)$ is uniquely determined, then S is a comprehensive information system, and the information will be considered completely; however, for other many information systems, we can not determine the attribute's values, for example, there exists missing data or some data is not unique, the fuzzy is hard to determine, so $a(x)$ will not be unique, that is $a(x) \in V_a$, where V_a is numerical range of the all objects from the attribute a, then S is an incomplete information system, and the information is incomplete.

When evaluating concept after the summary of dams and related water conservancy projects, Ames Wescoat has pointed out that the evaluation system should contain the attributes of comprehensive, integrated, long – term, cumulative and adaptive. Comprehensive evaluation of water conservancy project should use comprehensive data series which include the influence of environment, society, economy, public relations and related other water conservancy project. Integrity evaluation refers to analyzing the interaction of these different types of impact. Long – term evaluation that refers to the data used to assessment are monitored from the decades or even longer. Cumulative need is to analysis the regional water management structural measures and unstructured measures and the systematic relationship in water conservancy projects of the same basin. Adaptive evaluation requires assessment timely and adjustment decisions related in the changing external environment and social conditions.

Seen by the above requirements of post – evaluation system. Assessment focused on the socio-economic and environmental impact of the projects. Content covers a wide range and data involves more complex. From the actual data to predict data, sources of information and signal transduction pathways, data information reliability etc exist a huge difference. Thus hydraulic engineering post – evaluation system is incomplete information system. The analysis from the perspective of the policy makers was mainly caused by imperfect means of systematic observation, random factors and the limited nature of the system of complex behavior patterns. Specific performance is as follows.

2.2.1 The incompletion of information on time dimension
The information used in the ex – post evaluation process has strong temporal properties, especially the time – validity property. The pre – evaluation has made detailed analyses and predictions for the hydro – project, but the external conditions of the project have changed at the time of post – evaluation. The length of time series (real data) for the evaluation to the performance of hydro – project is short (only 5 to 10 years in general), and for the lack of a completed database for the ex – post evaluation of hydro – project construction. There is no related parameters or data series from horizontal comparison to be used for reference, thus the information needed for ex – post evaluation is very limited. On the other hand, due to historical reasons, the process of hydro – project construction is usually not standardized or under procedure, and in many projects the pre – valuations have not been conducted, not even a complete design or completion and acceptance report. The opinion of "emphasizing construction and omitting management" and

not attaching importance to the accumulation of practical information has resulted in the absence, missing or scattering of statistical information in many projects, which induce the incomplete information problem for assessment analysis.

2.2.2　The incompletion of information on spatial dimension

The subsystems of study on the ex – post evaluation are rather considerable, involving to the natural information, solid things information, biological and social information in the project place. Knowledge classification has many levels, below each level of classification there are a number of grade two and grade three class information representations. For example, former articles put emphasis of benefit evaluation theory on economic benefit, with external economic benefit analysis. Scholars are of the view that the benefit of water conservancy project should include economic benefit, social benefit and environmental benefit, and the ex – post evaluation should be comprehensive evaluation of the three aspects of benefit. The social benefit analysis system need to research on information of study area, such as population, employment, per capita, per capita consumption, industrial structure, social culture and so on , and with these multivariate data analysis to analysis the social system in the information flow problems. Because of so many involving factors, there are large amount of benefit data and information about flood control, power generation, water supply, comprehensive management, and are less other external benefit data with considerable incomplete information. While experts solve these benefit index information with less accuracy, even with contradictory nature, coupled with the limitation of the valuers' understanding on the problems and their own lack of knowledge or time and patience, the complexity of social environment, all these make the comprehensive benefit evaluation face a lot of incomplete information.

2.2.3　The incompletion of information on model dimension

The knowledge framework of ex – post evaluation should set comprehensive and specific index system reasonably, making the target indexing and quantification as far as possible, i. e. , most of the objectives proposed have clear indicators, and can be quantified, measured and compared (including historical comparison and industry comparison). The characteristic of the post – evaluation application model is that select the resource datum for a certain target from the pluralistic, multi – level and conflicting resource information datum, form each individual conclusion of the post – evaluation through the comprehensive evaluation, and finally all the individual form the ultimate evaluation conclusion. The weight and index have the nature of measurement, hysteresis, asymmetry, and interference, relativity and stripping, so need data processing for the large and all series of information resources (such as numerical, text, voice and image, etc.). Complicated problem is simplified, and individual is abstracted for general, so the original information of the model parameters must be complex, because the "save effect law" and the limitation of the material, the model parameters are rather incomplete.

2.2.4　The incompletion of information on personnel dimension

Based on the known information, when ex – post evaluation personnel generalize or classify things, there will be a difference in an accurate grasp of the development things and identifying information model, as a result of the different understanding of "grey" things mechanism or motion law. At the same time, because of recognition subject which is interfered by all kinds of noise, information dissemination source which is in disorder, the interference of other information(equal to noise), and the limitation of receiving information ability, it will influence the real degree of information. As a result, from information source to information receiver is a complex transmission process, inevitably, there will be different sides, different degrees of distortion rather than "information remain unchanged in the transmission process" of informatics ideal condition. In the level of transforming them into "social information" or "human information" , we call part distortion information as "incomplete information" .

Post – modern – country governing theory emphasizes cooperative game or win – win of the power and the multiple decentralized management. There are a lot of incomplete information problems in water conservancy project construction, coming up with the suitable industrial ex − post evaluation incentive mechanism is helpful to reducing the cost of searching information and improving enterprise operation efficiency.

3   The incentive mechanism design ideas of ex − post evaluation

Mechanism design has a special meaning, it is aimed at the principal, and how the optimal mechanism should be designed to achieve their goals by the principal of the contractual relationship with the agent. Analysis from the display principle, the design focuses on how to show real personal information system of the agent, and its direct impact to the agent of the order of specific activities and results. Later evaluation mechanisms should belong to the confusion of the oversight mechanisms and incentives. Especially, it is in the presence of multi − level principal − agent system, and it especially need to devise a mechanism to guarantee orderly and efficient running of incentive activities to curb the opportunism of the agents and to promote the efficient allocation of resources of the government, thus to maximize promote social welfare.

Water conservancy project influences national economy and people's livelihood. In a joint − stock enterprise, the board represents as capital are consistent, and the government decision − making should base on the integration of environmental and socio − economic. In the process of water conservancy project construction and management, the relationship between the government and enterprises is a principal − agent relationship. The goal of water administrative department is to pursue the coordinated development of economic, ecological and social benefits and the maximization of overall benefit. There exists information asymmetry between the two. Enterprises compared to the government, hold more accurate information about the enterprise itself. Due to the differences of benefit goals between the government and enterprises and the concealment of action and information, it is prone to occur moral hazard problem. Therefore, the water administrative department must design an optimal incentive mechanism, to induce enterprises to take the initiative to reveal their own information from their own interests in the post − evaluation stage, and to strive to take actions to improve the efficiency of resource utilization and reduce bad external effects of the project, so as to achieve maximization of benefits of economy , the society and environment.

The drive function from the entruster such as the water conservancy departments is performed in the public management. The Evaluation Agency takes on the role of intermediary in the ex − post evaluation. According to the view of Pro R. Selten, the deliver information system should be strengthened, and the effective signal mechanism designed, which can lead the agent release real information, and acts based on the entruster's goal. The design thought of the incentive mechanism should be as following: on the foundation of grasping the existing delegate − agency relationship clearly, seek for appropriate incentives and build the positive correlation relationship in the aspects of benefits and goals between the entruster and the agent in accord with their respective characteristics. In order to build the relationship, the incomplete contract should be made up, and the most effective method is to build a control mechanism based on the results between them before the agent acts, namely confirming the reasonable mechanism to share risk and benefit.

In the relationship, one effective incentive mechanism should meet three basic conditions: building favorable external environments which is fit to the mechanism; building scientific evaluation system about the agents' performance, which can be the basis of incentive; at last, choosing the motivational pattern.

3.1   The system of open information

Establish the ex − post evaluation information open system, by the use of the reputation incentive principle, standardize the behavior of project construction and management unit. Effective incentive is the combination of dominant and recessive incentive, and this conclusion is the foundation of a dynamic incentive mechanism. Recessive incentive mechanism hypothesis the game between the principal and the agent is multiple and dynamic, under such a long − term contract, in addition to the incentive mechanism, "time" itself may solve the agency problem, Rander

(Rander, 1981) proves Pareto optimal risk sharing and incentive can realize by using repeated game model, based on the opportunism's "bad guys" hypothesis, economists Kreps, through the establishment of economic models, proves if repeated games between participants are enough, the expected profit of cooperation in the future will exceeds the loss of being deceived. Therefore, both the game began to want to establish a cooperative reputation (even if he is cooperative in essence), in order to get the cooperation of expected return. The participator will destroy his reputation all at once in order to gain more short – term benefits only before the end of a period in the game. Kreps at the same time also pointed out that the establishment of the reputation does not need to keep long – term trade relationship. As long as a party exists for a long time, the other party can observe their long – term behavior. Reputation can play a role.

3.1.1 Model analysis

Assume that in the market mechanism, the agent's reputation is decided by their own business performance in ex – post evaluation, namely:

$$R = r(f(e,x)), \tag{1}$$

where, e is on behalf of the efforts level of the agent; x is on behalf of the agent's ability; $f(e,x)$ is the business performance function, which is decided by e and x.

In the long – term cooperation expectations, the agent's reputation income can be expressed as:

$$R = \alpha r(e_x,x) + \gamma r(e_r) \tag{2}$$

where, α and γ are on behalf of the coefficients of the agent's efforts in the pursuit of the current performance and future development respectively; e_x is on behalf of the efforts level of the agent for the current consideration; e_r is on behalf of the efforts level of the agent for the long – term social reputation in order to get more income; $e = e_x + e_\gamma$.

Assume that the agent money income is expressed as:

$$W = \beta f(e_x,e_r,x) + p + b \tag{3}$$

where, β is the coefficient of risk income; p is the fixed income of the agent; b is on behalf of the agent's gains for control power. The agent's efforts cost consists of two components, and the function is defined as:

$$C_{(e)} = C_{(e_x+e_r)}, \mathrm{d}c/\mathrm{d}r \tag{4}$$

Then agent's utility function being:

$$U = W + R - C = \beta f(e_x,e_r,x) + p + b + \alpha r(e_x,x) + \gamma r(e_r) - C(e_x+e_r) \tag{5}$$

Asking about the effort cost of a derivative on both sides of the equation, we can reach a conclusion that:

$$\mathrm{d}U/\mathrm{d}e = \beta \mathrm{d}f/\mathrm{d}e + \mathrm{d}p/\mathrm{d}e + \mathrm{d}b/\mathrm{d}e + \alpha \mathrm{d}r(e_x,x)/\mathrm{d}e + \gamma \mathrm{d}r(e_r)/\mathrm{d}e - \mathrm{d}c/\mathrm{d}e \tag{6}$$

When the agent utility reaches the maximum, namely $\mathrm{d}U/\mathrm{d}e = 0$, we can conclude that:

$$\beta \mathrm{d}f/\mathrm{d}e + \mathrm{d}p/\mathrm{d}e + \mathrm{d}b/\mathrm{d}e + \alpha \mathrm{d}r(e_x,x)/\mathrm{d}e + \gamma \mathrm{d}r(e_r)/\mathrm{d}e = \mathrm{d}c/\mathrm{d}e \tag{7}$$

According to the results deduced, we can derive that the condition for agent utility reaches the maximization is that the marginal monetary income and marginal reputation are equal to the marginal cost and that reputation incentive is also a kind of incentive mode. Li Chunqi and Shi Lei (2001) propose that the marginal utility of money income exists decreasing trend. So the role that it plays in certain conditions is limited, especially when the money income reaches to a level in which the effect of income may be over the substitution effect. Therefore, simple increases in monetary income may reduce the enthusiasm of the agent. Combined with the conclusion by Li Heng (2002) we can conclude that the entrepreneur's demand is not just money income, especially when money income reaches a certain satisfaction, in which the other need should be considered in the demand function. So when the currency plays incentive effect in codominance, more attention should be paid to reputation incentive at the same time.

3.1.2 Establishment of the post evaluation information public system by reputation incentive mechanisms

So, for construction and management of water conservancy projects, the action of reputation incentive mechanism is good reputation in the market to enhance the bargaining power or market competition ability, so as to give its benefits. As well, for enterprise operators and project manager, the reputation incentive mechanism is a kind of final incentive because it is a person that the highest level needs that pursuit good reputation and gets others think highly of and respect based on Mathlow's need theory of management.

Thus evaluation mechanism should increase the result of evaluation information diffusion and make public more transparent, additional information issued special ways and link in order to improve market reputation both the construction and management enterprise, and inspire the enthusiasm for work. Measures, such as show good department or enterprise, propaganda throw the newspaper, can be taken. We can also improve the reputation degree of development throw the evaluation results into the industry organizations and professional network information database.

On the other hand, excellent or success criteria need to be divided, according to the results of the post – evaluation, to praise technological innovator or excellent construction managers, units outstanding title, excellent personal title, awarded a certificate, documented evaluation of historical archives can be designed to give appropriate spiritual rewards. For example, the construction unit of credit rating, build construction unit historical performance information system, for the new generation of project evaluation to provide the necessary information, make construction unit through the norms to fulfill the generation task, to establish the credibility of their own.

Water conservancy project post evaluation for the industry level should be carrying out regularly, and many times after the valuation point should be set in the water conservancy project life cycle, for example, to carry out special study on the post evaluation for the management of water conservancy project operation, using the "time" to solve the agency problem. Be careful that there was a close correlation between the "Reputation" quality and impartial, objective evaluation of project management performance (output).

3.1.3　Feedback timeliness

Using the principle of timely incentive, the incentive effect can be expressed as the following functional relationship:

$$I = f[\,1/(t_0 - t)\,] \tag{8}$$

where, t_0 represents the time that managers promised to reward, t is the actual reward time. According to the functional relation $\Delta t = t - t_0$, the more is Δt approaches to the infinitesimal, the better does the incentive effects. With the latency time becomes longer, the value of $f(x)$ is going to decline, then the poorer incentive effect will be obtained. Therefore, once drawn evaluate message in the post – evaluation, reward or punishment should be timely given to the responsibility unit. The best way is to build water conservancy project ex – post evaluation expert system, timely release the evaluation results, and operate in accordance with the systems and regulations.

3.2　The design of punishment system

In the way of management incentives, positive incentives have the problem of diminishing incentive effect and rising costs, but negative incentives can compensate for this defect. Negative incentives is to punish employees who have the undesired behavior that goes against the goals of the organization so that this does not recur. Kahneman·Daniel (2002), Nobel economist winner of Israel's psychological economist, by the way of cognitive psychology research to study human decision – making behavior, put forward that people in the decision – making, will have its own "reference point", in the vicinity of the "reference point", the value damage (disutility) caused by a certain amount of loss brought greater value meet than the same amount of profit, the key problem is that these two is not symmetrical, showing that in the use of the same amount of stimulin. Deprivation has a larger stimulus than giving. Shao Jianping and Cao Lingyan (2003) pointed out that in the relative to positive incentives, negative incentives are more effective and lasting incentive. Huang Linde etc (2004) think that the negative incentives have the effect of additional positive incentives, to give warning, correction and education bad behavior and producing pressure, and it also can stimulate the initiative to achieve organizational goals. So, it can emphasize the rationality and importance.

The negative incentive mechanism of water project ex – post evaluation is missing currently, just that the system of ex – post evaluation is positioned at the level of the "lessons learned". The index of ex – post evaluation includes analysis of the degree of difference, the ex – post evaluation of engineering construction, mthe ex – post evaluation of financial, the ex – post evaluation of national economy and the ex – post evaluation of engineering management all contain the index of various deviation rate, but being lack of the abnormal accountability and disciplinary mechanisms which are based on the industry average standard, so we should strengthen the negative incentives in the system of ex – post evaluation, combined with the positive incentives.

First of all , "the negative incentive" is based on a set of effective evaluation appraisal system, which requires clear appraisal index, clear responsibility main body and quantitative index of appraisal, at the same time, we should strengthen the negative incentive mechanism and avoid formal evaluation. According to the comparison of relative performance, we can set up hierarchical level last warning measures. Through the comprehensive evaluation of the success of the analysis conclusion, we can also forbid the construction management unit (whose project construction management degree of success is the lowest on the historical record) to taking charge of the water conservancy project. In addition, we should also add some light disciplinary means such as issuing warnings, negligence criticism and warning, reminding the parties concerned to correct mistakes timely. The competent department or the evaluation sponsors can use some methods such as economic sanctions, functionary disciplinary action, stands mechanism punish the one whose evaluation conclusion is poor. The correlation and design of the disciplinary level and responsibility classification system need further discussion and research.

4 Conclusions

From a specific point of view of understanding, post evaluation of water conservancy project system should be a continuum of oversight mechanism and excitation mechanism. However, post evaluation of water conservancy project at present still mainly focus on the dominant economic benefits information and a comprehensive evaluation which is based on the comparative analysis of the presence or absence. This caused a significant reduce of stimulation. This chapter analyzes the principal – agent problem in water conservancy project, and optimal incentives were proposed in the post evaluation system to maximize the social welfare in water conservancy administrative department. There are several ways to design incentive mechanism of post evaluation of water conservancy project by utilizing economic analysis model of incentive theory and managing incentive theories: ① the need to adapt to the trend of the decentralization system evolution; ② the transformation from affiliation to the contractual relationship; ③the transformation from the process control to the results control; ④the transformation from "Rules – based" to a "results – oriented". With the establishment of scientific post evaluation mechanism and appropriate performance indicator system, the comprehensive study of managers' performance evaluation should be conducted from the perspective of the comparison between external and horizontal performance of the project.

References

Shannon C E. A Mathematical Theory of Communication I[J]. Bell System Technical Journal. 1948,(27):379 – 423.

Norbert Wiener. Wiener Writings[M]. Shanghai: Shanghai Translation Publishing House, 1978.

Deng Julong. Gray System Theory Course [M]. Wuhan: Huazhong University of Science and Technology Press, 1990.

James Wescoat. Ex – post Evaluation of Dams and Related Water Project[R]. http://www. dams. org/docs/kbase/contrib/obt183. pdf.

Zhang Z J, Zhang S B, Feng T H. Calculation of Flood Control Benefit for Economic Post _ evaluation for Taolinkou Reservoir Project[J]. China Rural Water and Hydropower, 2002

(8):6 – 7.

Feng T H. Features of Post – economy Evaluation in Water – saving Irrigation Project[J]. Journal of North China Institute of Water Conservancy and Hydroelectric Power (Social Sciences Edition) ,2006(2):19 – 21.

Ministry of Water Resources of the People's Republic of China. SL 72—94 Regulation for Economic Evaluation of Water Conservancy Construction projects[S]. 1996.

Liu Ya. Agency and Supervision Mechanism [D]. Wuhan: Wuhan University, 2004.

Effect Research of River Networks Water Fluidity Enhancing Engineering Measures Based on the 2D Numerical Model

Yang Fuxiang[1] , *Pan Jie*[2] , *LI Dongfeng*[3] and *Zhang Hongwu*[4]

1. Henan Vocational College of Water Conservancy and Environment, Zhengzhou, 450008, China
2. Yuyao Municipal Bureau of Water Resources, Yuyao, 315400, China
3. Zhejiang University of Water Resources and Electric Power, Hangzhou, 310018, China
4. State Key Laboratory of Hydro-science and Engineering, Department of Hydraulic Engineering, Tsinghua University, Beijing, 100084, China

Abstract: One of the effect measures of improving water flow fluidity and solving city water pollution is water diversion from outside of an urban area. Because of the different boundary condition, quantity of flow distribution and velocity is different for every river. The velocity of some river channels is even much smaller. In order to enhance the fluidity of those weak rivers with small velocity, to set up a series of sluice gates and to control gates by closing and opening in some need. Thus water quantity is redistributed and the velocity is increased. For the need of investigating the effect of setting up and closing the gates, 2D unsteady river networks flow finite element model is established and verified. Bases on the model, the velocity size and orientation of the every river and water levels are calculated at the condition of whether setting up gates. These numerical researches show that the water flow quantity and velocity of flow distribution are both increased significantly for the near of river. The maximum velocity increment 0. 23 m/s, the maximum banked up water level is 0. 014 m, the maximum water lowering is 0. 05 m, this engineering effect is obvious and this provided quantity data for the projects plan and design.

Key words: river network, gates control engineering, effect, finite element method; hydrodynamics

1   Introduction

In most cities, river networks is criss-crossing, water environment pollution is critical and complex problems. One of the effect measures of improving water flow fluidity and solving city water pollution is to implement water diversion from outside of an urban area.

Because of the different boundary condition, quantity of flow distribution and velocity is different for every river. The velocity of some river channels is even much smaller. In theory, in order to enhance the fluidity of those weak and small velocity rivers, a series of sluice gates and controlling gates by closing and opening in some need is set up. Thus water quantity will be redistributed and the velocity is also increased in the near of the gates. In fact, how to know quantity the effect of whether setting up gates is important for the plan and design of urban water environment. In order to investigate quantity, depth-averaged planar 2D shallow water finite element method Numerical Model is good method.

2   Numerical simulation basic theory methodology

2. 1   The governing equations and deterministic conditions (Li Dongfeng, 1999)

Using depth-averaged planar 2-D shallow water equations as the governing equations for computation.

The equations of continuity:

$$\frac{\partial z}{\partial t} + \frac{\partial hu}{\partial x} + \frac{\partial hv}{\partial y} = 0$$

The equations of motion:

$$\frac{\partial u}{\partial t} + u\frac{\partial u}{\partial x} + v\frac{\partial u}{\partial y} + g\frac{\partial z}{\partial x} + \frac{gn^2 u \sqrt{u^2 + v^2}}{H^{4/3}} - fv - \varepsilon(\frac{\partial^2 u}{\partial x^2} + \frac{\partial^2 u}{\partial y^2}) = 0$$

$$\frac{\partial v}{\partial t} + u\frac{\partial v}{\partial y} + v\frac{\partial v}{\partial y} + g\frac{\partial z}{\partial y} + \frac{gn^2 v \sqrt{u^2 + v^2}}{H^{4/3}} - fu - \varepsilon(\frac{\delta^2 u}{\partial x^2} + \frac{\delta^2 v}{\partial y^2}) = 0$$

where, u, v is x, y direction components of depth averaged velocity; z, h is water level (or tidal level) and depth; g is acceleration due to gravity; ε is turbulent viscosity coefficient; C is Checy's coefficient, C is calculated by Checy's formulation:

$$C = \frac{1}{n}R^{1/6}$$

where, n is Manning roughness coefficient; f is Coriolis force coefficient, $f = 2\omega\sin\Phi$, ω is rotation angular velocity of earth, Φ is the latitude of computed reach.

2.2 The deterministic conditions

The deterministic conditions involve boundary conditions and initial conditions.

2.2.1 Boundary conditions

Boundary conditions include opening boundary and closing boundary. The former opening boundary is inlet and outlet water boundary, which is govened by inlet flow quantity process and outlet water levels process for model. The latter closing boundary is land boundary and the normal velocity is treated as zero for model.

2.2.2 Initial conditions

The initial water level and the initial velocity are given by measured tidal level or given value zero, initial conditions do not affect the precision of computed result. At initial time, tidal level, velocity and other varibles are given.

3 The finite element solution of the equations and verification of model

3.1 Numerical procedure

According to the above formulation, finite element analysis is carried out first, then composing each element and the overall finite element equations can be obtained.

3.2 Solution of finite element model

The predictor-corrector method is used to estimate the iteration equations.

3 Verification of the model

The comparison of calculated data and filed shows that all these numerical simulation results, the water level and velocity process, are well agreement with the field ones. The detailed verification of the model is shown in reference (Li Dongfeng, 2009)

4 The analysis of calcualting results

4.1 Calcualting river channel networks

As an example, the Cixi city of Zhejinag Province, of which the river channel networks is shown in Fig. 1.

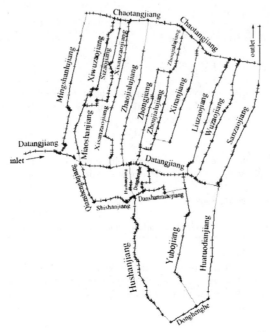

Fig. 1 Cixi urban river channel networks and boundary

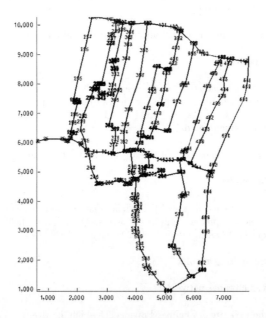

Fig. 2 Initial points and grids

4. 2   Calcualting river channel networks initial and calculated points and grids

River channel networks initial and calculated points and grids are shown in Fig. 2 and Fig. 3.

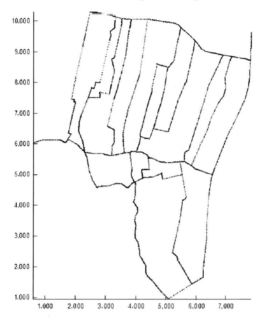

Fig. 3   Calculated points and grids

4. 3   Calculating boundary conditions

According to the advices the calculating boundary conditions is that at the inlet flow quantity is 18 m³/s and at outlet water levels is 1. 95 m. The closing boundary is land boundary and the normal velocity is treated as zero for model.

In order to conform quantitatively the effective and the effect of whether setting up gates, a project is calculated and researched, one case is not set up with sluice gate at the inlet of Mingshanlujiang River, the other case is set up with gate and close gate.

5   Results and Analysis

According to the above condition, the finite element software is used to calculate and simulate the water flow two cases of whether setting up sluice gates, the main hydraulic factors is water levels, deep-average velocity, water depth, the distribution of every river is shown from Fig. 4 to Fig. 9, these results are analyzed below.

5. 1   Water levels increment

5. 1. 1   Water levels calculated results of not setting up sluice gate

In the condition of not setting up sluice gate, water levels calculated Results is shown in Figure 4, the numerical is the water level elevation of key points

5. 1. 2   Water levels calculated results of setting up sluice gate

In the condition of setting up sluice gate, there is no water flow into the Mingshanlujiang

River, the water levels calculated results is shown in Fig. 5, the numerical is the water level elevation of key points.

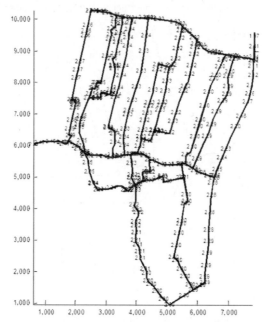

Fig. 4   Water levels no gate at Mingshanlujiang River

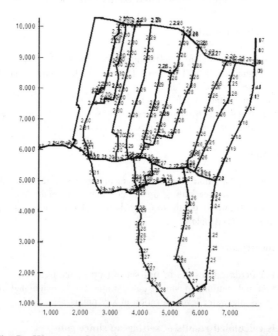

Fig. 5   Water levels setting gate at Mingshanlujiang River

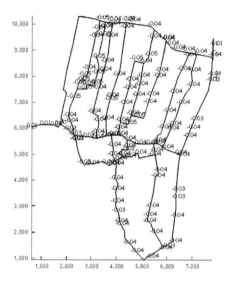

Fig. 6 Water levels increment of key points

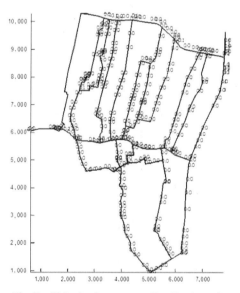

Fig. 7 Velocity increment distribution of *x*
direction (east and west)

5.1.3 Water levels increment

Water levels increment is the key indicator of investgating the effective and the effect of setting up sluice gate, the results of key points of water levels increment are show in Fig. 6.

The negative value means the fall of the water level, the positive value means the rise of the water level. The graph Fig. 6 shows that in the upper reach of the gate the maximum dammed water rising height is 0.014 m, the rest is that of 0.005 ~ 0.014 m, in the lower reach of the gate the maximum water fall height is 0.05 m. the rest is that of 0.01 ~ 0.05 m.

5.2 River network velocity increment

5.2.1 Velocity increment distribution in the x direction (east and west)

In the condition of setting up sluice gate, the velocity increment in the X direction is shown in Fig. 7.

the numerical of key points shows that the main increment ocuurs in Datangjiang River and Chaotangjiang River, the maximum velocity increment is respectively 0.26 m/s and 0.15 m/s.

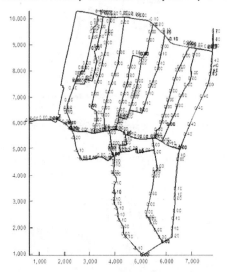

Fig. 8　Velocity increment distribution of y direction (south and north)

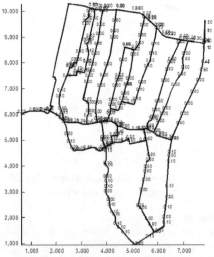

Fig. 9　Overall flow velocity increment distribution

5.2.2 Velocity size increment distribution in the y direction (south and north)

In the condition of setting up sluice gate, the velocity increment in the y direction is shown in Fig. 8, the numerical of key points shows that the main increment ocuurs in many branches river, the maximum velocity increment is respectively from 0.01 m/s and to 0.18 m/s (except nearby the outlet).

In the condition of setting up sluice gate, there is no water flow into the Mingshanlujiang River, the Water levels calculated Results are shown in Fig. 5, the numerical is the water level elevation of key points.

5.2.3 Overall flow velocity increment distribution

In the condition of setting up sluice gate, the overall flow velocity increment is shown in Fig. 9, the numerical of key points shows that the increment ocuurs in many branches river, the maximum velocity increment is respectively from 0.01 m/s and to 0.23 m/s (except nearby the outlet).

5.3 River networks water fluidity effect research of engineering measures (setting up gates)

It is setting up sluice gate at the inlet of Mingshanlujiang River that redistributes the water quantity of entering into all branches rivers, the quantity of branch river increases. In the same condition, velocity increase correspondingly, then the water levels fall in all branches river. Only in the inlet reach Datangjiang River networks, the water level rises 0.05 ~ 0.11 m.

Overall, setting up gate at proper location makes almost all river velocity incease, and enhence water fluidity, lower the water levels. This makes for not only the city water enviroment improvement but also the city flood control. Only in the short length of up reach the gate, the water levels rise, some measures can be made to lower the local water levels, such as dredged channel or widen the river.

6 Conclusions

These above analysis remarks that the established and verified planar 2D unsteady river networks flow finite element model is effective, setting up gate at proper location makes almost all river velocity incease, and enhence water fluidity, lower the water levels. This makes for not only the city water enviroment improvement but also the city flood control. Only in the short length of up reach the gate, the water levels rise, some measures can be made to lower the local water levels, such as dredged channel or widen the river.

Acknowledgements

This research was supported by the National Natural Science Foundation of China (No. 51039003) and the 2011 Projects of the Water Resources Bureau of Zhejiang Province (RC11092011), the 2009 Zhejiang Provincial Education Bureau Key Projects (No. Z200909405), the Zhejiang Provincial Education Science Plan Office Project [(2009) 12, No. SCG220] and the "325" Talent Training Program of Water Resources Department of Zhejiang Province.

* All authors contributed equally to this work.

References

Li Dongfeng, Zhang Hongwu, Zhang Junhua. Finite Element Method of Yellow River and Sediment Movement [J]. Journal of Sediment Research, 1999 (4): 59-63.

Li Dongfeng, Zhang Hongwu, Zhong Deyu. 2 - D Mathematical Model for Flow and Sediment Transport in the Estuary of Yellow River [J]. Journal of Hydraulic Engineering, 2004 (6): 1-7.

Li Dongfeng, Zhang Hongwu, Zhong Deyu. Numerical Simulation and Analysis on Tidal Current and Sediment Silting Process in Yellow River Estuary [J]. Journal of Hydraulic Engineering, 2004(11): 74 – 80.

Zou Bing, Li Dongfeng, Zhong Deyu. Effect of Training Levees on Movement of Water Flow and Sediment in Yellow River Estuary [J]. Journal of Hydraulic Engineering, 2006, 37(7): 880 – 885.

Li Dongfeng, Zhang Hongwu, Chen Bin. Numerical Simulation Analysis of Bridge Construction Flood Flow Effect based on Two Dimensional Finite Element Model [C]// Proceedings of 2010 International Conference on Modern Hydraulic Engineering. Xi'an: London Science Publishing, 2010: 295 – 298.

Li Dongfeng, Zhang Hongwu, Chen Bin. Dammed Water Level Comparison of Widening Bridge Piers Based on 2D FEM Numerical Simulation and Empirical Formula [C]// Proceedings of 2010 International Conference on Modern Hydraulic Engineering. Xi'an: London Science Publishing, 2010: 255 – 258.

Li Dongfeng, et al. Numerical Simulation and Analysis of Hydraulic Characteristic of Cixi Rural River Networks N Technical Report [R]: Report of ZJWCHC, 2009.

Li Dongfeng, Mao Lianming, Chen Dongyun, et al. 2D Numerical Calculation Analysis on River Water Flow Control [J]. Journal of Zhejiang Water Conservancy and Hydropower College, 2011, 23(1): 9 – 12.

Variation of the Reach-scale Channel Geometry in the Braided Reach of LYR after the Operation of XLD Reservoir

Li Xiaojuan, *Xia Junqiang* and *Zhang Xiaolei*

State Key Laboratory of Water Resources and Hydropower Engineering Science,
Wuhan University, Wuhan, 430072, China

Abstract: The Lower Yellow River (LYR) has experienced continuous channel degradation since the operation of water impoundment and sediment detention of the Xiaolangdi (XLD) Reservoir, and therefore significant variation of the channel geometry has occurred in the LYR. The channel evolution of the braided reach plays a key role in the channel adjustment of the LYR. In this paper, measured profiles of 28 sedimentation sections were collected in the braided reach of the LYR over the period from 1999 to 2010, and the reach-scale characteristic parameters under the bankfull stage were estimated using a composite method which integrates the geometric mean based on the log-transformation with the weighted average based on the distance between two consecutive sections. Recent variation of the reach-scale channel geometry in the braided reach was investigated. In the BaiHeZhen-Huayuankou (BHZ-HYK) reach, the reach-scale bankfull width increased from 1,055 m in 1999 to 1,375 m in 2010, with the corresponding bankfull depth increasing from 1.45 m to 3.60 m, which led to an increase of 3,421 m^2 in the corresponding bankfull area; In the Huayuankou-Jiahetan (HYK-JHT) reach, the reach-scale bankfull width increased from 1,136 m to 1,846 m, and the bankfull depth increased by 1.17 m, which led to an increase of 3,136 m^2 in the corresponding bankfull area; In the Jiahetan-Gaocun(JHT-GC) reach, the reach-scale bankfull width increased from 611 m to 823 m, with the bankfull depth increasing from 1.97 m to 3.73 m, and the corresponding bankfull area was increased by 1,864 m^2. In the braided reach, the reach-scale bankfull area increased by about 200% as the bankfull width increased by 43% and the bankfull depth was doubled. In the braided reach, the reach-scale regime coefficient decreased by 40% from 19.9 to 11.4, and therefore, the channel adjustment in the braided reach was characterized mainly by channel incision, also accompanied by significant channel widening.

Key words: channel geometry, braided reach, reach-scale channel geometry, lower Yellow River, channel incision, channel widening

1 Introduction

The XLD Reservoir is a major project in a heavy sediment-laden river, which is located at the key point to connect the upstream and the lower river to control the water and sediment entering the LYR. It can regulate the floods in the lower reach and trap the sediment with its storage capacity for sedimentation. Since October 1999, the reservoir has been put into use for the period of water impoundment and sediment detention, continuous channel degradation has occurred in the LYR, and it is most notable that the scour amount in the braided reach (282 km) accounts for 76% of the total one in the LYR. Along with the bed incision, the channel geometry in the braided reach made various adjustments. The variation of cross-sectional profiles manifests that during the process of the main channel degradation and widening, the discharging capacity increases, and the regime coefficient decreases to different extent (Shang et al., 2008). As the bankfull width and depth increase, the corresponding bankfull area increases, and the discharging capacity of the main channel in the LYR reach increases consequently. In the HYK-GC reach, the bankfull discharge is greater than 5,300 m^3/s. With the degradation and widening of the main channel, the water level under the same flood discharge is 1.2 m lower than the average before the water-sediment regulation (Wan et al., 2010), and the water level under the discharge of 2,000 m^3/s in the HYK-JHT reach has dropped 1. 8 m. As for the longitudinal channel profile, the longitudinal slope turns to flatten, the concavity values increases slightly. As the scouring in the main channel continues, the bed material of the main channel in the LYR reach become obviously armored, the lateral channel stability increases.

A natural alluvial river causes adjustment of cross-sectional profile to response the incoming flow and sediment conditions, the hydraulic geometry relations of stable equilibrium in alluvial rivers is presented by adjusting their channel form to an equilibrium state at the bankfull stage (Liu et al. , 2012), thus the bankfull width, depth and area are used to describe the characteristics of the main channel geometry. Because of the complicated cross-sectional profiles in the LYR, the characteristics of cross-sectional geometry change significantly along the braided reach, the variation of channel geometry of a specified cross section can hardly represent the adjustment that the whole reach made, thus, it is necessary to adopt the reach-scale conception. Most of existing studies concentrate on the variation of cross-sectional geometry.

Wu et al. (2007) conducted the research into the observed bankfull discharge at HYK and other four hydrometric stations, and the bankfull discharge in response to the incoming discharge and sediment load, the accumulated effects of the incoming discharge and sediment load in consecutive years are explained by the analysis on the correlations between the bankfull discharge and moving average discharge as well as moving average incoming sediment coefficient, with a formula for calculating the bankfull discharge being proposed. Liang et al. (2005) used the observed hydrological data to analyze the correlations between the reach-scale bankfull channel geometry and the discharge in consecutive years, and pointed out that the correlations between the reach-scale channel geometry and the former 3 ~ 5 a is the best. The method that Liang et al. (2005) used to estimate the reach-scale channel geometry is the direct geometric mean with the assumption that the distance between the cross sections are even, which is not reasonable when applying to the LYR. Considering the distance between the observed cross sections in LYR is unequal, Xia et al. (2008) introduced the weighted average based on the distance between two consecutive sections method to estimate the reach-scale channel geometry characteristics and the reach-scale bankfull discharges to analyze the processes of the recent channel adjustment in the LYR.

In this paper, the reach-scale bankfull width, depth and area in the braided reach of the Lower Yellow River are estimate based on the observed hydrological data from 1999 to 2010 using the composite method which integrates the geometric mean based on the log-transformation with the weighted average based on the distance between two consecutive sections method, and by comparison with the typical cross-sectional geometry, the variation of channel geometry in the braided reach in LYR is investigated since the Xiaolangdi Reservoir's operation.

2　Methods for describing the channel geometry

There is a quantitative relationship between the cross section profiles as well as the longitudinal profile and the inflow situation (including incoming discharge and sediment concentration) when the alluvial river is in regime, i. e. the hydraulic geometry. The channel width, depth and velocity measured at different discharges at a given cross section are referred to as at-a-station hydraulic geometry; channel properties geometry describes bankfull conditions for different cross sections located in the same fluvial system, which is referred to as downstream hydraulic geometry (Julien, 2002). The at-a-station hydraulic geometry describes channel properties at the given cross section while the downstream hydraulic geometry reflects the channel geometry along the river. Analysis of variation of the channel geometry in the reach will obtain the downstream hydraulic geometry.

2. 1　Analysis of cross-sectional geometry

The cross section is quite wide and shallow in the braided reach of the LYR, with the distance between the levees of more than 20 km, and the area of the floodplain makes up more than 80% of the total channel area (Xia et al. , 2010). Fig. 1 shows the cross-sectional profile at Babao after the flood season of 2010, as a typical cross-sectional profile in the braided reach. It can be seen from Fig. 1 that the total width of the section is 11 km with the main channel of about 1 km. However, the main channel is still the main course for discharging the floods. Thus, the channel geometry of a cross section in the LYR can be described by its main-channel geometry, which can reflect the trend of the channel adjustment. When the water stage reaches the floodplain level, the hydraulic geometry parameters such as the bankfull width, depth and area, et al. are referred to as bankfull hydraulic geometry. Due to the high velocity and active sediment transport capacity, the bed-forming action is strong at the bankfull discharge, and the bankfull hydraulic geometry is thus used widely as important indicators of main channel geometry and conveyance capacity in the researches of fluvial processes.

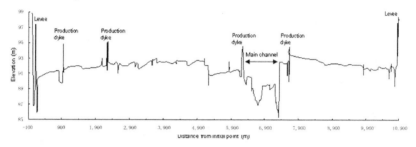

Fig. 1   Typical cross-sectional profile in the braided reach of the LYR

2. 2   Analysis of reach-scale channel geometry

The reach-scale channel geometry can be calculated by the arithmetic mean or the geometric mean. If the arithmetic mean is used, the product of the estimated reach-scale bankfull width, depth and velocity does not always equals the corresponding bankfull discharge. Harman et al. (2008) proposed the geometric mean based on the log-transformation method to avoid the complication. The continuity condition can be guaranteed when Harman's method is used to calculate the reach-averaged characteristic variables in a certain reach at the bankfull discharge.

The distance between the observed cross sections in LYR is often unequal, and the distance range in the braided reach is 5. 43 ~ 21. 34 km. With the effect of the uneven distance between sections considered, Xia et al. (2010) suggested the composite method which integrates the geometric mean based on the log-transformation with the weighted average based on the distance between two consecutive sections to calculate the reach-averaged characteristic variables. In the reach with a length of L, the number of the observed cross-sections of N, the distance between two consecutive sections $(i, i+1)$ is $(x_{i+1} - x_i)$, and the reach-scale variables can be expressed as follows:

$$\overline{B}_{bf} = \exp\left[\frac{1}{2L} \sum_{i=1}^{N-1} (\ln B_{bf}^{i+1} + \ln B_{bf}^{i})(x_{i+1} - x_i) \right] \tag{1}$$

$$\overline{H}_{bf} = \exp\left[\frac{1}{2L} \sum_{i=1}^{N-1} (\ln H_{bf}^{i+1} + \ln H_{bf}^{i})(x_{i+1} - x_i) \right] \tag{2}$$

$$\overline{U}_{bf} = \exp\left[\frac{1}{2L} \sum_{i=1}^{N-1} (\ln U_{bf}^{i+1} + \ln U_{bf}^{i})(x_{i+1} - x_i) \right] \tag{3}$$

$$\overline{Q}_{bf} = \exp\left[\frac{1}{2L} \sum_{i=1}^{N-1} (x_{i+1} - x_i)(\ln Q_{bf}^{i+1} + \ln B_{bf}^{i}) \right]$$
$$= \overline{B}_{bf} \cdot \overline{H}_{bf} \cdot \overline{U}_{bf} \tag{4}$$

where, $\overline{B}_{bf}$, $\overline{H}_{bf}$, are the reach-averaged bankfull water width and depth, respectively; $\overline{U}_{bf}$, $\overline{Q}_{bf}$ are the reach-scale velocity and discharge, respectively; and B_{bf}^{i} denotes the water surface width at the bankfull stage for the i th section.

This method can guarantee the flow continuity and reflect the effect of the uneven distance between the observed cross sections. The calculation accuracy of this method is closely related to the number of the observed cross-sections in the research reach, and thus this method can be used in practice in the LYR.

3 Variation of typical cross-sectional geometry

The main channel geometry is closely related to the flow rate, sediment concentration, flood duration, process of the discharge from upstream reservoir and the channel boundary conditions. The incoming water and sediment volumes in the LYR have been changed since the operation of the XLD Reservoir (Chen et al., 2008). The total amount of incoming water and sediment volumes are smaller, as compared with the values in the 1990s; The frequency of flood occurrence and the magnitude of peak flow decreased, and the incoming sediment volume also decreased significantly; The incoming water volume in the non-flood season accounted for greater than 50% of the annual water volume in the LYR, while the incoming sediment volume in the flood season occupied the majority of the annual sediment volume, with finer sediment particles released because the coarse and medium sediment particles were stored in the XLD Reservoir.

The variations of the cross-sectional bankfull channel geometry is shown in the Fig. 2 ~ Fig. 5, the adjustment trend of the cross-sectional bankfull area, width, depth, and regime coefficient can be observed. The HYK section is located 131.9 km downstream of the XLD Reservoir, the water width had been widened from 1,480 m to 2,382 m, during the period from 1999 to 2002 (Fig. 3), with an increase of 61%, and with the mean increase value of 300 m/a. The rapid widening rate of the main channel was due to significant bank erosion. After 2002, the main channel width increased relatively slowly, varying at the magnitude of about 2,500 m, and reached 2,561 m in 2010. The bankfull depth was doubled over the first 3 a, increasing from 0.83 m to 1.68 m, and then increased gradually with a slight decrease in 2008. Correspondingly, the bankfull area increased rapidly over the early five years and then mildly during the following years.

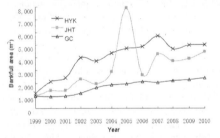

Fig. 2 Variations in bankfull area at typical sections

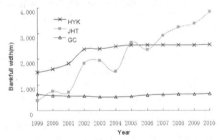

Fig. 3 Variations in bankfull width at typical sections

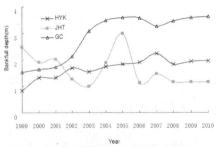

Fig. 4 Variations in bankfull depth at typical sections

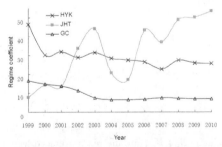

Fig. 5 Variations in regime coefficient at typical sections

A continuous increase of cross-sectional bankfull width has occurred at the section of JHT. The bankfull width in 2010 was 3,860 m, as greater as 8 times the value in 1999 (Fig. 3). At JHT, the bankfull depth decreased by 53%, while the bankfull width increased greatly. Thus, the cross-sectional bankfull area still increased over the 11 a, which led to a larger flood discharging capacity. Unlike the HYK and GC section, the regime coefficients of JHT increased with a fluctuation, which indicated that the section had a trend of becoming more wide and shallow. The variation of the cross-sectional bankfull geometry at JHT shows that the degradation and aggradation process at a special section in the LYR was rather remarkable, and however, the variation of a characteristic parameter at a section could not represent the change of the whole reach. Thus, the variation of the bankfull cross-sectional geometry is not enough to represent the adjustment of the braided reach.

The ratio of width to depth is generally used to characterize the channel geometry. When a cross section is widening and deepening at the same time, the regime coefficient as the specific ratio is used to justify the variation trend. GC section has changed mildly in the bankfull geometry, and the bankfull width changed little, and the bankfull area became larger with the increase of bankfull depth. The bankfull width was kept unchanged and the depth increased substantially, which resulted in a decrease of the regime coefficient from 16.6 to 7.1 (Fig. 5).

4 Variation of reach-scale channel geometry

In this study, the braided reach between BHZ and GC sections was divided further into three subreaches, the BHZ-HYK reach (consisting of 11 sections), the HYK-JHT reach (consisting of 10 sections) and the JHT-GC reach (consisting of 7 sections) The reach-scale bankfull characteristic parameters for each subreach and the whole braided reach were calculated, respectively. The variation of the reach-scale bankfull channel geometry can reflect the adjustment trend of the whole cross-sectional geometry perfectly. The variation of the reach-scale channel bankfull geometry in each subreach is closely related to the channel characteristics and locations. The variation process of the channel geometry in the braided reach in clear water scouring period can be obtained through the calculation of the cross-sectional characteristic parameters of the 28 sections from BHZ to GC, which can reflects the feature of each subreach synthetically.

The reach-scale bankfull width increased from 1,055 m to 1,375 m in the BHZ-HYK reach (Fig. 6), with an increase of less 30%, while it increased by 43% in average in the braided reach. The low increase of the reach-scale bankfull width in the BHZ-HYK reach was affected by the mountainside of Mount Mang, which restricted the bank erosion in this reach. Meanwhile, the river bed has been deepened from 1.45 m to 3.60 m, with an increase rate of 149% more than the average value of the braided reach. Because of continuous scouring and deposition in response to the incoming flow and sediment conditions, the bankfull depth increased much more as the bankfull width was constrained over the period of clear water scouring. As the bankfull width and depth both increased, the flood discharging capacity increased greatly, due to the increase of the bankfull area from 1,528 m^2 to 4,948 m^2.

The bankfull width in HYK-JHT reach increased markedly from 1,136 m to 1,846 m, with an increase of 62.5%, which was much higher than the average value of 43% in the braided reach. Especially in the years of 2002 and 2003, the main-channel widened by 133 m and 205 m, respectively. The hydrological data in flood season shows that in these two years the average discharge in the flood season was twice the one in 2001. This means the main channel was heavily scoured when the incoming discharge increased, and much of the scouring was due to the widening of the main channel by bank failure. The bankfull depth increased from 1.38 m to 2.55 m (Fig. 7), with an increase of 84%, which was less than the average value of the braided reach. With the widening and deepening, the reach-scale bankfull area increased from 1,570 m^2 to 4,706 m^2, tripling over the past 11 hydrological years.

The reach-scale bankfull width and depth in the JHT-GC reach increased slightly among the three subreaches, with an increase rate less than the average of the braided reach. The farer away the study reach is from the dam site, the less the bankfull area increases. The bankfull area in the

BHZ-HYK reach increased by 224% while the value in the JHT-GC reach increased by 155%.

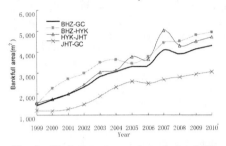

Fig. 6 Variations in reach-scale bankfull area of different reaches

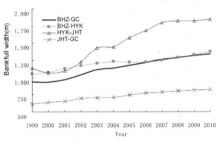

Fig. 7 Variations in reach-scale bankfull width of different reaches

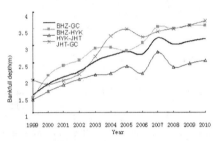

Fig. 8 Variations in reach-scale bankfull depth of different reaches

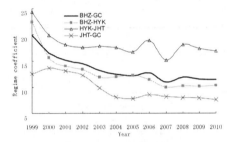

Fig. 9 Variations in reach-scale regime coefficients of different reaches

The data listed in Tab. 1 shows the change of reach-scale characteristic parameters in different subreaches. Obviously, the reach-scale bankfull depth increased more than the bankfull width, and the regime coefficient of each subreach reduced, especially in the BHZ-HYK reach. It can be concluded that the channel adjustment in the braided reach was characterized mainly by channel incision, also by channel widening. Tab. 1 also shows the reach-scale characteristic parameters changed rapidly in the early 5 ~ 6 a of the operation of the XLD Reservoir; after the year of 2005, the variation rate became mild.

Tab. 1 Change of reach-scale characteristic parameters in different reaches

	BHZ – HYK	HYK – JHT	JHT – GC	BHZ – GC
Bankfull width B(m)	1,055→1,375 (30% ↑)	1,136→1,846 (63% ↑)	611→823 (35% ↑)	941→1,341 (43% ↑)
Bankfull depth H(m)	1.45→3.60 (149% ↑)	1.38→2.55 (84% ↑)	1.97→3.73 (89% ↑)	1.54→3.2 (108% ↑)
Bankfull area A(m^2)	1,528→4,948 (224% ↑)	1,570→4,706 (200% ↑)	1,206→3,070 (155% ↑)	1,451→4,293 (196% ↑)
Regime coefficient ζ(m$^{-1/2}$)	22.4→10.3 (54% ↓)	24.4→16.9 (31% ↓)	12.5→7.7 (39% ↓)	19.9→11.4 (42% ↓)

5 Conclusions

It has been more than 10 years since the operation of the XLD Reservoir, due to the operation of water impoundment and sediment detention, the channel geometry downstream of the dam was affected significantly. In this paper, both the cross-sectional and reach-scale bankfull geometries are analyzed through the calculation of the observed data.

(1) The reach-scale bankfull geometry is more representative to describe the variation of the reach channel as compared with the cross-sectional bankfull channel geometry. The composite method which integrates the geometric mean based on the log-transformation with the weighted average based on the distance between two consecutive sections is rather practical in estimating reach- averaged bankfull characteristic parameters in the LYR.

(2) The variation of individual cross-sectional channel geometry may either not agree with the one of the study reach. It can still be used to indicate the adjustment tendency of the braided reach, the bankfull width, depth and area of the HYK, JHT and GC sections all increased, characterized by channel narrowing with a decreasing regime coefficient.

(3) The variation tendency and magnitude of the reach-scale channel bankfull geometry in the three subreaches are closely related to the channel characteristics and locations. The reach-averaged bankfull width increased by 43% and the bankfull depth was doubled; the bankfull area was nearly tripled with the discharging capacity being increased notably. The regime coefficient dropped by 40% , thus, the channel incision is more significant than channel widening.

References

Shang H X, Shen G Q, Li G X. Analysis on Scour and Fill Effects of the Lower Yellow River in Initial Debris Retaining of the Xiaolangdi Reservoir[J]. Yellow River, 2008, 11(11): 24 – 26.

Wan Q, Jiang EH, Zhang L Z. Characteristics of Flood Routing of the Lower Yellow River since the operation[J]. Yellow River, 2010, 32(7), 23 – 24.

Liu X F, Huang H Q, Deng C Y. Mathematical – physical Analysis of Stable Equilibrium Condition and Channel Form in Alluvial Rivers[J]. Journal of Sediment Research ,2012, (01):14 – 22.

Wu B S, Xia J Q, Zhang Y F. Response of Bankfull Discharge to Variation of Flow Discharge and Sedimentload in Lower Reaches of Yellow River[J]. Journal of Hydraulic Engineering, 38 (7): 886 – 892.

Liang Z Y, Yang L F, Feng P L. Relation of Channel Geometry to the Water and Sediment Rate for the Lower Yellow River[J]. Journal of hydroelectric Engineering, 06(06): 68 – 71.

Xia J Q, Wu B S, Wang Y P. Processes and Characteristics of Recent Channel Adjustment in the lower Yellow River[J]. Advances in Water Science 2008, 19(3):301 – 308.

Julien P Y. River Mechanics[M].Cambridge: Cambridge University Press, 2002.

Xia J Q, et al. Estimating the Bankfull Discharge in the Lower Yellow River and Analysis of its variation Processes[J]. Journal of Sediment Research, 2010, (2): 6 – 14.

De Rose, Ronald C, Michael J, et al. Downstream Hydraulic Geometry of Rivers in Victoria, Australia[J]. Geomorphology, 2008, 99(1 – 4):302 – 316.

Chen J G, et al. Initial Operation of Xiaolangdi Reservoir and Responses of Sedimentation in the Lower Yellow River[J] Journal of Sediment Research, 2008 (5): 1 – 8.

The Regional Differentiation of the CO_2 Emission Structure in Region along the Lower Yellow River[①]

Zhang Jinping[1,2] , *Qin Yaochen*[1] and *Wang Lujuan*[3]

1. Center for Yellow River Civilization and Sustainable Development / College of Environment and Planning, Henan University, Kaifeng, 475001, China
2. College of Environment and Planning, Liaocheng University, Liaocheng, 252059, China
3. College of Chemistry and Chemical Engineering, Henan University, Kaifeng, 4750004, China

Abstract: In China, it is important for the accounting, structural evolution and regional differences analyzing of the city-level carbon emissions for rational utilization of geographical potential in regions to inhibit the rapid growth of carbon emissions in the short term by technological learning and knowledge flows. In this study, CO_2 emissions from fossil fuel and cement industrial processes have been accounted in the region along the Lower Yellow River based on the municipal administrative units from 2000 to 2009, along with the accounting in the two sub-zones and 21 cities. Study shows that, the total CO_2 emissions at different spatial scales shows a typical 'S-type' growth characteristic, but there are significant spatial and temporal differences in the emission structures. Generally, emissions from the cement industrial processes have a rapid growth for the rapid industrialization in some cities. Optimization of industrial structure in a few cities has promoted the proportion of emissions from the cement industrial production to decline. Different evolutions in emission structure at different scales reflect features of the CO_2 emission structure under the pushing of rapid industrialization and urbanization.

Key words: CO_2 emissions at the municipal-scale, CO_2 emission structure, regional differences, region along the Lower Yellow River

1 Introduction

The large imbalance of China's regional economic development leads to a significant regional differences of carbon emissions . However, the domestic researches of carbon emissions are most of accounting and analysis for carbon emissions from energy at the national and provincial level with municipal-scale carbon emissions analysis having not got the attention it deserve. From the perspective of decision makers, carbon reduction of the municipal area is essential for China to achieve the carbon emission targets by 2020. The national and provincial level researches, however, may cover up the spatial and temporal variation characteristics of carbon emissions at the municipal and smaller scale, therefore, are not conducive to the final decomposition of China's carbon emission reduction targets and developing targeted emission reduction policies. Region along the lower Yellow River is one of the major grain producing areas in China, and is the binding region of "The Yellow River Delta High-efficiency Economic Zone" and "Zhongyuan Economic Zone". In this region, economic development differences of 21 cities and the contradiction between the rapid industrialization and environmental protection are very significant. We will account CO_2 emissions of cities along the lower Yellow River from 2000 to 2009, and analyze the spatial and temporal variations of CO_2 emissions and the structure of different levels of the whole region, the two sub-zones and cities in our study. It is important not only for the transformation of regional economic growth and low-carbon development, but also for the making of regionally balanced carbon reduction

① Foundation: Under the auspices of National Natural Science Foundation of China (NO. 41171438), Major Program of National Natural Science Foundation of China(NO. 2012CB955800), Major Program of Humanities and Social Sciences Key Research Base of Ministry of Education of China (NO. 10JJDZONGHE 015) and National Natural Science Foundation of China (NO. 41001359) policies at the larger-scales.

2 Study area

Taking "Old Mengjin" as the dividing point of the middle and lower reaches of the Yellow River, "region along the lower Yellow River" is defined by the scope of municipal area covered by the lower Yellow River irrigation area. The region's total land area is 1.639×10^5 km^2, involving a total of 21 prefecture-level cities in three provinces of Henan, Shandong and Hebei, and its agricultural production condition is the best in the Huang-Huai-Hai Plain. We define the intersection of region along the lower Yellow River and Zhongyuan Economic Zone as Zhongyuan zone, and the rest of the region as Delta zone. The industrialization processes of the two zones are rapidly advancing, and will become a powerful engine driving economic growth of region of the middle and lower Yellow River. However, there are large regional differences about the key economic development indicators of cities along the lower Yellow River and the sub-zones, such as GDP per capita, the level of urbanization and the proportion of industrial added value, which make regional carbon emissions growth and spatial variation based on the basic spatial unit of cities present different characteristics from large-scale study.

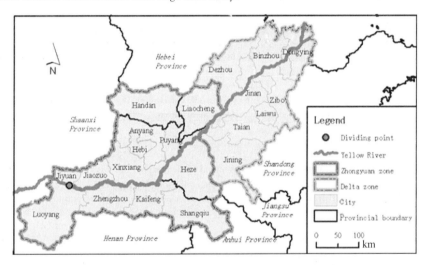

Fig. 1   Definition of the region along the Lower Yellow River

3 Data and methods

In this paper, we calculate CO_2 emissions from fossil fuels and cement industrial process for 21 cities along the lower Yellow River from 2000 to 2009. The data comes mainly from Statistical Yearbook of the three provinces of Shandong, Henan and Hebei and that of 21 cities from 2001 to 2010, and national economic and social development statistical bulletin of the corresponding year. GDP and industrial added value were translated into comparable prices of 2005.

3.1 CO_2 emissions accounting

3.1.1 CO_2 emissions from fossil fuels

CO_2 emissions of 20 cities affiliated with Shandong and Henan province from 2005 to 2009 and that of Handan city from 2007 to 2009 are calculated by Eq. (1) according to city-level energy consumption per unit of GDP (Tons of standard coal per ten thousand Yuan) in the Statistical

Yearbooks. CO_2 emissions in other years are estimated by the provincial energy consumption data. In the estimation method, Dhakal estimated the city's energy consumption in accordance with China's urban population proportion in their respective provinces. CO_2 emissions from the urban industrial energy consumption, however, accounted for at least 70 per cent of that from the total energy consumption. Therefore estimation based solely on the population proportion is unscientific. We decompose the total energy consumption as four parts of energy consumptions of livelihood, primary industry, secondary industry and tertiary industry according to the energy balance sheets of the three provinces, and estimate each part of the energy consumption by the city's total population and regional Gross Domestic Product of the three industries. Then the city-level energy consumption can be drawn by adding up the four parts' energy consumptions (ten thousand tons of standard coal).

$$C_E = E_Q \cdot G_k \cdot K \tag{1}$$

where, C_E is CO_2 emissions from fossil energy, E_q is energy consumption per unit of GDP, G_k is regional Gross Domestic Product, and K is emission factor of standard coal converted into CO_2 emissions.

3.1.2 CO_2 emissions from cement production process

CO_2 emissions from cement production process come mainly from the production of the raw materials of cement, i.e. cement clinker. According to the conclusions of the IPCC (2006), the emission factor for cement clinker was 0.52. In the case of interregional transporting data of cement clinker unable to get, we estimate CO_2 emissions from cement production (C_C) in accordance with the following formula.

$$C_C = q \cdot c \cdot e_f \tag{2}$$

where, q is the total output of cement, c is cement clinker content regardless of its type and is set as 75 percent, and e_f is CO_2 emission factor for cement clinker.

3.2 Uncertainty analysis

The uncertainty of calculation results is mainly affected by the fossil fuels data and CO_2 emission factors. Fossil energy consumption data of some years are converted by energy consumptions of the four provincial industries, population and three industry outputs of 21 cities. Therefore, the calculated total energy consumption is influenced by comprehensive factors of population, scale and structure of regional GDP. The impact of energy structure on CO_2 emissions from fossil fuels is failed to consider due to the current data limitations. CO_2 emission factors are set as constant during the study period, and can not consider their changes due to technical advances. Similarly, CO_2 emissions calculation of cement can not reflect the differences in production technology in various cities. The quality of statistical data and statistical standards of different years are also the uncertainty factors.

4 Results and analysis

4.1 Changes in CO_2 emissions

CO_2 emissions of the region along the Lower Yellow River increased by 3.04 times from 2000 to 2009 (Fig. 2). In 2006, the change rate of industrial output of Zhongyuan zone has caught up with and surpassed that of Delta zone and led to a bigger annual growth rate of CO_2 emissions of Zhongyuan zone, which became the main driving force of CO_2 emissions in the whole region during the "Eleventh Five-Year" in China. At the city-level, CO_2 emissions in 21 cities increased year by year (Fig. 3) and they were 3.64×10^6 t to 44.27×10^6 in 2000, while in 2009 they were 12.38×10^6 to 104.12×10^6. The growth of CO_2 emissions in the whole has accelerated since 2002, but slowed down after 2006. CO_2 emissions shows the "S-type" growth characteristics with time, but the regional differences are significant.

Handan is a heavy industrial city and its CO_2 emissions cardinality is very big, where CO_2 emissions have had a substantial rise because high energy-consuming industries, such as iron and steel and building materials developed rapidly in recent years. CO_2 emissions in cities of Jinan, Zibo, Zhengzhou and Jining, are mainly driven by industry. Although the cardinality was slightly lower, the growth rates accelerated from 2002 to 2006, and then were effectively controlled. The five cities are located in "the first camp". The cardinalities of CO_2 emissions in ten cities including Luoyang and Anyang were even lower, but maintained a rapid growth since 2002. They are located in "the second camp". Six cities including Shangqiu and Puyang belong to "the third camp" of CO_2 emissions for their features of low emissions and slow growth rate.

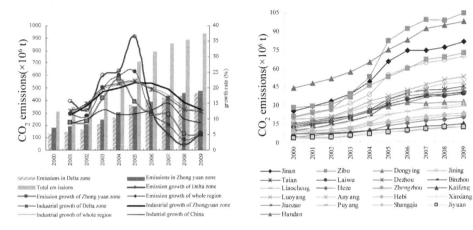

Fig. 2 Changes of CO_2 emissions and the growth rates in the whole region and two subzones

Fig. 3 Changes of CO_2 emissions in 21 cities

21 city's infrastructure construction and investment in fixed assets have sped up noticeably from 2002 to 2005 driven by the national investment-oriented economic growth, which resulted in the accelerated emission growth. Changes of CO_2 emissions in general are consistent with conclusions of studies at larger scales, however, effect of rapid development of secondary industry and investment increase are more prominent in cities with better industrial base in the first and second camp. The unbalanced nature demonstrated by the overheating of macro-investment in 21 cities leads to significantly increased inter-city differences in total CO_2 emissions after 2002.

4.2 Changes of CO_2 emission structure

Select five time points of 2000, 2002, 2004, 2006 and 2009 to analyze the evolution of CO_2 emission structure of the region along the Yellow River (Fig. 4). At the city-level, Fossil fuels are the main sources of CO_2 emissions and account for at least 79.6%. However, there are large regional differences in the proportion of CO_2 emissions from cement production process. Cities in Delta zone experienced a "decrease - increase - decrease" or "increase - decrease" process, while most cities in Zhongyuan zone experienced an overall "decrease - increase" process. In cities with strong cement industry base, such as Zhengzhou, Hebi, Xinxiang, Zibo and Jinan, the proportion of emissions from cement production is close to or more than 10% in most years. Nonetheless, emissions from cement production in Jinan and Zibo dropped significantly from 2007 to 2009 by eliminating backward production capacity (Jinan dropped to less than half of 2006, whereas the first place of Zibo was replaced by Zhengzhou). Urban and rural construction scale in Heze increased year by year, which drove the rapid development of cement industry. The proportion of emissions from cement production in Heze was overall rising. The emissions amounted to 2.784×10^6 t and accounted for 13.5% of the total emissions from cement.

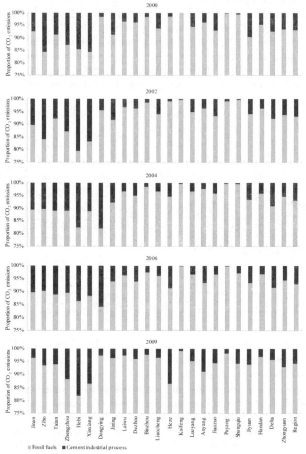

Fig. 4 CO_2 emission structure in the region along the Lower Yellow River

Strategies like internal structural optimization of the secondary industry and development of the tertiary industry in cities of Delta zone are the direct causes of the obvious change in CO_2 emission structure after 2005. In the typical city Jinan, proportion of the secondary industry increased from 42% in 2000 to 46% in 2006 and decreased to 43 percent in 2009. At the same time, proportion of added value of non-metallic mineral products industry (the cement industry is included) in the industrial added value increased from 6% to 8% and decreased to 5%, which partly shows that the upgrading of industrial structure has a more significant impact on emission structural changes. The proportions of heavy industry in GDP in most cities of Zhongyuan zone increased more rapidly compared with cities in Delta zone. The rapidly advancing of urban and rural construction also significantly improved the proportion of emissions from cement production. Thus, at a zone scale, the evolution of CO_2 emission structure of the two zones is opposite. The proportion of emissions from cement industry in Delta zone firstly increased to the maximum of 8.9% in 2004 and then decreased to the minimum of 4.3% in 2009, while that in Zhongyuan zone firstly decreased to the minimum of 5.3% in 2004 and then increased to the maximum of 7.0% in 2009. At the regional scale, the proportion of emissions from cement industry was almost unchanged before 2006, but there was a decline after 2006. Analysis on city-level CO_2 emission structure reveals important temporal and spatial variations concealed by large-scale study, i. e. the inter-city differences of the proportion of emissions from cement industry are very large (changes from 0.7% of Kaifeng to 18. 2% of Hebi), and the emission structure has a clear scale effect for the city, zone and region level.

5 Conclusions and discussions

From 2000 to 2009, CO_2 emissions in region along the Lower Yellow River, Delta zone, Zhongyuan zone and 21 cities shows typical "S-type" growth characteristics, but there are significant differences in evolution of CO_2 emission structure at different scales. The inter-city differences of CO_2 emissions are obviously enlarged due to China's investment-oriented economic growth mode from 2002 to 2005, so that the 21 cities have been separated into the three "camps" by the total emissions and their growth rates. Fossil fuels are the main source of CO_2 emissions, but there are large regional differences in the proportion of CO_2 emissions from cement industrial process.

The proportion of emissions from cement production overall increased in cities of Delta zone before 2005, and then gradually decreased; however, it significantly increased in most cities of Zhongyuan zone for the expanding urban and rural construction scale after 2005. During the "Eleventh Five-Year" period, the proportion of emissions from cement industrial production decreased rapidly in cities such as Jinan and Zibo, where carbon-lowering effect of the optimization of industrial structure and the technology upgrades within the secondary industry is very significant. The emission structure has a clear scale effect for the city, zone and regional level. Different evolutions in emission structure at different scales reflect features of the CO_2 emission structure under the pushing of rapid industrialization and urbanization. Zhongyuan zone is especially typical for its accelerated industrialization process, in which the ratio of emissions from cement industry gradually increased, what's more, the growth rate of CO_2 emissions has surpassed Delta zone since 2006 to become the leading force of CO_2 emission changes in the whole region.

References

Fleisher B, Li H Z, Zhao M Q. Human Capital, Economic Growth, and Regional Inequality in China[J]. Journal of Development Economics, 2010, 92 (2): 215 – 231.

Yi W J, Zou L L, Guo J, et al. How Can China Reach its CO_2 Intensity Reduction Targets by 2020? A Regional Allocation Based on Equity and Development [J]. Energy Policy, 2011, 39 (5): 2407 – 2415.

Chen W Y, Wu Z X, He J K, et al. Carbon Emission Control Strategies for China: A Comparative Study with Partial and General Equilibrium Versions of the China MARKAL Model [J]. Energy, 2007, 32(1): 59 – 72.

Auffhammer M, Carson R T. Forecasting the Path of China's CO_2 Emissions using Province – level Information [J]. Journal of Environmental Economics and Management, 2008, 55 (3): 229 – 247.

Zhao R Q, Huang X J, Zhong T Y. Research on Carbon Emission Intensity and Carbon Footprint of Different Industrial Spaces in China [J]. Acta Geographica Sinica, 2010, 65 (9): 1048 – 1057.

Guo Y Q, Zheng J Y, Ge Q s. study on the Primary Energy – related Carbon Dioxide Emissions in China[J]. Geographical Research, 2010, 29(6): 1027 – 1036.

Dhakal S. Urban Energy Use and Carbon Emissions from Cities in China and Policy Implications [J]. Energy Policy, 2009, 37 (10): 4208 – 4219.

Zhang J P, Qin Y C, Zhang Y, et al. Measurement of Urban CO_2 Emission Structure and Low – carbon Standard – A Case Study for Beijing, Tianjin, Shanghai and Chongqing city[J]. Scientia Geographica Sinica, 2010, 30(6): 874 – 879.

Research on Index Frame of Drought Risk in Typical Arid Area of Northwest Liaoning Province

Sun Tao, *Fu Jun' e* and *Huang Shifeng*

China Institute of Water Resources and Hydropower Research, Beijing, 100048, China

Abstract: Drought occurs frequently in recent years with a lot of economic and social losses. The build of drought risk index frame, which is one of the effective ways assisting drought risk management of agriculture, helps to quantify and standardize drought risk evaluations. Based on a series of data related with drought on northwest Liaoning province, drought characteristic is analyzed. Depending on the form mechanism of drought risk, hazard, exposure, vulnerability, ability of preventing and decreasing drought disaster are considered to select typical indicators which describe the influences to drought well. Index frame of agricultural drought risk is then set up to supply scientific solutions for decreasing, rescuing and preventing drought disasters.

Key words: Northwest Liaoning province, drought, risk, index system

1 Introduction

Agricultural droughts occur frequently and cause severe losses in Liaoning province. Northwestern Liaoning areas (NWLN) suffered drought disasters for 8 times at different degree from 1999 ~ 2006, particularly in 2009 where encountered catastrophic summer drought never seen for 50 years and brought significant agricultural losses. NWLN includes four main cities in western Liaoning province, Tieling city, Kangping and Faku county of Shenyang. "Drought happens nine times within ten years" is usually used to describe the drought situation of NWLN. Actually it is uncommon that there is no drought disaster now days. Statistics shows that there had occurred consecutive drought of two years for 14 times, three years for 4 times. The one from 1999 to 2001 caused severe cutoff of rivers and dry up of reservoirs. 2007 and 2009 saw the catastrophic drought disasters in history with huge amount of agricultural and economic losses.

There are a variety of reasons that cause agricultural drought in NWLN. High temperature, less precipitation and intensive evaporation are direct reasons for drought situation. Low soil moisture, little runoff of rivers, limited water storage in reservoirs (though large capacity) and ever-increasing water usage are all factors led to high frequency and easiness of drought in this region. At the same time, limited land resources, simplex plant structure and the problem between water supply and water demand will also increase the probability of drought disasters .

2 Analysis of indicators that affect agricultural drought

Risk analysis of drought disaster is assistant measure for decision support of agricultural drought relief. The build of risk analysis system aims to quantify and regulate the work of agricultural drought risk evaluation.

There are a lot of research achievements both home and abroad. Scientists from different intellectual discipline may have different opinion on drought. For agriculture, drought is characterized with low precipitation within certain time, space and cause reduction of crop yield. Low precipitation and degree of crop failure should be important factors of drought evaluation. Since the occurrence and development of agricultural drought have complex mechanism, a variety of natural or human factors such as hydrologic condition, meteorological condition soil, plant, regulation of cultivation, may play important role in the process of agricultural drought disasters .

2.1 Precipitation

Undoubtedly, rainfall is the most important natural factors. In areas where groundwater level is low and with limited irrigation facilities, abnormal of precipitation may cause water shortage for crops. Usually, precipitation anomaly percentage and rainless days are used as main indicators.

2.1.1 Analysis of precipitation during growing season

Precipitation (P) during June to August occupies some 60% ~ 70% of the year especially in the late 2 months, peak value occurs in July. It can be concluded that the precipitation in NWLN distributes unevenly and varied with seasons. Corn is the dominant plant in study area with few paddy rice, soybean and wheat planted in part areas. Crop often be sowed in April and harvested in September. Shortage of rainfall may affects the growth of plants and cause decrease of crop outputs or failure of yields. Considering the water demand of crops, analysis of precipitation during April to September is meaningful.

Fig. 1 shows the inter-annual variation of precipitation of the six typical stations in NWLN. Rainfall fluctuates from 100 mm to 950 mm yearly with obvious trend, though there is no evidence showing long term changes.

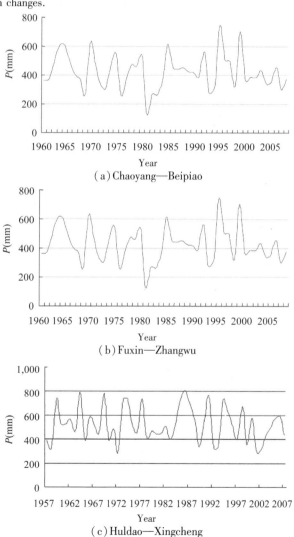

(a) Chaoyang—Beipiao

(b) Fuxin—Zhangwu

(c) Huldao—Xingcheng

Fig. 1   Inter-annual variation of P in typical stations of NWLN

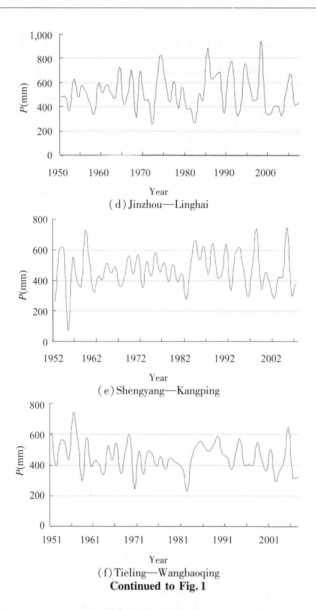

(d) Jinzhou—Linghai

(e) Shengyang—Kangping

(f) Tieling—Wangbaoqing

Continued to Fig. 1

2.1.2 Analysis of percentage of precipitation anomaly

In order to reflect the influence of rainfall, percentage of precipitation anomaly (PA) versus percentage of drought disaster suffered area (DR) is analyzed. Fig. 2 (b) shows better correlation with R^2 0.844,2 which refers to the close relation of agriculture to precipitation. Precipitation is the main source of agricultural plants. Drought suffered area depends on the quantity of rainfall in growing seasons. In addition, it seems that degree of drought affected by rainfall is deferent to different county. One thing is that the percentage of irrigated area and rainfall depended area are different. The other reason is the distribution of rainfall in the growing period, which affects the degree of drought, is very important. Shortage of rainfall in key growing period may cause more losses than it occurs at other time. Xingcheng, Huludao city suffered severe drought in 2005 with crop losses of 2.25×10^7 kg because of the rainfall in July (key time for crop) is 75 mm, 100 mm less than ever, though the annual rainfall is 572.7 mm which is larger than average level.

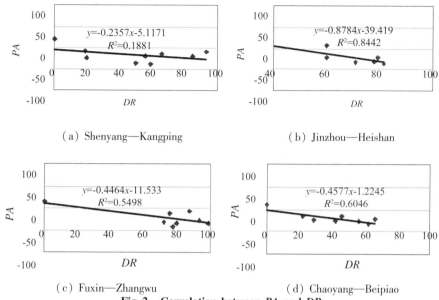

(a) Shenyang—Kangping   (b) Jinzhou—Heishan

(c) Fuxin—Zhangwu   (d) Chaoyang—Beipiao

Fig. 2   Correlation between *PA* and *DR*

2.2   Temperature

Undoubtedly, with the rise of the temperature the evaporation will be increased and the soil moisture content, which is the index of agriculture drought, will be lower. 2001 is the third drought year since it happened in 1999, causing serious disasters. Temperature in May 2001 is higher than the top value of corresponding period since 1949. 2007 saw severe summer drought (June to July) in NWLN with the temperature is 1.3 ℃ higher than average value. Autumn drought in 2009 coincides with high temperature in August, above 32 ℃ in most of areas of NWLN.

2.3   Evaporation

Evaporation data of 28 observation stations from 1934 to 2000 are analyzed. Changes of average evaporation monthly and drought index are studied to explain the effect of evaporation in NWLN.

2.3.1   Changes of average evaporation monthly

Six observation stations of evaporation are selected from 6 administrative regions to analyze monthly average evaporations. Fig. 3 shows that the evaporation varies greatly within a year and distributes unevenly. April to July evaporation occupies about 50% of the year, especially during May to June. Peak value of evaporation occurs in May. It is worth mentioning that May is the key period of seeding. However, the difference between rainfall and precipitation in May is the largest within a year. Statistics shows that the average precipitation is 40 mm, but the evaporation is 236 mm to 312 mm. Soil moisture content become worse affected by the difference value and sprouts are difficult to grow out. Drought disasters occurred frequently during the spring season in NWLN.

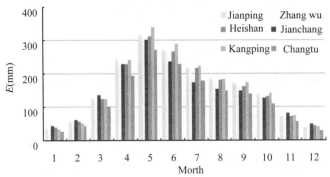

Fig. 3　Average evaporation monthly of represented counties in NWLN

2.3.2　Drought index (R)

Drought index R, defined as the ratio of evaporation and precipitation yearly, is usually used to describe the degree of drought disasters.

$$R = E_0/P \tag{1}$$

E_0 stands for capacity of yearly evaporation and P for yearly precipitation. E_0 is often substituted by water evaporation of E_{601}. R has close relation with meteorological factors. The little the R value and the more the precipitation, the more humid the climate is. On the contrary, the capacity of evaporation exceeds that of precipitation which will increase the probability of drought disasters.

R has close relation with percentage of drought occurred areas. The higher the R data, the larger the drought disaster area would be, vice versa (Fig. 4).

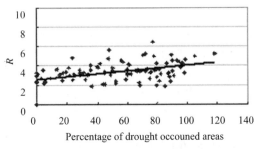

Fig. 4　Relation of drought index with percentage of drought occurred areas

2.4　Soil moisture

Soil moisture content is affected by precipitation, evaporation and temperature. It is not the direct reason of drought but it can increase the risk probability of drought disaster if the soil content is rather low before drought. Soil moisture content is below 60% which is lower than corresponding level in NWLN, June, 2007.

2.5　River runoff and storage of reservoir

River runoff and storage of reservoir are flags of drought disasters. Less precipitation and higher evaporation will cause the decrease of runoff water into river and eventually the storage of reservoir. Storage of reservoir can adjust the distribution of water resources and reduce the risk of drought disasters.

During the drought period of 2007, the precipitation and runoff of river are all decrease greatly. Rainfall in June is 52.4 mm, 42.8% less than average level, causing sharp decline of runoff in this area. Runoff value is one fourth of average value observed in Tieling hydro station, Liaohe river. Chaoyang station, Dalinghe river, saw extremely low runoff, 1/10 that of average value (Tab. 1). For the 23 large scale reservoirs in Liaoning province, water quantity stored decreased by 18% compared with the same period of 2005.

Tab. 1 Monthly runoff of Chaoyang hydro station, Dalinghe river, 2007

January	February	March	April	May	June	July	Auguest	September	October	November	December	Sum
4.59	2.15	8.82	3.79	4.22	1.14	12.04	26.76	6.36	0.81	2.15	17.76	90.6

2.6 Topography

For the same kind of crop, landform can also have different effect on the degree of drought risk. The small the gradient of landform, runoff within the soil will be slow. Runoff lost would be little and there will be more water left and be used by plant. Otherwise, large part of rainfall would be run away which will decrease the ability of plant to catch water and increase the probability of agricultural drought disaster.

2.7 Soil type

Risk probability of drought disaster varies with landform as well as soil type. Considering the main types of soil in NWLN, ability of water retention and fertilizer conservation, the sequence of drought risk from high to low is as follows: sandy soil, brown soil, cinnamon soil, meadow soil and black soil.

2.8 Groundwater

Groundwater is very important to the agriculture of NWLN. It occupies more than 50% of the total irrigation water used. In some counties, Kangping as an example, agricultural irrigation relies on groundwater completely. Groundwater level has some relations with surface water but the trend is not the same. Properties, retardation, recharge and discharge relations with surface water are different with surface water. Groundwater of different type or formation would increase the risk of drought.

2.9 Capacity building of drought relief

It can be concluded that precipitation, temperature and evaporation are main meteorological factors that affect the drought of agriculture. Irrigation level, production level and management level, in the type of project measures and non-project measures, reflect the ability of anti-drought and drought relief. Building and prefect of water conservancy projects such as water storage, water pumping, water diversion and water transfer projects can enhance the ability of regional drought control directly. Non-project measures include the draft of drought fighting plans, the improvement of funds for drought, pattern change of cultivation, adjustment of agricultural structure, the improvement of agricultural foundation and the improvement of peasant's skills.

3 Index frame of agricultural drought risk

The risk of natural disaster is decided by hazard of disaster causing factor (H), exposure (E) and vulnerability (V) of disaster affected body, capacity of drought relief (C). H represents the variation of disaster. The larger the value of H and the higher the frequency, the more serious the loss would be. E includes people or property, e. g. staff, livestock, house, crops etc., which may be threatened by affected factors. The more the people or property affected, the more the loss will be. *V* means the degree that disaster bearing body may be affected. *C* is the capacity that governments or local people have used to take certain measures to reduce drought risks. The stronger the capacity is, the lower the risk.

There are a variety of factors that affect agricultural drought. Precipitation, temperature and wind are meteorological conditions. Soil, landform, hydrology are underlying conditions. Crops and water conservancy projects are the third type of influence factors. Frame of agricultural drought risk is designed considering the above conditions in NWLN from the four main factors (*H*, *E*, *V* and *C*), each consists of specific sub-factors (Fig. 5).

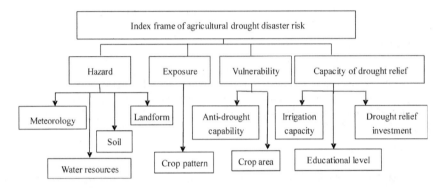

Fig. 5   Index frame of agricultural drought risk

4   Risk indicators and index system of agricultural drought disaster

Tab. 2 lists the main risk indicators of agricultural drought disaster. The index system consists of factors, indicators and sub-indicators is showing as follows.

Tab. 2   Index system of drought disaster risk evaluation

Factor	Indicator	Sub-indicator
Hazard (*H*)	Meteorology	I_{H1} Precipitation (Apr. ~ Sep.) (mm)
		I_{H2} Rainless days (Apr. ~ Sep.) (d)
		I_{H3} Evaporation (Apr. ~ Sep.) (mm)
		I_{H4} Anomaly of precipitation (%)
		I_{H5} Anomaly of temperature (%)
		I_{H6} Drought index (%)
	Water resource	I_{H7} Natural runoff of rivers (m^3)
		I_{H8} Water storage (m^3)
		I_{H9} Anomaly of groundwater depth (%)
	Soil	I_{H10} Relative moisture content (%)
		I_{H11} Soil type
	Landform	I_{H12} Landform type
		I_{H13} Gradient
Exposure (*E*)	Crop	I_{E1} Seeding area (hm^2)
	Crop	I_{E2} Ratio of drought prone area (%)
Vulnerability(*V*)	Anti-drought capability	I_{V1} Unit yield of crop (kg/ha)
		I_{V2} Ratio of water supply and demand (%)

Continued to Tab. 2

Factor	Indicator	Sub-indicator
Capacity of drought relief(C)	Irrigation	I_{C1} Ratio of irrigated land (%)
		I_{C2} Anti-drought facilities
	Fund	I_{C3} Anti-drought investment (yuan)
		I_{C4} Average per capita income (yuan)
		I_{C5} Number of skilled technician
	Education	I_{C6} Ratio of educated people (%)
	Regulations	I_{C7} Prefect degree of anti-drought plan (%)

5　Conclusions

A series of data, including meteorological, hydrological and geological data are analyzed to form the index system of drought risk evaluation in NWLN. Designed system consists of 24 sub-indicators, belongs to 4 kinds of factors respectively. It may well be used in the drought fighting campaign of NWLN to reduce the risk of drought disasters and help to decrease the loss of crops. Researched achievements would supply useful clues to indication system formation of related disasters in the nearby regions.

Acknowledgements

This work was financially supported by the Special Fund Projects (YAOJI 1256) and (YAOJI 1244) of China Institute of Water Resources and Hydropower Research. A lot of information and data are provided by the Public Welfare Project (201101061) of Ministry of Water Resources (MWR). Thanks for all the help and necessary work condition by Remote Sensing Center, MWR.

References

Fu G F, Meng Q J. Water Resources & Hydropower of Northeast China. Vol. 3 (2002), p. 25

Zong Y F, Tao L, Zhang Y L, et al. Journal of Anhui agri. Sci. Vol. 36 (2008), p. 11954

Xu X Y, Chen X J. Journal of Hohai University (Natural Sciences). Vol. 4 (2001), p. 56

Yang S Y, Hao A H, Ning D T. Journal. of Safety and Envi. Vol. 2 (2004), p. 79

Zhu H X, Li X, B. C. Li. Heilongjiang Meteorology. Vol. 2 (2009), p. 16

Studies on the Effect of Cooling Water Drainage of the Rabigh Power Plant in Saudi Arabia to the Nearby Wharf

Wu Xiufeng[1,2] and *Wu Fusheng*[1,2]

1. Key Labaoratory of Navigation Structurese, Nanjing Hydraulic Research Institute, Nanjing, 210029, China
2. State Key Labaoratory of Hydrology-Water Resources and Hydraulic Engineering, Nanjing Hydraulic Research Institute, Nanjing, 210029, China

Abstract: This paper is the hydraulic model test on energy dissipation and erosion control at the drainage oultlet of the 2 × 660 MW fuel power generating units of the Rabigh Power Plant in Saudi Arabia. The scale of the normal hydraulic model is 1 :40. The study is depend on the gravity similarity criterion. According to the test results: in the design scheme, open channel water flows into the deep sea and is fully dissipated. The flow near the outfall and the wharf is steady under different conditions. The speed of flow is lower than the start-up speed of natural sand. And the beach nearby the outfall and the wharf was not washout when cooling water was discharged from the drainage open channel to the sea. The part movable bed model was designed according to similarity theory of sand start-up speed. The beach nearby the discharging channel and the wharf has not had the washout destruction according to the scouring test. Along the beach of the drainage open channel and the wharf of the Rabigh Power Plant is affected by the waves. Therefore, the washout protection question nearby the drainage open channel and the wharf must be considered in the wave-silt model test at the same time.

Key words: Rabigh Power Plant in Saudi Arabia, the normal hydraulic model, wharf, incipient motion of sediment, move bed scouring

1　Preface

The existing Rabigh Power Plant (SEC, for short called the "existing plant") located on the Red Sea east beach of the Kingdom of Saudi Arabia, is approximately 150 km north of Jeddah city. The cooling water for the plant operation is taken from Red Sea through cooling water intake channels directly and discharged back into the sea through two outfall channels.

The Rabigh Independent Power Provider (hereinafter known as "IPP") will be installed adjacent to the existing Rabigh Power Plant to the south and will have 2(two) units of 660 MW oil-fired engines. The project site is in an open area. Its terrain slopes from east to west and the ground elevation is between about 1.2 ~ 5.0 m. The plant will cover about 96 km² with about 1.2 km from east to west and about 0.8 km from south to north. It will meet the needs of building and constructing of the plant. This project intends to use the cooling water system of modular direct supply means including seawater desulfurization system water supplying drainages.

The cooling water for the two units engines of this IPP stage will be supplied by one pump house, in which stalled 5(five) CW pumps for the condensers (4 in use and 1 reserved), 5(five) FGD pumps(4 in use and 1 reserved), 10 rotary raking machine filters and gratings respectively (10 coarser grid machines, 10 finer grid machines and 2 clean-up machines), 4 rinse pumps. The cooling water of each unit engine will be sent to the main plant A directly with a DN3800 pressure pipe, and then divided into two DN2600 steel pipes into the condenser. Through the condenser, it runs to the siphon drainage wells, water aeration tank desulfurization and eventually discharged to the Red Sea. The cooling water for 2 × 660 MW units condensers is up to 59.01 m³/s (intake water temperature 32 ℃) supplied by five pumps with a capacity of 15.47 m³/s each. Cooling water throughout condenser is about 39 ℃, 13.89 m³/s of them will be used for FGD with production water at 44 ℃. 55.13 m³/s will be used for aeration basin directly at 32 ℃. 2.64 m³/s for other usages. The total discharge into the sea is about 116.78 m³/s, and its temperature increase is 4.3 ℃.

The bottom elevation of intake window is -21.20 m, and the upper edge of intake window elevation is -15.0 m. The cooling water will discharge into the sea offshore through the outfall channel lined with precast concrete slabs on the bottom and rock revetment on the side slopes. These features of the channel lining were not reproduced in the hydraulic model as they do not impact the flow near the jetty.

1.1 Bathymetry

The Red Sea is a seawater inlet of the Indian Ocean, lying between Africa and Asia. The name of the sea may signify the seasonal blooms of the red-colored Trichodesmium erythraeum near the water's surface. The connection to the ocean is in the south through the Mandeb strait and the Gulf of Aden. In the north, connected to the Suez Canal and the Mediterranean, there is the Gulf of Aqaba, and the Gulf of Suez. The Red Sea has a surface area of about 4.5×10^5 km^2. It is about 2,100 km long and averaged 290 km wide, at its widest point, is 306 km wide. It has a maximum depth of 2,922 m in the central median trench, and an average depth of 558 m. The coast at Rabigh is made up of a narrow reef which is at about 0 m LAT for about 100 ~ 400 m from the coast, and then falls rapidly to about 80 m below LAT. About 2 km offshore of Rabigh there is a large reef (approximately 10 km long and up to 4 km wide). The channel between the offshore and the coast is up to 80 m deep at Rabigh but rapidly drops several hundred meters just to the south of the power plant. The recent survey of the bathymetry around the IPP site confirms that here the reef top is flat over a width of about 400 m.

1.2 Tide

The tidal range in most parts of Red Sea is small, and the largest will almost up to 0.5 m. At Rabigh, mean spring low water is about 0.5 m above LAT. There are seasonal mean sea level variations of up to 0.5 m (0.2 m above average in December and January, 0.3 m below in August and September), which means that at certain times of the year some reef tops may become exposed. It is also noted from the survey results that there are short-term variations in mean sea level of the order 0.1 m over a period of several days.

1.3 Currents

According to the sea currents and water temperature survey of HR Wallingford for the existing Rabigh power plant stage six thermal pollution study in 2008, the main results are: speeds are low (< 20 cm/s); Predominately the current is toward the south; current is not significantly influenced by the tide; peak currents occur usually in the early afternoon, when winds from the north-west are strongest; Currents appear to be uniform along the coast near Rabigh PP; the variation of currents at 2 m below the surface is probably associated with diurnal variations in the wind and that the tidal currents are negligible .

2 The goal and contents of the test and research

For the security of the drainage project, based on the model tests, hydraulic similarity theory and similitude criteria as well as the actual and excavated landforms, we want to make clear the drainage capacity of the port, the flow velocity distribution of the watercourse and the scouring effect near the water port. Still, we analysis the scour area under different operating conditions, and raise proper protection measurements, so as to make sure that every circulating water drainage port works and the wharf safely and economically.

The main contents of the test and research are as follow:

(1) Studying the velocity distribution near the outfall and the wharf on the condition of three different tide sea level under the designed cooling water discharge.

(2) Carry on the discharging tests of the open channel outfall energy dissipation and scouring.

（3）Studying the scouring condition along the sea bank near the drainage port and the wharf through part movable bed scouring test.

（4）According to the results of the model tests, determine reasonable type of drainage port, raise measurements to reduce scouring and assess the work.

3　Physical model design and test conditions

3.1　Model similarity scale

In order to reproduce successfully the prototype of water flow and sediment motion, the model has to satisfy the relevant simulation conditions. The flow motion is designed to the gravity similarity, and the sand motion is to the washout similarity. This model experiment lays emphases on the study of reduction water velocity nearby the project, and water flush and protection to the beach around the drainage port. It is proper to use the normal model. Synthesizing the factors of study goal, technological specification and location etc, we use the geometry ratio $\lambda_L = 40$.

3.1.1　Water flow motion similarity

（1）Geometry similarity condition

$$\lambda_L = \frac{L_p}{L_m} = 40, \qquad \lambda_H = \frac{H_p}{H_m} = 40,$$

（2）Current flow motion similarity condition

①Gravity similarity　　　　　　　$\lambda_V = \lambda_H^{1/2} = 6.32$

②Resistance similarity　　　　　　$\lambda_n = \lambda_H^{2/3}/\lambda_L^{1/2} = 1.85$

③Flow continuity similarity　　　$\lambda_{tl} = \lambda_L/\lambda_H^{1/2} = 6.32$

$$\lambda_Q = \lambda_L\lambda_H^{2/3} = 10,119.29$$

In the paper, fresh water was used instead of sea water. In this drainage port sediment physical model, the effect of buoyant plume of cooling water was not concerned.

3.1.2　Sediment kinematic similarity

There are many similitude criteria for the sediment movement. As for the characteristics of this subject, the sea bed washout similarity is the main consideration. The ratio of model sand flow velocity to the natural sand is up to the following formula: $\lambda_{V_0} = \lambda_V = \lambda_H^{1/2} = 6.32$.

Therefore, according to the geometry and other similitude criteria, the ratios of similarity are as shown in Tab. 1.

Tab. 1　General physical model similarity scales

Name	Symbol	Calculated Value	Adopted Value
Horizontal scale	λ_L	—	40
Vertical scale	λ_H	—	40
Velocity ratio	λ_V	6.32	6.32
Roughness coefficient	λ_n	1.85	1.85
Flow time scale	λ_{tl}	6.32	6.32
Flow capacity ratio	λ_Q	10,119.29	10,119.29
The model sand particle size	λ_d	—	According to starting similar conversion

3.2　Model scope

The area of the open channel outfall model of Saudi Arab Rabigh Power Plant is 246.6 m² (length × width = 18 m × 13.7 m). The drainage channel and the wharf are on the sea beach. The model imitates the beach nearby the cooling water discharging channel and the deep sea. It is about 550 m along the bank line and 400 m outside the drainage port in the deep sea area. The width of the beach including the channel and the wharf is about 300 m. The drainage open channel and the wharf and steep sea bank is shown in Fig. 1.

Fig. 1   Model of the drainage open channel, the wharf and steep sea bank

3.3   Physical model construction

The model is made according to a 1 : 2,000 topographic map supplied by the designing institute. It takes section method. The plane error is within 1 cm, and the altitude error is within 0.1 mm.

The extent of the model movable bed test is based on the fixed bed water flow test, including the greatest changes of the flow field caused by the project. It meets the demand of the sea bed terrain continuity change.

3.4   Test conditions and measurements

3.4.1   Test conditions

According to the technical requirement document of the designing institute, the overall cooling water discharge of the project in this period 2 ×660 MW is 116.78 m³/s. The characteristic water level is accord to the drawings provided by the design unit. As the tide changes small, the drainage flow distribution at low tide level is suitable by LAT tide level approximately. The three different tide sea level of the model test are as follow (MSL basic surface) : design high tide sea level : 0.44 m; design average tide sea level : 0.00 m; design low tide sea level : -0.54 m.

3.4.2   Measurements

(1) Three dimensional Doppler Acoustic Velocimeter ADV.

The ADV used in this paper is produced by Nortek AS companny. Its sampling frequency is 25 Hz, the velocity measuring range is 0.5 ~ 400 cm/s, it has higher measurement accuracy.

The measuring reliability is depended on two aspects: ① Grading analysis of water sample, the grain size is between 5 ~ 50 μm arriving at 97% , it indicates that the clay particle is basic in the water. The follow character of the particle is good. ② During the measuring time, two parameters will be shown in the interface of the software: SNR and Correlation Coefficient. The record data is true and effective when SNR number is greater than 15 and the Correlation Coefficient is at 70% ~ 100% .

(2) The discharge control.

The sharp rectangle weir is applied to control the water discharge in the test. The discharge is calculated by the T. Rehbock's formula: $Q = (1.782 + 0.24 \dfrac{h}{P})BH^{2/3}$, in which, P is the height of the weir, B is the width of the weir, and h is depth of water on the weir ($H = h + 0.001, 1$ m).

At the same time, the camera and spike particle were used to observe the flow pattern. The sea water level was controlled by the tail gate. The water level was measured by exploring tube, its accuracy is 0.1 mm.

4 Test of the design scheme(HUTA design scheme)

For decreasing the effect of drainage hot water to the water taking as far as possible, the outfall was changed to the direction of the wharf. The plane arrangement of the design scheme compare with the design scheme is shown in Fig. 2.

The velocity distribution is shown in Fig. 3 ~ Fig. 5, as water was discharged from the drainage cannel in the condition of different out sea level, which are the design high tide level, the design average tide level and the design low tide level.

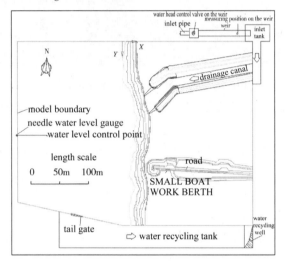

Fig. 2 Layout diagram of model in design scheme

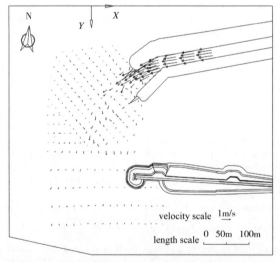

Fig. 3   Distribution of time-averaged velocity on high tide sea level (the design scheme)

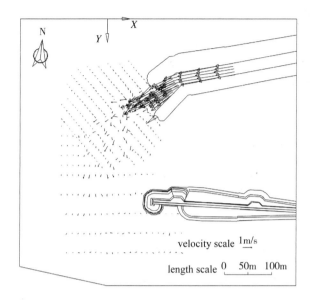

**Fig. 4   Distribution of time-averaged velocity on
average tide sea level (the design scheme)**

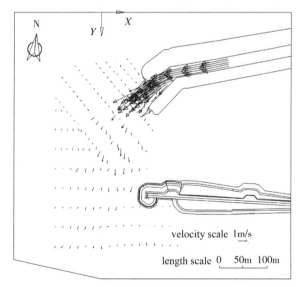

**Fig. 5   Distribution of time-averaged velocity on low
tide sea level (the design scheme)**

It can be seen in the design scheme that the flow along the wharf is steady in different tide level, because the drainage water is jumped into the deep sea (the depth of sea water is 30 ~ 50 m.) at the outfall and the flow energy was dissipated fully in the deep sea water. The velocity value is less than 0. 28 m/s nearby the wharf, and its transverse velocity is less than 0. 03 m/s. And the demands of the design scheme are satisfied. It also indicates that the higher surface velocities are all grouped near the northern dyke and the higher surface velocities persist beyond the end of the designed channel. The surface velocities are less than 2. 20 m/s. The end of the designed channel is the reef. It is stable for long time. But some appropriate protection to the bottom of the channel and the sea bed beyond the end of the channel maybe good for the drainage project.

5　Drainage port and wharf movable bed wash-out model test

5. 1　Selection of model sand

5. 1. 1　Natural sand starting speed

The pertinent data indicates that the nature sand median diameter of the Power Plant nearby the sea bed is 0. 322 mm. At present, there are many starting speed of flow formula, but there is not big difference to the natural sand computed result within certain particle size scope. The natural sand is coarse in this work. Compared to the others, Г. И. Шамов's formula is suitable both to the natural sand and model sand. So it was adopted for the natural sand.

$$V_c = 1.14 \sqrt{\frac{\gamma_s - \gamma}{\gamma} gd} \left(\frac{h}{d}\right)^{1/6}$$

where, V_c is starting up speed; h is water depth; d is sand diameter; γ_s is specific gravity of sand; γ is specific gravity of water; g is the acceleration of gravity, which is 9. 81 m²/s.

Based on the above formula, the natural sand starting flow speed is 0. 497 ~ 0. 671 m/s under the water depth 1. 0 ~ 6. 0 m, shown as Tab. 2.

Tab. 2　Natural sand and model sand starting speed contrast data

Natural sand $d_{50} = 0.322$ mm, $\gamma_s = 2.65$ t/m³		Model sand $d_{50} = 0.33$ mm, $\gamma_s = 1.16$ t/m³		Starting velocity ratio
Water depth H(m)	Starting speed (m/s)	Water depth H(cm)	Starting speed (m/s)	Starting speed (m/s)
1	0. 497	2. 5	0. 079	6. 31
2	0. 558	5. 0	0. 088	6. 31
3	0. 597	7. 5	0. 095	6. 31
4	0. 627	10. 0	0. 099	6. 31
5	0. 650	12. 5	0. 103	6. 31
6	0. 671	15. 0	0. 106	6. 31

5. 1. 2　Model sand choice

There are many types of sand type, the light quality material is generally used. At present, the main model sand is phenolic molding powder, plastic sand, wooden powder and coal scraps. As the drainage port testing model is big, and the starting speed of the sand is big too, finally wooden powder was used as the model sand pattern. Its median diameter is $d_{50} = 0.33$ mm.

5. 2　The laying scope of the model sand

The movable bed scope is designed according to the velocity distribution test. The layout scope of the movable bed is shown as Fig. 6. It covers about 440 m along the sea bank near the open channel outfall and the wharf. The width is approximately 100 m of the beach from the steep bank edge.

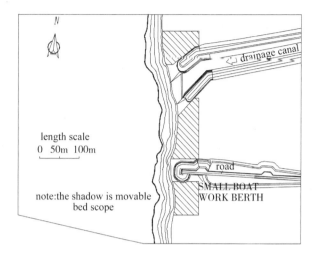

Fig. 6　Layout scope of movable bed

5.3　Local scouring test of the sea bank near the outfall and the wharf

According to the technical requirement document of the designing institute, the local scouring test condition is as follow: the cooling water discharge is 116.78 m^3/s, the three different tide sea levels are: high tide sea level: 0.44 m; average tide sea level: 0.00 m; low tide sea level: -0.54 m.

In the scouring test, according to the order: high tide sea level, average tide sea level, low tide sea level, carries on 2 h continuously for each and the total is 6 h, water discharge is the same, 116.78 m^3/s. Testing results are shown in Fig. 7.

Fig. 7　Result of scouring test

According to the test results of the velocity distribution, nearby the outfall and the wharf the velocity of flow is 0.03 ~ 0.28 m/s, is lower than the start-up speed of natural sand (the start-up speed of natural sand is shown in Tab. 2). The drainage open channel emissions cooling water can not cause the beach nearby the wharf washout. May know from local scouring test result of the discharging water channel, the beach nearby the outfall and the wharf has not had the washout destruction.

6   Conclusions and suggestions

Conclusions obtained from the above test results of fixed bed and local moveable bed:

(1) In the design scheme of the drainage open channel, open channel water flow into the deep sea and is fully dissipated. The flow near the outfall and the wharf is steady under different conditions. And the surface flow velocity is less than 0. 28 m/s. And the velocity in the transverse direction is less than 0. 03 m/s along the wharf, meet the requirements of the project.

(2) According to the test results, nearby the wharf the speed of flow is 0. 03 ~ 0. 25 m/s. It is lower than the silt start-up speed. The water was discharged from the drainage open channel to the sea. And the beach nearby the wharf was not washout. May know from local scouring test result, the beach nearby the wharf had not had the washout destruction.

(3) Along the beach of the drainage open channel and the wharf of Rabigh Power Plant is mainly affected by the waves. Therefore, the washout protection question nearby the drainage open channel and the wharf must be considered in the wave-silt model test.

References

Rabigh Power Plant Stage VI. Cooling Water Thermal Discharge Studies [R]. HR Wallingford Report EX5561 – 02, June 2007.

Hui Yujia, Wang Guixian. Model Experiment in River Engineering [M]. Beijing: China Water Conservancy and Electric Press,1999.

Wuhan Institute of Hydraulic and Electric Engineering. River Silt Engineering [M]. Beijing: China Water Conservancy and Electric Press,1980.

Wu Fusheng, Tong Zhongshan. Test Report on Water Taking Area Sediment Physical Model of Vietnam Guangning Power Plant 4 × 300 MW Power Units Project [R]. Nanjing: Nanjing Hydraulic Science Research Institute, National Key Laboratory of Hydrology – Water Resources and Hydraulic Engineering, 2006.

Tong Zhongshan, Wu Fusheng. Test Report on Water Taking Area Sediment Physical Model of Vietnam Haifang Power Plant 4 × 300 MW Power Units Project [R]. Nanjing: Nanjing Hydraulic Science Research Institute, National Key Laboratory of Hydrology – Water Resources and Hydraulic Engineering, 2006.

Environment Impact Analysis of Household's Heating Energy Consumption in Kaifeng City[①]

Zhang Yan[1,2], *Qin Yaochen*[2], *Zhang Lijun*[2] and *Lu Chaojun*[2]

1. College of Environment Science and Tourism Nanyang Normal University, Nanyang, 473061, China

2. College of Environment and Planning, Henan University, Kaifeng, 475001, China

Abstract: To explore the carbon emissions characteristics and impact factors of household's winter heating energy consumption in the cities of the Yellow River Basin, households in Kaifeng city was as an example to be researched. Based on the household life style, large random sample method was used to research household's winter heating energy consumption. The features of per capita CO_2 emissions were analyzed. This analysis was carried out according to family size, income, housing size, respectively. And the impact factors were analyzed by using multiple stepwise regression method. The results are as following: ①family income, housing area, heating form, and heating period length are significant factors of per capita heating CO_2 emissions from household's winter energy consumption in Kaifeng City; ②per capita heating CO_2 emissions from household's winter energy consumption can be regulated by changing heating mode, reducing heating period length, controlling housing size and so on.

Key words: carbon emission, household, heating energy consumption, influence factor

1 Introduction

With the acceleration of urbanization and improvement of living standard, household's heating energy consumption has been increasing. This led to more pressure on environment. Household's heating in winter is a basic consumer demand in north cold area, and CO_2 emissions from household's winter heating energy consumption occupies above 38% of that from direct energy consumption. This was caused attention from many scholars. The problems facing to the central heating systems development of cities were analyzed by Liu. The impact factors were analyzed by Sun based on statistic data. Households' winter heating energy consumption in Beijing was studied by Sun based on household research. But the character and impact factors of urban households' winter heating energy consumption were not studied deeply. In order to clear-out the character and impact factors of carbon emission from urban households' winter heating energy consumption, households' winter heating energy consumption in Kaifeng city were researched. It was good for finding out the regulation and adjusting the carbon emission from urban households energy consumption.

2 Methodology

2.1 Theoretical frame

According to lifestyle theory, it is considered the heating lifestyle includes household characteristics (income, family size, housing size, housing ownership, the age of household head), heating style (including heating mode, heating period length). It is different of home heating with the heating energy type, energy consumption strength, energy heating period length, heating frequency difference among households. And this makes the difference of per capita CO_2

① Foundation: National Natural Science Foundation of China, NO. 41171438; Major Program of National Natural science Foundation of China, NO. 2012CB955800; Major Program of Humanities and Social Sciences Key Research Base of Ministry of Education of China, NO. 10JJDZONGHE 015

emissions from households' heating energy consumption. It is estimated per capita CO_2 emissions from households' heating energy consumption is influenced by heating lifestyle. Accordingly, the microscopic research model is established, as is shown in Fig. 1.

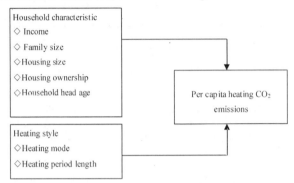

Fig. 1   Analysis framework

2.2   Calculation method of heating carbon emissions

Per capita CO_2 emissions from households' heating energy consumption were calculated by the method of energy carbon conversion coefficient. The computational formula is as following:

$$C = \sum E_i \times \eta_i \times K/P \tag{1}$$

where, E_i is the quantity of energy consumption; η_i is coefficient of the energy converted into standard coal; K is coefficient of standard coal converted into CO_2; P is the number of household members.

2.3   Analytical procedure

The influential factors of households' carbon emissions from heating energy consumption in Kaifeng were analyzed by using multiple regression method. The independent variables are introduced to stepwise multiple regression model one by one by testing on the dependent variable significantly affect. Only important independent variables are into the equation, and those unimportant, irrelevant are to be exclude. Thus there are only independent variables having significant role in dependent variable in the regression equation. Increase any of these factors, no significant regression effect improvements. The regression equation has strong explanatory power. Per capita CO_2 emissions from households' heating energy consumption in Kaifeng city were for the dependent variables, factors of household characteristics and heating energy forms were for the independent variables, household heating energy. The analysis was carried out by SPSS 16.0.

2.4   Study area

Kaifeng city locates 113°51'51"E ~ 115°15'42"E, N4°11'43"N ~ 35°11'43"N, belonging to warm temperate zone continental monsoon climate, with distinct four seasons. moderate rainfall, and average temperature, 14 ℃. By the end of 2010, Kaifeng area is 95 km^2, urban population was of 949,000, the per capita GDP reached 19,893 yuan, urban per capita disposable income was 13,695 yuan.

2.5   Data

Households' heating energy consumption research was carried out from October to November in 2011 in Kaifeng city. According to large sample of random method, 1,000 questionnaires were extended and 792 valid questionnaires were received. The questionnaire survey content consists of urban household's characteristics (including income, family size, housing size, housing ownership, the age of household head), heating style (including heating mode, heating period length), data of CO_2 emission from heating. Based on survey data, it was established "households' carbon emissions from heating energy consumption in Kaifeng" database by Microsoft excel 2003.

3 Results

3.1 Characteristic of heating carbon emissions

3.1.1 Income and per capita heating CO_2 emissions

It is presented in Fig. 2 that the quantity of heating CO_2 emissions from households with the annual income of less than 50,000 yuan is 279 kg per capita, lower than the city average level of 424 kg per capita. Per capita heating CO_2 emissions increase with the increasing of household annual income. When household's annual income is up to 150,000 ~ 200,000 yuan, per capita heating CO_2 emissions reaches 914 kg per capita, the highest value, on average, 3.3 times the lowest level. But when the income is over 200,000 yuan, the per capita heating CO_2 emissions are down to 567 kg per capita. The relationship between per capita heating CO_2 emissions and income complies with the inverted U-shaped trend.

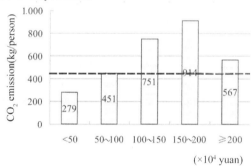

Fig. 2   Per capita heating CO_2 emissions from household with different income levels

3.1.2 Family size and per capita heating CO_2 emissions

It is presented in Fig. 3 that the quantity of heating CO_2 emissions from households with the family size of above 6 is 182 kg per capita, lower than the city average level of 424 kg per capita. Per capita heating CO_2 emissions increase with the decreasing of family size. When family size is less than 2, per capita heating CO_2 emissions reaches peach with the level of 482 kg per capita, which is 2.6 times that of above 6.

3.1.3 Housing size and per capita heating CO_2 emissions

It is presented in Fig. 4 that the quantity of heating CO_2 emissions from households with the housing size of less than 80 m^2 is 236 kg per capita, which is the lowest level in the city. Per capita heating CO_2 emissions increase with the increasing of housing size. It ups to average level with housing size of 100 ~ 120 m^2. Per capita heating CO_2 emissions achieve the highest level, 750 kg per capita, on average, which is 3.2 times the amount of the lowest level.

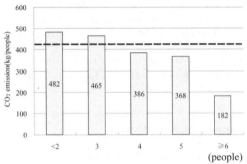

Fig. 3 Per capita heating CO₂ emissions from household with different family size

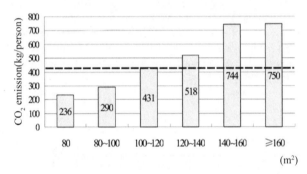

Fig. 4 Per capita heating CO₂ emissions from household with different housing size

3.2 Estimating the impact factors

Through the establishment of a multiple regression model, the per capita household heating CO_2 emissions and household characteristics of CO_2 emissions per capita heating energy and heating energy life form of the variable relationship between the return of objects to be interpreted, the explanatory variables for the family size, annual household income, housing area, housing ownership, the age of household head, heating forms, heating can be of length. By four times multiple regression, the result (Tab. 1) are as following:

(1) Determinants of heating CO_2 emission from households in Kaifeng city are heating period length, heating mode, income, and housing size. Heating period length and heating mode have greater influence than income and housing size.

(2) Heating mode is highly correlated with heating CO_2 emissions. There are many kinds of heating modes for home in Kaifeng City: central heating by municipality, coal boiler heating in residential area, heating using natural gas from residential area, geothermal heating from residential area, households heating by electrical , natural gas, coal and so on. The heating modes are different in energy type, operation mode, heating range, heating temperature. Carbon emissions vary due to the different types of heating mode. High-carbon heating mode is an important factor of stimulating urban households' carbon emissions.

(3) Per capita household heating CO_2 emissions have positive correlation with heating period length, income, housing size. Per capita CO_2 emissions increase with the increasing of heating period length, income and housing size.

Tab. 1 **Multiple stepwise regression result of determinants of heating CO_2 emission from households in Kaifeng city**
(Dependent variable: ln heating CO_2 emission per capita)

Model	Unstandardized Coefficients		Standardized coefficients	t	Sig.	R^2
	B	Std. Error	Beta			
1 (Constant)	1.948	0.084		23.279	0.000	
Heating period length	0.137	0.028	0.369	4.981	0.000	0.369
2 (Constant)	1.768	0.091		19.384	0.000	
Heating period length	0.118	0.027	0.318	4.421	0.000	
Heating mode	0.075	0.018	0.292	4.059	0.000	0.468
3 (Constant)	1.618	0.102		15.935	0.000	
Heating period length	0.110	0.026	0.295	4.193	0.000	
Heating mode	0.075	0.018	0.293	4.183	0.000	
Income	0.031	0.010	0.212	3.055	0.000	0.513
4 (Constant)	1.598	0.102		15.935	0.000	
Heating period length	0.106	0.025	0.291	4.193	0.000	
Heating mode	0.072	0.017	0.285	4.183	0.000	
Income	0.028	0.009	0.202	3.055	0.000	
Housing size	0.012	0.007	0.156	2.671	0.000	0.523

4　Conclusions

The characteristics of heating CO_2 emissions from households in Kaifeng city were quantitatively analyzed. The influential factors of carbon emissions were analyzed by using multiple regression method. The results are as following:

(1) Heating CO_2 emissions from households in Kaifeng city vary with the income, housing size, family size, and show certain regularity. The relationship between per capita heating CO_2 emissions and income complies with the inverted U-shaped trend. Per capita heating CO_2 emissions increase with the decreasing of family size and the increasing of housing size.

(2) Determinants of heating CO_2 emission from households in Kaifeng city are heating period length, heating mode, income, and housing size. Heating period length and heating mode have greater influence than income and housing size. Per capita heating CO_2 emissions from household's energy consumption can be regulated by changing heating mode, reducing heating period length, controlling housing size and so on.

References

Ye H, Pa L Y, Chen F, et al. Direct Carbon Emission from Urban Residential Energy Consumption: a Case Study of Xiamen, China [J]. Acta Ecologica Sinica, 2010, 30(14): 3802 – 3811.

Zhang Y, Qin Y C, Yan W Y, et al. Urban Types and Impact Factors on Carbon Emissions from Direct energy Consumption of Residents in China [J]. Geographical Research, 2012, 31 (2):345 – 356.

Zhang Y, Qin Y C. Research on the Difference in Carbon Emissions Due to Direct Household Energy Consumption in the Same Urban Residential Community [J]. Ecological Economy, 2012 (4): 42 – 46.

Sun J. Researching and Analyzing of Resident Energy Consumption in Typical Cities [D]. Shanghai:Tongji University, 2009.

Chen L, Fang X Q, Li S 3, et al. Comparisons of Energy Consumption between Cold Regions in China and the Europe and America [J]. Journal of Natural Resources, 2011, 26 (7): 1258 – 1268.

Yu X P, Fu X Z, Huang G D, et al. Analysis on Impacts of Dividing Air – conditioning/Heating Periods on Limited Value energy Consumption for Residential Building in Hot summer and Cold Winter Area [J]. Building Science,2007, 23(8): 27 – 31.

Fan Y F. Research on Recommended Value of Index for District Heating in China [J]. Building Science,1989, 5(3): 17 – 23.

Liu J, Huo Z Y, Yin H C. Problems Facing to the Central Heating Systems Development of Cities [J]. Energy Conservation and Environmental Protection, 2008 (4): 33 – 35.

Chen L, Li S 3, Fang X Q, et al. Influence Factor Analysis of Urban Residential Heating Energy Consumption in Severe Cold and Cold Regions in China—A Case of Jilin Province [J]. Scientia Geographica Sinica, 2009, 29(2): 212 – 216.

Experimental Study on Sand Wave Evolution in Flume[①]

Wan Qiang, *Cao Yongtao* and *Ma Tao*

Yellow River Institute of Hydraulic Research, YRCC
Key Laboratory of Yellow River Sediment Research, MWR, Zhengzhou, 450003, China

Abstract: It is observed that the river bed of coal ash evolves from smooth bed to sand ripples in the flume model test. The change law of model sand wave scale is studied and the relationship between the height of the ripples and the water power parametric is discussed. It is analyzed how the sand wave and bed form changes under different water and sediment process condition. So the model river bed resistance, the bed load transport and the effect on the test measurement by the ripples would be researched based on this foundation.

Key words: flume experiment, sand wave evolution, water power parameter, bed form

The interaction of flow and sediment can generate kinds of bed forms in rivers. Different sand wave scale presents on the bed with the current intensity changes when the water flows on the smooth river bed. The change of sand wave scale affects the channel resistance and the bed load transport observably. At present, the cause of ripple formation is not uniform conclusion. Yalin M. S considers that the sand wave is a result of flow turbulence action. Kennedy J. F. thinks that the cause of the sand wave formation is the unstable interface of two kinds of fluid when they move relatively. BaiYuehuan and Andreas Maleherek regard that the cause of ripple evolution is the laminar flow instability or the near wall flow layer Coherent Structure dynamic. In the physical model experiment, the analysis on the sand wave evolution is very significant to the law of similitude and tests precision and so on.

1 Experiment on sand wave evolution in flume

The test flume was built following the test requirement and useable area. The flume length is 30 m with the width of 1 m and the height of 0.5 m. The discharge can be adjusted with the electromagnetic flow meter and the sediment concentration is controlled by the orifices box. The model sediment is the coal ash produced by Zhengzhou Thermal Power Plant. $D50$ is 0.035 mm and the density is 2.1 t/m^3. The test initial boundary condition is the approximate U pattern cross section with the depth of 6 cm. The initial bed slope is 1.5‰. There are 15 levels discharge in the test and every level duration time is 2 h. The process is divided into 5 stages successively: low water stage, water rising stage, flood stage, the second low water stage and the second flood stage. The sediment concentration changes with the flow process (Tab. 1).

Tab. 1   Groups of sand wave evolution experiment plan

Stage	Parameters	Group one	Group two	Group three	Group four	Group five
Low water stage	Discharge(L/s)	1.4	2.5	3.6	—	—
	Concentration(kg/m^3)	0	5.6	5.6	—	—
	Duration(h)	2	2	2	—	—
Water rising stage	Discharge(L/s)	5	6.5	7.9	9.4	10.8
	Concentration(kg/m^3)	5.6	8.3	8.3	8.3	13.9
	Duration(h)	2	2	2	2	2

① Project supported by the Ministry of Water Resources Public Welfare Industry Research Project. (No_ 201101009)

Continued to Tab. 1

Stage	Parameters	Group one	Group two	Group three	Group four	Group five
Flood stage	Discharge(L/s)	12. 2	13. 6	15. 1	16. 5	17. 9
	Concentration(kg/m^3)	13. 9	13. 9	13. 9	13. 9	13. 9
	Duration(h)	2	2	2	2	2
Low water stage	Discharge(L/s)	3. 6	—	—	—	—
	Concentration(kg/m^3)	5. 6	—	—	—	—
	Duration(h)	2	—	—	—	—
Flood stage	discharge(L/s)	21. 5	—	—	—	—
	Concentration(kg/m^3)	13. 9	—	—	—	—
	Duration(h)	2	—	—	—	—

The test content: ① measurement of flow factors (including velocity and flow depth). The duration time of measurement is about 30 min in every discharge level, the measurement of 6 times are carried out. The bed slope is controlled in 1. 5‰ and the water level are recorded along the flume;② sand wave characteristics measurement. 20 ripples of continuous distribution are selected in the test section. The length width and height of these samples are measured in every discharge level; ③ particle size analysis. The bed load sample particle is analyzed in every discharge level.

2   Analysis on the test result

2. 1   Sand ripples scale characteristics

In the test, the sand wave scale increases in the low water and water rising stage, but the size increases at first then decreases gradually in the flood stage, and the size increases again in the second low water stage. The ripple height maximum is 2. 95 cm at 10. 8 L/s, and the maximum of ripple length and width are 17. 4 cm and 17. 35 cm at 13. 6 L/s separately, Fig. 1, Fig. 2 and Fig. 3. In addition, the fine particle in the bed load is taken away by the current for the continue scour. The average grain size of bed load increases with the rising discharge. The suspend load drops and silts on the bed and the bed load average grain diameter decreases, Fig. 4.

Fig. 1   Wave length with the test duration 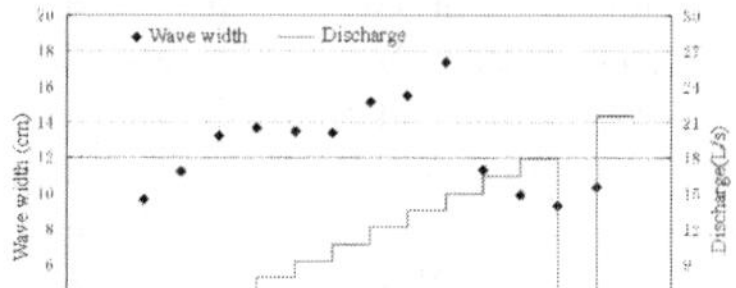   Fig. 2   Wave width with the test duration

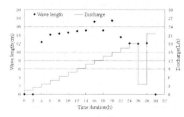

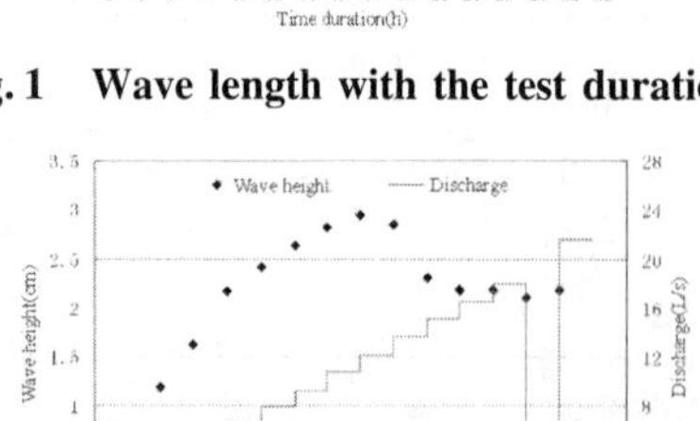

Fig. 3   Change of ripple height
with the test duration

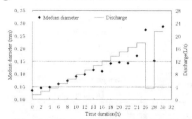

Fig. 4   Change of grain size with
the test duration

2.2 Relationship between the ripple height and the typical water power parameter

In order to research the mechanism of the sand wave change with the flow condition, three water power parameters are chosen which are closely related to the ripple and it is analyzed that the ripple size retroacts followed the discharge process by the parameters variation.

2.2.1 Relationship between the ripple height and the Shields number.

It is generally acknowledged that the shields number is the decisive dimensionless mechanical parameter $\theta = \tau_0 / (\gamma_s^l - \gamma^l) D$. Where τ_0 is the river bed shear force, γ_s is the volume weight of sediment, γ is the water volume weight, and D is the bed load median diameter. It reflects the ratio of bed load starting force to the moving resistance. The shields number is bigger, and the sediment mobility is more powerful.

According to the result of the flume test, the shields number appears increase at first and then reduces with the increasing discharge, Fig. 5. The maximum is 0.488 at 10.8 L/s. It is because that the bed load is coarsening with the growing discharge and the sediment resistance is increasing. The ripple height is increasing with the growing Shields number, Fig. 6. It reflects that the sand wave height change trend is consistent with the bed load mobility.

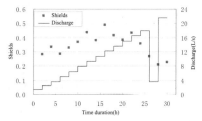

Fig. 5   Change of Shields number with duration

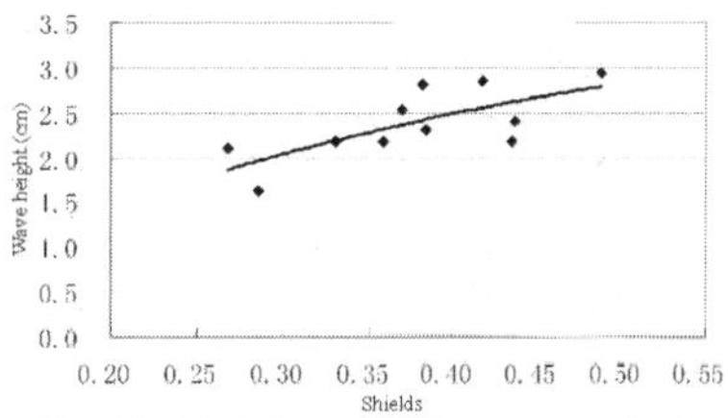

Fig. 6   Relationship between ripple height and the Shields number

2.2.2 Relationship between the ripple height and the sand grain Reynolds number

The sand wave evolution is related to the near wall flow layer and turbulent motion. The sand grain Reynolds number $R_* = BU_* D/V$ reflects directly the ratio of the grain height to the viscous sublayer thickness (where U_* is friction velocity, and V is coefficient of viscosity). It could reflect indirectly the ratio of sediment starting power to viscous force. The sand grain Reynolds number is a dimensionless mechanics parameter too. Based on the flume experiment, Fig. 7, the sand grain Reynolds number increases with growing discharge and it is consistent with the bed load diameter change trend. The ripple height appears increase at first then reduction with the growing sand grain Reynolds number Fig. 8.

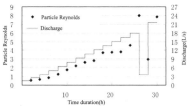

Fig. 7   Change of Reynolds number with duration

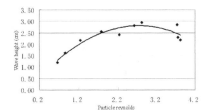

Fig. 8   Relationship between ripple height and Reynolds number

2.2.3 Relationship between the sand wave height and Froude number

The Froude number ($Fr = v/(gh)$) is the inertance to gravity ratio at specific cross section in open channel flow. It determines the flow regime and it is an important mechanic parameter which determines the sand wave evolution. The Froude number change trend followed the discharge is not obvious. However the Froude number increases rapidly in the lower water stage, Fig. 9. The ripple height appears downtrend with the growing Froude number, Fig. 10.

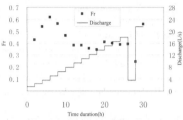

Fig. 9　Change of Froude
number with duration

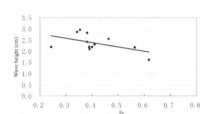

Fig. 10　Relationship between ripple
height and Froude number

3　Conclusions

The bed ripples evolution process is discussed quantitatively by the flume test, it is concluded:

(1) Based on the flume test, the bed presents the change from smooth bed to two dimensions ripple and three dimensions ripple. The sand ripples scale and stability increases with the bed form evolution. The ripple scales increases at first then decreases with the increase of discharge.

(2) With the increase of discharge, the flume bed is coarsened and the bed load grain size increases continuously.

(3) The relationship between typical water power parameters and the sand ripples height and the flow condition is studied, it is concluded that Shields number presents increase at first then decrease with the increase of discharge and the ripple height rises with the increase of Shields number. The sand grain Reynolds number presents increase with the increase of discharge, and the ripple height increases at first and then decreases with the increase of sand grain Reynolds number. The Froude number decreases after the ripples have been formed and the ripple height decreases with the increase of Froude number.

References

Qian Ning, Wan Zhaohui. Mechanics of Sediment Transport [M]. Beijing: Beijing Science Press, 1983.

Yalin M S. Mechanics of Sediment Transport[M]. Pergamon Press, 1972;290.

Kennedy J F. The formation of Sediment Ripples, Dunes, Antidunes, AnnualRev. Flu. Meeh. 1969(1):147 – 168.

Bai Yuehuan, AndereaS Maleherek. A linear Theory for Disturbanee of Coherent Structure and Mechanism of Sand Waves in Open Channel Flow[J]. Journal. Sediment Researeh, 2001, 16(2):234 – 243.

Wan Xinkui, Shao Xuejun, Li Danxun. The Basis of River Mechanics [M]. Beijing:Press of China Water Conservancy &Hydroelectricity, 2002.

Study on Deposition Morphology of Xiaolangdi Reservoir

Wang Ting, *Ma Huaibao*, *Zhang Junhua* and *Li Ping*

Yellow River Institute of Hydraulic Research, YRCC
Key Laboratory of Yellow River Sediment Research, MWR, Zhengzhou, 450003, China

Abstract: By October 2010, the mainstream delta vertex of Xiaolangdi Reservoir had moved to cross section HH12 which is 18.75 km from the dam, and the amount of sediment reached $2.822,5 \times 10^9$ m³. The results show that, under the same amount of sediment and water storage, compared with cone deposition morphology, the backwater length of delta deposition morphology is obviously shorter, the water lever before dam is obviously lower. The delta deposition morphology is obviously superior to cone deposition morphology in the aspects of trapping the coarse sand and discharging the fine sand, reducing deposition, optimizing process of outlet water and sediment and so on. The delta deposition morphology is beneficial to shape density current, and the sediment ejection effect of density current is superior to backwater free flow. Therefore, we suggest timely sediment ejection should be carried out at present so as to delay the transformation from delta deposition morphology into cone deposition morphology as long as possible. Studying deposition morphology can supply some technical supports for later operation of Xiaolangdi Reservoir.

Key words: Xiaolangdi Reservoir, delta deposition morphology, cone deposition morphology, density current, free flow

Xiaolangdi Reservoir is a comprehensive utilization project and its development mission was definitely oriented to flood control (including ice prevention) and sedimentation reduction, and balanced water supply, irrigation and hydropower generation. It is about 130 km from Xiaolangdi dam to Sanmenxia dam. The controlled basin areas of Xiaolangdi Reservoir are 694,000 km² and accounts for 92.3% of basin areas of Yellow River. The highest operation water level is 275 m. The initial storage capacity is 12.75×10^9 m³, and the long – term effective storage capacity is 5.1×10^9 m³. Xiaolangdi Reservoir started its build of main project in September 1994, river closure in October 1997, impoundment in October 1999 and formal use in May 2000.

1 Deposition morphology of mainstream since Xiaolangdi Reservoir operation

In the initial operation period of Xiaolangdi Reservoir, the deposition morphology of mainstream was a cone. Because of sediment deposition, by October 2000, the obvious delta continent surface reach, foreslope reach and deposition reach before dam had came into being, the deposition morphology had transformed into a delta, and the vertex was 69.39 km from dam and the elevation of it was 225.22 m. With the continuous deposition in reservoir, the elevation of delta continent surface uplifts gradually, and the delta vertex moves gradually downstream to the dam. The Fig. 1 shows the profiles of Xiaolangdi Reservoir.

By October 2010, the delta vertex had moved to cross section HH12 which is 18.75 km from dam, the elevation of delta vertex was 215.61 m, and the amount of sediment reached $2.822,5 \times 10^9$ m³. However, the elevation of deposition before dam was only 188.9 m. According to the report on research of initial operational mode of Xiaolangdi Reservoir, the amount of sediment had exceeded the cut off value of initial sediment retaining period and later sediment retaining period, however the delta vertex did not move to dam. In the whole, the deposition position leaned to upstream reach, the amount of sediment below elevation 205 m was lower than the designed, and there was still $0.175,8 \times 10^9$ m³ storage capacity under elevation 205 m. If we extend delta continent surface to dam according to slope in October 2010, the storage capacity of sediment retaining was still 0.4×10^9 m³. This means that the operation of Xiaolangdi Reservoir is about to

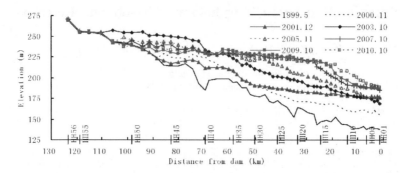

Fig. 1 Longitudinal thalweg profile of mainstream in Xiaolang di Reservoir

transform into the operation of later sediment retaining period. Because mainstream deposition morphology has great influence on the comprehensive benefits such as sediment retaining, sediment ejection, hydropower generation and so on, therefore, it is necessary to study deposition morphology during later sediment retaining period.

2　Reservoir deposition morphology and operation

In order to study the influence of deposition morphology on reservoir operation, we preliminarily study storage capacity eigenvalues of cone deposition morphology with amount of sediment of $2.822,5 \times 10^9$ m^3 (in October 2010), and compare with delta deposition morphology at present (Tab. 1). We can see that, compared with cone deposition morphology, in the same water level, the water storage of delta deposition morphology is more bigger and the backwater length is more shorter. The research on the law of reservoir sediment transport based on measured data and entity model test shows, in the same amount of sediment and water storage, the delta deposition morphology which can keep more storage capacity near dam is superior than cone deposition morphology in the aspects of trapping the coarse sand and discharging the fine sand, reducing deposition, optimizing the process of outlet water and sediment and so on.

Tab. 1　Storage capacity eigenvalues of different deposition morphology

Elevation(m)	Storage capacity($\times 10^8$ m^3)		Backwater length(km)	
	Delta	Cone	Delta	Cone
215	4.730	3.367	19	27
220	7.158	6.291	34	42
225	10.503	10.206	54	57

2.1　The delta deposition morphology is beneficial to reservoir sediment ejection and deposition reduction

According to the operation regulations of Xiaolangdi hydro – junction in the initial sediment retaining, the reservoir operational mode will transform from operational mode of water storage and sediment retaining and water and sediment regulation into operational mode of multi – year sediment regulation and man – made precipitation washout at right occasion. Namely, that is the operational mode of the combination of water and sediment regulation under general water and sediment conditions and man – made precipitation washout under larger flood condition.

In the process of water and sediment regulation under general water and sediment conditions, the reservoir is always in water storage state, and the water storage is between 0.2×10^9 m^3 and regulating storage capacity. Obviously, the backwater end of delta deposition morphology is more near to dam, and more beneficial to shaping density current to sediment ejection, and the sediment ejection effect of density current is superior to backwater free flow.

2.1.1 Delta deposition morphology is beneficial to plunging of density current

A lot of researches show that the water depth of density current plunging point in Xiaolangdi Reservoir can also be calculated by following formula.

$$h_0 = \left[\frac{1}{0.6\eta_g g} \frac{Q^2}{B^2} \right]^{1/3} \tag{1}$$

Han Qiwei thinks that after density current plunges and moves a certain distance, the flow patterns transform into uniform flow and the water depth can expressed as follows:

$$h'_n = \frac{Q}{V'B} = \left(\frac{\lambda'}{8\eta_g g} \frac{Q^2}{J_0 B^2} \right)^{1/3} \tag{2}$$

where, η_g is correction coefficient of gravity; $\eta_g g$ is effective acceleration of gravity; Q is flow; B is average width; J_0 is bottom slope of reservoir; λ' is drag coefficient of density current and taken as 0.025.

If water depth h'_n of uniform flow formed by density current is less than h_0, namely $h'_n < h_0$, then the plunging is successful. Otherwise, the water depth of density current will exceed the water depth of surface clear water, and the density current will float upward and disappear.

When water depth h'_n is equal to h_0, correspondingly, the critical bottom slope of reservoir is $0.001,875$, that is, $J_{0,c} = J_0 = 0.001,875$. In general, for density current, except meeting plunging condition Eq. (1), bottom slope also should be bigger than $J_{0,c}$. Therefore, once cone deposition morphology is formed, and then density current is difficult to shape.

2.1.2 Sediment ejection effect of density current is superior to backwater free flow

Under water storage condition, in backwater region, there are two kind of flow patterns of sediment transport, that is free flow and density current, among them the formula of sediment ejection of backwater free flow is as follows:

$$\eta = a\lg Z + b \tag{3}$$

where, η is sediment delivery rate, $Z = \left(\dfrac{V}{Q_{out}} \dfrac{Q_{in}}{Q_{out}} \right)$ backwater index; V is water storage volume during calculation interval; Q_{in}, Q_{out} is reservoir inflow and outflow; a, b is constants, respectively.

During the period of water and sediment regulation of 2006 to 2008, there was no sediment transport of backwater free flow in continent surface reach of delta. We choose the incoming water and sediment process and water storage condition in the same period, suppose that sediment ejection pattern is backwater free flow, calculate sediment ejection amount by Eq. (3), and compare the calculated values with the practically measured values of density current(Tab. 2). It can be seen that the sediment ejection effect of density current is superior to backwater free flow.

Tab. 2 Comparison of sediment ejection of backwater free flow and density current

Year	Period (month. day)	Moving distance of density current(km)	Reservoir inflowing sediment amount ($\times 10^8$ m^3)	Reservoir outflow sediment amount ($\times 10^8$ m^3)	
				Calculated values	Measured values
2006	6.25~6.28	44.13	0.230	0.052	0.071
2007	6.26~7.2	30.65	0.613	0.161	0.234
	7.29~8.8	–	0.834	0.153	0.426
2008	6.28~7.3	24.43	0.741	0.157	0.458

In addition, in the process of preflood water and sediment regulation, the delta deposition morphology is more beneficial to scouring delta continent sediment to shape density current. After density current moves to the front of the dam, the sediment particles suspended in flow is very fine and the concentration is very high, and the turbid water reservoir settle very slowly. By using this characteristic, according to incoming water and sediment condition and the law of sediment transport of Lower Yellow River, the discharge tunnels at different elevation are opened so as to optimize output water and sand combination.

2.2　The delta deposition morphology is beneficial to trapping the coarse sand

The longitudinal slope of delta foreslope reach is more than 10 times of cone deposition morphology. When water level before dam is low than delta vertex, if the uplifted height of water level is equal, then the increase of backwater length of delta deposition morphology is much less than cone deposition morphology. If deposition morphology of reservoir is a cone, except that coarse sand deposits in the river reach of backwater zone, a large amount of fine sand will deposit along the downstream. Comparatively speaking, after plunging of density current, the moving distance of density current is relatively shorter, and therefore the sediment delivery rate of fine sand is relatively bigger.

2.3　The delta deposition morphology is beneficial to the effective utilization of tributary storage capacity

When tributary is in the downstream river reach of main river deposition delta vertex, the deposition pattern of tributary is flowing backward of density current, the channel – mouth bar is difficult to form, tributary storage capacity can be used to water and sediment regulation. If deposition morphology is a cone, in a long low water period, the sand bar is easy to form at estuary of some tributary, storage capacity under the elevation of sand bar is difficult to effectively use in some period. Zhenshui River is the biggest tributary of Xiaolangdi Reservoir, and also the most difficult to flow backward. In October, 2011, when the delta vertex just moved to the estuary of which, the sand bar of which quickly formed and the height was more than 10m. Therefore, keeping delta deposition morphology as long as possible is beneficial to using storage capacity of several big tributaries nearby dam.

3　Study of operational mode of keeping delta deposition morphology

Nearby the vertex of deposition delta, the coarse sand in sediment – laden flow sorts and deposits, and the flow carrying fine sand forms density current and moves towards dam. Therefore, the bed material nearby dam, mostly being fine sand, is viscous deposit. When the deposits are still in the unconsolidated state, it can be seen as Bingham fluid and described by rheological equation as $\tau = \tau_b + \eta \dfrac{du}{dy}$. When the shear stress of the deposits along a certain sliding surface exceeds its yield shear stress, the slump will happen and storage capacity will recovery. During the period of physical model tests of Xiaolangdi Reservoir, when channel suffers from retrogressive erosion and water level drops, under the combined action of gravity and seepage water pressure, the unconsolidated deposits in saturation state on both bank beaches will lost stability and slump into the channel(Fig. 2).

In the process of reservoir operation, meeting proper flood process, by controlling water level, retrogressive erosion can be shaped in river reach before dam and delta continent surface. By scouring of density current deposition reach before dam and erosion recession of delta, the storage capacity downstream the delta vertex can be recovered. At the same times, in the process of moving toward dam, the sediment scoured from delta will sort once again, the finer sand will be discharged from reservoir and is easy to transport in the Lower Yellow River. Therefore, during the later period of sediment retention, the delta deposition morphology(Fig. 3(a)), not cone deposition morphology (Fig. 3(b)), is kept as long as possible.

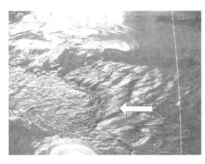

Fig. 2　Channel retrogressive erosion and beach slump in physical model tests of Xiaolangdi Reservoir

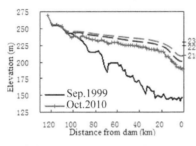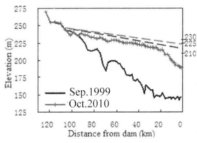

（a）Delta deposition morphology　　　（b）Cone deposition morphology

Fig. 3　Schematic diagram of deposition surface uplift of Xiaolangdi Reservoir

4　Conclusions

By October 2010, the delta vertex had moved to cross section HH12 which is 18.75 km from dam, the delta vertex was at elevation of 215.61 m, and the amount of sediment reached 2.822,5 × 10⁹ m³. The research show that, in the same condition of amount of sediment and water storage, the backwater length of delta deposition morphology is shorter than cone deposition morphology and the sediment ejection effect of density current is superior to backwater free flow. In the process of reservoir operation at present, we suggest carrying out sediment ejection timely and trying to delay transformation from delta deposition morphology to cone deposition morphology, so as to reduce reservoir sedimentation effectively, adjust bed material composition, optimize outlet water and sediment process, at the same time, enhance operation flexibility and regulation capacity of Xiaolangdi Reservoir.

Acknowledgements

The work was supported by public welfare industry research special fund of the Ministry of Water Resources (MWR) (No. 200901015, No. 200801024), the National Natural Science Foundation of China(NSFC) (NO. 51179072) and central – level nonprofit research institute fund (HKY – JBYW – 2012 – 14)

References

Wang Ting, Chen Shukui, Ma Huaibao, et. al. Distribution of Deposition in Xiaolangdi Reservoir [J]. Sediment Research, 2011(5).

Wang Ting, Ma Huaibao. Characteristics of Water and Sediment and Scour and Silting Evolution of

Xiaolangdi Reservoir in Early Sediment Retaining Period [R]. Zhengzhou: Yellow River Institute of Hydraulic Research, YRCC, 2011.

Zhang Junhua, Chen Shukui, Li Shuxia, et al. Sediment Research of Xiaolangdi Reservoir in Early Sediment Retaining Period[M]. Zhengzhou: Yellow River Conservancy Press 2007.

Li Shuxia, Zhang Junhua, Chen Shukui, et. al. Formation of Density Current by Means of Reasonable Operation of Reservoir in Xiaolangdi Project [J]. Journal of Hydraulic Engineering,2006,37(5):567 – 572.

Li Shuxia, Zhang Junhua, Chen Shukui, et al. Study on Density Current in Xiaolangdi Reservoir [C]//The Ninth International Symposium on River Sedimentation. Beijing: Tsinghua University Press, 2004.

Han Qiwei. Reservoir Sedimentation[M]. Beijing: Science Press, 2003.

Lin Xiushan, LI Jingzong. The Engineering Planning of the Planning and Design Series of Xiaolangdi project of the Yellow River [M]. Zhengzhou: Yellow River Conservancy Press, 2006.

Li Tao, Zhang Junhua, Chen Shukui, et al. Special test on Precipitation Washout Operation of Xiaolangdi Reservoir [R]. Zhengzhou: Yellow River Institute of Hydraulic Research, YRCC,2008.

The Foundation Treatment Research of Hekoucun CFRD

Xing Jianying[1], *Guo Qifeng*[1], *Yan Shi*[2] and *Zhu Qingshuai*[1]

1. Yellow River Engineering Consulting Co. , Ltd. , Zhengzhou, 450003, China
2. Hekoucun Reservoir Project Construction Bureau of Henan Province, Jiyuan, 454650, China

Abstract: The plinth of CFRD generally rests on a firm, non – erodible rock surface which can be grouted in a normal way, however, due to the limitation of topographic and geological condition, when plinth rests on overburden, the problem of deformation resulted from overburden's low Young Modulus could be a key problem for the project is whether success or not. This article, based on the practical situation of Hekoucun Project, analyses the relevant treatment method for the thick overburden and applies high pressure spray piles to solve the deformation problem in the key area. It is proved from inspecting data obtained from the treated foundation that the average Young Modulus has been enhanced greatly, and the result of 3D finite element analysis shows the aim of reducing the dam deformation and lessening the joint displacement has been achieved. This treatment method for thick overburden not only solves the displacement problem efficiently, but also avoids the mass excavation, reduces investment, shortens constructing time. Practice and theory proves it is the beneficial quest and try for ecological treatment of deep and thick alluvial deposit and the sustainable development in the basin.

Key words: CFRD, plinth, foundation treatment, high pressure spray pile

1 Introduction

Hekoucun Project lies in the lower reaches of the Qinhe River which is the first level branch of Yellow River. It controls 9,223 km² drainage area which accounts for 68.2% of the total catchment area of the Qinhe River, and accounts for 34% of the non – project section between Xiaolangdi and Huayuankou. The development objective of this project is mainly flood control and water supply, with considerations of irrigation, electricity generation, and eco – environment improvement. This project is a large scale 2, consisting of dam, discharge tunnel, spillway and diversion power generation system. Its design flood standard is at 500 year frequency, and 2,000 years frequency for check flood standard. Its normal water level is 275.00 m, design flood level (check flood level) is 285.43 m, with total capacity of 3.17×10^8 m³.

The dam is CFRD with its riverbed plinth resting on thick overburden and other plinth on firm rock, the maximum dam height is 122.5 m, the upstream slope is 1:1.5, and downstream comprehensive slope is 1:1.68 with access road on it. Concrete cutoff wall was designed as the anti – seepage measure for riverbed overburden, 4 – m – width linking plate for connecting the concrete cutoff wall with the riverbed plinth.

2 Research objective

It is recommended in Chinese current standards that plinth shall rest on firm, non – erodible rock surface which can be grouted in a normal way, however, due to the limitation of geological condition, a great quantity of excavation usually cannot be avoided to meet these requirements, which not only prolong the construction period, increase the project costs, but also cause some environmental problems, which goes against sustainable development of the basin where the project located, so it is vitally necessary to seek a suitable method for foundation treatment and study it.

The depth of the overburden of the dam foundation is up to 41.87 m, in addition, the foundation includes several clay interlayers and sand lens bodies which are with different continuity. Though the method of the plinth resting on the rock has comparatively mature construction technique and has more engineering practices, the mass excavation will result in a series problems such as higher cofferdam, longer discharge tunnel and etc. The result of comprehensive analysis shows that the scheme of plinth resting on the rock has 10% more investment than that on overburden, and prolongs the construction time.

The method of plinth resting on the overburden has been proved to be feasible by several projects in China, the foundation of which all meet the requirements after treatment. The foundation of Hekoucun Dam is deeper and more complicated than others, so it must be treated carefully before the plinth constructed. In order to seek the method which is economical, effective and has less influence on construction period, this research has been carried on.

3 Research method

This research mainly incorporate with the geological conditions and treatment methods of Hekoucun Dam foundation.

3.1 Foundation geological conditions

The foundation overburden mainly is fluvial and alluvial layer. The general depth is about 30 m, and maximum depth is up to 41.87 m. The lithology is sand – gravel layer including some boulder and soil, with 4 unwell continuous clay interlayers and 19 sand lens bodies.

3.1.1 Composition of the riverbed overburden

3.1.1.1 Sand – gravel layer

The sand – gravel layer including boulder is divided into three parts, the gravel's lithology is mainly dolomite and limestone with medium corrosion round, poor sorting. Its average dry density is 2.05 g/cm^3, porosity ratio is 0.327, specific gravity is 2.72. Longitudinal wave velocity is between 1,020 m/s and 1,460 m/s, shear wave velocity is between 298 m/s and 766 m/s, dynamic Poisson's ratio is between 0.43 ~ 0.46, dynamic modulus is 420 ~ 1,220 MPa, shear modulus is 160 ~ 1,100 MPa.

The permeability coefficient of the sand – gravel layer is 1 ~ 106 m/d, most of them between 40 and 60 m/d, the general rule is that the permeability of the middle is larger than that of the both sides.

3.1.1.2 Clay interlayer

According to analysis from quantity of drilling data of the riverbed, there are 4 clay interlayer with different continuity in the foundation overburden, the top surface elevation of the 4 layers are 173 m, 162 m, 152 m and 148 m respectively, the general thickness is 0.5 ~ 6.6 m. The length of the layers along the river is 350 ~ 800 m, and is a critical element for the stability and deformation of the foundation. The lithology is yellowish gray, heavy silty loam, has a well natural consolidation.

3.1.1.3 Sand lens body

The sand lens body is generally distributed at the convex bank, which is about 30 ~ 60 m in length, 10 ~ 20 m in width, 0.2 ~ 5 m in thickness. Most of them are lie in the upper 15m. Its lithology is silty – fine sand, well graded and good compaction.

3.1.2 Engineering geological problems of deep overburden

3.1.2.1 The anti – slide stability of clay interlayer

According to the test data, the internal friction angle of clay interlayer is 19° ~ 23°, however, that of sand lens body and sand gravel layer is up to 27.5° ~ 39.5°, therefore, the clay interlayer controls the stability of foundation.

3.1.2.2　Uneven settlement

By the relevant geological data, the accumulative thickness of clay interlayers is up to 5 ~ 20 m, which accounts for nearly 1/6 ~ 1/2 of the total overburden, and its compression coefficient is 0.1 ~ 0.2 MPa^{-1}, while the compression coefficient is 0.01 ~ 0.06 MPa^{-1} for sand gravel layer.

The three materials mentioned above have different compaction in addition to the uneven spatial distribution. The uneven settlement is inevitable when dam constructed on this foundation.

3.2　The design of dam foundation treatment

3.2.1　Anti – sliding stability analysis

To meet the requirements of the design standard, 2D stability calculation have been carried out to determine the foundation stability, such as simplified Bishop Method and Morgenstern method. The computation result shows that the design slope meets the stability requirement in all conditions.

3.2.2　Finite element analysis of dam

In order to control the dam deformation in construction stage and operation stage, and to assure the dam safety, 3D finite element analysis has been conduct using the geological data of foundation to reveal the deformation characteristic of dam body and foundation, especially for anti – seepage system. The results of it are as follows:

(1) The settlement at the completion time and the operation time is 93 cm and 100 cm respectively, and upstream displacement is 14 cm and – 1 cm respectively, downstream displacement is 25 cm and 32 cm respectively. The maximum major principal stress of dam body and foundation at the two stages are 2.671 MPa and 1.978 MPa respectively, and the representative value of the stress level is 0.5.

(2) The axial displacement of face slabs is toward the valley, the axial displacement of face slabs on the both river sides are 5.9 cm (2.93 cm) and 4.6 cm (2.09 cm) during operating period (construction period) respectively. Due to the riverbed plinth rests on the overburden, the maximum deflection of the face slab bottom are bigger than that of the plinth resting on rock, the value is up to 36.2 cm(18.1 cm), and 39.6 cm(14.6 cm) for the middle of the face slabs (about 2/3 height of the dam).

(3) Whether in construction period or operating period, the deformation of all kinds of joints are bigger than other projects due to the overburden. In operating period, the maximum displacement of the top of cutoff wall is 20.9 cm toward downstream, and for linking plate and plinth they are 32.64 cm and 18.78 cm toward downstream separately. The relative deformation of impervious system are all small except for the relative settlement between linking plate and cutoff wall, and the value of it is up to 5.22 cm.

From the result of the finite element analysis we can see, the settlement displacement of dam body and deflection of face slab has little differences with other projects. However, the relative settlement between linking plate and plinth is slightly larger than the present capacity of the joint seal. For the safety of the project, some strengthening measure for the foundation should be taken.

3.2.3　Treatment method for the dam foundation

3.2.3.1　Principle of the treatment

Based on the result of calculation and analysis, the aim of reducing the foundation settlement and enhancing the capacity of resistance to deformation, the dam foundation need to be strengthened. The criterias for strengthen are about the maximum settlement, displacement value and stress. To be specified, the settlement of dam should be less than 1% of the dam height, the displacement of all kinds of shear, opening and settlement should be less than 3 cm, and stress should be less than the allowable stress or can be solved by reinforcement.

3.2.3.2　Comparison of treatment schemes

When the foundation needs to be treated due to its comparative soft, larger deformation or uneven displacement, there are several following treatment methods at present according to the composition and physic – mechanical properties of the overburden:

(1)Substitution.

Substitution method is a treatment method that replaces the soft part that cannot meet the requirement in the foundation with the well graded soil or rock, and then compact it. It is the best method for small shallow foundation. For the foundation of Hekoucun Dam, it will be just like the excavation scheme and will lead to more investment and longer construction period, so it is not suitable for this project.

(2)Dynamic compaction.

Dynamic compaction method is a method that is used to increase the density of soil deposits. The process involves of dropping a heavy weight repeatedly on the ground at regularly spaced intervals. This method is suitable for old fills and granular soils. For the foundation of Hekoucun Dam, because its foundation is sand gravel layer containing some boulders and the limited influence depth of the method, it therefore is not suitable for this project too.

(3)High pressure spray.

It is the treatment method that suit for loose permeable foundation, like soil, sand soil, gravel, and pebble foundation. The mission of the treatment for Hekoucun Dam foundation is enhancing the entirely modulus, reducing the deformation of dam body and anti – seepage system due to the soft foundation. Fluid and cement flow with high pressure is injected into foundation through grout pipe system, in this process, cement flow fills the foundation pore and consolidates, the integrity of the foundation is enhanced, therefore, it is the most suitable method for the Hekoucun Dam foundation.

3.2.3.3　Treatment design

According to the relevant standards, the foundation that is within the range of 0.3 ~ 0.5 dam height from the plinth to downstream should has low – compression. The dam height of Hekoucun Project is 122.5 m, based on the 3D finite element analysis result, the foundation between cutoff wall and plinth is the key area that controls the displacement of all kinds of joints.

Through analysis and comparison, high pressure spray piles have been applied to treat the 50 m area from the cutoff wall. 5 rows of piles with row spacing being 2 m have been arranged near the cutoff wall, to meet the requirement of deformation transition, the spacing is adding to 2.0 m, 3.0 m, 4.0 m, 5.0 m, 6.0 m toward downstream gradually, the length of all piles is 20 m (or to rock), pile diameter is 1.2 m, about 630 piles are arranged in the treatment area.

The technique requirements of high pressure piles are: Portland cement or ordinary Portland cement, strength grade is 42.5 MPa. The diameter of piles should be larger than 1.2 m, and 28 d compressive strength should be larger than 3.0 MPa.

3.3　Test and construction of the high pressure spray piles

To confirm the adaptability and effect of the high pressure spray piles in Hekoucun Dam foundation, test has been conduct in this area according to relevant standard. Test area has been selected in the strengthen area, includes all kinds of strata configuration which will be encountered in the construction, the test area is about 520 m². The number of test piles is equal to 8% of the total number, 50 piles in all, the interval between piles is same to the arrangement of the design. The aim of the test is to check the rationality of grouting parameter and the pile effect in these strata.

Before and after the test, static test, pressure meter test and cross holes wave velocity test have been conducted to the natural foundation and treated foundation. Due to the limitation of different spacing of piles, the result of static test might be not fully reasonable. On the basis of the test result, the following conclusion has been drawn.

(1)Bearing capacity of foundation: before the treatment, the value is generally 500 ~ 600 kPa at the elevation 175 m, and the value at the same place is about 990 ~ 1,100 kPa after treatment, increased nearly 200%.

(2) Young Modulus of foundation: the value is 40MPa and 46. 1 ~ 154. 1 MPa respectively, correspond to before and after treatment, increased 15% ~ 300%.

(3) The result of different interval: the enhanced effect of spacing 2 m × 2 m is more distinct than the others, up to 200% ~ 300% increase.

From the result of grouting test, the Young Modulus of treated foundation is all enhanced distinctly. From the result of excavating test, the piling result is good, the diameter all meet the design requirement. Therefore, according to the adjusted grouting parameter, piles have been constructed as the design requirements. For part of them that cannot continue due to the boulder encountered, different measures were taken according to its position and importance.

4   Research results

It is proved that all the physical and mechanics parameters have variously increased in composite foundation by multi – test, especially for bearing capacity and Young Modulus. Overall, the high pressure spray piles reduce the uneven characteristic of the riverbed and enhance the bearing capacity and Young Modulus. So the project reaches the designers' aim.

According to the result of test, the calculation parameters of 3D finite element have been adjusted, and then recomputation has been performed. The results show the shear displacement between plinth and linking plate decreased by 50%, the uneven settlement between linking plate and cutoff wall decreased by about 30%, the opening displacement between linking plate and cutoff wall decreased by about 60%, than before treatment. All these proved that the treat measurement is appropriate and successful.

The following experiences have been gathered from the practice of Hekoucun Project and can be refer for other projects.

(1) When grouting in the sand – gravel layer, due to it has large permeability in addition to ground water influence, some problems will be encountered, like too much leakage, cement flow don't return at orifice and pile shape is not regular, the measures such as dewatering and reduce the flow should be taken during being constructed.

(2) Engineering geological property should be studied before construction when this measure will be taken, and the grouting parameters, construction working procedure and holes spacing should be adjusted during construction while incorporating with the specific geological condition, to achieve the aim of economic and efficient.

(3) It is necessary tot perform a field trial before large – scale construction.

(4) Appropriate ways and methods should be selected to check the treat effect according to the engineering geological property of overburden. Some detection means recommended in the specification might not be suitable for all the geological conditions, the effect should be determined by multiply means, mainly relying on static test, wave velocity test and pressure meter test, supplemented by excavation test. Only one or two methods may be not reflected the reality exactly.

5   Research conclusions

At present, high pressure spray piles usually service as an anti – seepage measures in hydraulic projects, combined with the composition of Hekoucun Dam foundation and technical and economic comparison, it is the first time to be serviced as a strengthen measure in CFRD foundation treatment. According to the implementation situation and detection result, combined with the result of 3D finite element analysis, this measure for thick overburden not only can solve the displacement problem efficiently, but also avoids the mass excavation. Practice and theory proves it is the beneficial quest and try for ecological treatment of deep and thick overburden and the sustainable development in the basin.

362

References

SL 228—98 Design code for Concrete Face Rockfill Dams [S]. Beijing:China WaterPower Press, 1999.

DL/T 5016—2011 Design Specification for Concrete Face Rockfill Dams [S]. Beijing: China Electric Power Press,2011.

DL/T 5200—2004 Technical Specification of Jet Grouting for Hrdropower and Water Resources Projects[S]. Beijing:China Electric Power Press,2005.

Jiang Suyang,Guo Qifeng. Design and Foundation Treatment of CFRD on Deep Overburden [M]. Beijing:China WaterPower Press,2011.

Industrial CO_2 Emission Structure and Evolution Mechanisms in Zhengzhou—Kaifeng Metropolitan Area

Zhang Lijun[1], *Qin Yaochen*[1], *Zhang Jinping*[1] and *Lu Chaojun*[1,2]

1. Key Research Institute of Yellow River Civilization and Sustainable Development/College of Environment and Planning, Henan University, Kaifeng, 475004, China
2. Department of Energy and Environment, Chinese Research Academy of Environmental Sciences, Beijing, 100012, China

Abstract: Sources of CO_2 emissions from metropolitan areas are divided into six categories, consisting of agricultural, industrial, commercial and residential energy consumption, as well as construction and transportation energy consumption. Industrial CO_2 emissions are further subdivided into 34 kinds of industrial energy consumption. Based on the 2000 ~ 2009 energy consumption data, a CO_2 estimation model to account for CO_2 emissions in scope 1 (only refer to energy combustion) and scope 2 was constructed in this paper. Evolution mechanisms of CO_2 emissions of 2000 and 2009 of six units and 34 kinds of industries were also researched by using LMDI methods. The conclusions can be drawn as follows: ① the changes of CO_2 emissions show different sectoral and regional characteristics in various stages. ② The changes of CO_2 emissions took place in the interaction of factors including intensity (technical) effects, structural effects and scale effects of economies. The positive driving factors for CO_2 emissions increase are the exogenous economic increasing pattern of economic scale and element input; the negative driving factors are intensity of carbon emission and density of work labour. T ehe labour efficiency and the variation of industrial structure affect the CO_2 emission bidirectionally.

Key words: low-carbon city and economy, industrial carbon emissions, LMDI analysis, Zhengzhou-Kaifeng metropolitan area

1 Introduction

The conflict between the great demand of energy in the process of rapid urbanization and industrialization and the CO_2-abate pressure under the condition of global warming has become one of the most prominent foci in man-land territorial system. In essence, urbanization and industrialization's impact on environment changed largely through transition of economic development patterns and adjustment of industrial structure.

To explore the relationship between economic development and environmental damage, industrial CO_2 emission and evolution mechanisms in cities has become academic focus. Various approaches have been used to explore the interaction mechanisms between economic development and CO_2 emission and regional differences from some respects, such as economy and population size, energy intensity and structure, industrial structure and emission factors, foreign investment and total factor productivity, economic development mode, etc. The methods mainly include LMDI analysis, IO analysis, Granger analysis and Malmquist index based on DEA approach.

The role of industrial structure received extensive attention among different perspectives, and was considered as an effective way to achieve low carbon economic development. However, many studies have found that the contribution of industrial structure to carbon emissions change was not large, and only relying on economic structural adjustment to reduce CO_2 emissions for China was not a valid political choice.

There seemed to be contradictions between the theory and the empirical evidence. Thus, some scholars concerned about the importance of economic development patterns' impact on carbon emissions from perspectives of exports, consumption and investment. However, those were laws in national level, applied defectively in local scale. Douglas production function characterizes the different types of economic development from micro perspectives, such as exogenous capital-intensive and labor-intensive economic development, endogenous technological progress, to some extent compensating for this deficiency.

Zhengzhou and Kaifeng are the new and the old core cities in urban systems of China's Central Plains respectively. The integrated metropolitan area formed by two cities' space docking, economic and social sectors convergence, will be a strong growth pole to lead the development of Central Plains Economic Region. Thus the urbanization and industrialization development has special needs for low-carbon city for Zhengzhou and Kaifeng.

The article built CO_2 emissions calculation model, assumed constant returns to scale, and considered the impact of different economic development patterns and industrial structure on the driving mechanism in CO_2 emissions change, which not only were important to clarify the relationship between economic development patterns and industrial structure adjustment and enrich low-carbon development theory, but also provided practical guidance for the development of low-carbon industry in Zhengzhou and Kaifeng metropolitan area.

2 Study area and Methodology

2.1 Study area

Nutrition and energy structures in urban centers with "inverted pyramid" style determine the low-carbon eco-city development should be supported by the areas in which they are. The urban sprawl area (built-up area and expansion area) covered by urban carbon cycle system is not always adjacent to urban carbon footprint area. Maybe it is located hundreds of kilometers away.

The study area includes not only the core area of urban district, also the close area of contact layer in Zhengzhou and Kaifeng, which is integrated counties (cities) with closely socio-economic links, covering two urban districts under jurisdiction of Zhengzhou City and Kaifeng City, Xinzheng City, Xinmi City, Gongyi City, Xingyang City, Dengfeng City, Zhongmou County, Weishi County, Tongxu County, Lankao County, Qi County and Kaifeng County.

2.2 Variables and data

This study mainly analyzed CO_2 emission changes referring scope 1 (only fuel combustion) and scope 2 in Zhengzhou and Kaifeng metropolitan area from 2000 to 2009. The sources of CO_2 emissions in metropolitan area were divided into energy consumption of 6 sectors, including agriculture, industry, construction, transportation (i.e., transportation storage and communications sector), commerce (combination of wholesale, retail and accommodation, catering and other industries) and consumer consumption.

The continuing growth of industrial energy consumption was the main factors to drive CO_2 emission growth. Hence, CO_2 emissions structure of 37 kinds of industries was further detailed analyzed. In order to avoid double counting, production and supply of electricity, heat industry, petroleum processing and coking industry and gas production and supply industry were removed from the total industrial CO_2 emissions accounting.

The main variables were as follows: energy consumption from 6 kinds of sectors within 34 types of industries, 19 kinds of emission factors of energy, population amount, household income levels, as well as increment, number of employees and fixed assets of various industries in Zhengzhou and Kaifeng metropolitan area.

Part of energy consumption data was derived from estimation, emission factors were calculated by the average low calorific value and carbon content, and other data were all from Statistical Yearbook in Henan, Zhengzhou and Kaifeng. To eliminate the effect of price changes, added value, fixed assets and household income were expressed on 2000 comparable prices.

2.3 Methodology

CO$_2$ emissions are estimated as follows.

$$C = C_f + C_{eh} \tag{1}$$

$$C_f = 44/12 \times 1,000 \sum_m E_{fm} \times NCV_{fm} \times F_{fm} \tag{2}$$

$$C_{eh} = E_e \times F_e + E_h \times F_h \tag{3}$$

where, C is total CO$_2$ emissions, 10^4 t; C_f is fossil-fuels-related CO$_2$ emissions, 10^4 t; C_{eh} is electricity and heat-related CO$_2$ emissions, 10^4 t; E_{fm} is the No. m (m = 1, 2, $\cdots$, 17) kind of fossil energy consumption, 10^4 t; NCV_{fm} is the average low calorific value for the No. m kind of fossil energy, J/g; F_{fm} is the carbon content for the No. m kind of fossil energy, kg/GJ; E_e and E_h are electric power consumption and heat consumption, 10^4 t respectively; and F_e and F_h are CO$_2$ emission coefficients of electricity and heat respectively.

Drawing on diversified industrial structure coefficient, diversified coefficient of CO$_2$ emissions (*CSD*) is built up to explain the status and trends in CO$_2$ emissions' evolution. The basic formula is expressed as follows.

$$CSD = \sum (C_1/C_1, C_2/C_1, C_3/C_1, C_4/C_1, C_5/C_1, C_6/C_1) \tag{4}$$

where, C_1, C_2, $\cdots$, C_6 respectively represents CO$_2$ emissions from 6 sectors of agriculture, industry, construction, transportation, commerce and residential consumer; The higher *CSD* means the greater dependence on carbon-based energy.

Zhengzhou—Kaifeng metropolitan area consumed much coal, accounting for 80%, which is dertermined by the resource endowments in Henan Province. Thus, this article did not consider the impact of the energy structure. 6 sectors were further divided into production sectors and consumption sectors, of which the latter referred to residential sector, and the former were other five sectors. Total CO$_2$ emissions in six sectors can also be expressed as follows.

$$C = \sum_n C_n/V_n \times V_n/V_T \times V_T/P \times P + C_6/THI \times AHI \times HN = \sum_n I_n \times S_n \times R \times P + HI \times AHI \times HN \tag{5}$$

where, C_n is CO$_2$ emissions of the nth (n = 1,2, $\cdots$, 5) sector (10^6 t); V_n is added value (10^8 yuan) of the nth sector; V_T is total increase value of the nth sector, 10^8 yuan; P is total population; 10^4; C_6 is CO$_2$ emissions of consumer; *THI* is total household income, 10^4 yuan; *AHI* is average annual family income (yuan per household); and *HN* is the number of households. Production sectors focus on the impact of carbon emission intensity (I_n), industrial structure (S_n), economic level (R) and population (P) on CO$_2$ emissions, while residential sector highlights household carbon emissions intensity (*HI*), household income (*AHI*), the number of household (*HN*). Eq. (1) and Eq. (5) separately represent the bottom-up and top-down CO$_2$ emissions accounting methods and the results are consistent.

The industrial sector has an important role in CO$_2$ emissions. Industrial development is closely related to factor inputs. The paper explored the influence on CO$_2$ emissions change of intensity (technology) effects, scale effects, structural effects and economic development patterns. Total CO$_2$ emissions (C_i) from 34 kinds of industries can be expressed as follows from the perspective of production structure.

$$C_i = \sum_k C_k/V_k \times V_k/V \times V/L \times L/K \times K = \sum_k I_k \times S_k \times LP \times LI \times K \tag{6}$$

In addition, C_i can also be expressed in form of employment structure. That is:

$$C_i' = \sum_k C_k'/V_k' \times V_k'/L_k' \times L_k'/L \times L'/K' \times K' = \sum_k I_k' \times LP_k' \times S_k' \times LI' \times K' \tag{7}$$

where, C_k is CO$_2$ emissions of the kth (k = 1, 2$\cdots$, 34) industry, 10^4 t; V_k is added value of the kth industry, 10^8 yuan; V is total increase value of industrial sectors, 10^8 yuan; L is the number of employees in industrial sectors, and K is gross fixed assets in industrial sectors, 10^8 yuan. The meanings between I and S are the same with that are mentioned above. *LP* represents labor productivity and *LI* represents labor intensive.

The CO_2 emissions variation of six sectors ΔC can be expressed as follows based on LMDI method:

$$\Delta C = C^t - C^0 = \Delta I + \Delta S + \Delta R + \Delta P + \Delta HI + \Delta AHI + \Delta HN \qquad (8)$$

The CO_2 emissions variation of industrial sectors ΔC_i can be expressed as follows:

$$\Delta C_i = C_i{}^t - C_i{}^0 = \Delta iI + \Delta iS + \Delta iLP + \Delta iLI + \Delta iK \qquad (9)$$

3 Results and analysis

3.1 Evolution of CO_2 emissions structure

3.1.1 Evolution of CO_2 emissions structure for 6 sectors

Evolution of CO_2 emissions structure gradually accelerated in Zhengzhou-Kaifeng metropolitan area from 2000 to 2009, demonstrating significant regional differences (Fig. 1). Diversified coefficient of CO_2 emission structure firstly increased and then decreased with an overall upward trend in Zhengzhou-Kaifeng metropolitan area. The evolution level and speed of CO_2 emissions structure of Zhengzhou is much higher than that of Kaifeng.

Most CO_2 was emitted from industries in 6 sectors with the highest share from 81.3% to 86.5% of total emissions. Household consumption-related CO_2 emissions were the second, with only a small share from 6.2% to 7.8%. Commercial CO_2 emissions ranked the third with rapid growth. Construction and transport industry discharged relatively small amount of CO_2 emissions in 2000, and became the two fastest growing sectors after 2004 with the average annual growth rate of 16.5% and 11.5% respectively. Agricultural CO_2 emissions accounted the minimum with the slowest rate of change.

As for Zhengzhou and Kaifeng, the share of industrial CO_2 emissions of Zhengzhou was higher than that of Kaifeng, but the increment amounts and speed of Kaifeng were greater. Household consumption-related CO_2 emissions were with much smaller proportion in Zhengzhou than that of Kaifeng, and the share of Zhengzhou showed a rising trend. The CO_2 emissions amount and speed of construction, transport and commerce of Zhengzhou were much greater than that of Kaifeng.

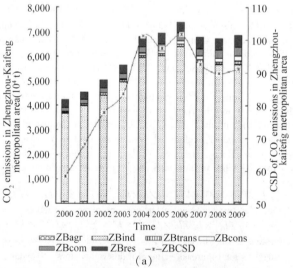

(a)

Fig. 1 CO_2 emissions structure of six departments

in Zhengzhou—Kaifeng metropolitan area

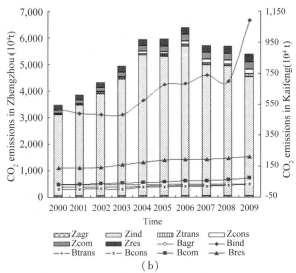

(b)

Note: ZB represents Zhengzhou—Kaifeng metropolitan area,
Z represents Zhengzhou, and B represents Kaifeng.
Continued to Fig. 1

3.1.2 Evolution of CO_2 emissions structure for industries

The cumulative contribution rate of first 9 industries was more than 80%, according to sequence of the proportion of various industrial CO_2 emissions in Zhengzhou – Kaifeng Metropolitan area and Zhengzhou and Kaifeng from 2000 to 2009. The combination relations among 9 industries of high-CO_2 emissions, to some extent, reflected CO_2 emissions structure in industrial sectors. (Fig. 2).

Industrial CO_2 emissions showed the following features in Zhengzhou – Kaifeng metropolitan area (Fig. 2(a)). First of all, there existed some difference in the rate and the trend of industrial CO_2 emissions change. CO_2 emissions substantially declined in non-ferrous metal smelting industry, coal mining industry and chemical raw materials and chemicals manufacturing industry, while increased a little in non-metallic mineral smelting industry. Furthermore, high-emissions industries consisted of heavy industry, chemical industry and large-scale light industry, and heavy and chemical industries accounted for a larger proportion.

Zhengzhou and Kaifeng showed their respective geographical features with different resource endowments, industrial development history and development stages (Fig. 2(b)). Firstly, primacy ratio of industrial CO_2 emissions had big difference. During 2000 to 2009 period, primate industry (chemical raw materials and chemical products industry) occupied a share between 30.7% and 60.7% and the proportion of other industries was all less than 1% in Kaifeng; while primate industry (non-ferrous metal smelting industry) was with a share between 35.5% and 39.6%, and the secondary industry (non-metallic mineral products industry) was still with a larger proportion which varied between 23.3% and 32.6%. Secondly, industry sectors were endowed with obvious regional characteristics. The major components of high-emission industries in Zhengzhou are resource-based heavy industries whilst in Kaifeng, they are chemical and light industries.

3.2 Evolution Mechanism of CO_2 emissions change

3.2.1 Evolution mechanism of CO_2 emissions change for 6 sectors

A series of driving factors influencing CO_2 emissions were explored from perspectives of

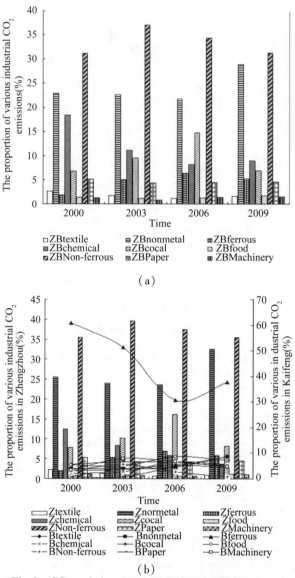

Fig. 2　CO$_2$ emissions structure of major industries in
Zhengzhou – Kaifeng metropolitan area

production and consumption of socio-economic activities. It was concluded that the change of CO$_2$ emissions took place in the interaction of factors including intensity effects, structural effects and scale effects of economies (Fig. 3(a)).

Scale effects of economies significantly accelerated CO$_2$ emissions with a contribution of 49. 1% , while carbon emissions intensity effectively restrained CO$_2$ emissions with a contribution of 32. 8%. Industrial structure adjustment played a bidirectional role, resulting in an increment of CO$_2$ emissions during 2002 to 2008 and a decrease soon after. Continuing increment in population size caused the rise of CO$_2$ emissions, but it was only a smaller contribution. Residential carbon

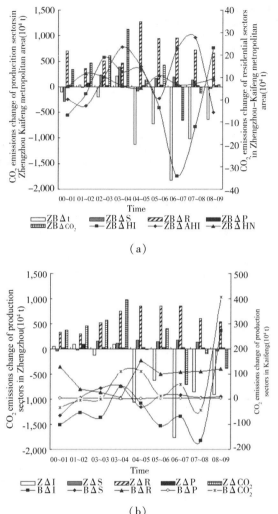

**Fig. 3 Driving forces for CO$_2$ emissions change in
Zhengzhou—Kaifeng metropolitan area**

emissions intensity, average annual family income and the number of households had smaller impact on CO$_2$ emissions change with the contribution of less than 1%.

Overall, the production sector's contribution to CO$_2$ emissions change was much larger than the consumer sector's, but the gap was gradually narrowing. CO$_2$ emissions from residential consumptions had similar trends in Zhengzhou and Kaifeng, synchronously with the change in Zhengzhou – Kaifeng metropolitan area.

However, CO$_2$ emissions change, change trends and evolution mechanism were different (Fig. 3(b)). The value of CO$_2$ emissions variation in Zhengzhou was higher compared to that in Kaifeng with larger variation range. The intensity and the patterns of driving factors were distinctly different from each other in Zhengzhou and Kaifeng. Some factors exerted a more tremendous influence on Kaifeng, such as carbon intensity of industries and residents, industrial structure, as well as the family income level and the change in the number of families. The variation of economics of scale effects exerted a more tremendous influence in Zhengzhou with a contribution to 51.1%, compared to Kaifeng with a contribution to 36.1%. The influence of variation of carbon intensity on CO_2 emissions change firstly increased and then decreased in Zhengzhou. on the contrary, the change direction was just on the opposite in Kaifeng with a substantial increase in CO_2 emissions especially after 2008.

3.2.2 Evolution mechanism of CO_2 emissions change for industries

Industrial CO_2 change mechanism from 2000 to 2009 was explored triennially based on some effects including carbon emissions intensity, labor productivity, interactions between factor inputs and lagged effects of fixed capital (Fig. 4). Meanwhile, added value and employment were both used to illustrate industrial structure but its calculation was basically the same (Fig. 4(a)). Production structure was chosen to analyse evolution mechanism of CO_2 emissions change for various industries in this article.

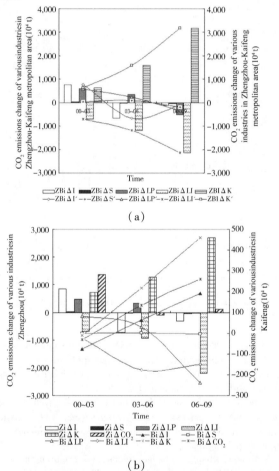

(a)

(b)

Fig. 4 Driving forces for industrial CO_2 emissions change in Zhengzhou—Kaifeng metropolitan area

The changes of industrial CO_2 emissions took place in the interaction of factors including technical effects, structural effects and scale effects of economies. Technological progress, including energy production and using technology, caused a decrease in carbon emission intensity and an increment in labor productivity, dramatically restraining the increase of CO_2 emissions. Exogenous pattern of economic development by inputting factors such as the labor intensity reduction and the capital inputs increase respectively became important factors to suppress and promote the change of CO_2 emissions. Among them, average contribution of carbon emission intensity and labor intensity were respectively -30.9% and -15.8%. Labor productivity and industrial structure promoted CO_2 emissions to change from positive to negative, but the contribution of industrial structure was not significant, and the average contribution was about 0.8%. Capital investment with an average contribution of 39.2% had much greater influence.

The patterns and the intensity of driving factors for CO_2 emissions change had differences between Zhengzhou and Kaifeng (Fig. 4(b)). Zhengzhou was in the transition from the mid- to late industrialization when capital investment's positive effect and labor-intensity's negative effect made a great difference due to heavy industrialization. Kaifeng was in the transition to mid-industrialization with solid foundation of light industry, when the gradual increase of labor productivity led to a clearly negative effect and the carbon emissions change had relatively small sensitivity to the labor intensity change.

In general, the rate of capital investment increase was relatively slow, with a downward trend in carbon intensity and labor intensity in Zhengzhou. While capital investment and its increase rate were relatively greater and labor productivity played a more prominent role in inhibiting CO_2 emissions in Kaifeng.

4 Results and analysis

(1) The change of CO_2 emissions shows different sectoral and regional characteristics in various stages due to the adjustment to CO_2 emissions structure in Zheng – Bian metropolitan area. Total CO_2 emissions increased in the beginning and then reduced, of which industry ranks the first. Residential CO_2 emissions can not be ignored. CO_2 emissions from commerce, construction and transportation are from a small starting base, but the growth rate is quite faster.

9 kinds of high-emission industries discharge about 80% CO_2, characterized in heavy chemical industry and large-scale light industry, which are key industries to mitigate CO_2 emission. There existed great difference in potential regulation and sensitivity, in consequence of great disparity in total CO_2 emissions and the level and rate of industrial emissions structure evolution.

(2) The change of CO_2 emissions took place in the interaction of factors including intensity (technical) effects, structural effects and scale effects of economies. The main reasons for CO_2 emissions increase are the rapid growth of economies of scale and the exogenous pattern of economic development by inputting factors. The residential sector has less contribution than production sectors to CO_2 emissions, while the gap is gradually narrowing.

The change of carbon emissions intensity and labor intensity plays an obvious prohibitive role and the change of labor productivity and industrial structure plays bidirectional role in CO_2 emissions. The contribution from labor productivity and industrial structure to CO_2 emissions is greater for 6 kinds of sectors compared to 34 kinds of industries.

The contribution of population size is small and the increase in capital investment is an important factor to promote the change in CO_2 emissions. Factor inputs make a great difference in the patterns and the intensity of CO_2 emissions change because of Zhengzhou and Kaifeng at different stages of economic development.

(3) Industrial structure adjustment has been an important means to reduce CO_2 emissions for academics and politicians. However, the LMDI analysis showed that the contribution of industrial structure adjustment was much smaller than the contribution of economy scale and economic development patterns. The following reasons may be explained for that.

Firstly, industrial restructuring is a long process, and has a lag effect, while economy scale and economic development patterns can quickly change CO_2 emissions. Secondly, the high-emissions industries mainly concentrate in a few industries which are often subject to regional development policy with a slow change. Finally, the change in economy scale or economic development patterns is partly caused by industrial structure adjustment reducing the contribution of industrial restructuring. Longer-period study and other methods will be used to explain the impact of industrial structure adjustment on the change of CO_2 emissions in further research.

Acknowledgements

This work is supported by the National Natural Science Foundation of China (NO. 41171438), the Major Program of National Natural Science Foundation of China (NO. 2012CB955800) and the Major Program of Humanities and Social Sciences Key Research Base of Ministry of Education of China (NO. 10JJDZONGHE 015).

References

Liu W D, Zhang L, Wang L M. A Sketch Map of Low-carbon Economic Development in China [J]. Geographical Research, 2010, 29(5): 778 –788.

Liu L C, Fan Y, Wu G, et al. Using LMDI Method to Analyze the Change of China's Industrial CO_2 Emissions From Final Fuel Use: an Empirical Analysis[J]. Energy Policy, 2007, 35 (11):5892 –5900.

Tarancón MorAn M A, del Río Gonzáleza P. A Combined Input – output and Sensitivity Analysis Approach to Analyse Sector Linkages and CO_2 Emissions[J]. Energy Economics, 2007, 29 (3): 578 –597.

Pao H T, Tsai C M. Multivariate Granger Causality between CO_2 Emissions, Energy Consumption, FDI (Foreign Direct Investment) and GDP (Gross Domestic Product): Evidence from a Panel of BRIC (Brazil, Russian Federation, India, and China) countries[J]. Energy, 2011, 36 (1): 685 –693.

Zhou P, Ang B W, Han J Y. Total Factor Carbon Emission Performance: A Malmquist Index Analysis[J]. Energy Economics, 2010, 32(1): 194 –201.

Zhang Y G. Economic Development Pattern Change Impact on China's Carbon Intensity[J]. Economic Research Journal, 2010 (4): 120 –133.

Tunc G I, Türüt – Asik S, Akbostanci E. A Decomposition Analysis of CO_2 Emissions From Energy use: Turkish case[J]. Energy Policy, 2009, 37(11): 4689 –4699.

Wei B Y, Fang X Q, Wang Y, et al. Estimation of Carbon Emissions Embodied in Export in the View of Final Demands for China[J]. Scientia Geographica Sinica, 2009, 29(5): 634 –640.

Folke C, Jansson A, Larsson J, et al. Ecosystem Appropriation by Cities[J]. Ambio, 1997, 26 (3): 167 –172.

Wang G X, Cai J M. Research on the Method of Quantifying the Spatial Boundary of Metropolitan Regions[J]. Economic Geography, 2008, 28(2): 191 –195.

IPCC/OECD. 2006 IPCC Guidelines for National Greenhouse Gas Inventories//Eggleston H S, Buendia L, Miwa K, et al. Prepared by the National Greenhouse Gas Inventories Programme. Japan, IGES. 2006.

National Reform and Development Commission. A Notice on the Determination of China's Regional Grid Baseline Emission Factor. Clean Development Mechanism in China. 2008 – 07 – 18. http://cdm. ccchina. gov. cn/ web/ NewsInfo. asp? NewsId =3239.

Ang B W. The LMDI Approach to Decomposition Analysis: a Practical Guide[J]. Energy Policy, 2005, 33(7): 867 –871.

Study on Features of Vertical Distribution of Suspended Sediment Concentration in a Sediment – laden Reservoir

Zhang Chao, *Zhang Cuiping*, *Jiang Naiqian* and *Yi Xiaoyan*

Yellow River Institute of Hydraulic Research, Zhengzhou, 450003, China

Abstract: The paper is intended to show the research results about the features of the vertical and longitudinal changes in suspended sediment concentration and longitudinal changes in particle size grading in the reservoir of Sanmenxia on the Yellow River in China and quantitative relationships between point and average concentrations, by using the measured sediment concentration and particle size grading data. First, there are two forms of vertical concentration distributions: one with the concentration increasing from the surface all the way down to the bed bottom and the other with the concentration increasing from the surface to somewhere above the bed bottom where it reaches the maximum. Second, the concentration at the relative flow depth of 0. 4 is found equal to the vertical average, just like the velocity at the point equal to the vertical average. Third, concentration and particle size decrease from upstream to downstream in the reservoir when the flow retained is made due to smaller flow releasing capacity of the tunnels while they increase in the period of drawdown. The research findings will be helpful in analyzing sediment distribution and its variations in sediment – laden reservoirs, and flushing deposits during the drawdown.

Key words: sediment concentration, vertical distribution, particle size grading, reservoir

1　Introduction

Considering the fact that suspended sediment transport is a crucial factor in the fluvial processes, it is one of the important research topics about suspended sediment transport to investigate the law governing the distribution of sediment concentrations along a vertical.

Concentration along a vertical in a natural river often increases from surface to the bed bottom while it, together with particle size grading, changes in more complicated forms in reservoirs due to dam operating, which leads to different flow states from those in the river. Study on the sediment vertical distribution and its longitudinal change in a reservoir on a sediment – laden river will help river professionals not only to plan and design the structures for discharging sediment from a reservoir to be built, but also to provide technical support in flushing sediment particles from an existing reservoir to extend its lifespan.

The Xiaolangdi dam is located on a crucial site in controlling flow and sediment processes into the lower Yellow river. One of the important problems is how to operate the dam so that the released flow and sediment can cause its lower reach to be eroded or be less deposited. Considering the similar flow and sediment transport properties between the Xiaolangdi and Sanmenxia reservoirs, it is possible to refer to and apply the law governing the vertical distribution of sediment concentrations in the Sanmenxia reservoir onto the Xiaolangdi reservoir so as to determine the sediment transport rates through the sediment discharging tunnels for the expected flow rate and concentration values.

2　Features of vertical distributions of suspended sediment

To study the features of vertical distributions of suspended sediment, we use the data measured in the years from 1963 to 1964 at Tongguan, Taian, Beicun, Maojin, and Shijiatan stations, which are located within the Sanmenxia reservoir. The data covers nearly all the frequent fluid states such as free flow, retarded flow, return flow from the floodplains, and includes nearly all the frequent dam operating modes such as water impounding for icing control, power generating at low hydraulic head, flood making by dam operating, free flow discharging, flushing sediment from the reservoir by drawdown.

In the years when the dam was not restructured, the flow releasing remained to be made through the 12 deep – lying flow tunnels as before, which had a small total flow conveyance capacity of 3,084 m³/s for the dam – hell water level of 315 m, therefore there was sedimentation due to flow retaining for larger incoming flow rates. Especially in 1964 which saw a large flood with a large sediment load carried, severe flow retained in the reservoir was made even all the 12 tunnels were completely opened with dam – heel flow level varying 306. 0 ~ 325. 9 m. To the contrary, in Nov. ~ Dec. , 1964, free flow flushed more sediment, causing erosion to the reservoir channel bed.

Based on data analysis, it is found that there exist two forms of vertical sediment concentration distributions: One with the concentration increasing from the surface to the bottom and the other with the concentration increasing from the surface to the maximum somewhere above the bed bottom (Fig. 1).

Analysis of measured data shows that when sediment concentrations takes the first form, the concentration at the bottom can be as large as twice that at the surface for a high average concentration but not so significant for a low average concentration. Besides, the maximum concentration is frequently found to takes place at the relative depth of 0. 2 with the second form of the concentration distribution.

3 Relationship between point and average concentrations

Turbulence diffusion theory, which is found more applicable explaining the suspended sediment transport than other theories, is widely applied in the research on sediment – related problems. With Flick's Law and various formulas for velocity distribution, vertical concentration distribution formulas can be established. Analysis of much data show there is a strong correlation between average concentration (or velocity) over the vertical and the concentration (or velocity) at a certain point, which is verified by the data measured at the deep and main channel flow at the Beicun hydrometric station in 1963 ~ 1964.

3.1 Point velocity vs. average velocity

Lots of relevant research results show that even though velocity distribution in a sediment – laden flow takes the same form of velocity formula as that in a clear flow, the former is much different from the latter in that the Karman constant (k) is not 0. 40 any more in the former. Therefore the key problem with velocity distribution in the deep and main channel flow is how to find a law governing the change of k. Qian Ning, et al. (1983), argue that k decreases as the concentration (or its gradient) increases, that compared with velocity distribution along a vertical in the main flow zone as the criterion especially in term of its slope, the velocity distribution near the bottom doesn't agree with well the logarithm formula, and that the discrepancy increases as concentration increases.

The logarithmic formula of velocity distribution is given as follows (Th. von. Karman – L. Prandtle, 1930):

$$\frac{u_{max} - u}{u_*} = \frac{1}{k}\ln\frac{h}{y} \tag{1}$$

where, u_{max} is the maximum surface velocity; u is the velocity at a distance of y up from the bottom; μ_* is the frictional velocity; and k is the Karman constant.

Analysis of measured velocity data shows that point velocity at the relative flow depth ($\eta = 0. 4$) is nearly equal to the average u_{pj} (Fig. 2), which has been applied in the hydrometric measurement of velocity.

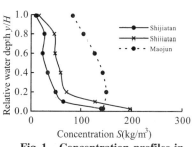

Fig. 1 **Concentration profiles in Sanmenxia Reservoir**

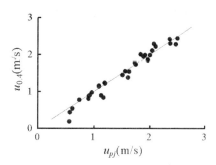

Fig. 2 **Relationship u_{pj} of with $u_{0.4}$**

Calculate the average Karman constant k_{pj}, by using the logarithm – formed formula and velocity data measured at different points along a vertical. Then plot k_{pj} vs. S_{pj}, with the latter denoting the mean average concentration (Fig. 3). Please note the data in the figure doesn't include the data measured during density current venting. From Fig. 1 it is clear that k_{pj} decreases as S_{pj} increases. k_{pj} varies between 0.057 ~ 0.349, and that k_{pj} tend to be small at the starting of flushing by drawdown.

Estimate k values corresponding to various relative depths by using Eq. (1). Plot k vs. k_{pj} to find k at the relative depth of 0.4 equal to the average k_{pj} (Fig. 4), which corresponds to the relationship between point velocity and the average velocity.

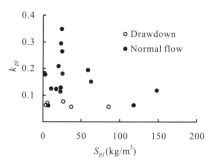

Fig. 3 **Relationslip of S_{pj} with $k_{0.4}$**

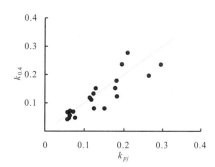

Fig. 4 **Relationship k_{pj} of with $k_{0.4}$**

3.2 Point concentration vs. average concentration

Rouse formula is a general formula for concentration distribution along a vertical based on the diffusion theory, and is given as in Eq. (2)

$$S/S_a = [(h/y - 1)/(h/a - 1)]^Z \qquad (2)$$

where, S is the concentration at a distance (y) up from the bottom, h is the flow depth, Z is a kind of suspension index, $Z = \omega/ku_*$.

Analysis of measured data leads to a conclusion that concentration distribution at the Beicun hydrometric station can be well represented by the Rouse formula with its z varying between 0.052 ~ 0.423.

Analysis of the relationship of point concentration vs. the average concentration (Fig. 5) leads to a conclusion that the concentration at the relative depth of 0.4 is equal to the average, i.e., $S_{pj} = S_{0.4}$, which is similar to the case with velocity. Note: there is a data point standing far away from the correlation line, which was the measured on 21 Aug. 1964 when sediment was being discharged from the reservoir by density current venting which made a different concentration distribution from that by open – channel turbulent sediment venting.

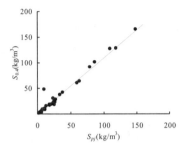

Fig. 5   Relationship of S_{pj} with $S_{0.4}$

4   Longitudinal change in concentration distribution

4. 1   Sediment venting during flow retained

In the flood season(July ~ Oct.) in 1964 when the Sanmenxia dam was operating with its flow retained due to lower flow conveyance capacity of the tunnels than the inflow rates, the concentration distribution along a vertical is shown in Fig. 6. On 7 Aug. 1964 the flow rate and concentration measured at Tongguan was 5,460 m³/s and 51 kg/m³ respectively while the released flow rate and concentration released from the dam was 4,030 m³/s and 20 kg/m³ respectively with the flow level measured at Shijiatan station, which is immediately upstream of the dam (dam heel) at 320. 11 m,which indicates clearly how severe the flow and sediment transport retaining were at that time. At the Taian station, 72. 6 km upstream of the Shijiatan station, backwater lead to longitudinal decreases in flow velocity and concentration which is evidenced by the decreases in the average concentration measured at the verticals in the deep and main channel from 50 kg/m³ to 26 kg/m³ and changes in concentration distribution along a vertical, which is evidenced by the decreases by 51 kg/m³, 24 kg/m³, 23 kg/m³ at the relative depths of 0. 2, 0. 4, and 0. 8 respectively, indicating the decrease values decreases gradually from bottom to surface.

Therefore, while the dam operation was made such that flow retaining was made, flow slowed down, suspended sediment particles was falling to the bottom and the average concentration was decreasing longitudinally from upstream to downstream, with much larger decrease at the bottom than at the surface.

From the sediment particle size grading curves measured at the stations on 7 Aug. 1964 when the dam operation was in a mode of flow retained (Fig. 7) , it can be seen that flow retained made the particles become finer longitudinally from upstream to downstream with the fraction between 0. 01 mm and 0. 05 mm decreased by larger margins.

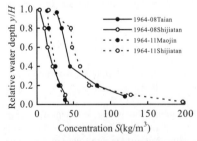

Fig. 6   Vertionl distribution of concontration

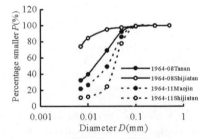

Fig. 7   Particle size grading curve

4.2 Erosion during drawdown

In the post – flood season in 1964 when the dam operation was in a mode of drawdown, the concentration distribution is shown in Fig. 6. On 2 Nov. 1964 the concentration measured at Tongguan, which is at the entrance to the reservoir, was 21.5 kg/m^3, while the concentration released from the dam was 55.5 kg/m^3, with the flow level at the dam heel (i.e., at the Shijiatan station) at 314.57 m. Concentration measured at the deep and main channel flow increased from 21 kg/m^3 at the Maojin station, 13.9 km upstream of the Shijiatan station, to 53 kg/m^3 at the Shijiatan station. In the meanwhile the concentrations measured the relative depths of 0.2, 0.4, and 0.8 increased by 42 kg/m^3, 35 kg/m^3 and 31 kg/m^3, respectively, indicating the increases decrease from the bottom to the surface gradually. Similarly the concentrations measured the relative depths of 0.2, 0.4, and 0.8 on 4 Apr. 1964 increased by 27 kg/m^3, 15 kg/m^3 and 13 kg/m^3, respectively.

Therefore, while the Sanmenxia dam operation was in a mode of drawdown, concentration increased longitudinally in the downstream direction, with the increased value decreasing from the bottom to the surface gradually.

Fig. 7 shows that drawdown made the bed eroded, and particles coarser with the fraction between 0.01 ~ 0.05 mm increasing by larger margins, which is evidenced by the data measured on 2 Nov. 1964.

5 Conclusions

Analysis of the data of the suspended sediment concentration and particle size grading measured in the Sanmenxia Reservoir leads to the conclusions:

(1) When the dam operation makes drawdown or flow retained, there are two forms of concentration distributions along a vertical: one with the concentration increasing from the surface all the way down to the bottom while the other with the concentration increasing from the surface all the way down to somewhere above the bottom where it reaches the maximum. And the first form is most frequently seen.

(2) The average concentration over a vertical is found to be nearly equal to the concentration at the relative depth of 0.4, which is similar to the property of velocity. The concentration distribution along a vertical can be well expressed with the Rouse formula with the suspension index varying between 0.052 ~ 0.423.

(3) Longitudinal change in concentration is subject to the dam operation. When the dam is operated such that flow retained is formed, concentration decreases in the downstream direction with the decreased value the largest at the bottom and the smallest at the surface. When the dam is operated such that bed is eroded by drawdown, concentration increases in the downstream direction with the increased value the largest at the bottom and the smallest at the surface.

(4) Longitudinal change in the suspended particle size grading in the reservoir depends on the dam operation. When the dam is operated such that flow retained is formed, sediment particles become finer in the downstream direction. When the dam is operated such that bed is eroded by drawdown, sediment particles become coarser in the downstream direction.

Acknowledgements
This research was funded by the National Technologies Research and Development Program in the twelfth Five year Plan(2012BAB02B02).

References

Qian Ning, Wan Zhaohui. Sediment Dynamics[M]. Beijing: Science Press, 1983. (in Chinese)
Qian Ning, Zhang Ren, Zhou Zhide. River Morphological Changes [M]. Beijing: Science

Press,1987.

Zhang Ruijin. River Sediment Dynamics (Version 2) [M]. Beijing: Water Resources and Hydropower Press. 1989.

Th. von. Karman. Selected Works of Th. Von. Karmon[C].

Hui Yujia. Analysis of vertical distributions of velocity and sediment concentration in the Yangtze River and the Yellow River[J]. Journal of Hydraulic Engineering, 1996(2) :11 − 17.

The Research for Spatial and Temporal Differences of China's Carbon Emissions and the Situation of Low Carbon Economy

Wang Xi and *Miao Rui*

College of Environment and Planning, Henan University, Kaifeng, 475004, China

Abstract: Carbon emissions intensity is the most direct and efficient index for the development of low carbon economy development. The results show that the carbon emissions and the per capita carbon emissions of various regions are growing gradually, but the trend of growth has been slowed, which will be continued in the future. The spatial and temporal differences of the carbon emission intensity are obvious. The carbon emission intensity of the eastern region is lower significantly than the middle and western regions. The research indicated that carbon emission intensity was closely related to the level of economic development and industrial structure. At the same time, the carbon emission intensity of China was reduced gradually. From that we can foresee, with the raising development level of economy and the improvement of the industrial structure, the carbon emission intensity will be reduced continually. The situation of low carbon economy development can be analyzed through the relationship between the carbon emission intensity and the speed of economic development. The study results showed that the low carbon development situations of the most of the provinces in the eastern region of China are better than others, which belongs to the mode of low carbon emission and high economic growth; Otherwise, the low carbon development situations of the western regions are relatively poor, which belongs to the mode of high emissions growth and high economic growth or the mode of high emissions growth and low economic growth; The situation of low carbon economy development are general in the middle regions.

Key words: carbon emissions, spatial and temporal differences, carbon emissions intensity, low carbon economy

1 Introduction

At present, the low carbon economy has become a global economic hot topic and many countries are developing low carbon economy in the world. England put forward the concept for "low carbon" firstly and actively advocates low carbon economy. Presently, many countries have carried out low carbon economy through various ways with local characteristics, put forward their own development objectives, made their relevant planning and have obtained the certain effect. Facing the urgency of world to develop low carbon economy, China has already taken active measures to create the conditions for low carbon economy development.

The change of carbon emissions is the most direct index to reflect the low carbon economy development. China is a huge geographical country. Natural resources distribution, industrial structure and the level of economic development have distinct spatial difference, so the carbon emissions and carbon emission intensity are different significantly (Yue Chao et al., 2010; Zhao Yuntai et al., 2011). In order to realize the carbon reduction target, the target must be decomposed in every region effectively. At the same time, the mode and path for low carbon economy development must be made according to the various regional situation. Based on the calculation analysis of primary energy from 2000 to 2009 in China, spatial difference and dynamic change of China were analyzed to provide the necessary technical basis for making Carbon emissions goal setting and regulation. At the same time, the situation of low carbon economic development for various regions were analyzed through the relationships between the economic growth and carbon emissions intensity change to provide necessary theoretical reference to control the low carbon economy developmen.

2　Data sources and processing

2.1　Energy data sources

The key to this study is the calculation of carbon emissions for the various regions in different time. In this study, the main carbon emissions data came from china energy statistical year book from 2001 to 2010, and the carbon emissions were calculated according to the carbon emissions coefficients of kinds of energy. In addition, in order to facilitate analysis of carbon emissions over a large area space change, the carbon emissions for eastern, central and western areas in China were calculated. The division of the eastern, central and western areas is consistent with the National Bureau of Statistics. The western area include ten provinces (cities and autonomous) regions such as Chongqing, Sichuan, Guizhou, Yunnan, Tibet, Shaanxi, Gansu, Ningxia, Qinghai, Xinjiang. The central area include nine provinces (cities and autonomous) regions such as Shanxi, Inner Mongolia, Jilin, Heilongjiang, Anhui, Jiangxi, Henan, Hubei, Hunan. The eastern area include eleven provinces, cities and autonomous regions such as Beijing, Tianjin, Hebei, Liaoning, Shanghai, Jiangsu, Zhejiang, Fujian, Shandong, Guangdong and Hainan. A final note about this paperis that Taiwan and Tibet's carbon emissions were not calculated because of the constraints for data collection.

2.2　The carbon emissions coefficients of all kinds of energy

At home and abroad, many scholars (Wang Zheng et al. , 2012) have estimated and argued for the carbon dioxide production of different units energy (i. e. transform coefficient). The transform coefficient released by IPCC was used in this paper, and the specific can be seen in Tab. 1.

Tab. 1　The carbon emissions coefficients of all kinds of energy

Energy type	Carbon emission coefficient ($\times 10^4$)	Energy type	Carbon emission coefficient ($\times 10^4$)
Raw coal	0.755,9	Fuel oil	0.618,5
Washed clean coal	0.755,9	Other petroleum products	0.585,7
Coke	0.855,0	Liquefied petroleum gas	0.504,2
Other coking products	0.644,9	Natural gas	0.448,2
Crude oil	0.585,7	Coke oven gas	0.354,8
Gasoline	0.553,8	Refinery dry gas	0.460,2
Kerosene	0.571,4	Other gas	0.354,8
Diesel	0.592,1		

2.3　Other data

The other data except energy came from China and provinces (municipalities, autonomous regions) statistical year book from 2001 to 2010. In order to make data of different years comparable, the economic statistical data in this paper were all taken 2000 as the benchmark period, and other years data were transferred to constant price as 2000.

3　The calculation results and analysis

3.1　The results calculated for carbon emissions

The carbon emissions of provinces (municipalities, autonomous regions) were calculated according to all kinds of energy consumption of various provinces (municipalities, autonomous regions) from 2000 to 2009 and carbon emissions coefficients. The specific results are in Tab. 2.

Tab. 2 Energy consumption carbon emissions of all provinces, municipalities and autonomous regions in 2000 ~ 2009

(Unit: ×10⁴ t)

Province	2000	2001	2002	2003	2004	2005	2006	2007	2008	2009	Average growth(%)
Beijing	10,331	10,752	11,226	11,737	12,813	13,766	14,719	15,669	15,774	16,380	5.25
Tianjin	6,965	7,275	7,534	8,015	9,216	10,183	11,219	12,322	13,371	14,644	8.61
Hebei	27,911	25,905	28,889	33,598	43,248	49,451	54,333	58,798	60,634	63,369	9.54
Shanxi	16,773	19,864	23,282	25,895	28,049	31,785	35,147	38,893	39,079	38,830	9.78
Inner Mongolia	8,848	10,154	11,368	13,008	19,003	24,098	27,973	31,853	35,152	38,251	17.66
Liaoning	26,565	26,565	26,423	28,542	32,594	33,932	37,362	41,244	44,379	47,646	6.71
Jilin	9,390	9,630	10,852	12,443	13,967	13,251	14,729	16,347	18,003	19,191	8.27
Heilongjiang	15,372	15,050	14,968	16,738	18,612	20,069	21,766	23,377	24,879	26,093	6.06
Shanghai	13,710	14,504	15,255	16,698	18,462	20,505	22,127	24,108	25,447	25,846	7.30
Jiangsu	21,471	22,140	23,955	27,573	34,034	42,798	47,469	52,223	55,425	59,107	11.91
Zhejiang	16,355	16,279	18,413	21,253	26,986	29,995	32,955	36,209	37,661	38,808	10.08
Anhui	12,163	12,759	13,253	14,564	15,000	16,219	17,624	19,294	20,755	22,177	6.90
Fujian	8,634	7,885	8,701	9,785	13,584	15,311	17,022	18,915	20,577	22,229	11.08
Jiangxi	6,245	5,806	6,355	7,566	9,508	10,685	11,618	12,596	13,420	14,491	9.80
Shandong	28,325	24,818	27,543	32,494	48,922	60,236	66,711	72,737	76,211	80,824	12.36
Henan	19,742	20,552	21,447	23,838	32,594	36,459	40,467	44,470	47,308	49,240	10.69
Hubei	15,629	15,088	16,736	19,059	22,736	25,134	27,545	30,273	32,022	34,174	9.08
Hunan	10,148	11,523	12,577	13,866	18,945	24,205	26,378	28,991	30,802	33,234	14.09
Guangdong	23,553	25,376	28,308	32,656	37,920	44,677	49,787	55,387	58,526	61,462	11.25
Guangxi	6,655	6,654	7,434	8,529	10,479	12,137	13,438	14,951	16,197	17,638	11.44
Hainan	1,197	1,296	1,560	1,984	1,851	2,050	2,295	2,635	2,830	3,073	11.05
Chongqing	6,053	7,519	7,988	7,409	9,149	12,323	13,384	14,826	16,136	17,525	12.54
Sichuan	16,249	16,977	18,722	22,946	26,675	29,458	32,375	35,436	37,757	40,690	10.74
Guizhou	10,667	11,064	11,144	13,816	15,010	14,064	15,388	16,952	17,660	18,863	6.54
Yunan	8,647	8,701	9,820	10,348	12,988	15,018	16,505	17,782	18,724	20,024	9.78
Shaanxi	6,809	8,120	9,257	9,852	11,906	13,889	15,281	16,890	18,492	20,053	12.75
Gansu	7,508	7,242	7,524	8,788	9,742	10,889	11,824	12,737	13,328	13,666	6.88
Qinghai	2,237	2,318	2,540	2,800	3,401	4,164	4,745	5,223	5,682	5,854	11.28
Ningxia	2,940	4,187	4,753	6,477	5,789	6,322	7,055	7,672	8,051	8,445	12.44
Xinjiang	8,296	8,716	9,030	10,064	12,240	13,728	15,076	16,394	17,624	18,761	9.49
Total	375,386	384,721	416,855	472,341	575,423	656,801	724,317	795,205	841,906	890,588	10.08

3. 2　Analysis for carbon emissions of the temporal and spatial distribution in China

3. 2. 1　Analysis for spatial and temporal distribution of the total carbon emissions

In order to analyze the difference of carbon emissions about spatial distribution in China, taking the provincial units for regional unit, based on Arcmap, the carbon emissions were analyzed about temporal change, and the results can be seen in Fig. 1 and Fig. 2. From Fig. 1 we can see, China's carbon emissions, larger located mainly in the middle eastern region which population is more, economic development level is higher regions, and carbon emissions of the northwest region is relatively low.

Fig. 1　The spatial distribution of carbon emissions and its variation

Fig. 2　The spatial difference carbon emissions growth rate in China

As we can see from Fig. 2, the average speed of national carbon emissions growth is 10. 07% during the period 2000 ~ 2009. In Inner Mongolia, Shaanxi, Chongqing, Sichuan, Qinghai, Guangxi western provinces and Shandong, Jiangsu, Fujian and Guangdong Provinces , the speed of economic development are higher, but in central and northeast it is relatively low, which is lower than average growth; According to the change of time , the carbon emissions growth rate is divided into two stages in China from 2001 to 2009. From 2001 to 2004 , in eastern, western, central regions the carbon emissions growth rate increases year by year, of which in the western region get the highest in 2003 and other areas get the peak value in 2004. This is mainly because China's economic is developing, but most of which is production industry, whose energy mostly come from fossil fuels such as coal for the source . Before 2004, new energy resources and low carbon industry did not get popularity, and there is not a very effective policy to reduce carbon emissions to limit the production industry carbon emissions.

In order to macro analysis, we can put all carbon emissions data of the eastern, central and western regions corresponding provinces (municipalities, autonomous regions) together and calculate the carbon emissions and their growth situation changes of the east, middle and west three regions separately and the results are shown in Fig. 3 and Fig. 4. As we can see from the graph, the eastern carbon emissions growth rate is higher than the national average growth rate.

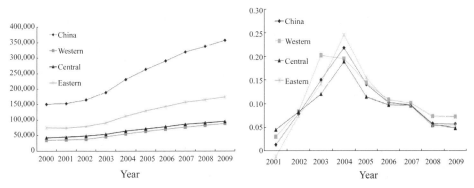

Fig. 3   The carbon emissions of national and the eastern, central and western areas ($\times 10^4$ t)

Fig. 4   China's eastern, central and western carbon emissions growth rate in 2000 ~ 2009

3.2.2   The analysis of the spatial and temporal variation of per capita carbon emissions

Carbon emissions are related to the level of economic development, the population, and other factors. In order to analysis per capita carbon emissions in per area, according to the population of provinces (municipalities, autonomous regions) in 2001 ~ 2010, calculate the per capita carbon emissions (t/a) in each area from 2000 ~ 2009, and selected the data of the per capital carbon emissions in each area from 2000 to 2009 and do the space analysis. the results are shown in Fig. 5 and Fig. 6.

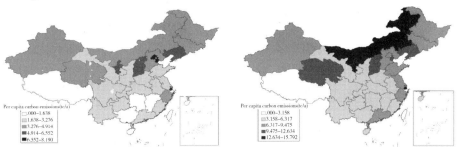

Fig. 5   The per capita emissions (t/a) in 2000

Fig. 6   The per capita carbon emissions (t/a) in 2009

From the space analysis of the per capita carbon emissions, we can see that the per capita carbon emissions in the northern , the eastern and the northwest regions is higher, including Xinjiang, Qinghai, Inner Mongolia, Ningxia autonomous region, Gansu, Shanxi, Hebei, Beijing, Tianjin, Liaoning, Shandong, Jiangsu, Zhejiang, Shanghai, Heilongjiang, Jilin provinces, municipalities, autonomous regions per capita carbon emissions where the per capital carbon emissions is higher , which is related to the level of economic development, industrial structure factors in these areas.

3.2.3   The spatial and temporal variation analysis of carbon emissions intensity

Carbon emissions intensity refers to carbon emissions that the unit GDP generated, whose indicator can measure accurately energy utilization efficiency and carbon production ability.

According to the carbon emissions, GDP data of each provinces (municipalities, autonomous regions) from 2000 to 2009, we can calculate carbon emissions intensity of provinces (municipalities, autonomous regions). The results are shown in Tab. 3 and Fig. 7. Select carbon emissions intensity from 2000 to 2009 for spatial contrast, and the results can be seen on Fig. 8 and Fig. 9.

Tab. 3　The change of carbon emissions intensity in each province (municipalities, autonomous regions)

(Unit: $\times 10^4$ yuan)

Province	2000	2001	2002	2003	2004	2005	2006	2007	2008	2009
Beijing	4.17	3.90	3.69	3.48	3.36	3.22	3.05	2.83	2.61	2.46
Tianjin	4.25	3.96	3.64	3.38	3.36	3.23	3.10	2.95	2.75	2.58
Hebei	5.48	4.68	4.76	4.97	5.68	5.73	5.55	5.32	4.99	4.74
Shanxi	10.20	11.15	11.70	11.42	10.84	10.83	10.61	10.13	9.38	8.85
Inner Mongolia	6.32	6.62	6.61	6.47	7.92	8.11	7.91	7.55	7.08	6.59
Liaoning	5.69	5.22	4.71	4.57	4.62	4.27	4.12	3.95	3.75	3.56
Jilin	5.16	4.84	4.98	5.18	5.18	4.39	4.24	4.05	3.85	3.61
Heilongjiang	4.73	4.23	3.82	3.87	3.86	3.73	3.60	3.46	3.29	3.10
Shanghai	3.01	2.89	2.74	2.68	2.61	2.60	2.49	2.36	2.27	2.13
Jiangsu	2.50	2.34	2.27	2.30	2.47	2.71	2.62	2.51	2.36	2.24
Zhejiang	2.71	2.44	2.45	2.48	2.75	2.71	2.61	2.50	2.37	2.24
Anhui	4.00	3.88	3.70	3.72	3.41	3.32	3.20	3.07	2.93	2.78
Fujian	2.20	1.85	1.84	1.86	2.30	2.32	2.25	2.17	2.09	2.01
Jiangxi	3.12	2.66	2.64	2.78	3.09	3.08	2.98	2.85	2.68	2.56
Shandong	3.32	2.64	2.63	2.72	3.56	3.81	3.68	3.51	3.28	3.10
Henan	3.84	3.67	3.50	3.51	4.22	4.13	4.01	3.84	3.65	3.42
Hubei	3.65	3.23	3.29	3.42	3.67	3.62	3.50	3.36	3.13	2.95
Hunan	2.75	2.86	2.87	2.88	3.52	4.01	3.87	3.70	3.45	3.27
Guangdong	2.44	2.40	2.40	2.42	2.46	2.54	2.47	2.39	2.29	2.19
Guangxi	3.25	3.00	3.03	3.16	3.47	3.55	3.46	3.35	3.21	3.07
Hainan	2.31	2.30	2.53	2.91	2.46	2.46	2.44	2.42	2.35	2.29
Chongqing	3.81	4.34	4.18	3.48	3.83	4.62	4.46	4.26	4.05	3.83
Sichuan	4.05	3.88	3.87	4.24	4.37	4.29	4.15	3.97	3.81	3.59
Guizhou	10.74	10.23	9.45	10.64	10.37	8.62	8.36	8.02	7.51	7.20
Yunan	4.42	4.18	4.36	4.23	4.76	5.05	4.98	4.78	4.55	4.34
Shaanxi	4.10	4.48	4.66	4.47	4.78	4.91	4.74	4.52	4.26	4.06
Gansu	7.64	6.73	6.39	6.78	6.78	6.78	6.60	6.33	6.02	5.59
Qinghai	8.49	7.85	7.66	7.53	8.15	8.89	8.94	8.67	8.31	7.78
Ningxia	11.07	14.32	14.75	17.92	14.42	14.20	14.06	13.57	12.64	11.85
Xinjiang	6.08	5.91	5.66	5.70	6.24	6.31	6.24	6.05	5.86	5.77
Total	3.79	3.55	3.49	3.53	3.80	3.84	3.72	3.57	3.38	3.20

Fig. 7 Carbon emissions intensity distribution and variation of space (t/10⁴ yuan)

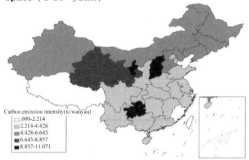

Fig. 8 The carbon intensity (t/10⁴ yuan) of each area in 2000

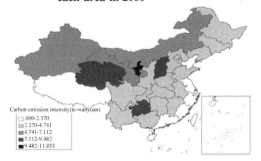

Fig. 9 The carbon intensity (t/10⁴ yuan) of each area in 2009

From the Tab. 3 we can see, carbon emissions intensity on the whole is reduced constantly in China, the national average lowed from 3. 79 t/10⁴ yuan in 2000 to 3. 20 t/10⁴ yuan in 2009, which was according with the basic requirements for a low carbon economy development. The difference between China's spatial carbon emissions intensity is significant . In 2009 China's carbon emissions intensity for the average is 3. 20 t/10⁴ yuan, and the standard deviation is 2. 31 t/10⁴ yuan, and the variation coefficient was 1. 38. The carbon emissions intensity is higher in Ningxia, Shanxi, Guizhou, Qinghai, Inner Mongolia, Gansu, Xinjiang, which are located in China's northwest and southwest and are the base of main energy source production; Carbon emissions intensity is lower in Fujian, Guangdong, Zhejiang, Jiangsu, Shanghai and other province (city), which are located in China's eastern regions, where economic development level is higher, and they are also the most important economic centre in China (Liu Zhixiong, 2011; Wang Huitong and Wang Miaoping, 2011).

In order to analysis the spatial differences of carbon emissions intensity comprehensively, calculated respectively carbon intensity according to the east, middle and west three regions and the results can be seen from Fig. 10. As we can be seen from the graph, the carbon emissions intensity in western is the highest, and the second is in central, and the lowest is in eastern. But in general, The carbon emissions intensity of three main parts are in decreasing gradually trends.

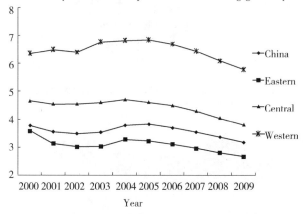

Fig. 10   Carbon emissions intensity variations (t/10^4 yuan) in east, middle and west areas

4   The analysis for a low carbon economy development trend in China

From the above analysis, we can see that China's gross domestic product increased year by year, and carbon emissions intensity reduced year by year between 2000 and 2009, which suggests that with economic growth and the progress of science and technology, each region has been focusing on a low carbon economy development, and made success at first. But China is a large country, the difference between the change of carbon intensity and economic development speed is bigger, and the difference of low carbon economy development situation in regions are significant. Many scholars at home and abroad use different methods to study low carbon economy development situation from different angles (Wang Huitong and Wang Miaoping, 2011; Liu Zhu et al., 2011). According to the general requirements for low carbon economy development, combination with the actual situation of China, the author holds that the state which is the high growth of economy and carbon emissions will exist for a long time to some extent, but carbon emissions intensity will gradually decrease. We can compare the regional economic development speed and the change for carbon emissions intensity and get the low carbon economic development situation in different areas. Here through the analysis of the average carbon emissions intensity and the relative relations between growth rates in GDP all over the world between 2000 and 2009 to study the low carbon economy development situation. Use the average carbon intensity as the x axis in 2000 ~ 2009, and use annual growth rate in GDP as the y axis in 2000 ~ 2009. Make the scatter-plot chart all over the country. According to the average of the strength of the national carbon emissions and average in GDP growth for the standards respectively between 2000 and 2009, low carbon economy development trend in China is divided into four modes: high emissions low growth, low emission and high growth, high growth and low emission, low emissions and low growth, the specific can be seen from Fig. 11.

We can see from Fig. 11, the eastern are the low emission high growth model except for Fujian

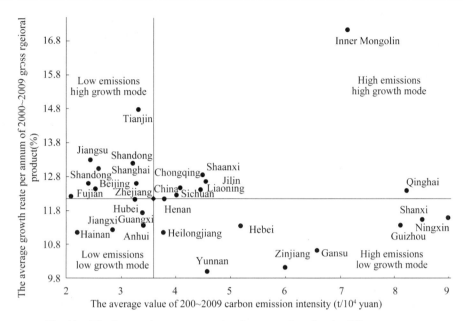

Fig. 11　The low carbon economy development situation in different areas

and Hubei, Hainan, Liaoning. This is because in the east region economic development level is higher, and the capital-intensive industries and technology-intensive industries is the main in the industrial structure. And with the superior location, we can make the industrial structure optimization and adjustment in time. In general the low carbon economy development situation is better; And the western region is more high emissions low growth and high emissions high growth model, this is because in the western region with high energy consumption industries, economic development level is relatively low and energy consumption is higher, but the technical level is relatively backward, so the result of carbon intensity is bigger. In general, the low carbon economy development situation is poorer. And in the central region is low emissions and low growth pattern, this is because the regional economic development speed is relatively slow, and the level of economic development are technology accumulation are not good. In general, the low carbon economy development trend is on average.

5　Conclusions and suggestions

We can see the following points from analysis carbon emissions between 2000 and 2009.

(1) Since China is in a period of rapid growth of economy and society, the carbon emissions and the per capita carbon emission have been increasing. After 2005, China's carbon emissions' growth speed slows down, this is related to China's response to the climate change and measures we take like clean production, low carbon life and other policies; We can foresee, carbon emissions and the per capita carbon emissions will still continue to grow in the following years in China, but the growth speed will gradually slow down.

(2) Spatial and temporal change of China's carbon emissions intensity is obvious. From the view of space , the eastern carbon emissions strength is lower than the central, it indicates that carbon emissions intensity and the level of economic development are closely related to industrial structure; looking from time, carbon emission intensity in every area of China is gradually reduced over time; and we can foresee, with the raise of economic development level and improvement of industrial structure, China will continue to reduce carbon emissions intensity.

(3) Through the analysis of the relationship between carbon emissions intensity and velocity of

economic development, we can find that most of the eastern provinces low carbon development situation is good, belonging to the low emission and high growth mode; the low carbon economy development of the western area are relatively poor, belonging to high emissions and low growth mode; and in most parts of central area, low carbon economy development situation are general.

(4) China should actively take various measures to respond to the international climate change and the international climate talks. Improving the industrial structure, improving energy efficiency, developing clean energy and zero carbon energy are China's important ways to develop low carbon. At the same time, we should draw on the experience of low carbon economy development of Europe and America and other developed countries, the government as the leading factor, to develop, introduce and popularize the low carbon technologies, and be more solid to promote the development of low carbon economy (Wang Zheng and Zhu Yongbin, 2010; Zhao Xia, Zhu Lin and Wang Sheng, 2010).

References

Yue Chao, et al. Provincial Carbon Emissions and Carbon Intensity in China from 1995 to 2007 [J]. Acta Scientiarum Naturalium Universitatis Pekinensis, 2010,46(4):510 – 516.

Zhao Yuntai, et al. Spatial Pattern Evolution of Carbon Emission Intensity from Energy Consumption in China[J]. Environmental Science, 2011,32(11):3145 – 3152.

Wang Zheng, et al. Prediction on Beijing s, Tianjin s and Hebei s Carbon Emission [J]. Geography and Geo – Information Science, 2012, 28(1):84 – 89.

Liu Zhixiong. Research on the Relation Between Energy Consumption, Economic Growth and Carbon Emissions in China[J]. Research on Coal Economic, 2011, 31(4):37 – 41.

Wang Huitong, Wang Miaoping. Analysis of Carbon Emission Spatiotemporal Patterns and Grey Incidence of Factors Influencing Carbon Emission in 30 Provinces in China [J]. China Population,Resources and Environment, 2011,21(7):140 – 145.

Liu Zhu, et al. Low – carbon City s Quantitative Assessment Indicator Framework based on Decoupling model[J]. China Population,Resources and Environment, 2011, 21(4):19 – 24

Wang Zheng, Zhu Yongbin. Study on the Status of Carbon Emission in Provincial Scale of China and Countermeasures for Reducing its Emission [J]. Strategy and Decisions, 2010, 23 (2):110 – 114.

Zhao Xia, Zhu Lin, Wang Sheng. Lessons from European Union Greenhouse Gas Emission Trading Scheme[J]. Environmental Protection Science, 2010,36(1):57 – 59.

Study on Carbon Storage and Carbon Density of Crops in Kaifeng[①]

Lu Fengxian[1,2] , *Zhao Liqin*[3] , *Ning Xiaoju*[1] ,
Yang Shuichuan[1] , *Yan Weiyang*[1] and *Wang Xi*[1]

1. College of Environment and Planning, Henan University, Kaifeng, 475004, China
2. Institute of Policy and Management, Chinese Academy of Sciences,
Beijing, 100190, China
3. College of Geography and Resources Science, Sichuan Normal University,
Chengdu, 610000, China

Abstract: Sequestration activities can help prevent global climate change by enhancing carbon storage in crops and soils, preserving existing crops and soil carbon, and by reducing emissions of CO_2, methane (CH_4) and nitrous oxide (N_2O). Carbon storage and carbon density of crops in Kaifeng (known as a land of crops in China) were calculated with vegetation biomass method based on the statistical data of crop yield and cultivate area from 2000 to 2009. The conclusions are as following: ① the total carbon storage of crops increased by 21.2% from 2004 to 2009; ② in 2009, Qixian county had the highest carbon storage of 5 counties in Kaifeng since it was determined by crops yield and cultivate area; ③ the proportion of wheat and corn in total carbon storage was high form 2004 to 2009, accounting for 53% and 36% respectively. As a result, in order to increase both agricultural production and carbon storage, the following countermeasures have been put forward: enhancement of infrastructure construction, improvement of agricultural production conditions, promotion of new agricultural technology and adjustment of crop management system.

Key words: carbon storage, carbon density, crop, Kaifeng

The global warming and climate change owing to the rising of emission on CO_2, methane (CH_4) and nitrous oxide (N_2O) are obvious, which have attracted more attention from governments, scientists and public. The international community generally believes that the increase of carbon fixation and storage in terrestrial ecosystem can effectively retard the rising of CO_2 concentration. There are forest, grassland, farmland, natural wetland ecological four subsystems in the land ecological system, the farmland ecological system is an important component of the terrestrial ecosystem for crops could absorbing and fixing atmospheric CO_2 by own photosynthesis, and providing the survival food for human. At present, studies on farmland ecological system is mainly concentrated in the soil organic carbon, the studies of carbon storage of crops mainly focus on national or provincial scale, the county scale is little. In addition, the study on carbon emission and carbon storage in Henan Province also less.

Kaifeng was the center of seven dynasties in history, and now is the main producing areas of China's grain core and the core built-up areas of the Central Plains Economic Region. In this paper, we analysis the changes of carbon stocks in crops and explore the characteristics and spatial distribution of the carbon density based on the data of Kaifeng agriculture such as the main crops' output, planting area. The aim is to increase carbon storage of the land ecological system and slower the growth rate of CO_2 concentration in atmospheric, and improve the planting of crop of Kaifeng, provide a reference for the development of agriculture.

① Funding: Major Program of National Natural Science Foundation of China (NO. 2012CB955804); National Natural Science Foundation of China (NO. 41171438); Natural Science Foundation of the Education Department of Henan Province (NO. 2010A170001)

1 Study area and research methods

The Kaifeng city is divided into urban areas and the five counties (Qixian County, Tongxu County, Weishi County, Kaifeng County, Lankao County) , which is a national commodity grain and wheat production base , and it is also the region of wheat, cotton , peanuts , soybeans producing base, whose planting structure has a strong representative and could reflect the characteristics of the crops in central China.

In cropland ecosystems, the main influence factors of soil organic carbon fixed, transformation and release is farmland management practices and soil properties, and other factors existing such as land use, plant varieties, climate change. So it need a comprehensive opinion on considering the interaction of many factors for the sake of evaluate the soil carbon fixed potential and adjustment measures. We could estimate the carbon storage of crops by means of the vegetation biomass carbon content of wheat, corn, soybeans, peanuts, cotton, and so on. The calculation formula is :

$$T_i = \sum_{i=1}^{n} C_i \times (1 - W_i) \times (1 + R_i) \times Y_i / H_i \tag{1}$$

$$D_i = T_i / A_i \tag{2}$$

where , T_i is total carbon content of the crop type $i(t)$; C_i is the carbon content of the crop type i ; W_i is the water content of crop type i ; R_i is root-shoot ratio of type i , below-ground biomass and aboveground biomass ratio; Y_i is economic yield of the crop type i , harvest yield(t) ; H_i is i kind of crops economic coefficient the ratio of economic yield and biological yield, the root-shoot ratio of different crops, economic coefficient and carbon content see Tab. 1 , take 15% as the average water content of the crop; D_i is the carbon density of crops at i , kg/hm^2 ; A_i is the total sown area of crop type i , hm^2.

If the A_i is the region's arable land , the D_i reflects the ability of crop carbon sequestration of the regional unit area. The data on the main crop yields, the area of sowing and the common cultivated land come from Kaifeng City Statistical Yearbook and five counties statistics from 2005 to 2009.

Tab. 1 The root-shoot ratio, carbon content, economic coefficient of different crops

Crop	Root-shoot ratio	Carbon content	Economic coefficient
Wheat	0.48	0.485,3	0.35
Corn	0.44	0.470,9	0.5
Soybeans	0.92	0.439,9	0.435
Peanuts	0.38	0.45	0.43
Cotton	0.19	0.45	0.383

Note : Economic coefficient come from related literature.

2 Results and analysis

2.1 Time variation of crop carbon storage

2.1.1 Kaifeng major crops carbon reserves inter-annual change in 2004 ~ 2009

In 2004 ~ 2009, there is a steadily growing tendency for the crops carbon stocks of Kaifeng, the carbon storage from 4,388,600 t increased to 5,569,700 t (Fig. 1). Simultaneously, the crop yields from 2,071,593 t increased to 2,856,986 t, an increase of 27.5%, the amount of the scwing area from 561,720 hm^2 increased to 593,160 hm^2, a rise of 5.3%. The results indicate that although there is a little change in the amount of the main crop planting area, but the crops yield per unit of area is increased and carbon storage also got uninterruptedly rising owing to the improvements of the agricultural technique and the change of varieties crops. These measurements include the introduction of new varieties of products, the subsidy policy for the thoroughbred varieties, the fertilization technology by testing the characteristic of soil, the wheat accurate seeding and semi-accurate seeding technology, the water-saving and high-yield cultivation technology, the corn "one-increase & four-change" technology, and so on. On the other hand, carbon stocks has been increasing with the total increase of 21. 2% from 2004 to 2009, it is benefit from an increased rainfall in this period, but little inter-annual fluctuations, making the Kaifeng crop carbon storage stability change.

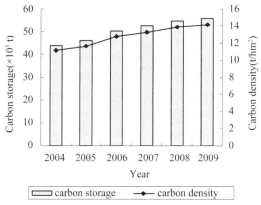

Fig. 1   The change trends of carbon storage and carbon density in 2004 ~ 2009

Kaifeng crops carbon density demonstrated a stable growth. It increased by 21. 3% from 11, 126 kg/hm^2 in 2004 to 14,135 kg/hm^2 in 2009 . This shows that the ability of crops fixing carbon of Kaifeng increased year by year.

2. 1. 2   The change of carbon storage and carbon density of major crops in Kaifeng

As shown in Fig. 2, during the period of 2004 ~ 2009, the carbon stocks of wheat was significantly higher than other crops, and it increased year by year, from 2,424,577 t to 2,981,829 t, a growth of 18. 7% ; carbon storage of peanuts also displayed an increased trend with each passing year, from 242,100 t to 477,020 t, a growth of 49. 3%; corn carbon stocks showed a rising trend ,a growth of 18. 7% ; the carbon stocks of soybeans, cotton was significantly lower than wheat , and the trend was increased at first and then decreased. About cotton, the planting area is only 66,910 hm^2 and the production is 62,909 t in 2009. The squeeze acreage of cotton planting due to fluctuations in cotton prices in recent years, the expansion area of grain and vegetables, and the lower cotton technology. So the total production of cotton annual is fluctuating, and the carbon storage is reduced.

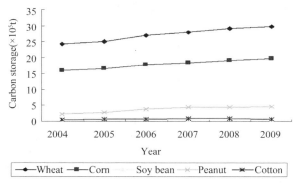

Fig. 2   The change trends of carbon storage of the major crops of Kaifeng in 2004 ~ 2009

As shown in the Fig. 3, about the carbon density of the major crops of Kaifeng from 2004 to 2009, wheat has a gradually increasing, but the rate is not large, the value is 12.3% ; peanut from the 3,012 kg/hm^2 increased to 4,986 kg/hm^2, an increase of 39.6% ; the carbon density of corn, soybeans, and cotton has a fluctuating phenomenon, in which corn carbon density fluctuates from 16,024 kg/hm^2 to 16,936 kg/hm^2. Carbon density also reflects the ability of the crop fixing carbon, and carbon sequestration capacity of corn is the highest, followed by wheat.

According to the composition analysis of crop average carbon storage of Kaifeng in the past six years(Fig. 4), it is known that the carbon stocks proportion of wheat in the five kinds of crops is very prominent, and the contributions of different crops to carbon stocks are as following: wheat (53%), corn (36%), peanuts (8%), cotton (2%), and soybean (1%).

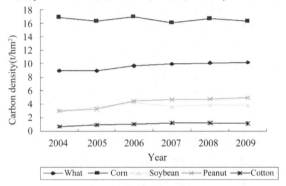

Fig. 3   The change trends of carbon density of the major crops of Kaifeng in 2004 ~ 2009

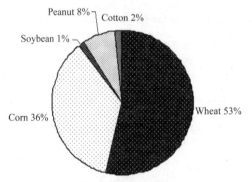

Fig. 4   The composition of crops carbon storage of Kaifeng

2.2   The spatial distribution of crop carbon storage and carbon density

Form the Tab. 2, it is obvious that the carbon storage of Qixian County is the largest in 2009 by the influence of cultivated land area, crop yields and other factors. In addition, carbon storage of Weishi County and Kaifeng County is the second largest, and the urban district has the smallest carbon storage. Due to its good natural conditions, Qixian County is a developed agricultural area and has a long-term cultivation of grain, cotton, and better farming foundation. On the other hand, QiXian County is gradually advancing to modern agriculture, including the new agriculture technology promotion, new varieties of grains introduction and cultivation and other measures. Therefore, the Qixian county has a higher crop yield and the carbon stocks is the highest in

KaiFeng. This arable land of Kaifeng County is only 19. 2% of the whole city, but the carbon stocks has accounted for 21. 6%. The reasons of this fact are as follows: superior natural conditions, sufficient sunshine, fertile soil, and some measures such as, optimization the structure of varieties and the promotion of new agricultural technology, so the per unit area yield of grain is high, these factors make Kaifeng County crops higher carbon storage. The carbon storage of urban district is the smallest by larger proportion of industrial land and the minimum of arable land.

Tab. 2   Kaifeng and the 5 - counties crop carbon storage and carbon density in 2009

Variables	Urban district	Qixian county	Tongxu county	Weishi county	Kaifeng county	Lankao county
Carbon storage (t)	151,396	993,579	577,733	913,618	940,574	778,840
Carbon density (kg/hm^2)	6,303	11,227	10,864	10,812	12,410	11,448

According to the distribution of carbon density in the counties and urban of Kaifeng in 2009 (Fig. 5), the Kaifeng carbon density distribution crops is uneven, it increases gradually from east to west. Kaifeng County has the highest carbon density and reach up to 12,410 kg/hm^2, the urban district has the lowest value of 6,303 kg/hm^2.

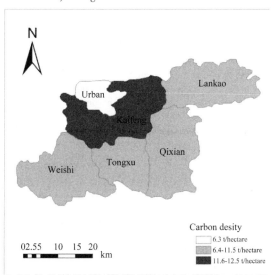

Fig. 5   The spatial distribution of carbon density of Kaifeng in 2009

2. 3   The structure of carbon storage of the entire Kaifeng city in 2009

Fig. 6 shows that in each county, the proportion of wheat carbon reserves is higher and more than 65. 0 %, the proportion of corn is followed, and the proportion of soybeans, cotton is relatively low. Among that, the proportion of wheat carbon storage has reached 71.2% in Tongxu County, far more than other counties. The proportion of corn and wheat carbon storage of Qixian County is also higher than other counties. It indicates that the main crops of the city and five counties are wheat and corn, but there is some difference in the structure of carbon stocks in each county because of the natural resource and planting structure.

3   Conclusions and discussions

We could draw some conclusions by estimating the crop carbon storage of Kaifeng from 2004 to 2009: ①the total carbon storage of Kaifeng increases stably with time, owing to the rising of crop

394

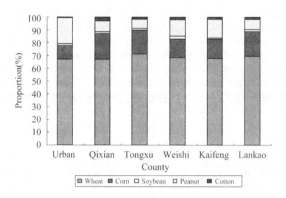

Fig. 6   The composition of crops carbon storage
of the Kaifeng region

yields and sown area. ②In the period of 2004 ~ 2009 , wheat and cotton take a higher proportion at the average carbon storage composition of Kaifeng's crops. So carbon storage and crop planting structure have a strong correlation. ③The crops carbon density of Kaifeng in southwest is relatively low, the spatial differences is significant.

In order to improve the carbon storage of corps through increasing the crop output of Kaifeng, retard the speed of CO^2 increase, some measures can be taken as follows: ①we could enhance the resistance of corps against nature disaster and reduce the dependence of corps on nature condition through enlarging input of agriculture basic facilities and improving the condition of agriculture manufacture. It may bring in the increasing of crops output during normal years and the stability of crops yield in disastrous years. ② It is necessary that take some measures to strengthening popularization on agriculture science and technology and farmland management, such as import excellent varieties for enhancing the adaptability of crops; fertilizing according to prescription and reduce the use of chemical fertilizers; improving the multiple crop indexes along with the reasonable rotations and the use of chemical fertilizer. These measures can significantly increase crop yield, thus enhance the carbon storage of the crops. ③ Adjusting the cropping system. Encouraging farmers to plant those corps which may be adept in absorbing carbon according to the local soil in the case of the grain production has no change. Take Wei shi County as a case, there are Shuangji River, Dugong River, Jialu River and Cornwall ditch flowing through the Weishi territory, and the cultivated area ranked the second in the whole city, which reaches to 21.4%. Local government may improve the carbon storage of corps by way of optimizing the planting structure and devoting more efforts to wheat and cotton planting.

Several standpoints might be discussed: ① the agro-ecosystems is not only a huge carbon sinks library, but also a source of GHG emissions. In order to reflect the carbon surplus situation of agricultural ecological system completely and extend the carbon sink of plants, the quantitative analysis of crops carbon storage should be combined with the study of crops carbon emission. ②We could increase crop yields by Agricultural technology and superior strains, thereby, the development of agriculture can contribute to carbon absorb and sequestration. ③ The point that we should only pay attention to the carbon sink of straw because of the process of consumption agricultural products engender carbon emissions is not really: it is no doubt about this view from a wide range of carbon cycle, but it's unfair when the producer and the consumer are not the same subject. So we should take some carbon compensation measures for the region of agricultural production. ④The compensating system as the core content, should be united with ecological compensation, which may urge green agriculture and low carbon development. More studies on carbon finance theory and practice also should be deepen and spread.

References

Huang Yao, Zhou Guangsheng, Wu Jinshui, et al. Chinese Terrestrial Ecosystem Carbon Budget Model[M]. Beijing: Science Press, 2008.

Tang HuaJun, Qiu JianJun. Estimations of Soil Organic Carbon Storage in Cropland of China Based on DNDC Model [J]. Geoderma, 2006, 134(1-2):200-206.

Liang Er, Cai Dianxiong, Dai Kuai, et al. Changes in Soil Organic Carbon in Croplands of China: II Estimation of Soil Carbon Sequestration Potentials[J]. Soil and Fertilizer Sciences in China, 2010(6):87-91.

Wang Shufang, Wang Xiaoke, Ouyang Zhiyun. Modeling Soil Organic Carbon Storage and Dynamics in Croplands of the Upperstream Watershed of Miyun Reservoir in North China[J]. Ecology and Environmental Sciences, 2009, 18(5):1923-1928.

Xu Xinwang, Pan Genxing, Sun Xiuli, et al. Changing Characteristics and Sequestration of Farmland Topsoil Organic Carbon in Guichi County Anhui Province [J]. Journal of Agro-Environment Science, 2009, 28(12):2551-2558.

Wang Shaoqiang, Yu Guirui. Spatial Characteristics of Soil Organic Carbon Storage in China's Croplands [J]. Pedosphere, 2005, 15(4):417-423.

Lu Chunxia, Xie Gaodi, Xiao Yu, et al. Carbon Fixation by Farmland Ecosystems in China and Their Spatial and Temporal Characteristics [J]. Chinese Journal of Eco-Agriculture, 2005, 13(3):35-37.

YU Guirui, Li Xuanran. Carbon Storage and its Spatial Pattern of Terrestrial Ecosystem in China [J]. Journal of Resources and Ecology, 2010, 1(2):97-109.

Jin Shulan, Yang Fangying. Carbon Stock Estimation and Analysis of Crops in Jiangxi [J]. Guangdong Agricultural Sciences, 2011(2): 216-218.

Zhao Rongxin, Qin Mingzhou. Temporospatial Variation of Partial Carbon Source/sink of Farmland Ecosystem in Coastal China [J]. Journal of Ecology and Rural Environment, 2009, 23(2): 1-6, 11.

Li Qingyun, Fan Wei, Yu Xinxiao, et al. Carbon Storage of Poplar-crop Ecosystem in Eastern Henan Plain [J]. Chinese Journal of Applied Ecology, 2010, 21(3):613-618.

Guang Zengyun. Study on Forest Biomass Carbon Storage in Henan Province [J]. Areal Research and Development, 2007, 26(1):76-79.

Zhang Jian, Luo Guisheng, Wang Xiaoguo, et al. Carbon Stock Estimation and Sequestration Potential of Crops in the Upper Yangtze River Basin [J]. Southwest China Journal of Agricultural Sciences, 2009, 22(2):402-408.

Wu Zhengfeng, Cheng Bo, Wang Caibin, et al. Effect of Continuous Cropping on Peanut Seedling Physiological Characteristics and Pod Yield [J]. Journal of Peanut Science, 2006, 35(1): 29-23.

Zhang Debian, Li Kexing, Yu Qianhua, et al. Analysis of the Similarities and Differences in Temperature and Precipitation Change in Recent 47 Years in Luoyang and Kaifeng City [J]. Meteorological and Environmental Sciences, 2009, 32(S1): 193-197.

Study on Carbon Lock – in in the Middle and Lower Yellow River Regions[①]

Lu Chaojun[1,2] , *Qin Yaochen*[1] , *Zhang Lijun*[1] and *Zhang Yan*[1]

1. Key Research Institute of Yellow River Civilization and Sustainable Development/College of Environment and Planning, Henan University, Kaifeng, 475004, China
2. Department of Energy and Environment, Chinese Research Academy of Environmental Sciences, Beijing, 100012, China

Abstract: With the increasing development of industrialization and urbanization, the energy-environment-resources pressure appeared. It's necessary for the regions to change to low-carbon economy pattern. Based on the concept of carbon lock-in and 1995 ~ 2010 energy consumption data, the regional characters about carbon emission pattern were studied. Carbon emissions per unit of GDP and per capita of the regions increased, Meanwhile, the evolution law of regional carbon emissions was also revealed by using Theil index. Then estimate whether the regions were in the carbon lock-in status or not. The results showed that the carbon emissions in the regions increased year by year, carbon emissions per unit of GDP was stable with a slight decline, carbon emissions per capita was rising. The difference of regional carbon emissions changed a lot, while that of regional carbon emission intensity decreased over time. So there is carbon lock-in in the middle and lower Yellow River.
Key words: the middle and lower Yellow River regions, evolution laws, carbon lock-in, regional distinction

1 Introduction

"Carbon lock-in" mentioned in the low-carbon economy is developed from the lock-in technology, and it means that technology highly dependent on fossil energy has taken a dominant role from the industrial revolution. The combination of political, society and carbon lock-in hinder the development alternative technologies, then keep the industrial economy in carbon lock-in status, especially in the fossil fuel energy system. The lock-in roots in different levels like technology, industry, regime and enterprises . In the study of China, carbon lock-in existed in the sector of power generation, automotive consumption and construction. The determining factor is technology. To unlock the situation, in addition to choose and spread the low-carbon technology as well as make it grow, the present regime should be broken down . Besides, a law was found that there was an inverted U-shaped curves between economy and carbon emissions. There are 3 peaks about carbon emissions, carbon emission intensity and carbon emission per capita, which the peak time differs from countries or regions . So the carbon lock-in status will be maintained until the 3 peaks have been stepped over.

As we all know, the proportion of carbon-based energy (coal, oil, natural gas) is up to 87% of the total energy. China is undergoing rapid industrialization and urbanization and resources shortage makes coal as the main primary energy in the future for a long period of time. Although China has already passed the peak of carbon emission intensity, the other 2 peaks have not come . In order to comply with the global trend of low-carbon development, implement the national greenhouse gas emission reduction objectives, China is facing severe carbon emission reduction

① Under the auspices of National Natural Science Foundation of China (NO. 41171438)
 Under the auspices of Major Program of National Natural Science Foundation of China (NO. 2012CB955800)
 Under the auspices of Major Program of Humanities and Social Sciences Key Research Base of Ministry of Education of China (NO. 10JJDZONGHE 015)

pressure. The middle and lower Yellow River regions are China's major coal production bases and has characteristics of the high-carbon. Its energy structure is dominated by coal. Meanwhile, the regions are relatively fragile of the ecological environment. It is important for the regions to coordinate the relationship between the rigid constraint of economic development and resources and the environment by shifting high-carbon economy to low-carbon economy.

2 Methodology

2.1 Carbon emission accounting

The middle and lower Yellow River regions (Shaanxi, Shanxi, Henan, Hebei, Shandong) were the research object in this paper. Data in this paper were from *China Statistical Yearbook* (1996 ~ 2011) and *China Energy Statistical Yearbook* (1997 ~ 2011). GDP was calculated at comparable price in 1995. Reference Method provided by IPCC was used to account carbon emission from primary energy consumption by the following equation:

$$E = \sum C_i \cdot EF_i \tag{1}$$

where, E_i is the carbon emissions from primary energy consumption, t-CO_2; i is the energy type, e. g. crude coal, gasoline and natural gas; C_i is the energy consumed in energy type i, tec; EF_i is the carbon emission coefficient of energy type i, t-CO_2/tce, EF_i is from *China Sustainsble Development Strategy Report 2009* (Tab. 1).

Tab. 1 Coefficient of CO_2 emissions of different sources of energy

(Unit: t · CO_2/tec)

	Coal	Coke	Natural gas	Crude oil	Fuel oil	Gasoline	Kerosene	Diesel oil	Electricity
Coefficient	0.747,6	0.112,8	0.447,9	0.585,4	0.617,6	0.553,2	0.341,6	0.591,3	2.213,2

2.2 Theil index

Variation Coefficient, Gini Coefficient and Theil Index are commonly used to study regional differences. In this paper Theil Index is used to discriminate the carbon emission differences among the middle and lower Yellow River regions. The equation is as follows:

$$T = \sum_{i=1}^{n} (C_i/C) \cdot \lg \frac{C_i/C}{GDP_i/GDP} \tag{2}$$

where, n is the number of the provinces involved in the calculation; GDP_i is GDP of each province; GDP is regional gross; C_i is provincial carbon emissions or carbon emission intensity; C is regional carbon emissions or carbon emission intensity.

The higher the Theil Index is, the greater the intra-regional difference will be.

3 Results

3.1 Regional carbon emission characteristic

On existing research results on the pattern of regional CO_2 emission in China showed that the scale of carbon emissions was divided into four main categories, light, general, heavy and extra heavy. Accounting results showed the spatial pattern in the middle and lower Yellow River regions greatly changed from 1952 to 2010, experiencing four categories in turn (Tab. 2). In 2010, carbon emissions in a descending order as: Shandong, Hebei, Henan, Shanxi, Shaanxi.

In 1995 ~ 2010, the carbon emissions of each province in the middle and lower Yellow River regions increased year by year. Growth rate fluctuated in the inter-annual (Fig. 1). The average annual growth rates of carbon emissions were 9.03% (Shaanxi), 5.41% (Shanxi), 8.32% (Henan), 7.18% (Hebei) and 9.53% (Shandong). The rate peak emerged respectively in 2002

(Shanxi), 2004(Shaanxi and Henan), 2005(Hebei and Shandong). Fig. 2(a) shows carbon emissions per unit of GDP trends. The curve of Shanxi declined significantly while others increased from 2003 to 2006. Besides, Fig. 2(b) shows carbon emissions per capita trends. The curves were on the rise overall. Three curves represented Henan, Hebei and Shaanxi increased steadily. Another two ones increased sharply in 2001 ~ 2007. The curve of Shanxi fluctuated in 2007 ~ 2010. The average annual growth rates of carbon emissions per capita were 8. 62% (Shaanxi), 4. 43% (Shanxi), 8. 14% (Henan), 6. 43% (Hebei) and 8. 86% (Shandong).

Tab. 2   The changing patterns of CO_2 emission in middle and lower Yellow River regions in 1952 ~ 2010

Year	1952	1980	1990	1995	2000	2005	2010
Shaanxi	●	●●	●●	●●	●●○	●●●	●●●●
Shanxi	●	●○	●●○	●●●	●●●	●●●●	●●●●
Henan	●	●●	●●○	●●●	●●●	●●●●	●●●●
Hebei	●	●○	●●○	●●●	●●●	●●●●	●●●●
Shandong	●	●●○	●●●	●●●	●●●	●●●●	●●●●

Note: ●Light(≤999 × 10^4 t/a) ●○General II(1,000 × 10^4 ~ 1,999 × 10^4 t/a) ●●General I(2,000 × 10^4 ~ 2,999 × 10^4 t/a).

●●○Heavy II(3,000 × 10^4 ~ 5,999 × 10^4 t/a) ●●●Heavy I(6,000 × 10^4 ~ 9,999 × 10^4 t/a) ● ●●●Extra HeavyI(≥10,000 × 10^4 t/a)

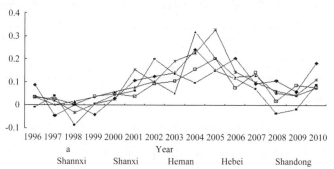

Fig. 1   The change rate of carbon emissions in Middle and Lower yellow River regions

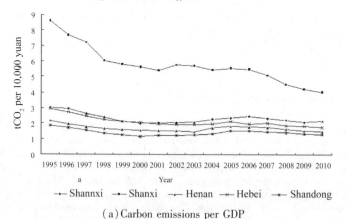

(a)Carbon emissions per GDP

Fig. 2   Carbon emissions intensity in Middle and Lower Yellow River regions in 1995 ~ 2010

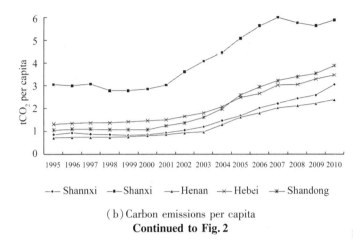

—•— Shannxi —■—Shanxi —▲—Henan —×—Hebei —*—Shandong

(b) Carbon emissions per capita
Continued to Fig. 2

3. 2 Regional Distinction

In 1995 ~ 2010, the regional distinction in the middle and lower Yellow River regions were changed(Fig. 3 (a)). Carbon emissions Theil Index waved in the period of 1995 ~ 1999 and 2004 ~ 2010. The Theil Index curve rose slowly between the two periods. The figure showed that maximum index presented in 2010 while the minimum in 2005, and average was 0. 109. It indicated that the greatest regional distinction was in 2010. On the other hand, carbon emissions intensity Theil Index changed quite gentle overall(Fig. 3 ~ (b)). The most significant year was 1995. After that the index declined in three year. Until 2006, it didn't stop rising. But in the short-term decline through 2005 and 2008, it rebounded in 2009 and 2010. As a whole, the inter-regional differences was decreasing over time. Carbon emissions Theil Index and intensity index was basically consistent with the trends of the regions' actual carbon emissions and carbon intensity.

3. 3 Carbon emissions evolution

According to the three inverted-U curves for carbon emissions, the evolution of carbon emissions can be divided into four stages . Phase I is before carbon emissions intensity peak. Phase II is from carbon emissions intensity peak to carbon emissions per capita peak. Phase III is from carbon emissions per capita peak to carbon emissions peak. Phase IV is carbon emissions steady decline phase. In the last phase, regional economic development and carbon emissions will be in a strong decoupling status, and low-carbon transition will be done.

Based on the analysis above, the middle and lower Yellow River regions have crossed the peak of carbon emissions intensity, but carbon emissions per capita and carbon emissions are still in the rising. Therefore, the regions are undergoing Phase II (Fig. 4).

As is mentioned before, a region will last in a carbon lock-in status until it crosses three peaks of carbon emissions. So do the middle and lower Yellow River Regions. In spite of the same phase, 5 provinces (Shaanxi, Shanxi, Hebei, Henan, Shandong) are being through, their different economic development and GDP per capita will make them reach the diverse peak of carbon emissions as well as vary in carbon lock-in period.

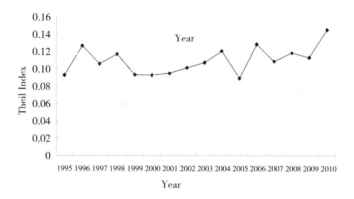

(a) Carbon emissions Theil Index

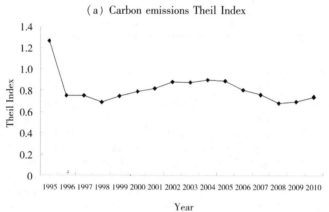

(b) Carbon emissions intensity Theil Index

Fig. 3 Carbon emissions Theil Index in the Middle and Lower Yellow River regions in 1995 ~ 2010

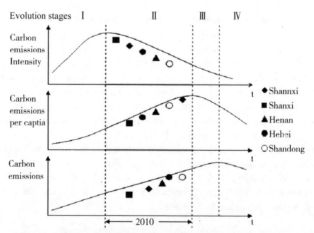

Fig. 4 Carbon emissions evolution in the Middle and Lower Yellow River regions

4　Conclusions and discussions

The middle and lower Yellow River regions (Shaanxi, Shanxi, Henan, Hebei, Shandong) were the research object in this paper. The regional carbon emissions evolution laws were revealed by analyzing the pattern change and regional differences of their carbon emissions, carbon emissions per capita and carbon emissions intensity. Then estimate whether the regions were in the carbon lock-in status or not.

(1) The middle and lower Yellow River regions' spatial changing pattern of carbon Emissions was as same as previous research. The spatial pattern in the regions greatly changed from 1952 to 2010, experiencing four categories (light, general, heavy and extra heavy) in turn. Shandong Province topped carbon emissions list and Shaanxi Province ranked the last one. The results also showed that the carbon emissions in the regions increased year by year, carbon emissions per unit of GDP was stable with a slight decline, carbon emissions per capita was rising. The waved carbon emissions Theil Index indicated that the difference of regional carbon emissions changed a lot, while that of regional carbon emission intensity decreased gently over time which mean the regional distinction was narrowing.

(2) The regions were undergoing Phase Ⅱ (period from carbon emissions intensity peak to carbon emissions per capita peak). Their different economic development and GDP per capita will make them reach the diverse peak of carbon emissions as well as vary in carbon lock-in period. They will maintain carbon lock-in status before they cross the peak of carbon emissions per capita and carbon emissions.

(3) The regions are China's important energy production and export bases which are rich in coal resources. They are in a period of rapid urbanization and industrialization with features of sharp population growth and industrial structure optimization and upgrading. At the same time, the regions face the resources constraints in the future development. The coal-dominated energy consumption structure is hard to change in the short time and it will make the regions in the carbon lock-in status for a long time. Although 5 main provinces in the regions are through the same carbon emissions evolution phase, the regional difference will be widen or narrowed by industry and energy structure adjustment as well as energy conservation and emission reduction strategy implement. Meanwhile, low-carbon transition of each province will in different pace. And further study should be done to quantitatively determine when the regions will be unlocked.

References

Zhuang Guiyang. How Will China Move Towards A Low Carbon Economy [M]. Beijing: China Meteorological Press,2007.

Kemp R. Technology and the Transition to Environmental Sustainability [J]. Futures, 1994,26 (10): 1023 – 1046.

Messner F. Material Substitution and Path Dependence: Empirical Evidence on the Substitution of Copper for Aluminum [J]. Ecological Economics, 2002(42): 259 – 271.

Unruh G C. Understanding Carbon Lock – in [J]. Energy Policy,2000, 28(12): 817 – 830.

Unruh G C. Globalizing Carbon Lock – in [J]. Energy Policy,2006, 34(10): 1185 – 1197.

Unruh G C. Escaping Carbon Lock – in [J]. Energy Policy, 2002, 30(4): 317 – 325.

Yang Lingping, Lv Tao. Analysis of Carbon Lock – in in China and Unlocking Strategies [J]. Journal of Industrial Technological Economics, 2011 (4): 151 – 157.

Xie Laihui. Carbon Lock – in, Unlocking and Low Carbon Development[J]. Opening Herald,2009 (5): 8 – 14.

The Sustainable Development Strategy Research Team of Chinese Academy of Sciences. China Sustainable Development Strategy Report 2009 [C]// China's Approach towards a Low Carbon Future. Beijing: Science Press,2009.

Zhang Lei, Huang Yuanxi, Li Yanmei, et al. An Investigation on Spatial Changing Pattern of CO_2 Emissions in China [J]. Resources Science, 2010,32(2): 211 – 217.

Research on the Role of Microbial Bacteria in Inhibiting Algae[①]

Shi Wei, Cui Wenyan, Hou Siyan, Wang Liming,
Zhou Xushen, Zhang Hui and *Lin Chao*

Haihe River Water Resources Commission of Ministry of Water Resources,
Tianjin, 300170, China

Abstract: Photosynthetic bacteria and EM bacteria, the most widely used microbial bacteria in the area of environmental protection, are effective in disposing sewage and inhibiting algae. But there are few studies on which one can achieve better effects and higher economic profits. To figure out the optimal bacteria and dosage for algae removal, two lab experiments are designed. Firstly, the experiments compare the algae exhibiting effects of photosynthetic bacteria and EM bacteria and then compare the algae exhibiting effect by different bacteria dosages. The result shows that photosynthetic bacteria are superior to EM bacteria in exhibiting alga and the optimal concentration of photosynthetic bacteria is 0.05%. At last, this paper compares the economic benefits of photosynthetic bacteria and EM bacteria to demonstrate that photosynthetic bacteria have higher application values than EM bacteria in exhibiting algae.

Key words: photosynthetic bacteria, EM bacteria, algae exhibiting effect

In recent years, the algal bloom is no longer a term special for the water body in South China. It may also happen in northern areas. Studies reveal that some tiny organisms in water play an important role in the biodegradation of algae and relevant toxic byproducts. Now it is found that many myxobacteria, blue-green algae bacteriophage and funguses can split the nurse cells of algae or damage a particular cell structure. The microorganisms released most frequently are photosynthetic bacteria and effective microorganisms.

Photosynthetic bacteria are a type of beneficial bacteria being able to photosynthesize. This microorganism is very important in nature. Photosynthetic bacteria can absorb and transform the phosphorus in the eutrophic water, decompose and release nitrogen and transform organics rapidly into nutrients which can be absorbed by aquatic organisms. Photosynthetic bacteria can also exhibit algae well. Studies show that photosynthetic bacteria can improve water quality, promote water transparency, increase the content of DO and beneficial algae and suppress algal bloom.

Effective microorganisms are microorganisms cultivated through special fermentation processes of mixing selected aerobic and facultative aerobic microorganisms, including eighty species of beneficial microorganisms in ten genera among seven types of microorganisms such as photosynthetic bacteria, lactic acid bacteria, saccharomycetes, bacillus, acetic acid bacteria, bifidobacterium and actinomycetes. EM bacteria are widely applied to industry, agriculture, animal husbandry and environmental protection. As for environmental protection, EM bacteria can promote the decomposition of organic pollutant, decrease BOD_5 and COD, purify water, improve the purifying capacity of the sewage disposal system, reduce the sludge and cost, compete with algae with nutrition, inhibit the growth and breeding of algae and decrease the possibility of cyanobacterial bloom.

Both photosynthetic bacteria and EM bacteria can play a role in disposing sewage and inhibiting algae, but which one can lead to better effect and high lab experiments. At first it compares the algae removal effect of photosynthetic bacteria and EM bacteria, selects out the optimal bacteria, and then determines the best use concentration and at last compares the economic profit of photosynthetic bacteria and EM bacteria.

①   The study is financially supported by the Commonweal Projects Specific for Scientific Research of the Ministry of Water Conservancy of China (Grant No. 201101018).

1 Comparison of the algae removal effect of photosynthetic bacteria and EM bacteria

1.1 Experimental design

The sample water is from a fish pond, DO 3.8 mg/L and pH 6.6. The dominating alga is green alga. The algae density is 7.23 × 109 per L. The experiment has seven processes which are EM 0.1%, EM 0.3%, EM 0.4%, EM 0.5%, photosynthetic bacteria 0.3%, photosynthetic bacteria 0.5% (all are volume ratio) and the control group. Every process has two parallels. Pour 4,000 mL sample water into a 5,000 mL beaker. BG11 nutrient solution (excluding photosynthetic bacteria 0.5%) is added in case of loss of nutrition in algae. Microcystis aeruginosa is inoculated in the sample water until the density of the microcystic aeruginosa reaches up to 107 per L. Then photosynthetic bacteria and EM bacteria are added according to the bacteria dosage. Natural lighting and room temperature (25 ~ 29 ℃) are required. The experiment lasts eleven days. Absorbance, DO, pH, TP and TN are measured for every day.

1.2 Experimental method

Absorbance: 722 Spectrophotometer, wavelength 440 nm;
DO: Leici JPB – 607 Portable Dissolved Oxygen Meter;
pH: LC PHB – 4 pH Meter;
TP, TN: SEAL Autoanalyzer 3 Flow Injection Analyzer.

1.3 Experimental findings and analysis

From the third day on, photosynthetic bacteria seem to be significantly superior to EM bacteria. The sample water processed by photosynthetic bacteria is clearer, less stinky and more light-colored and has fewer bubbles than that processed by EM bacteria. All of the bottoms and walls are attached with green dirt. Till the seventh day, the difference among the processes becomes more significant. In view of the overall effect, photosynthetic bacteria are superior to EM bacteria. 3% photosynthetic bacteria are the best, and then 5% photosynthetic bacteria. Among the EM processes, the less the EM bacteria are used, the better the effect is. The sample water processed by 0.5% EM is the greenest and the transparency is the lowest. Till the ninth day, the stink of the sample water processed by photosynthetic bacteria all disappears. The sample water is clear and transparent. However, the sample water processed by EM bacteria is still stinky. The sample water is green and there are still some bubbles. The following part discusses from the perspectives of biological indicators and chemical indicators.

1.3.1 Changes of absorbance

Absorbance is positively correlated to algal density. What absorbance reflects is the quantity of algae. See Fig. 1 for the experimental results. It can be seen from the chart that except the water processed by 0.3% photosynthetic bacteria, the absorbance of the water processed by others all reaches up to the maximum on the fourth day, then decreases sharply until the eighth day, decreases slowly from the eighth day to the tenth day and then tends to be stable after the tenth day. This result is closely related to the growth characteristics of microcystis aeruginosa which grows rapidly during the first four days, presenting an obvious exponential growth trend, then steps into the decline period and after about ten days grows slowly. At the beginning of the experiment, the activity and the algae control effect of all bacteria should also be the strongest. However, because the growth speed of algae is far higher than the speed of bacteria in inhibiting algae, during the first four day, the algae still presents a significant growth trend. After four days, the algae step into the decline period and the growth speed decreases greatly. The two (the growth speed of algae and the speed of bacteria in inhibiting algae) shall not strike a dynamic balance until the tenth day.

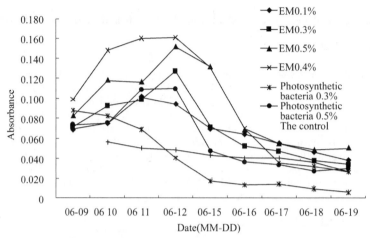

Fig. 1 **Changes of absorbance with time curve under different processes**

1.3.2 Changes of DO

Some research shows that photosynthetic bacteria can significantly increase dissolved oxygen content in the water. But some studies demonstrate that photosynthetic bacteria exert little influence on DO in the water. After eleven days' cultivation through different processes, DO tends to increase. The effect caused by 0.3% photosynthetic bacteria is the most evident, see Fig.2.

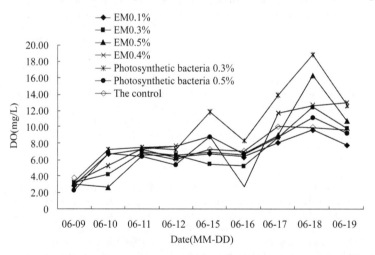

Fig. 2 **Changes of DO with time curve under different processes**

1.3.3 Changes of pH

See Fig.3 for the changes of pH with time curve under different processes. The changes of pH under different processes present the same trend. They all at first increase, then on the seventh day reach up to the maximum and later start to decrease. But generally it remains between eight and ten days. It shows that both photosynthetic bacteria and EM bacteria can stabilize pH. This conclusion is in accordance with the research achievements made by other researchers

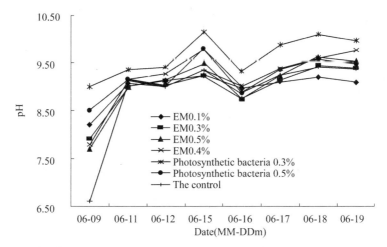

Fig. 3　Changes of pH with time curve under different processes

1.3.4　Changes of TN and TP

As shown in Fig. 4. changes of TN are significantly different between EM bacteria and photosynthetic bacteria. TN of water processed by EM 0. 1% , 0. 3% and 0. 5% decreases at first, then reaches to the minimum on the fourth day and then starts to increase. But it tends to increase on the whole. TN tends to change similarly in case of EM 0. 4% , the control group, photosynthetic bacteria 0. 3% and photosynthetic bacteria 0. 5%. TN at first decreases, then increases and later decreases again. It tends to decrease generally.

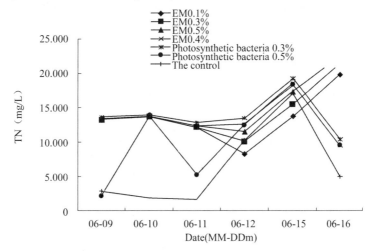

Fig. 4　Changes of TN with time curve under different processes

As shown in Fig. 5, the changes of TP under different processes tend to be the same (except photosynthetic bacteria 0. 5%). Generally TP tends to decrease. TP of water processed by EM presents two small peaks on the fourth day and the ninth day, but generally it still tends to decrease. TP under the process of photosynthetic bacteria 0. 3% demonstrates significant dephosphorization effect from the beginning of the experiment. During the entire experiment, TP removal rate reaches up to 83. 9% , followed by EM 0. 4% （70. 3%）, EM 0. 5% （53. 1%）and

EM 0.3% (39.0%). Therefore, as for EM, if the concentration is set, 0.4% should be the optimal for dephosphorization. TP under the process of photosynthetic bacteria 0.5% is 0.020 mg/L on the first day. During the experiment, TP does not decrease. On the contrary, it gradually increases. The reason may be that the process does not include any BG11 nutrient solution. For lack in nutrition, some algae die and then release phosphorus in the water.

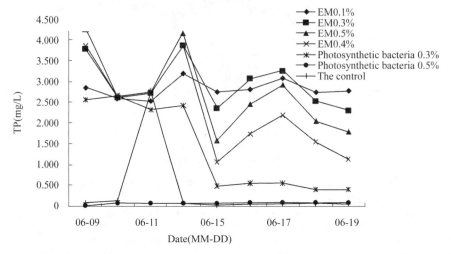

Fig. 5　Changes of TP with time curve under different processes

Many studies demonstrate that EM bacteria and photosynthetic bacteria show excellent effects in removing phosphorus and nitrogen. This research comes to a similar conclusion. But TN under the processes of EM 0.1%, 0.3% and 0.5% increases, which is against this conclusion. The reason may be that the concentration is not suitable. The TN removal rate under the process of EM 0.4% is 29.8%, which illustrates that 0.4% is a better choice for nitrogen removal.

Compared with EM, photosynthetic bacteria 0.5% demonstrates significant superiority from the seventh day on. The absorbance is 0.023 less than the smallest (0.1%) by EM, which demonstrates that photosynthetic bacteria 0.5% is better than all of EM processes in exhibiting algae. Photosynthetic bacteria 0.3% has better performance. From the first day till the end of the experiment, it always shows excellent effect in exhibiting algae. There is no growth peak which appears under other processes. So it denies the conclusion that the more the bacteria are added, the better the inhibiting effect is. There is no positive correlation between them. The reason is that the excessive microbial bacteria may promote the growth and reproduction of algae because of the addition of other nutrient substances. So there is an optimal dosage when microbial bacteria are used to inhibiting algae.

According to the above analysis, the experimental phenomenon, the water quality index and the changes of absorbance all demonstrate that photosynthetic bacteria have better performance than EM bacteria in inhibiting algae and purifying water.

2　The optimal concentration of photosynthetic bacteria

2.1　Experimental scheme

The above experiment illustrates that no matter for photosynthetic bacteria or EM bacteria, there is an optimal dosage. Some studies show that the optimal dosage of EM bacteria is 0.1% ~ 0.01% (volume ratio). The optimal dosage of photosynthetic bacteria is 0.02% ~ 0.1% (volume ratio). To verify these conclusions, this experiment consists of five processes which are the control group, EM 0.01%, EM 0.05%, photosynthetic bacteria 0.01% and photosynthetic bacteria

0. 05% . Every process has two parallels. Pour 4,000 mL deionized water into a 5,000 mL beaker and add BG11 nutrient solution in case that the algae may be in shortage of nutrition. Microcystis aeruginosa is inoculated in the sample water until the density of the microcystic aeruginosa reaches up to 1×10^7 per L. After the algae stay in the water for two days, photosynthetic bacteria and EM bacteria are added according to the bacteria dosage. Natural lighting and room temperature (25 ~ 30 ℃) are required. The experiment lasts 11 d. The density of algae is measured for every day.

2. 2　Indicator measuring method

Algae density: Blood cell counting method.

2. 3　Experimental results and analysis

On the third day, compared with the control group, the bacteria process has more sediment on the bottom and the sample water is clearer. The sample water processed by photosynthetic bacteria is clearer than that processed by EM and the color is lighter. As for the sample water processed by the same type of bacteria, 0. 05% dosage leads to better effect than 0. 01% dosage. There is more sediment on the bottom, the sample water is clearer and the color is lighter. From the fifth day on, the sample water processed by photosynthetic bacteria 0. 05% becomes clearer and clearer. And there are bubbles on the surface. The sample water processed by EM bacteria has the most bubbles. Till the eighth day, the color becomes lighter and the water becomes clearer. Especially the sample water processed by photosynthetic bacteria 0. 05% becomes transparent.

See Fig. 6 for the experimental results. Photosynthetic bacteria 0. 05% starts to perform well in exhibiting algae from the fourth day on. On the eighth day, the algae density is 0. But as for photosynthetic bacteria 0. 01% , the algae density starts to increase from the first day to the seventh day and reaches up to the maximum on the seventh day. And then it gradually decreases until the end of the experiment when the density is 7. 4 $\times 10^7$ per L. So we can come to an evident conclusion that photosynthetic bacteria 0. 05% performs far better than photosynthetic bacteria 0. 01% in exhibiting algae. In this experiment, the optimal concentration of photosynthetic bacteria for algae exhibition is 0. 05% . The law is similar to the two processes of EM bacteria. The algae density at first increases and reaches up to the maximum on the sixth day. And then it starts to decrease. But during the entire experiment, the difference between the two processes is not significant. The algae density of the control group is significantly lower than that of the two processes of EM for the third day (except the seventh day). There are two possible reasons. The first is the experimental error. The second is that the EM bacteria have complicated composition and are rich in nutrition. In addition to the high concentration, the algae may grow and multiply rapidly.

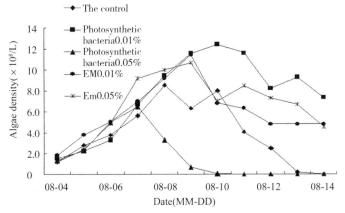

Fig. 6　Changes of algae density with time curve under different processes

3 Comparison of economic profits between photosynthetic bacteria and EM bacteria

The above experiment proves that photosynthetic bacteria are superior to EM bacteria in exhibiting algae and purifying water. And then what about their economic profits? If the calculation is based on 0.05%/dosage, the market price of EM is 10,000 ~ 20,000 yuan/t and the EM drug charge sewage is 5.0 ~ 10.0 yuan/m^3. However, the market price of photosynthetic bacteria is 4,000 ~ 6,000 yuan/t and the photosynthetic bacteria drug charge sewage is 2.0 ~ 3.0 yuan/m^3. And after the photosynthetic bacteria is added, we can greatly shorten the aeration time, decrease power consumption, reduce foul smell, decrease sludge and enhance nitrogen and phosphorus removal as well as algae exhibiting effect. Thus photosynthetic bacteria can lead to sound economic benefits and environmental benefits.

4 Conclusions and discussions

According to the above conclusion and analysis, photosynthetic bacteria are superior to EM bacteria both in exhibiting algae and economic benefits. So photosynthetic bacteria have better application values in purifying water and preventing water blooms. The optimal concentration of photosynthetic bacteria is 0.05% (volume ratio). In addition, we must pay attention to the problems such as adding mode and time when using photosynthetic bacteria, so as to make sure that we can make full use of them. We can also dilute photosynthetic bacteria with the water in the pond and then sprinkle it evenly in the pond. It can also be used together with zeolite powder mixture. Acid water is not favorable for the growth of photosynthetic bacteria. If the water is acid, add quick lime in the water and pH value shall increase. Then photosynthetic bacteria can be used too. Don't use it together with disinfectant and CuO_4. The effect shall be better if it is used during 10 p m and 4 a m on sunny days. Photosynthetic bacteria shall take effect in exhibiting water bloom after four days. Therefore they should be used three or four days before water bloom.

References

Sun Xuekai, Xu Chengbin, Ma Xiping. Photosynthetic Bacteria Characteristics and Applications in Each Industry [C]. Distinguished Papers in Annual Conference of Chinese Society for Environmental Sciences, 2007: 2133 – 2136.

Wang Yutang. Photosynthetic Bacteria and Applications in Aquaculture Industry (I) [J]. China Fishery, 2009 (3):48 – 51.

Fu Baorong, Cao Xiangyu, Leng Yang, et al. The Effect of Photosynthetic Bacteria on Aquaculture Water Quality and Aquatic Organism [J]. Biological Sciences, 2008(2): 102 – 106.

Lin Kunxia, Wang Xiaoquan, Hu Peiduo. Specific Function Light and Bacteria Suppressing Algal Bloom — Research on Photosynthetic Bacteria Inhibiting Algal Growth [J]. Environmental Science Survey, 2008, 27(1): 52 – 55.

Li Weijiong, Ni Yongzhen. Research on Application of EM Technique [M]. Beijing: China Agricultural University Press, 1998.

Chen Yuwei, Gao Xiyun, Chen Weimin, et al. Growth Traits of Microcystis in Taihu Lake and Preliminary Study on Isolation Pure Culture [J]. Journal of lake Science, 1999, 11(4): 351 – 355.

Xu Chengbin, Ma Xiping, Meng Xuelian, et al. Isolation Identification of Photosynthetic Bacteria and Application in River Crab Aquaculture [J]. Journal of Liaoning University (Natural Science Edition), 2009(1): 77 – 81.

Lin Dongnian, Ye Ning, Liu Xinghua, et al. The Effect of Photosynthetic Bacteria on Water Quality and Ecological in Tilapia Pond [J]. Research of Soil and Water Conservation, 2007 (2):207 – 212.

Meng Fanping, Li Kelin. Research on Degradation Ability of Effective Microorganisms to Organics in Living Sewage [J]. Journal of Central South Forestry University, 1997, 17(4): 8 – 13.

Shao Qing, Luo Wensheng. Application of EM on Batch Reactor to Deal with High Efficiency and Energy Saving Problem of Town Sewage[J]. China Rural Water and Hydropower, 2002(10): 27 30.

Qin Yunqi. Ion with Photosynthetic Bacteria to Remove Nitrogen in Eutrophic Water[J]. Shanxi Architecture, 2006, 32(1): 203 – 204.

Xiong Xiaojing, Cao Xiaoting. Application of EM Bacteria on Biological Sewage Treatment Process [J]. Environmental Hygiene Engineering, 2007, 15(3): 11 – 14.

Wang Lin, Wang Yingchun, Li Ji, et al. Indoor Laboratory Research on Microbial Agents Effect on Eutrophic Scenic Waters [J]. Journal of AgroEnvironment Science, 2007, 26 (1): 88 – 91.

Song Fengmin, Hu Shibin. Experiment for EM Bacteria Activated Sludge System Effect on Depth of Saponin Wastewater[J]. Industrial Water and Wastewater, 2008, 39(2): 39 – 41.

Yi Wenli, Wang Guodong, Liu Xuanwei, et al. Effect of Nitrogen and Phosphorus Ratio on Microcystis Aeruginosa Growth and Part of the Biochemical [J]. Journal of North West Agriculture and Forestry University (Natural Science Edition), 2005, 33(6): 151 – 154.

Zhao Ying, Zhang Yongchun. Interaction of Velocity and Temperature Effect on Microcystis Aeruginosa Growth[J]. Jiangsu Environment Science, 2008,21(1): 23 – 26.

Wang Ping, Wu Xiaofu, Li Kelin, et al. Research on Algae Exhibiting Effect of EM Bacteria[J]. Research of Environmental Sciences, 2004,17(3): 34 – 38.

Experimental Study of the Feature of Cohesive Sediment Movement with the Lower Energy Conditions

Wang Jiasheng [1], *Chen Li* [2], *Lu Jinyou* [1] and *He Shan* [1]

1. Changjiang River Scientific Research Institute, Wuhan, 430010, China
2. Water Resources and Hydropower Engineering Science,
Wuhan University State Key Laboratory, Wuhan, 430072, China

Abstract: Cohesive sediment is a major component of clay. Its erosion and movement has a significant impact on the soil environment. Moreover, cohesive sediment is an important part of the sediment of lakes and oceans, -moreover, its movement and transportation also have a crucial influence over dispersion of pollutants in water. Open channel flume experiments were carried out to study the characteristics of cohesive sediment movement; simultaneously the velocity and sediment concentration vertical profile was investigated. The experimental results indicate that under the conditions of suitable hydrodynamic velocity and sediment concentration, a special-shaped sediment flow, stratified movement, takes place in the flume. Its formation and features are closely related to fluid parameters and sediment concentration. Its velocity distribution shows a V-shaped pattern falling forward. At the corner of V, the maximum velocity is vertically located in the middle of the flume. The concentration distribution is vertically stratified into three discrete layers on whose interfaces the concentrations change suddenly. To form lamination movement, different hydrodynamic conditions are required under different sediment concentrations. When the concentration increases, the corresponding flow intensity also -rises up.

Key words: cohesive sediment, lamination movement, flocculation net structure, velocity distribution.

1 Introduction

Under the condition of strong flow, cohesive sediment is completely suspended and fine sediment is unsaturated. However, many observations on the regions of estuaries, ports and lake entrances indicate that under certain hydrodynamic conditions, sediment concentration, water depth and slope, the motion of cohesive sediment shows the characteristics of lamination. That means there is an obvious interface between clear and muddy waters, i. e. sudden change in sediment concentration in the vertical direction. Unfortunately, most of the studies are still primary observations; few experimental investigations have been carried out .

One of the most important topics in the sediment science is to study the vertical distribution of sediment concentration and to explore the laws of sediment motion and its interaction with water flow. Sediment motion is the interacting process of sediment depositing under gravity and water diffusing laterally, while the vertical distribution of sediment concentration is the result of the interacting process. By studying the process and its result, the laws that govern the interaction between fluid flow and sediment subsidence could be revealed.

In this paper, open channel flume experiments were performed to study the formation and features of lamination motion of cohesive sediments under different hydrodynamic conditions and sediment concentrations.

2 Experimental setup, methods, and factors

The experimental setup is shown in Fig. 1. The flume size is 20 cm in width, 25 cm in height, and 3.5 m in length. Its slope varies from 0 to 1%. The available maximum flow in the flume is 8 m^3/h. The cross-section used to measure the turbulence velocity is in the middle along the flume direction. The measured points were manually regulated and located. Flow velocity and turbulence were measured by a three-dimensional acoustic Doppler velometer. The oven-drying method was applied to measure the sediment concentration. A siphon sampler was used for sand sampling. In order to control the muddy fluid outflowing slowly and to get the required sand sample at the measured point, the height of the siphon outlet was set slightly lower than that of the measured point.

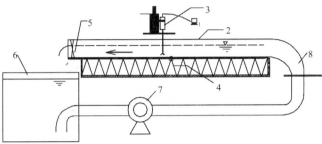

1—Computer; 2—Glass flume; 3—MicroADV; 4—Steel frame support;
5—Tail gate; 6—Reservoir; 7—Centrifugal pump; 8—Inlet

Fig. 1　Schematic of flume experiment system

Besides measuring flow velocity, sediment concentration and graduation, some parameters of fluid quality, such as its pH value, total saline concentration, the concentrations of Ca^{2+}, Cl^- and HCO_3^-, Ca^{2+} hardness, and Cl^- alkalinity, were also measured during the experiment. After the muddy fluid in the reservoir was mixed sufficiently by stirring to form uniform sediment concentration, it was pumped into the flume through electromagnetic flow meter, butterfly gate, and transmission pipe, before returning to the reservoir.

The parameters of muddy current flow in the flume are listed in Tab. 1. Tab. 2 shows the measurement results of the fluid quality in the experiments. The sediment $d_{50} = 4.5$ and sediment graduation used in the experiments is shown in Fig. 2.

Tab. 1　Parameters of flume experiment

Discharge (m³/h)	Depth(cm)	Re	Sediment concentration (g/L)	Temperature(℃)	pH
0.7~4.5	>15	>2,000	>1	15~25	7.5~8.5

Tab. 2　Ions concentration in experiment

pH	Ca²⁺ (mg/L)	Hardness (mg/L)	Cl⁻ (mg/L)	HCO₃⁻	Alkalinity (mmol/L)	Total saline Concentration (mg/L)
7.8	66.58	76.53	71.93	129.69	2.12	343.66

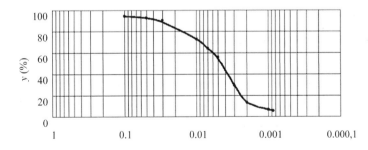

Particle diameter (mm)
y—Percentage of particles diameter less than particular value (%)

Fig. 2　Cumulative weight percentage frequency

3　Experimental results and analyses

The effects of hydrodynamic velocity and total sediment concentration on the motion properties of cohesive sediment were analyzed based on experimental results of the fluid velocity at the measured section and the sediment concentration in the vertical direction. The research was carried out - in two experimental conditions. Experiment 1 was performed under the same sediment concentration and different hydrodynamic intensities. Experiment 2 was conducted under the same hydrodynamic intensity and different sediment concentrations. Tab. 3 lists the experimental conditions in detail.

Tab. 3　List of experiment conditions

Experiment	Sediment concentration (g/L)	Hydrodynamic condition: Discharge (m³/h) [number#]					
Experiment 1	3. 5	3. 72 [11#]	2. 15 [12#]	1. 44 [13#]	0. 96 [14#]	0. 82 [15#]	0. 72 [16#]
	Discharge (m³/h)	Sediment concentration (g/L)					
Experiment 2	1. 45	1. 5 [21#]	2. 3 [22#]	3. 5 [23#]	3. 7 [24#]		
	2	3. 5 [25#]	10 [26#]				
	4	5 [27#]	10 [28#]				

3. 1　Vertical distribution of sediment concentration

Two groups of the experimental results of the vertical distribution of sediment concentration are given in Fig. 3 and Fig. 4, which corresponds to experiment Nos. 11# ~ 16# in Experiment 1 and experiment Nos. 21# ~ 28# in Experiment 2 respectively.

3. 2　Vertical distribution of muddy velocity

The vertical distributions of muddy velocity in the flume are shown in Fig. 5, among which experiment Nos. 11# and 12# were carried out in Experiment 1, while experimental Nos. 13# and 14# in Experiment 2.

3. 3　Analyses of experimental results

Based on the above experimental results, the sediment movement, under the conditions of certain water flow velocity and sediment concentrations, can be divided into three layers from the bottom to the water surface in the flume: the deposition layer on the bottom, sand-flowing layer in the middle, and the clear water layer on the surface. This kind of lamination movement is formed only in a specific flow velocity range. When the flow velocity is greater than the specific range, all the sediment is suspended, vertically and evenly distributed in the fluid body of the flume. In other words, the lamination cannot be formed; when the flow velocity is lower than the specific range, almost all the sediment is precipitated and the lamination disappears. Therefore, the lamination movement is a specific form of movement of cohesive sediment in a specific hydrodynamic velocity range.

As shown in Fig. 6 from Experiment 1, the vertical distributions of sediment concentration in curves 14 #, 15 #, and 16 # indicate that the lamination motion varies from its formation to disappearance with the flow velocity decreasing gradually. When the lamination is formed, the sediment concentrations of the middle sand-flowing layers of the three curves, which are at about 6

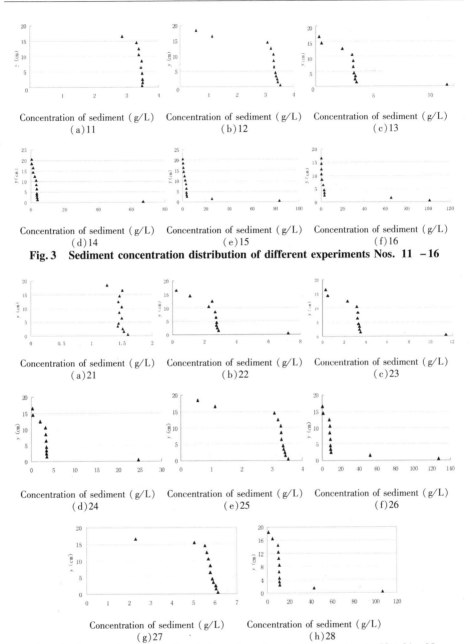

Fig. 3 Sediment concentration distribution of different experiments Nos. 11 – 16

Fig. 4 Sediment concentration distribution of different experiments Nos 21 ~ 28

cm from the bottom of the flume, are roughly the same at about 3.4 g/L. When the sediment concentration is higher than this number, the concentration at the deposition layer increases gradually with the decrease of the hydrodynamic velocity. The relationship of the layer thickness to fluid flow velocity, as shown in Fig. 7, also indicates that with the decrease of hydrodynamic velocity, the lamination movement first drops rapidly, then slowly, and at last dramatically.

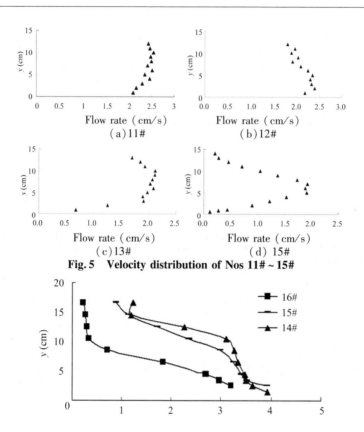

Fig. 5   Velocity distribution of Nos 11# ~ 15#

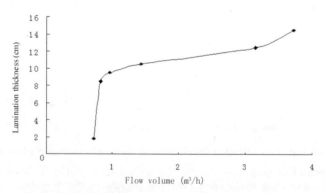

Sediment concentration（g/L）

Fig. 6   Comparison of sediment concentration distributions of 14# ~ 16#

Fig. 7   Relationship between hydrodynamic conditions and lamination thicknesses

Results of Experiment 2 show that the sediment concentration is a key factor to form the lamination movement. Under the given hydrodynamic velocity, the layered motion gradually becomes clear with the increasing sediment concentration. The lamination forms when the sediment concentration reaches a certain value, over which concentration, the layered form of motion is sustained and the sediment concentrations of the middle sand-flowing and bottom precipitation layers are increased.

The experimental results also indicate that the stronger the hydrodynamic velocity, the greater the optimum sediment concentration for the lamination movement is. Namely, the lamination is formed through the interaction of the current velocity with the sediment concentration.

The characteristics of velocity distribution in the flume are similar to that of density current. As shown in Figs. 5, after the formation of the lamination motion, the main flow with the maximum current velocity in the flume is located in the middle sediment-transporting layer, away from which the current velocities decrease gradually. Thus the flow velocity distribution in the flume shows a V-shaped pattern falling forward, typically as shown in Fig. 5. The characteristics of velocity distribution bear obvious resemblance to those of density current.

4 Formation mechanism of lamination motion

The formation, development, and disappearance of layered motion are affected by both hydrodynamic and gravitational interactions, resulted from flocculated and hindered settlings. When cohesive fine sediment forms flocculent masses, the settling velocity of aggregated masses is closely related to the suspended sediment concentration. Based on hydrostatic settlement, the settling velocity of cohesive fine sediment accelerates with the suspended load concentration increasing (so-called the flocculation settling); when the concentration reaches a certain value, flocculation net begins to form; after which time, when the concentration is increased continuously, the flocculation settling velocity will decrease (so-called the hindered settling). When the structure of flocculation net forms, the boundary of clear and muddy waters forms, i. e., the lamination movement emerges.

When the layered motion forms on the boundary of the upper layer and the middle layer, the internal waves occur due to the interaction between shear stress and buoyancy to which sediment is subjected. In the middle layer, sediment's vertical mixing will take place and flocculent masses diffuse upward towards the upper layer because of the collapse of the internal waves. Furthermore, owing to their potential difference, watery flocculent masses entering the upper layer at the lower concentration state will return to the previous positions after re-flocculation. Therefore, the upper layer remains stable under specific hydrodynamic and concentration conditions. Because the presence of floc network will make the settling velocity of sediment decrease, lower depositing velocity in the middle layer leads to the uniform distribution of concentration. While in the bottom layer, higher sediment concentration, stronger buoyancy, and greater sediment cohesiveness lead to the disappearance of the turbulence within the layer. Therefore, it is difficult to cause vertical exchange between the bottom and middle layers, whose boundary could prevent this situation. But with the current velocity increasing, turbulence will be activated again in the vicinity of the bottom layer, causing the enhancement of the re-suspending force of deposited matter. And finally, the lamination disappears.

The generation, development, and disappearance of flocculation net and lamination motion require specific gravitational and hydrodynamic conditions. The hydrodynamic velocity can not only promote, but also hinder the formation of flocculent network. In addition, cohesive sediment concentration, water quality, as well as complicated interactions between water and sediment all play very important roles. Accordingly, it is necessary to comprehensively consider various complex factors when studying the interaction between cohesive sediment and flow.

5 Conclusions

The experimental researches on sediment motion under the weak hydrodynamic velocity are presented in this article. The results indicate that the lamination movement occurs under the conditions of appropriate hydrodynamic velocity and sediment concentrations. The sediment concentration is vertically divided into three layers. The lowest concentration is in the upper layer, the highest in the bottom, and the medium in the middle. Among the layers, there are two boundaries where the sediment concentration changes discontinuously.

The occurrence of lamination movement is closely related to flow velocity and sediment concentration. Under a specific concentration, the formation of lamination motion requires a specific range of hydrodynamic parameters, beyond which (greater or smaller) the stratified motion will disappear. Corresponding to different fluid condition, different stratified thicknesses are formed with different sediment concentrations. While the lamination motion occurs, the velocity distribution on its cross section shows a V-shaped pattern falling forward, i. e. the maximum velocity is in the middle, over both sides of which, velocities decrease gradually towards the watery surface and the bottom of the flume.

Acknowledgements

This work was financially supported by Project supported by the National Natural Science Foundation of China (Grant No. 10802013) and the special fund for commonweal industries of Water Resources Ministry(200801004) .

References

Chinese Hydraulic Engineering Society. Sediment Handbook [M]. Beijing: China Environment Science Press, 1989.

Qian Ning, Wan Zhaohui. Sediment Movement Mechanics [M]. Beijing, Science Press, 1983.

Xie Jianheng. River Silt Engineering [M]. Beijing: Hydraulics Press,1981.

Xu Jianyi, Yuan Jianzhong. Study on the Fluid Mud in the Yahgtze Estuary [M]. Journal of Sediment Research, 2001(3): 74 – 81.

Shi Zhong, Ling Honglie. Vertical Profiles of Fine Suspension Concentration in the Changjiang Estuary [J]. Journal of Sediment Research, 1999(2): 59 – 64.

Li Jiufa, He Qing, Xu Haigen. The Fluid Mud Transportation Processes in Changjiang River Estuary[J]. Oceanology et Limnology Sinica, 2001(3): 302 – 310.

Dong Lixian. Mixing Process in Highly Turbid Waters of Jiaojiang River Estuary[J]. Oceanology et Limnology Sinica, 1998 (29): 535 – 541.

Li Yan, Xia Xiaoming, Dong Lixian. Distribution of Fluid Mud Layer in the Jiaojiang Estuary[J]. Acta Oceanologica Sinica, 1998(20): 72 ~ 82.

Xie Qinchun, Li Bogen. Distribution of Suspended Sediment Concentration and Development of Fluid Mud Layer in the Jiaojiang Estuary [J]. Acta Oceanologica Sinica, 1998(6): 58 – 69.

Validity Analysis and Downscale Research on GLDAS Temperature Data in the Weihe River Basin

Wang Mingcheng, *Yang Shengtian*, *Lv Yang* and *Song Wenlong*

School of Geography, Beijing Normal University, State Key Laboratory of
Remote Sensing Science, Beijing Key Laboratory of Environmental Remote Sensing and
Digital City, Beijing, 100875, China

Abstract: The temperature observation data of Huan County station and Changwu station in Weihe River Basin are used for the validity analysis of the temperature parameters provided by GLDAS products which are assimilated in NOAH mode on day, month and year scale. Based on the temperature vertical decline rate, GLDAS temperature data are downscaled with the surface elevation with the reversed distance weighted method. The spatial resolution of GLDAS temperature data are increased from 0.25 ℃ to 0.01 ℃, the data precision are improved, and the downscaled data highlight the influence of the terrain, thus the distributions of temperature are more reasonable. The result shows that the temperature parameters provided by GLDAS products which are assimilated in NOAH mode are able to reflect the distributions of temperature in Weihe River Basin, and the downscale method used in this article is feasible.

Key words: Weihe River Basin, GLDAS temperature data, validity analysis, downscale

1 Introduction

GLDAS (Rodell M, et al., 2004), namely Global Land Surface Data Assimilation System, obtain more accurate data mainly by driving land surface modes with observation data or real-time model assimilation. GLDAS provides four assimilation and analysis products in land surface mode, including Mosaic, NOAH, VIC and CLM. The GLDAS product assimilated in NOAH mode with temporal resolution of 3 h and spatial resolution of 25 km are used in this article, and 28 parameters including atmospheric pressure, temperature and wind velocity are provided.

Temperature is one of the most common meteorological input parameter in hydrological models. The site temperature observation data are used for the validity analysis of GLDAS temperature products, and the downscale method of GLDAS temperature data is discussed. This method is used in the typical arid and semi-arid area in China of— Weihe River Basin in China, which has important significance on the research of distributions of temperature and water cycle in Weihe River Basin.

2 Study area and data processing

2.1 Study area

Weihe River is the first major tributaries of the Yellow River, its length is 818 km, and it originated in Niaoshu Mountain in Weiyuan County, Gansu Province. It flows through 86 counties in Gansu, Shaanxi and Ningxia Province, and injects in the Yellow River in Tongguan County, Shaanxi Province (Wang Yanlin, et al., 2004). Weihe River Basin is located in the southeast of the Loess Plateau, covers an area of 134,800 km². The scope of the basin is mainly in the center of Shaanxi Province, and it is located in arid and semi-arid area, the climate there is continental monsoon. The average annual temperature of Weihe River Basin is 6 ℃ to 14 ℃, while the average annual rainfall is 450 mm to 700 mm, and rainfall is mostly in July to October. The rainfall decreases from southeast to northwest in the basin and it is unevenly distributed. The average annual evaporation capacity is 700 mm to 1,200 mm, and the average annual frost-free period is 120 d to 220 d.

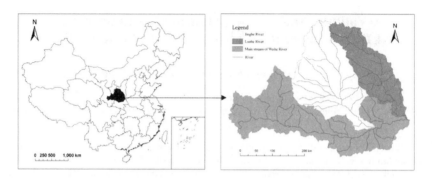

Fig. 1   The geographical position of Weihe River Basin

2. 2   Data processing

The GLDAS data are downloaded from "http://disc. sci. gsfc. nasa. gov/hydrology/data-holdings" "http://disc. sci. gsfc. nasa. gov/hydrology/data-holdings". The original format of GLDAS data is . grb. The temperature parameters are extracted by GLDAS data batch program compiled by IDL language, and they are converted to ENVI standard format data with grid size of $0.01° \times 0.01°$. The temperature data distribute as grid shape, and the grid size is $0.25° \times 0.25°$ degree, as shown in Fig. 2.

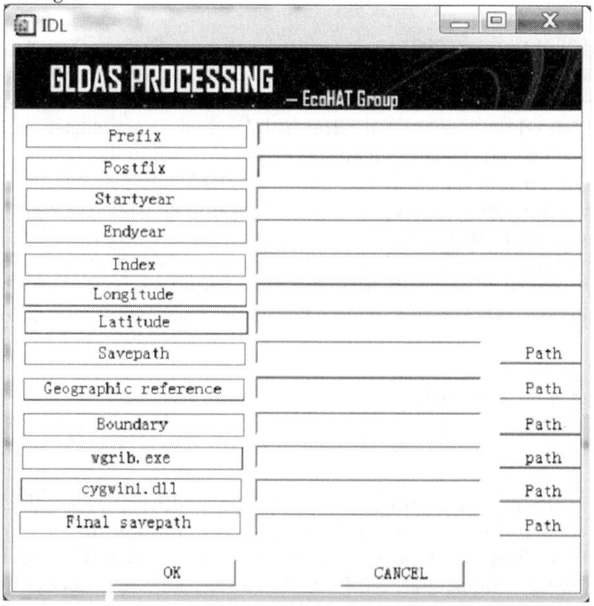

Fig. 2   GLDAS data batch program interface

The 1: 250,000 DEM from National Foundation Geographical Information System are chosen as surface elevation data (Zhai Liang, et al., 2006). The main information source is the national 1: 250,000 scale topographic map, the related elements including landform and water system are collected from the map, and raster data sets are formed after complex data processing. The data are stored as the shape of map block, the spatial resolution is 100 m. The data is the domestic authority and important surveying and mapping data with high accuracy. The sheets within Weihe River Basin are joined together by ARCGIS, and a surface elevation grid map with spatial resolution of 0.01° is formed after resample and clip process, as shown in Fig. 3.

Fig. 3   Elevation map of Weihe River Basin

3   GLDAS data validation

The GLDAS temperature values of Huan County station and Changwu station in Weihe River Basin in 2010 are selected, and China's ground climate data sets downloaded from China Meteorological Data Sharing Service System are used for the validity analysis of GLDAS temperature parameters on day, month and year scale.

As shown in Fig. 4, the average annual GLDAS temperature values are slightly greater than the observation data, but deviations are all within 20%. As shown in Fig. 5, the correlation coefficients between the mean monthly temperature and the observation data achieve 0.996 and 0.988 respectively, and the root-mean-square errors are 0.758 ℃ and 1.612 ℃ respective. The GLDAS temperature data from June, 2010 to August, 2010 are used for the validity analysis of GLDAS temperature parameters on day scale. As shown in Fig. 6, the correlation coefficients between the mean daily temperature and the observation data achieve 0.816 and 0.877 respectively, and the root-mean-square errors are 1.511 ℃ and 1.606 ℃ respectively.

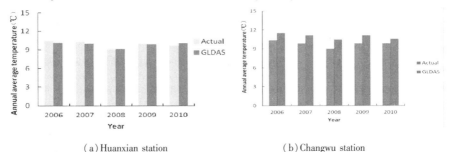

(a) Huanxian station   (b) Changwu station

Fig. 4   The comparison between GLDAS temperature and the measured temperature in Huanxian station and Changwu station in yearly scale

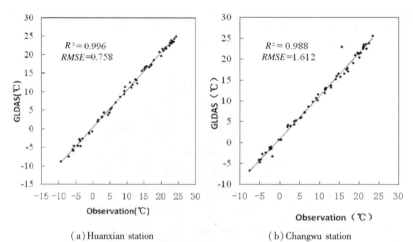

(a) Huanxian station (b) Changwu station

Fig. 5 The comparison between GLDAS temperature and the measured temperature in Huanxian station and Changwu station in monthly scale

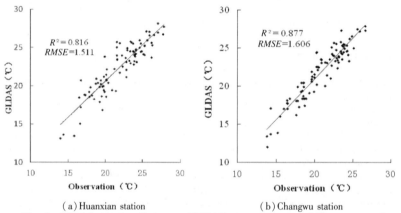

(a) Huanxian station (b) Changwu station

Fig. 6 The comparison between GLDAS temperature and the measured temperature in Huanxian station and Changwu station in daily scale

4 GLDAS data downscale

Assume that the GLDAS temperature values of the central point of each grid are the actual temperature of the point, based on the regional temperature distribution characteristics, the temperature values of the whole image are estimated with the surface elevation with the reversed distance weighted method. The specific steps are as follows.

(1) The temperature values of each grids are modified to the temperature values at sea level with the surface elevation based on the lapse rate of temperature — "Rising elevation 100 m, the temperature fell 0.6 ℃" (Liao Shunbao, et al. , 2003; Li Jun, et al. , 2006).

$$Tair = Gldas\_tair + 0.006H \qquad (1)$$

where, $Tair$ is the temperature values at sea level, K; $Gldas\_tair$ is the temperature value of each grid, K; H is the surface elevation, m.

(2) The temperature values at sea level of the central point of each grid are extracted to

interpolated to estimate the temperature values of the whole image. The reversed distance weighted method interpolates by weighted average using the distances between the sample points and the interpolation points as weight, shown as follows:

$$z_0 = (\sum_{i=1}^{n} \frac{z_i}{d_i^k}) / (\sum_{i=1}^{n} \frac{1}{d_i^k}) \tag{2}$$

where, z_0 are the values of interpolation points; z_i is the value of sample points; d_i is the distance between interpolation points and sample points; k is the power specified, in this article the value is 2.

(3) The temperature values at sea level after interpolating are modified to the temperature values in the actual height with the surface elevation using Eq. (1), and the spatial resolution is modified to 0.01°. Fig. 7 shows the comparison before and after interpolation.

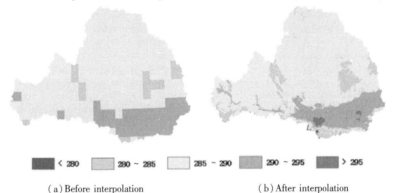

< 280	280 ~ 285	285 ~ 290	290 ~ 295	> 295

(a) Before interpolation   (b) After interpolation

Fig. 7   Comparison between before and after interpolation of GLDAS data (Unit: K)

The GLDAS temperature values after interpolation of Huan County station and Changwu station from June, 2010 to August, 2010 are used for the validity analysis of GLDAS temperature parameters on day scale. As shown in Fig. 8, the correlation coefficients between the mean daily temperature and the observation data achieve 0.817 and 0.881 respectively, and the root-mean-square errors are 1.511 ℃ and 1.313 ℃ respectively.

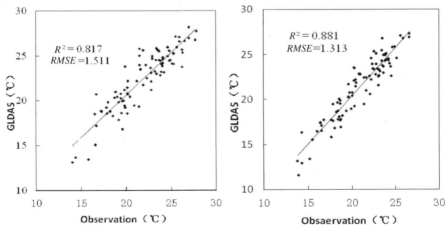

(a) Huanxian station   (b) Changwu station

Fig. 8   The comparison between downscaled GLDAS temperature and the measured temperature in Huanxian station and Changwu station in daily scale

The precision of downscaled GLDAS temperature data are improved, and the data highlight the influence of the terrain, thus the distributions of temperature are more reasonable. The result shows that the downscale method used in this article is feasible.

5   Conclusions

The temperature observation data of Huan County station and Changwu station in Weihe Basin are used for the validity analysis of the temperature parameters provided by GLDAS products which are assimilated in NOAH mode on day, month and year scale. The main conclusions are as follows:

(1) On the year scale, the average annual GLDAS temperature values are slightly greater than the observation data, but deviations are all within 20%. On the month scale, the correlation coefficients between the mean monthly temperature and the observation data achieve 0. 996 and 0. 988 respectively, and the root-mean-square errors are 0. 758 ℃ and 1. 612 ℃ respectively. On the day scale, the correlation coefficients between the mean daily temperature and the observation data achieve 0. 816 and 0. 877 respectively, and the root-mean-square errors are 1. 511 ℃ and 1. 606 ℃ respectively.

(2) Based on the temperature vertical decline rate, GLDAS temperature data are downscaled with the surface elevation with the reversed distance weighted method. The spatial resolution of GLDAS temperature data are increased from 0. 25° to 0. 01°, the data precision are improved, and the downscaled data highlight the influence of the terrain, thus the distributions of temperature are more reasonable.

(3) The GLDAS temperature values after interpolation of Huan County station and Changwu station from June, 2010 to August, 2010 are used for the validity analysis of GLDAS temperature parameters on day scale. the correlation coefficients between the mean daily temperature and the observation data achieve 0. 817 and 0. 881 respectively, and the root-mean-square errors are 1. 511 ℃ and 1. 313 ℃ respectively.

The result shows that the temperature parameters provided by GLDAS products which are assimilated in NOAH mode are able to reflect the distributions of temperature in Weihe Basin, and the downscale method used in this article is feasible.

References

Rodell M , et al. The Global Land Data Assimilation System[J], Bull. Am. Meteorol. Soc. , 2004, 85 : 381 – 394.

Wang Yanlin, Wang Wenke, Yang Zeyuan. Discussion on Eco – environmental Water Demand in Weihe River Basin of Shanxi Province [J]. Journal of Natural Resources, 2004, 19 (1) : 69 – 78.

Zhai Liang, Tang Xinming, Zhou Yi, et al. Contents Refinement of NFGIS Database[J]. Bulletin of Surveying and Mapping, 2006(1) : 47 – 48, 57.

Liao Shunbao, Li Zehui, You Songcai. Comparison on Methods Rasterization of Air Temperature Data[J]. Resources Science, 2003, 25(6) : 83 – 88.

Li Jun, You Songcai, Huang Jingfeng. Spatial Interpolation Method and Spatial Distribution Characteristics of Monthly Mean Temperature in China during 1961 ~ 2000 [J]. Ecology and Environment, 2006, 15(1) : 109 – 114.

Study on Decomposition of Wheat Straw and Function of Soil Fauna in Lower Reaches Irrigation District of the Yellow River[①]

Guo Zhifu[1,2] , *Song Bo*[3] , *Wang Guanshou*[1] and *Zhang Peipei*[4]

1. College of Environment and Planning, Henan University, Kaifeng, 475004, China
2. College of Environment and Planning, Shangqiu Normal University,
Shangqiu, 476000, China
3. Institute of Natural Resources and Environment, Henan University,
Kaifeng, 475004, China
4. School of Environment, Beijing Normal University, Beijing, 100875, China

Abstracts: Basied on the biodiversity function disturbed intensively by human behavior and farmer's low passivity of straw returning in agroecosystem, decomposition of wheat straw and function of soil fauna were detected by standard litterbag method in 107 d (from 15 June to 30 September, 2010) in the field with dry-cropping of lower reaches irrigation district of the Yellow River in Lankao county, Henan province. Different mesh-size litterbags (three mesh sizes, 0. 01 mm, 2 mm and 4 mm respectively) were used to control community composition of decomposers entering into the litterbags. The contributions of different soil fauna groups in decomposition of wheat straw can be found by considering the mass loss rates of straw decomposed by soil microorganism only (mesh size = 0.01 mm), soil microorganism + microfauna + mesofauna (mesh size = 2 mm), and soil microorganism + microfauna + mesofauna + macrosfauna(mesh size = 4 mm). Results showed that contribution rates of soil microorganism, microfauna + mesofauna, macrofauna for soil-returning straw treatment (SRT) are 1%, 15% and 84% respectively. And for crushing stock treatment (CST), contribution rates are 16%, 6% and 78% respectively. So, the higher biodiversity of soil fauna accelerates the mass loss of wheat straw. Functions of main decomposers, including soil microorganism, microfauna and mesofauna, were promoted in SRT plot comparing with CST. Results showed that wheat straw decayed rapidly in early decomposition stage (the first month after straw returning), and then mass of wheat straw lost slowly. At the end of study period, the mass loss rates of wheat straw are 86% and 67% respectively, for SRT and CST. The soil-returning straw treatment should be the priority returning straw method because that this method significantly improves material cycle efficiency in agroecosystem. The biodiversity protection of soil fauna should be concerned in system management because of significance of biodiversity function in agroecosystem.

Key words: decomposition of wheat straw, soil fauna, straw returning, lower reaches irrigation district of the Yellow River

1 Introduction

A large number of cellulose, lignin and nutrients, such as nitrogen and phosphorus, enrich in wheat straw. Nutrient release and decomposition of straw through returning can improve the physical condition and fertility of soil (Gong et al. , 2008). However, the usage of straw retuning is low because of straw utilization technologies. And burning straw in situ still exists in many districts in China. Burning straw not only causes serious problems of air quality and degradation of soil quality, but also accelerates global change because of a lot of carbon emissions (Norby et al. , 2001).

Soil fauna play vital roles in litter decomposition because they can consume litter, break up

---

① Foundation item: Under the auspices of Henan province education department natural science research item (No. 2010B170003)

organic material, and stimulate activities of microbes (Bradford et al. , 2002; Yin et al. , 2010). The previous studies showed that soil fauna could accelerate litter-decay rates (Smith and Bradford, 2003; Joo et al. , 2006). Most of the conclusions were established, which are derived from natural ecosystems, such as forest, grassland, and wetland ecosystem (Xu and Hirata, 2005; Yang and Zou, 2006; Song et al. , 2008; Trinder et al. , 2009). Additionally, biodiversity of soil fauna and role of earthworms has been concerned frequently (Zida et al. , 2011; Wolfarth et al. , 2011; Kabuyah et al. , 2012). But the function of biodiversity soil fauna is focused seldom in agroecosystem (Li et al. , 2006). Thus, it is necessary to study the decomposition functions of different soil fauna groups in agroecosystem.

The objective of this study is to investigate functions of different groups of soil fauna in straw decomposition and test weather higher biodiversity of soil fauna increases the decomposition of straw in agrosystem. Additionally, we try to find out effective methods of wheat straw returning in straw retuning to increase decomposition rate in lower reaches irrigation district of the Yellow River. To accomplish the objective, we evaluate the decomposition rates of wheat straw returning by different mesh-size litterbags and returning methods. While physical charactersistics of soil were analysized in decomposition state. Finally, the functions of soil fauna and the differences of decomposition rate in different returning treatments are discussed.

2 Materials and methods

2.1 Study site

The field site locates at lower reaches irrigation district of the Yellow River in Lankao county, Henan province, China ($34°12'$ N to $35°01'$ N and $113°52'$ E to $115°02'$ E). The research area has a warm temperate monsoon climate. The average temperature is $15°C$ and the mean precipitation is 678 mm. 65% of the precipitation is concentrate in June to August. The alluvial soil is the main soil type in the area (Soil Survey Office of Henan province). The continuous dry-cropping of wheat and corn is the dominant in the irrigation district, and the study area is one of major grain producing areas in China.

2.2 Litter decomposition

Litter decomposition rates in decomposing litter were evaluated using 14 cm × 12 cm litterbags with three mesh-size (0.01 mm, 2 mm, and 4 mm). The 0.01 mm mesh-size litterbags were made of nylon mesh. Only microfauna can enter into such mesh-size litterbags. The other two mesh-size litterbags were made of plastic. Microfauna and all mesofauna organisms can enter into 2 mm mesh-size litterbags. Microfauna, mesofauna and macrofauna can enter into 4 mm mesh-size litterbags. Microbe can enter into all mesh-size litterbags. So, the decomposition of litter in 4 mm mesh-size litterbag simulated the natural decomposition environment.

The standing dead straw of Triticum aestivum on early June 2010 were selected as the litter material in this study. The litter species represent the dominant food crop of lower reaches irrigation district of the Yellow River. The collected litter materials were air-dried, cut into 8 cm length pieces, and enclosed in the litterbags (4.71g air drying per litterbag, according to mass of straw return to field). Bag edges were sewed by nylon line.

Half of these litterbags were replaced in situ and anchored to the ground by short pieces of wire in each of six 5 m × 5 m blocks on 15 June 2010. And the other litterbags were replaced in situ and buried in six soil blocks below 10 cm. The former simulate straw decomposition by Crushing Stock Treatment (CST), and the latter simulate straw decomposition by Soil-Returning Straw Treatment (SRT) in field.

Six replicate litterbags of each treatment were retrieved semimonthly from the field from 15 June to 30 September 2010, and total 252 litterbags (2 treatment × 3 mesh-size × 6 replicates × 7 sampling date) were retrieved. Each litterbag was unearthed and gently slipped into a polythene bag without lifting to prevent loss of materials (Musvoto et al. , 2000). Macrofauna in the litter were picked out in the laboratory. Then the litter materials were placed in Tullgren extractors to remove the meso- and microfauna for 24 h. The litter materials in which soil fauna had been removed was rinsed carefully with distilled water to remove other extraneous organic material and soil particles, weighed after drying at 65℃ for 18 h, and calculated the mass percentages of the residues (Koukoura, 1998).

2.3 Physical characters of soil

The soil sample collected on every sampling time for measuring soil moisture content by oven drying method in laboratory. And the soil sample collected by cutting ring on every sampling time for determining the soil density (Institute of Soil Science, 1978). Soil temperature in 0, 5 cm, and 10 cm depth was measured by soil thermometer at 2PM on every sampling time.

2.4 Statistical analysis

Variance analysis was used to investigate the effects of mesh size litterbags on litter mass loss. Comparison of means (Ducan) was used to determine the variations of the differences litter mass loss between different treatments. The Data analyses were performed by SPSS (v.13.0).

3 Results

3.1 Characteristics of soil in plots

The values of soil temperature, moisture, and soil bulk density in CST and SRT plots are showed in Fig. 1. From Fig. 1, the differences of physical properties of soil in two plots showed that different ways of straw returning produce different environment through the process of straw decomposition. There are small differences of soil surface temperature at 2PM between two plots. The value of soil temperature in depth of 5 cm in SRT plot is slightly lower than that in CST plot (Fig. 1(a)). After raining or irrigation (30 June and 17 September), the value of soil moisture in CST plot is higher than that of SRT plot in short term, while in the dry period (15 July and 30 August), the results are opposite (Fig. 1(b)). Compared with CST plot, soil moisture in SRT plot varied slowly. The water content of soil in CST and SRT plots are 11.03% ~ 26.16% and 14.21% ~ 24.92% respectively. The values of soil bulk density in SRT plot are lower than those in CST plot in the whole study period (Fig. 1(c)). These suggest that the porosity, soil structure, and water retention capability of soil in SRT plot will be better than those in CST plot.

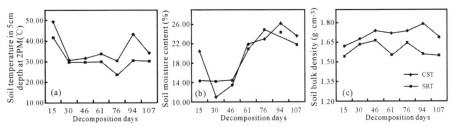

Fig. 1 Physical characters of soil in CST and SRT plots in study period

3.2 Mass loss of straw

Comparing the percents of straw mass remaining in SRT and CST plots (Fig. 2), the mass remaining of straw in former is lower than that in the latter on each retrieved sampling time. Furthermore, there are significant differences of the mass remaining in two plots for every mesh size litterbags (Tab. 1), which indicates that the decay rate of straw in SRT plot has been raised significantly in decomposition process.

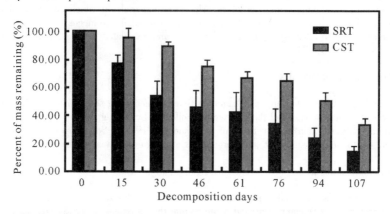

Fig. 2 Effects of different returning ways of straw on remaining mass.
Percent of mass remaining (mean ±1 SE, percentage of original)
in 4 mm mesh size litterbags on every retrieved time is shown

Tab. 1 Sig. p values of mean comparison between percent of mass remaining
in CST and SRT plots

Mesh size	Decomposition days						
	15	30	46	61	76	94	107
4 mm	0	0	0	0.01	0	0	0
2 mm	0	0	0	0.01	0	0.01	0
0.01 mm	0	0	0	0	0	0	0

The significant higher decomposition rate for SRT was contributed by the characteristics of microenvironment. This study has found that the value of soil bulk density in SRT plot is smaller than that in CST plot (Fig. 1). The changes of soil physical characteristics not only can directly strengthen the leaching process in straw decomposition, but also can improve the decomposing role of the decomposer through enriching soil fauna community, and then accelerate decompose of straw (Hobbie, 2003; Cleveland et al. , 2006; Ma et al. , 1999).

The mass loss percents of straw in 4 mm mesh litterbags for SRT and CST on every retrieved sampling time were shown in Tab. 2. In the first four weeks after straw returning, the mass loss percent in SRT plot is the highest in the whole study period, and that is higher than that in CST plot. Since then, the straw decayed slowly in SRT plot until the last two weeks. The percent of mass loss in CST changed similarly after the first four weeks delay period. At the end of decomposition period, the percents of mass loss for SRT and CST are 85. 77% and 66. 53% respectively.

Tab. 2 Mass loss percent of straw in 4 mm mesh litterbags

Decomposition days	15	30	46	61	76	94	107
SRT	22. 89	23. 18	8. 07	3. 40	8. 10	10. 83	22. 89
CST	4. 78	5. 80	14. 23	8. 07	2. 65	14. 19	4. 78

3.3 Decomposition contributions of different soil fauna groups

Mass remaining (air drying weight) of wheat straw in 4 mm, 2 mm and 0.01 mm mesh size litterbags on every retrieved time for SRT and CST was shown in Fig. 3. In the first 8 weeks after decomposition, the remaining weights of straw in 2 mm mesh litterbags were the highest in SRT plot. The higher mass loss in 4 mm mesh litterbags was found in last 4 weeks. At the end of this study, straw remains 0.67 g(4 mm), 0.71 g(2 mm), and 1.11 g (0.01 mm) respectively. In CST plot, the remaining weights of straw in 0.01 mm mesh litterbags are the lowest in the first eight weeks decomposition period. At the end of this study, straw remained 1.58 g(4 mm), 2.13 g(2 mm), and 2.28 g (0.01 mm) respectively.

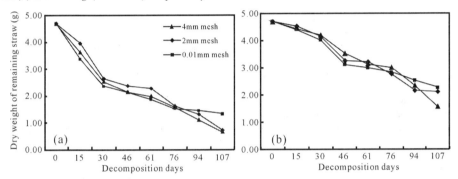

**Fig. 3 Mesh size effects on remaining mass of wheat straw in SRT plot (Fig. 2(a))
and CST plot (Fig. 2(b)). Mass remaining (air drying weight) in 4 mm,
2 mm and 0.01 mm mesh size litterbags on every retrieved time is shown**

The results of variance analysis on mass remaining percent of straw in different mesh size litterbags were shown in Tab. 3. For SRT, there is no significant difference between mass remaining in different mesh size litterbags in the middle decomposition period (15 ~ 94 d). At the end of decomposition period, the percents of mass remaining in 4 mm and 2 mm mesh litterbags are lower significantly than that in 0.01 mm mesh litterbags. And for CST, there is no significant difference between mass remaining in different mesh size litterbags in the former decomposition period (0 ~ 94 d). At the end of decomposition period, the percents of mass remaining in 4 mm mesh litterbags is lower significantly than those in 0.01 mm and 2 mm mesh litterbags.

**Tab. 3 Variance analysis on mass remaining percent of straw in different mesh
size litterbags**

Decomposition days		15	30	46	61	76	94	107
SRT	4 mm	77.11b	53.93a	45.39a	42.51a	34.36a	23.53a	14.23b
	2 mm	84.33a	56.48a	50.21a	48.24a	33.93a	27.92a	15.11b
	0.01 mm	71.30b	50.14a	48.30a	39.57a	32.17a	30.57a	27.71a
CST	4 mm	93.76a	89.42a	73.51a	65.90a	64.47a	50.28a	33.48b
	2 mm	95.12a	87.86a	69.51a	68.58a	59.09a	46.00a	45.19a
	0.01 mm	94.27a	85.71a	67.05a	64.84a	59.87a	54.00a	48.30a

Note: * Mean value is given; and the value with different letters in the same column mean significant difference among the treatments ($p < 0.05$).

In the whole study period, the total mass loss percent of straw for SRT in 4 mm, 2 mm, and 0.01 mm mesh litterbags are 85.77%, 84.89%, and 72.29% respectively. Therefore, the contributions of soil meso- and micro-fauna are 12.60% to decomposition (this group decomposition percent); the contributions of soil macro-fauna and microorganism are 0.88% and 72.29% respectively. And for CST, the total mass loss percent of straw in 4 mm, 2 mm, and 0.01 mm mesh litterbags are 66.53%, 54.81%, and 51.70% respectively. Therefore, the contributions of soil meso- and micro-fauna are 3.11% to decomposition (this group decomposition percent of total mass loss percent); the contributions of soil macro-fauna and microorganism are 11.71% and 51.70% respectively.

4　Discussions

4.1　Decomposition rate of wheat straw

The decomposition stage of wheat straw can divide into three periods for SRT (Tab. 2). In the first 4 weeks after wheat straw returning, straw decomposed rapidly, and the decay rate was above 1.5% per day. In the followed eight weeks, straw decomposed slowly, and the decay rate was less than 0.6% per day. In the last 2 weeks, the decay rate rose to 1.5% per day. Straw decayed rapidly in the first period benefited from the mass loss of readily degradable components in straw, such as carbohydrate and polysaccharide (Wardle et al., 2004). The followed period straw decomposed slowly because of resistant components to decomposition, such as cellulose, lignin, and Polyhenols, in litter (Rantalainen et al., 2004; Mayer et al., 2005). The colonizing and convergence of more soil fauna in litterbag increased the decay rate of straw in the last decomposition period (Ke et al. 1999; Li et al., 2006).

The change of decomposition rate of straw for CST is similar as that for SRT, but the rapid decomposition period is delayed. There are two reasons may make this difference clear. The one reason is colonizing time of soil fauna is longer in CST plot. And the other one is the lower moisture and higher temperature in litterbags (perceive in practice). The lower moisture and higher temperature can decrease decomposed rate (Huish et al., 1985; Drewnik, 2006).

4.2　Effects of microenvironment on soil fauna function

There are no continuous significant differences of decomposition rate through the three mesh size litterbags (Tab. 3). Although the decomposition rate rises with the more groups of soil fauna in litterbags at the end of study period, the decomposition rate in 4 mm mesh litterbags is not the highest on other retrieved sampling time (Fig. 2). This contrast result between decomposition rate and soil fauna diversity reveals the influence of microenvironment factor in process of straw decomposition. The litterbags control the body size of soil fauna which participate in decomposition process, and they construct the small scale microenvironment, at the same time. The more moisture in 0.01 mm mesh litterbags was discovered in practice. The characters in 0.01 mm mesh litterbags will increase mass loss of straw through leaching in early decomposition period (Knacker et al., 2003). So the contribution of soil fauna to straw decomposition evaluated by using litterbag method may be lower than that in real. And the variations of moisture status and temperature are related to soil fauna community (Bardgett and Cook, 1998). The influence of decomposer composing because of microenvironment in litterbags on decomposition process should be discussed in microcosm carefully.

5　Conclusions

The results of our study highlight that the contribution rate of soil microorganism to decomposition of wheat straw is 52% ~72%; and that of soil animal is 14% ~15%. In different retrieved sampling time, the advantage of soil fauna diversity will be weakened because of the influence of the microenvironment in litterbags. Compared with the soil returning straw treatment, the role of soil fauna is weak in the crushing stock treatment. The biodiversity protection of soil fauna should be concerned in system management because of significance of biodiversity function in agroecosystem.

During the study period, the decomposition rate of wheat straw is rapid at early stage; then it decreases continuously long time; and it increases in the last stage. In study area, the decomposition rate of wheat straw for SRT is significantly higher than that for CST. After 107 d, the accumulation percent of mass loss for SRT and CST are 85. 77% and 66. 53% respectively. Therefore, plowing after the straw returning should be the preferential soil tillage management in lower reaches irrigation district of the Yellow River. This method can increased the decay rate of straw and accelerate the nutrient cycling in agroecosystem.

Acknowledgements

We appreciate the assistance provided by Henan province education department natural science research item (No. 2010B170003) and gratefully acknowledge the Institute of Environment Resource in Henan University for chemical analysis.

References

Bardgett R D, Cook R. Functional Aspects of Soil Animal Diversity in Agricultural Grasslands[J]. Applied Soil Ecilogy,1998(10):263-276.

Bradford M A, Tordoff G M, Eggers T, et al. Microbiota, Fauna, and Mesh Size Interactions in Litter Decomposition[J]. Oikos,2002, 99(2): 317-323.

Cleveland C C, Reed T C, Townsend A R, Nutrient Regulation of Organic Matter Decomposition in A Tropical Rain Forest[J]. Ecology,2006,87(2):492-503.

Drewnik M, The Effect of Environmental Conditions on the Decomposition Rate of Cellulose in Mountain Soils[J]. Geoderma,2006,132(1-2): 116-130.

Gong C F, Wang J X, Chen X W, et al, The Present Situation and Technology Returning of Crop Straw[J]. Science and Technology Innovation Herald,2008(9): 251.

Hobbie S E. Interactions Between Litter Lignin and Soil Nitrogen Availability During Leaf Litter Decomposition in a Hawaiian Montane Forest[J]. Ecosystems,2000(3):484-494.

Huish S, Leonard M A, Anderson J M. Wetting and Drying Effects on Animal/microbial Mediated Nitrogen Mineralization and Mineral Element Losses from Deciduous Forest Litter and Raw Humus[J]. Pedobiologia,1985(28): 177-183.

Institute of Soil Science. Physical and Chemical Analysis of Soil [M]. Shanghai: Shanghai Scientific and Technological Press, 1978.

Joo S, Yim M H, Nakane K. Contribution of Microarthropods to the Decomposition of Needle Litter in a Japanese Cedar (Cryptomeria japonica D. Don) Plantation [J]. Forest Ecology and Management,2006, 234(1-3): 192-198.

Kabuyah R N T M, van Dongen B E, Bewsher A D, et al. Decomposition of Lignin in Wheat Straw in a Sand-dune Grassland[J]. Soil Biology and Biochemistry,2012, 45, 128-131.

Ke X, Zhao L J, Yin W Y. Succession in Communities of Soil Animals During Leaf Litter Decomposition in Cyclobalanopsis Glauca Forest[J]. Zoological Research,1999,20(3): 207-213.

Li Y L, Qiao Y H, Sun Z J, et al. The Eco-process of Agricultural Organic Matter Decomposition under Different Soil Conditions[J]. Acta Ecologica Sinica,2006,26(6): 1933-1939.

Ma L W, Peterson G A, Abuja I R, et al. Decomposition of Surface Crop Residues in Long-term Studies of Dry Land Agro Ecosystems[J]. Agronomy Journal,1999,91:401-409.

Mayer P M, Tunnell S J, Engle D M, et al. Invasive Grass Alters Litter Decomposition by Influencing Macrodetritivores[J]. Ecosystems,2005(8): 200-209.

Norby R J, Cotrufo M F, Ineson P, et al. Elevated CO_2, Litter Chemistry, and Decomposition: a Synthesis[J]. Oecologia,2001,127:153-165.

Rantalainen M L, Kontiola L, Haimi J, et al. Influence of Resource Quality on the Composition of Soil Decomposer Community in Fragmented and Continuous Habitat[J]. Soil Biology and

Biochemistry,2004,36, 1983-1996.

Smith V C, Bradford M A. Litter Quality Impacts on Grassland Litter Decomposition are Differently Dependent on Soil Fauna Across Time[J]. Applied Soil Ecology,2003,24(2): 197-203.

Soil Survey Office of Henan Province . Henan Soil[M]. Beijing: China Agriculture Press,2004.

Song B, Yin X Q, Zhang Y, et al. Dynamics and Relationships of Ca, Mg, Fe in Litter, Soil Fauna and Soil in Pinus Koraiensis-Broadleaf Mixed Forest [J]. Chinese Geographical Science,2008,18(3):284-290.

Trinder C J, Johnson D, Artz R R E. Litter Type, But Not Plant Cover, Regulates Initial Litter Decomposition and Fungal Community Structure in a Recolonising Cutover Peatland[J]. Soil Biology and Biochemistry,2009, 41:651-655.

Wardle D A, Bardgett R D, Klironomos J N, et al. Ecological Linkages Between Aboveground and Belowground Biota[J]. Science,2004,304: 1629-1633.

Wolfarth F, Schrader S, Oldenburg E, et al. Earthworms Promote the Reduction of Fusarium Biomass and Deoxynivalenol Content in Wheat Straw under Field Conditions[J]. Soil Biology and Biochemistry,2011,43(9): 1858-1865.

Xu Xiaoniu, Hirata E J. Decomposition Patterns of Leaf Litter of Seven Common Canopy Species in a Subtropical Forest: N and P Dynamics[J]. Plant and Soil,2005,273(1-2): 279-289.

Yang X D, Zou X M. Soil Fauna and Leaf Litter Decomposition in Tropical Rain Forest in Xishuangbanna, SW China: Effects of Mesh Size of Litterbags[J]. Journal of Plant Ecology, 2006,30 (5): 791-801.

Yin X Q, Song B, Dong W H et al. A Review on the Eco-geography of Soil Fauna in China[J]. Journal of Geographical Sciences,2010,20(3): 333-346.

Zida Z, Ouédraogo E, Mando A, et al. Termite and Earthworm Abundance and Taxonomic Richness under Long-term Conservation Soil Management in Saria, Burkina Faso, West Africa [J]. Applied Soil Ecology,2011, 51, 122-129.

Three Dimensional Large Eddy Simulations of Free Surface Curved Open-channel Flow

Bai Jing

Dept. of Hydraulic Engineering, the State Key Laboratory of Hydro-Science and Engineering, Tsinghua University, Beijing, 100084, China

Abstract: Secondary flow is very common natural phenomenon in channels with complex boundary boundaries such as curved open-channels. And it has very important effects on the main flow, then water level, bed shear stresses and sediment transport. Three dimensional numerical simulations of flow in a rectangular U-bend channel, using large-eddy simulation model (LES), is presented in this paper. Free surface level is captured dynamically by solving a 2D Poisson equation instead of a simple assumption of a horizontal rigid. Velocity profile and water level distributions, vortex locations are investigated via the LES model. The calculated velocities and water levels are compared with the experimental data to validate the presented numerical model. Generally the discrepancy between the numerical results and the experiment results are acceptable. The simulation results agree well with the experimental data indicating that the LES model is reliable and can be used to predict dynamically the water level, velocity distributions, secondary flow, and other flow characteristics of under complex boundaries in detail, and in the further study, the LES model can be used to calculate and predict the sediment transport with a sediment module.

Key words: open channel flow, water level, curved channel, large eddy simulation

1 Introduction

LES, RANS, and DNS are the three most important numerical model methods in computation fluid dynamics. DNS, due to the challenge of computation and storage load, is limited to low Reynolds number flow simulation. RANS, decomposing variables into time-average ones and impulse ones, obtains time-averaged flow information by Reynolds-averaged NS equations and turbulence models. In consequence, it is difficult to capture flow dynamically by RANS. In LES, the large energy-carrying eddies which contribute most to the turbulent transport are computed directly, whereas the influence of the small eddies are modeled by a sub-grid scale model. Small eddies are more universal, random, and isotropic, and this simplifies the development of sub-grid models. The advance of LES is that it can reflect the impulse information of flow field and simulate unsteady flow in a more reasonable way.

Flow in curved channels is of high interest in river engineering, and a lot of numerical investigations and simulations of such flows are very popular. Secondary flow development, shear stresses distribution and free water level of flows in meandering channels are investigated through numerical simulation. Most of the advancements, achieved through RANS, are limited to time-average values and analysis. And also there are some advancement obtained by LES. Breuer and Rodi (1996) gave a LES example regarding fully turbulent flow in a 180° duct. Booij (2003) presented a good reproduction of the main flow and the secondary flow of a 180° river bend. Stoesser et al. (2010) carried out calculations of flow and boundary shear stresses in a meandering channel, which were validated by experimental data of a physical model. In researches of Booij (2003) and Stoesser et al. (2010), the free surface was treated as a plane of symmetry where zero gradient conditions were applied for the variables parallel to the surface with the surface-normal variables set to zero. The free surface is treated as a horizontal rigid lid where free-slip conditions are applied. van Balen et al. (2009, 2010) computed curved channel flow over flat-bottom and topography, aiming at addressing second flows features and flow characteristics in detail, and also a horizontal rigid is adopted at the water surface.

In nearly all the LES achievements listed above about open-channel flow in curved channel, the horizontal rigid lid assumption is used for the representation free surface without other special treatments. The assumption of a rigid lid is justified as the longitudinal and transverse water-level slopes are negligibly small. It should be noted that in a refined simulation technique, such as LES, the errors may have an effect on the numerical results, especially in curved channel. In this paper, LES is introduced to compute the flow in curved channel flow. Using the approach of solving a 2D Poisson equation, the free water level is traced step by step. And instantaneous flow field is caught at the same time. The calculation results are reliable compared and validated by the experiment data.

2 Governing equations

Averaged over the control volumes formed by the numerical mesh, 3D filtered time-dependent incompressible dimensionless NS equations in the hydrodynamic module of LES model are given as follows.

$$\frac{\partial \overline{u_i}}{\partial x_i} = 0 \tag{1}$$

$$\frac{\partial \overline{u_i}}{\partial t} + \frac{\overline{\partial}}{\partial x_j}(\overline{u_i u_j}) = -\frac{\partial \overline{p}}{\partial x_i} + \frac{\partial}{\partial x_j}\left[\frac{1}{Re}\left(\frac{\partial \overline{u_i}}{\partial x_j} + \frac{\partial \overline{u_j}}{\partial x_i}\right)\right] - \frac{\partial \tau_{ij}}{\partial x_j} \tag{2}$$

where, ' $-$ ' denotes variables after being filtered, and it means the resolved variable; u_i ($i = 1$, 2, 3) are the dimensionless velocity components; p is dimensionless dynamic pressure; ν is the inverse of the Reynolds number, Re ; and the subgrid scale (SGS) stress τ_{ij} results from filtering of the non-linear convective fluxes.

This term reflects the influence of the subgrid scale turbulence structures on the large eddies. The SGS stress is calculated through an eddy viscosity relation as:

$$\tau_{ij} = -\nu_{\text{SGS}}\left(\frac{\partial \overline{u_i}}{\partial x_j} + \frac{\partial \overline{u_i}}{\partial x_i}\right) + \frac{1}{3}\partial_{ij}\tau_{kk} \tag{3}$$

where, the SGS viscosity ν_{SGS} is computed from the dynamic subgrid scale (SGS) model proposed by Germano et al. (1991).

According to Wu et al. (2000), Free water level $\overline{z_s}$ is capture by solving a 2D Poisson equation.

$$\frac{\partial^2 \overline{z_s}}{\partial x^2} + \frac{\partial^2 \overline{z_s}}{\partial y^2} = \frac{Q}{g} \tag{4}$$

In which

$$Q = \frac{\partial}{\partial t}\left(\frac{\partial \overline{U}}{\partial x} + \frac{\partial \overline{V}}{\partial y}\right) - \left(\frac{\partial \overline{U}}{\partial x}\right)^2 - 2\frac{\partial \overline{U}}{\partial y}\frac{\partial \overline{V}}{\partial x} - \left(\frac{\partial \overline{V}}{\partial y}\right)^2 - U\left(\frac{\partial^2 \overline{U}}{\partial x^2} + \frac{\partial^2 \overline{V}}{\partial x \partial y}\right) - V\left(\frac{\partial^2 \overline{U}}{\partial x \partial y} + \frac{\partial^2 \overline{V}}{\partial y}\right) +$$

$$\frac{1}{\rho}\left(\frac{\partial^2 \overline{T_{xx}}}{\partial x^2} + 2\frac{\partial^2 \overline{T_{xy}}}{\partial x \partial y} + \frac{\partial^2 \overline{T_{yy}}}{\partial y}\right) - \frac{1}{\rho}\frac{\partial}{\partial x}\left(\frac{\tau_{xb}}{h}\right) - \frac{1}{\rho}\frac{\partial}{\partial y}\left(\frac{\tau_{xb}}{h}\right) \tag{5}$$

where, ' $-$ ' denotes variables after being filtered; U is depth $-$ averaged velocity in x direction; V is depth $-$ averaged velocity in y direction; T_{xx}, T_{xy}, and T_{yy} are depth $-$ averaged turbulent stresses; τ_{xb} and τ_{yb} are bed shear stresses x in y and directions.

3 Numerical mothods

The code used for calculation is LESOCC2 (Large Eddy Simulation on Curvilinear Coordinates, 2nd edition), developed at the Karlsruhe Institute of Technology(Hinterberger et al., 2007). The equations are discretized with the finite volume method on curvilinear grids with a collocated variable arrangement. The convective fluxes and diffusive fluxes in momentum equations are approximated using the central scheme. A second-order accurate Runge-Kutta method is adopted for time discretization.

Boundary conditions (bcs) for the momentum and transport equations need to be provided at all domain boundaries, i. e. inlet, outlet, walls. First the bcs of the momentum equations: At the inlet, instantaneous velocities are specified directly for every time step. A convective boundary is specified at the outflow boundary. This non-reflective condition ensures vortices to pass the boundary without disturbing the upstream flow. The wall function of Werner and Wengle (1991) is adopted for the smooth sidewalls。

The method proposed by Stoesser (2010) is used to deal with the rough bottom. In the method, the bed topography is generated using random numbers with a normal distribution function. The mean bed elevation is known and the only parameter left is the standard deviation of the bed elevation. In this paper, the standard deviation is adjusted through the comparison of velocities between the calculation and the experiments. The bed roughness is considered by adding forcing terms to the source terms of discretized momentum equations, which aims to correct the velocities in roughness to zero.

4 Case study

The flume studied by Blanckaert (2010) consisted of straight inflow and outflow sections, together with a 193° curved section. The bed topography and water level were measured using acoustic limmimeters. For the purpose of avoiding high computational cost in computation, the inflow and outflow sections are shortened to 4 m long. And the geometry in computation (Fig. 1) is obtained by interpolation of a digitalized topographic map. As showed in Fig. 1, some big dunes are distributed in the flume. The sidewalls of the flume are hydraulically smooth. The average value of sand particles is about 2. 0 mm. Reynolds number is 68,600 based on the mean water depth $H = 0. 14$ m and the bulk velocity $U = 0. 49$ m/s. In computation the grid size is .

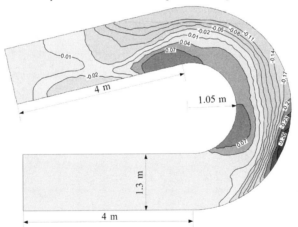

Fig. 1 Size and geometry of computational domain (m)

In Fig. 2, the color denotes the magnitude of dimensionless velocity by the bulk velocity U and the arrows denotes the direction of velocity. According to Fig. 2, because of the hydraulics of curved channel, there are two clockwise vortexes along the convex bank behind the dunes on the bed. In sediment-laded flow, deposit may happen in the recirculation zones, the velocities are very small compared with the main flow.

Velocity profiles in streamwise and transverse direction in the centerline are shown as Fig. 3. In the figures, the abscissa is the dimensionless velocity by the bulk velocity U, and the ordinate is the dimensionless distance from the bed by mean water depth. The bed level of the curved channel is not constant, and the flume averaged bed level is set as the reference level in Fig. 3. As the Figures show, there is good agreement between computation result and experiment result at most

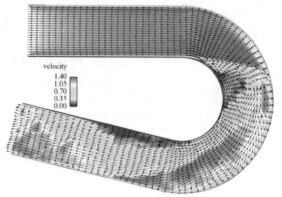

Fig. 2 Snapshot of instantaneous velocity field with low-pass filtering (m/s)

locations except a little discrepancy, such as at the surface of cross 30° (Fig. 3 (a)) and cross 180° (Fig. 3 (d)) , there are a few differences. The streamwise and transverse velocities are a little smaller than experiment data. At the bottom of cross 120° (Fig. 3 (c)) , the computed values of transverse velociy are a little bigger than experiment results. In general, the qualitative agreement between the experiment and the LES computations is rather good. And the comparisons indicate that LESOCC2 can be employed for curved channels in laboratory scale.

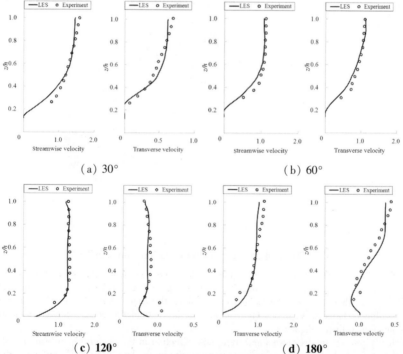

(a) 30°   (b) 60°

(c) 120°   (d) 180°

Fig. 3 Velocities in the centrelines of different cross sections

The calculated water levels at various cross-sections are compared with the experimental data as shown in Fig. 4. The abscissa is the distance from convex bank (the right bank along the direction of flow) , and the unit is meter. The ordinate is the water level, and the unit is centimeter.

Consistent with treatment in last section, flume averaged bed level is set as the reference level. The relative difference between both values is smaller than 0.5% except the smallest experiment data at some cross sections. The highest water level is predicted precisely and the errors mainly arise at the lowest water level at different cross sections. The max error is at cross section 90° and its value is smaller than 4%. The numerical results agree well with the experimental data generally. As shown in Fig. 4, the water level near the concave bank is higher than the water level near the convex bank, which is corresponding with the hydraulic characteristics of the curved channel.

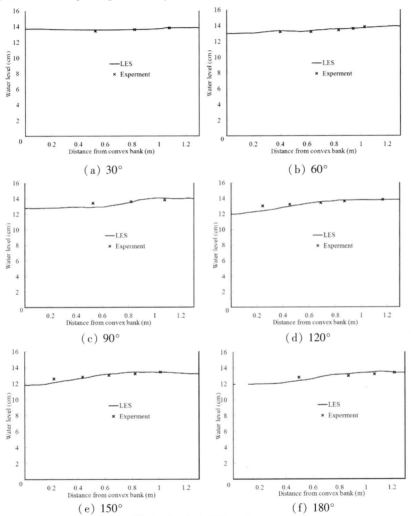

Fig. 4　Water level at different cross sections

5　Conclusions

In this paper, LES with a water level capturing module via solving a 2D Poisson equation is presented to compute the flow in curved channel flow. The velocity calculated distributions in the curved channel is in good agreement with the experimental results.

The detailed flow information in curved channel is also given by the LES model. In the curved channel, vortexes due to bed topography and hydraulics features are predicted. Also the water level changes from the convex bank to the concave bank are presented. The coincidence between the experimental data and the simulation shows that the LES model is practicable and credible. Further numerical simulations of sediment transport will be carried out in the future to investigate flow and sediment transport in the curved channel.

Acknowledgements

This work is part of a research program named Natural Science Foundation for the Youth (No. 10902061) funded by National Natural Science Foundation of China. And the calculations were supported by THPCC (Tsinghua High-performance Computing Center).

References

Breuer M, Rodi W. "Large-eddy Simulation of Complex Turbulent Flows of Practical Interest[J] Hirschel, E. H. (Ed.), Flow Simulation with High-Performance Computers II, Notes on Numerical Fluid Mechanics, 1996 (52): 258 – 274.

Blanckaert K. Topographic Steering, Flow Recirculation, Velocity Redistribution, and Bed Topography Sn sharp Meander Bends [J]. Water Resour. Res. , 2010:46.

Booij R. Measurements and Large Eddy Simulations of the Flows in Some Curved Flumes [J]. J. Turbul. ,2003: 4.

Germano M, Piomelli U, Moin P, et al. A Dynamic Subgrid – scale Eddy Viscosity Model [J]. Phys. Fluids, 1991 3(7): 1760 – 1765.

Hinterberger C, Frohlich J, Rodi W. Three – dimensional and Depth – averaged Large – eddy Simulations of Some Shallow Water Flows [J]. J. Hydraul. Eng. – ASCE, 2007, 133(8): 857 – 872.

Stoesser T. Physically Realistic Roughness Closure Scheme to Simulate Turbulent Channel Flow over Rough Beds within the Framework of LES [J]. J. Hydraul. Eng. – ASCE, 2010, 136 (10): 812 – 819.

Stoesser T, Ruether N, Olsen N R B. Calculation of Primary and Secondary Flow and Boundary Shear Stresses in a Meandering Channel [J]. Adv. Water Resour. , 2010, 33 (2): 158 – 170.

van Balen W, Uijttewaal W S J, Blanckaert K. Large – eddy Simulation of a Mildly Curved Open – Channel Flow [J]. J. Fluid Mech. , 2009, 630, 413 – 442.

van Balen W , Uijttewaal W S J, Blanckaert K. Large – eddy Simulation of a Curved Open – channel Flow Over Topography [J]. Phys. Fluids, 2010, 22(7).

Werner H, Wengle H. Large – eddy Simulation of Turbulent Flow over and Around a Cube in a Plate channel [C]// 8th Symposium on Turbulent Shear Flows, 155 – 168.

Wu W M, Rodi W, Wenka T. 3D Numerical Modeling of Flow and Sediment Transport in Open Channels[J]. J. Hydraul. Eng. – ASCE, 2000, 126(1): 4 – 15.